# Biomass Gasification, Pyrolysis, and Torrefaction

Practical Design and Theory

**Second Edition**

# Biomass Gasification, Pyrolysis, and Torrefaction

## Practical Design and Theory

**Second Edition**

**Prabir Basu**
Dalhousee University and
Greenfield Research Incorporated

AMSTERDAM • BOSTON • HEIDELBERG • LONDON
NEW YORK • OXFORD • PARIS • SAN DIEGO
SAN FRANCISCO • SINGAPORE • SYDNEY • TOKYO
Academic Press is an imprint of Elsevier

Academic Press is an imprint of Elsevier
32 Jamestown Road, London NW1 7BY, UK
525 B Street, Suite 1800, San Diego, CA 92101-4495, USA

First edition 2008
Second edition 2013


**British Library Cataloguing-in-Publication Data**
A catalog record for this book is available from the Library of Congress

**Library of Congress Cataloging-in-Publication Data**
A catalogue record for this book is available from the British Library

ISBN: 978-0-12-396488-5

For information on all Academic Press publications
visit our website at store.elsevier.com

Typeset by MPS Limited, Chennai, India
www.adi-mps.com

# Dedication

**This book is dedicated to one who taught me.**

***"We are the children of immortal bliss... Each Soul is potentially divine"***

# Contents

# Preface

The art of energy conversion of biomass is as old as our natural habitat. Such processes have been at work since the early days of vegetation on this planet. Flame leaping from forest fire is an example of "flaming pyrolysis." Trace of blue flame in a swamp is an example of methane gas formation through decomposition of biomass and its subsequent combustion in contact with air. Burning vegetation on ground to increase soil fertility is an example of biochar production. Human beings, however, learned to harness these processes much later.

Use of biomass for energy, though nearly as ancient as human civilization, did not rise at the same pace with industrialization because of the abundant supply and low prices of oil and natural gas. Only in the recent past has there been an upsurge in interest in biomass energy conversion, fueled by several factors:

- Interest in the reduction in greenhouse gas emissions as a result of energy production
- Push for independence from the less reliable supply and fluctuating prices of oil and gas
- Interest in renewable and locally available energy sources
- Rise in the price of oil and natural gas.

Several excellent books on coal gasification are available, but a limited few are available about biomass gasification and pyrolysis, and none on torrefaction. A large body of peer-reviewed literature on biomass gasification, pyrolysis, and torrefaction is available; some recent books on energy also include brief discussions on these topics. For example the previous edition (Biomass Gasification and Pyrolysis) of this book along with its Chinese and Italian versions presents a good treatment of these topics. There is yet a dearth of comprehensive publications specifically on torrefaction. For this reason, the previous book was revised and expanded with several new chapters on such new topics to develop the monograph.

Engineers, scientists, and operating personnel of biomass gasification, pyrolysis, or torrefaction plants clearly need such information from a single easy-to-access source. Better comprehension of the basics of biomass conversion could help an operator understand the workings of such plants, a design engineer to size the conversion reactors, and a planner to evaluate different conversion options. The present book was written to fill this important need. It attempts to mold available research results in an easy-to-use design methodology whenever possible. Additionally, it brings into focus new advanced processes such as supercritical water gasification and torrefaction of biomass.

This book comprises 13 chapters and a number of appendices, which include several tables that could be useful for the design of biomass conversion units and their components. Chapter 1 introduces readers to the art of different forms of biomass energy conversion and its present state of art. It also discusses the motivations for such conversion in the context of current energy scenario around the world. A brief introduction to economic issues around biomass utilization is available in Chapter 2. The properties of biomass, especially those relevant to gasification, torrefaction, and pyrolysis of biomass are included in Chapter 3. Chapter 4 discusses the principles of torrefaction and its recent developments. It also includes a simple method for design of a torrefaction plant. The basics of pyrolysis are included in Chapter 5 that discusses in some details on biochar, a new option for carbon sequestration and soil remediation.

Chapter 6 deals with an important practical problem of biomass gasification—the tar issue. This chapter provides information on the limits of tar content in product gas for specific applications. Chapter 7 concerns the basics of the gasification of biomass. It explains the gasification process and important chemical reactions that guide pyrolysis and gasification. Chapter 8 discusses design methodologies for gasifiers and presents some worked-out examples on design problems. Chapter 9 hydrothermal gasification of biomass, with specific reference to gasification in supercritical water. In recent past, there is much interest in partial substitution of coal with greenhouse gas (GHG) neutral biomass in existing coal-fired power plant. This near term cost effective means for reduction in GHG from coal-fired plants is presented in Chapter 10 that discusses the basics and different aspects of this new option of biomass energy conversion.

The production of chemicals and synthetic fuels is gaining importance, so Chapter 11 provides a brief outline of how some important chemicals and fuels are produced from biomass through gasification. Production of diesel and bio-gasoline along with Fischer–Tropsch synthesis process are also discussed briefly here. One of the common, but often neglected, problems in the design of biomass plant is the handling of biomass. Chapter 12 discusses issues related to this and provides guidelines for the design and selection of handling and feeding equipment. Chapter 13 presents a brief discussion of some commonly used analytical techniques for measurement of important properties of biomass that are essential for design of a biomass energy conversion system. Appendix A contains definitions of biomass, Appendix B lists physical constants and conversion units, and Appendix C includes several tables containing design data. Glossary presents definitions of some terms used commonly in the chemical and gasification industries.

# Acknowledgments

The author is greatly indebted to a large number of students, professional colleagues, and institutions who helped to revise numerous drafts of this edition of this book and provided permission for the use of published materials. Many students worked tirelessly to support the work on this book. Special efforts were made by S. Rao, B. Acharya, A. Dhungana, and A. Basu. My hope is that what is here will benefit at least some students and/or practicing professionals in making the world around us a little greener and more habitable.

Finally, this book would not have materialized without the constant encouragement of my wife, Rama Basu.

# About the Author

Dr. Prabir Basu, founding President of Greenfield Research Incorporated, a private research and development company in Canada that specializes in gasification and torrefaction, is an active researcher and designer of biomass energy conversion systems. Dr. Basu holds a position of professor in Mechanical Engineering Department and is head of Circulating Fluidized Bed Laboratory at Dalhousie University, Halifax. His current research interests include frontier areas, chemical looping gasification, torrefaction, and biomass cofiring among others.

Professor Basu is also the founder of the prestigious triennial International Conference series on Circulating Fluidized Beds, Fluidized Bed Systems Limited that specializes on design, training, and investigative services on fluidized-bed boilers.

Professor Basu has been working in the field of energy conversion and the environment for more than 30 years. Prior to joining the engineering faculty at Dalhousie University (formerly known as the Technical University of Nova Scotia), he worked with both a government research laboratory and a boiler manufacturing company.

Dr. Basu's passion for the transformation of research results into industrial practice is well known, as is his ongoing commitment to spreading advanced knowledge around the world. He has authored more than 200 research papers and 7 monographs in emerging areas of energy and environment, some of which have been translated into Chinese, Italian and Korean. He is well-known internationally for providing expert advices on circulating fluidized-bed boilers and conducting training courses on biomass conversion and fluidized bed boilers to industries and universities across the globe.

Chapter 1

# Introduction

The quest for renewable sustainable energy sources has given biomass a prominence it had lost during the industrial revolution after the discovery of coal. The share of biomass in meeting current world's primary energy mix is at a modest level of 10% (World Energy Council, 2010), but given the rising concern about global warming and sustainability, this share is very likely to rise. The most common use of biomass for energy is direct combustion, followed by gasification, carbonization, and pyrolysis. The production of transportation fuel from biomass through pyrolysis, trans-esterification, fermentation, and gasification-based synthesis is also gaining commercial importance. Carbonization that produces charcoal from biomass was widely practiced for extraction of iron from iron ore in ancient India and China (~4000 BCE). Charcoal is still being used in many parts of the world as a smokeless fuel as well as a medium for filtration of water or gas. Torrefaction (French word for "roasting"), a relatively new biomass conversion option, is similar to carbonization that produces solid fuels from biomass but has some important differences. In any case, this option is also attracting much attention especially in its near term application in co-firing biomass in coal-fired power plants and possibly for replacement of coke in metallurgy. This monograph deals primarily with three biomass conversions—gasification, pyrolysis, and torrefaction—which produce gas, liquid, and solids respectively from biomass.

Gasification is a chemical process that converts carbonaceous materials like biomass into useful convenient gaseous fuels or chemical feedstock. Pyrolysis, partial oxidation, and hydrogenation are related processes. Combustion also converts carbonaceous materials into product gases but with some important differences. For example, the product gas of combustion does not have any useful heating value, but the product gas from gasification does. Gasification packs energy into chemical bonds in the product while combustion releases it. Gasification takes place in reducing (oxygen-deficient) environments requiring heat, whereas combustion takes place in an oxidizing environment releasing heat.

The purpose of gasification or pyrolysis is not just energy conversion; production of chemical feedstock is also an important application. Nowadays, gasification is not restricted to solid hydrocarbons. Its feedstock

**Biomass Gasification, Pyrolysis and Torrefaction.**

includes liquid or even gases to produce more useful fuels. For example, partial oxidation of methane gas is widely used in production of synthetic gas, or *syngas*, which is a mixture of $H_2$ and CO.

Torrefaction (Chapter 4) is gaining prominence due to its attractive use in co-firing biomass (Chapter 10) in existing coal-fired power plants. Pyrolysis (Chapter 5), the pioneering technique behind the production of the first transportable clean liquid fuel *kerosene*, produces liquid fuels from biomass. In recent times, gasification of heavy oil residues into syngas has gained popularity for the production of lighter hydrocarbons. Many large gasification plants are now dedicated to the production of chemical feedstock from coal or other hydrocarbons. Hydrogenation, or hydrogasification, which involves adding hydrogen to the feed to produce fuel with a higher hydrogen-to-carbon (H/C) ratio, is also gaining popularity. Supercritical gasification (Chapter 9), a new option for gasification of very wet biomass, also has much potential.

This chapter introduces the above biomass conversion processes with a short description of the historical developments of gasification, its motivation, and its products. It also gives a brief introduction to the chemical reactions that are involved in important biomass conversion processes.

## 1.1 BIOMASS AND ITS PRODUCTS

Biomass is formed from living species like plants and animals—i.e., anything that is now alive or was alive a short time ago. It is formed as soon as a seed sprouts or an organism is born. Unlike fossil fuels, biomass does not take millions of years to develop. Plants use sunlight through photosynthesis to metabolize atmospheric carbon dioxide and water to grow. Animals in turn grow by taking in food from biomass. Unlike fossil fuels, biomass can reproduce, and for that reason, it is considered *renewable*. This is one of its major attractions as a source of energy or chemicals.

Every year, vast amounts of biomass grow through photosynthesis by absorbing $CO_2$ from the atmosphere. When it burns, it releases the carbon dioxide that the plants had absorbed from the atmosphere only recently (a few years to a few hours). Thus, the burning of biomass does not make any net addition to the earth's carbon dioxide levels. Such release also happens for fossil fuels. So, on a comparative basis, one may consider biomass "*carbon-neutral*," meaning there is no addition to the $CO_2$ inventory by the burning of biomass(see Section 1.3.2.1).

Of the vast amount of biomass in the earth, only 5% (13.5 billion metric tons) can be potentially mobilized to produce energy. Even this amount is large enough to provide about 26% of the world's energy consumption, which is equivalent to 6 billion tons of oil (IFP, 2007).

Biomass covers a wide spectrum: from tiny grass to massive trees, from small insects to large animal wastes, and the products derived from these. The principal types of harvested biomass are *cellulosic* (noncereal), *starch*, and *sugar* (cereal).

All parts of a harvested crop like corn plant are considered biomass, but its fruit (e.g., corn) is mainly starch while the rest of it is cellulosic. The cereal (namely corn) can produce ethanol through fermentation, but the cellulosic part of the corn plant requires a more involved process through gasification or hydrolysis.

Table 1.1 lists the two types of harvested biomass in food and nonfood categories, and indicates the potential conversion products from them. The division is important because the production of transport fuel (ethanol) from cereal, which is relatively easy and more established, is already being pursued commercially on a large scale. The use of such food stock for energy production, however, may not be sustainable as it diverts cereal from the traditional grain market to the energy market, with economic, social, and political consequences. Efforts are thus being made to produce more ethanol from nonfood resources like cellulosic materials so that the world's food supply is not strained by our quest for more energy.

### 1.1.1 Products of Biomass

Three types of primary fuels could be produced from biomass and are as follows:

1. *Liquid fuels* (ethanol, biodiesel, methanol, vegetable oil, and pyrolysis oil).
2. *Gaseous fuels* (biogas ($CH_4$, $CO_2$), producer gas (CO, $H_2$, $CH_4$, $CO_2$, $H_2$), syngas (CO, $H_2$), substitute natural gas ($CH_4$).
3. *Solid fuels* (charcoal, torrefied biomass, biocoke, biochar).

**TABLE 1.1** Sources of Biomass

| | | |
|---|---|---|
| Farm products | Corn, sugarcane, sugar beet, wheat, etc. | Produces ethanol |
| | Rapeseed, soybean, palm sunflower seed, Jatropha, etc. | Produces biodiesel |
| Lignocellulosic materials | Straw or cereal plants, husk, wood, scrap, slash, etc. | Can produce ethanol, bioliquid, and gas |

These biomass products find use in following four major types of industries:

1. Chemical industries for production of methanol, fertilizer, synthetic fiber, and other chemicals.
2. Energy industries for generation of heat and electricity.
3. Transportation industries for production of gasoline and diesel.
4. Environmental industries for capture of $CO_2$ and other pollutants.

The use of ethanol and biodiesel as transport fuels reduces the emission of $CO_2$ per unit of energy production. It also lessens our dependence on fossil fuels. Thus, biomass-based energy is not only renewable but also clean from the standpoint of greenhouse gas (GHG) emission, and so it can take the center stage on the global energy scene. However, this move is not new. Civilization began its energy use by burning biomass. Fossil fuels came much later, around AD 1600. Before the nineteenth century, wood (a biomass) was the primary source of the world's energy supply. Its large-scale use during the early Industrial Revolution caused so much deforestation in England that it affected industrial growth. As a result, from AD 1620 to AD 1720, iron production decreased from 180,000 to 80,000 tons per year (Higman and van der Burgt, 2008, p. 2). This situation changed with the discovery of coal, which began displacing wood for energy as well as for metallurgy.

#### *1.1.1.1 Chemicals Industries*

Theoretically, most chemicals produced from petroleum or natural gas can be produced from biomass as well. The two principal platforms for chemical production are sugar-based and syngas-based. The former involves sugars like glucose, fructose, xylose, arabinose, lactose, sucrose, and starch, while the latter involves CO and $H_2$.

The syngas platform synthesizes the hydrogen and carbon monoxide constituents of syngas into chemical building blocks (Chapter 11). Intermediate building blocks for different chemicals are numerous in this route. They include hydrogen, methanol, glycerol (C3), fumaric acid (C4), xylitol (C5), glucaric acid (C6), and gallic acid (Ar), to name a few (Werpy and Petersen, 2004). These intermediates are synthesized into a large number of chemicals for industries involving transportation, textiles, food, the environment, communications, health, housing, and recreation. Werpy and Petersen (2004) identified 12 intermediate chemical building blocks having the highest potential for commercial products.

#### *1.1.1.2 Energy Industries*

Biomass was probably the first on-demand source of energy that humans exploited. However, less than 22% of our primary energy demand is currently met by biomass or biomass-derived fuels. The position of biomass as a primary source of energy varies widely depending on the geographical and socioeconomic conditions. For example, it constitutes 90% of the primary energy source in Nepal but only 0.1% in the Middle Eastern countries. Cooking, although highly inefficient, is one of the most extensive uses of biomass in lesser-developed countries. Figure 1.1 shows a cooking stove still employed by millions in the rural areas using twigs or logs as fuel. A more efficient modern commercial use of biomass is in the production of steam for processing heat and generating electricity like the facility shown in Figure 1.2.

Heat and electricity are two forms of primary energy derived from biomass. The use of biomass for efficient energy production is presently on the rise in developed countries because of its carbon-neutral feature, while its use for cooking is declining because of a shortage of biomass in lesser-developed countries. Substitution of fossil fuel with biomass in existing plants is made simpler with the developments of the torrefaction process (Chapter 10).

### *1.1.1.3 Transport Industries*

Diesel and gasoline from crude petro-oil are widely used in modern transportation industries. Biomass can help substitute such petro-derived transport fuels with carbon-neutral alternatives. Ethanol, produced generally from sugarcane and corn, is used in gasoline (spark-ignition) engines, while biodiesel, produced

**FIGURE 1.1** A cooking stove using fire logs.

from vegetable oils such as rapeseed, is used in diesel (compression–ignition) engines.

Pyrolysis, fermentation, and mechanical extraction are three major means of production of transport fuel from biomass. Of these, the most widely used commercial method is fermentation, where sugar (sugarcane) or starch (corn) produces ethanol. The yeast helps ferment sugar or starch into ethanol and carbon dioxide. The production and refining of market grade ethanol, however, take a large amount of energy.

The mechanical means of extraction of vegetable oil from seeds like rapeseed has been practiced for thousands of years. Presently, oils like canola oil are refined with alcohol (trans-esterification) to produce methyl ester or biodiesel.

Liquid fuel may also be produced through pyrolysis that involves rapid heating of biomass in absence of air. The liquid product of pyrolysis is a precursor of bio-oil, which may be hydro-treated to produce "green diesel" or "green gasoline." At this time, ethanol and biodiesel dominate the world's biofuel market.

Gasification and anaerobic digestion can produce methane gas from biomass. Methane gas can then be used directly in some spark-ignition engines for transportation or converted into gasoline through methanol.

### *1.1.1.4 Environmental Industries*

Activated charcoal produced from biomass has major application in the pollution control industries. One of its extensive uses is in water filter. Activated charcoal impregnated with suitable chemicals like zinc chloride is very effective in removing mercury from flue gas from coal-fired power

**FIGURE 1.2** A modern fluidized-bed boiler firing varieties of biomass plant in Canada.

plants (Zeng et al., 2004). Biochar produced from biomass provides viable and less expensive means of sequestration of carbon dioxide. Biochar, produced through pyrolysis, can provide long-term sink for storage as atmospheric carbon dioxide in terrestrial ecosystems. Besides this it also helps in soil fertility and increased crop production (Lehmann et al., 2006). Thus biochar can retain the carbon naturally buried in ground instead of releasing it as $CO_2$ to the atmosphere. The potential annual biochar production from agricultural waste materials such as forest residues and urban wastes is 0.162 Pg/year (Lehmann et al., 2006). Life cycle analysis for stover and yard waste shows a negative $CO_2$ emission exceeding 800 kg/$CO_2$ equivalent per ton of dry feedstock (Roberts et al., 2010).

## 1.2 BIOMASS CONVERSION

Bulkiness, low energy density, and inconvenient form of biomass are major barriers to a rapid transition from fossil to biomass fuels. Unlike gas or liquid, biomass cannot be handled, stored, or transported easily. This provides a major motivation for the conversion of solid biomass into liquid and gaseous fuels, which are more energy dense and can be handled and stored with relative ease. This conversion can be achieved through one of two major routes (Figure 1.3): biochemical conversion (fermentation) and thermochemical conversion (pyrolysis, gasification). The inconvenience of bulkiness and other shortcomings of solid biomass are overcome to some extent through the production of more convenient cleaner solid fuel through carbonization and torrefaction.

Biochemical conversion is perhaps the most ancient means of biomass gasification. India and China produced methane gas for local energy needs by anaerobic microbial digestion of animal wastes. In modern times, most of the ethanol for automotive fuels is produced from corn using fermentation. Thermochemical conversion of biomass into gases came much later. Large-scale use of small biomass gasifiers began during the Second World War, when more than a million units were in use (Figure 1.4).

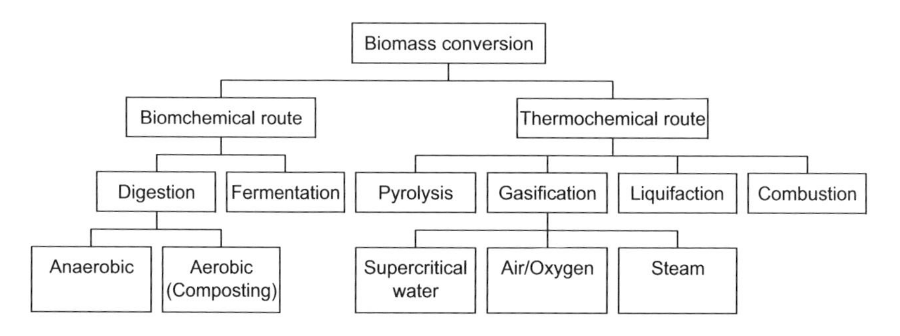


**FIGURE 1.3** Different options for conversion of biomass into fuel gases or chemicals.

Bus, Germany

**FIGURE 1.4** Bus with on-board gasifier during Second World War. Source*: http://www.wood-gas.com/history.htm.*

A brief description of the biochemical and thermochemical routes of biomass conversion is presented in the following sections.

### 1.2.1 Biochemical Conversion

In biochemical conversion, biomass molecules are broken down into smaller molecules by bacteria or enzymes (Figure 1.5). This process is much slower than thermochemical conversion process but does not require much external energy. The three principal routes for biochemical conversion are as follows:

1. Digestion (anaerobic and aerobic)
2. Fermentation
3. Enzymatic or acid hydrolysis.

The main products of anaerobic digestion are methane and carbon dioxide in addition to a solid residue. Bacteria take oxygen from the biomass itself instead of from ambient air.

Aerobic digestion, or composting, is also a biochemical breakdown of biomass, except that it takes place in the presence of oxygen. It uses different types of microorganisms that access oxygen from the air, producing carbon dioxide, heat, and a solid digestate.

In fermentation, part of the biomass is converted into sugars using acids or enzymes. The sugar is then converted into ethanol or other chemicals with the help of yeast. The lignin is not converted and is left either for combustion or for thermochemical conversion into chemicals. Unlike anaerobic digestion, the product of fermentation is liquid.

Fermentation of starch- and sugar-based feedstock (e.g., corn and sugarcane) into ethanol (Figure 1.5A) is a fully commercial process, but this is not

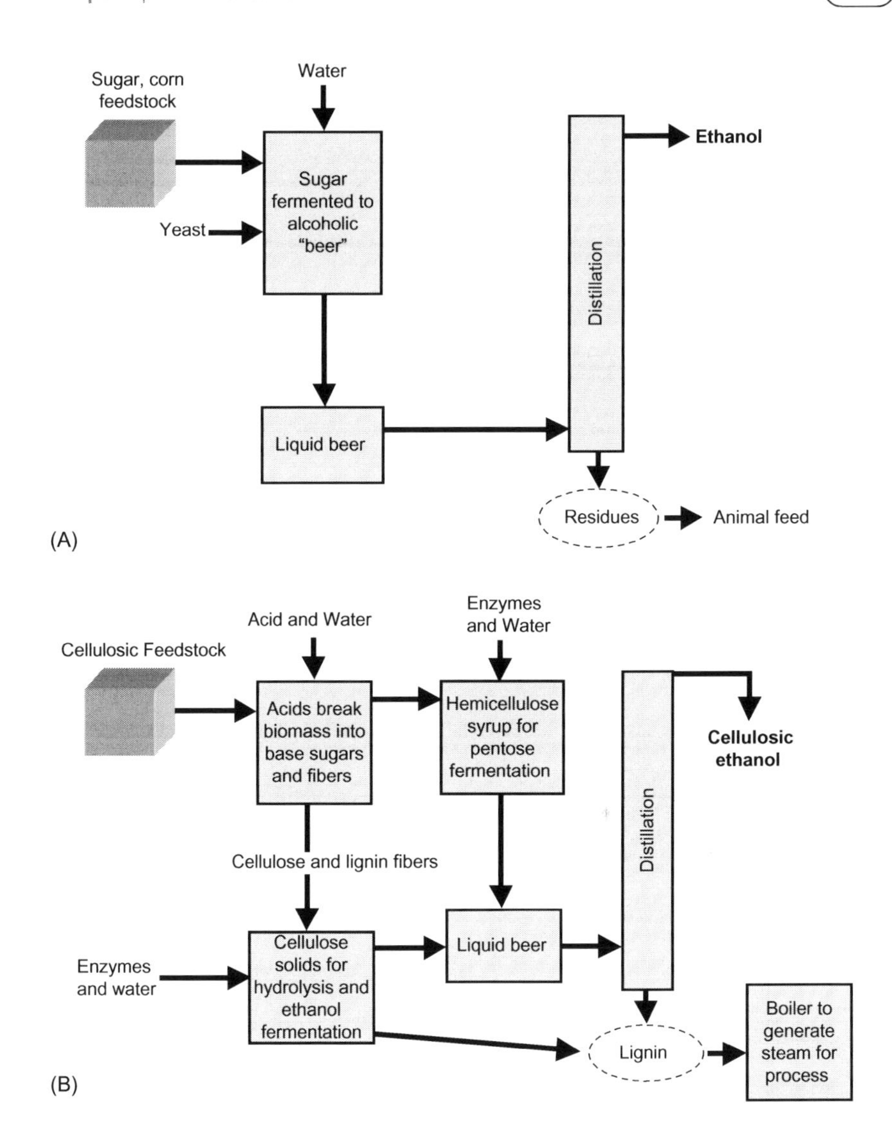


**FIGURE 1.5** Two biochemical routes for production of ethanol from sugar (noncellulosic) and cellulosic biomass: (A) Conversion of food-feedstock into ethanol and (B) conversion of cellulosic feedstock into ethanol.

the case with cellulosic biomass feedstock because of the expense and difficulty in breaking down (hydrolyzing) the materials into fermentable sugars. Lignocellulosic feedstock, like bagasse, requires hydrolysis pretreatment (acid, enzymatic, or hydrothermal) to break down the cellulose and hemicellulose into simple sugars needed by the yeast and bacteria for the fermentation process (Figure 1.5B). Acid hydrolysis technology is more mature than enzymatic hydrolysis technology though the latter could have a significant cost advantage.

## 1.2.2 Thermochemical Conversion

In thermochemical conversion, the entire biomass is converted into gases, which are then synthesized into the desired chemicals or used directly (Figure 1.6). The Fischer–Tropsch synthesis of syngas into liquid transport fuels is an example of thermochemical conversion. Production of thermal energy is the main driver for this conversion route that has five broad pathways:

1. Combustion
2. Carbonization/torrefaction
3. Pyrolysis
4. Gasification
5. Liquefaction.

Table 1.2 compares the above five thermochemical paths for biomass conversion. It also gives the typical range of their reaction temperatures.

Combustion involves high-temperature exothermic oxidation (in oxygen-rich ambience) to hot flue gas. Carbonization covers a broad range of processes by which the carbon content of organic materials is increased through thermochemical decomposition. In a more restrictive sense for biomass, it is a process for production of charcoal from biomass by slowly heating it to the carbonization temperature (500–900°C) in an oxygen-starved atmosphere. Torrefaction is a related process where biomass is heated instead to a lower temperature range of 200–300°C without or little contact with oxygen.

Unlike combustion, gasification involves chemical reactions in an oxygen-deficient environment producing product gases with heating values.

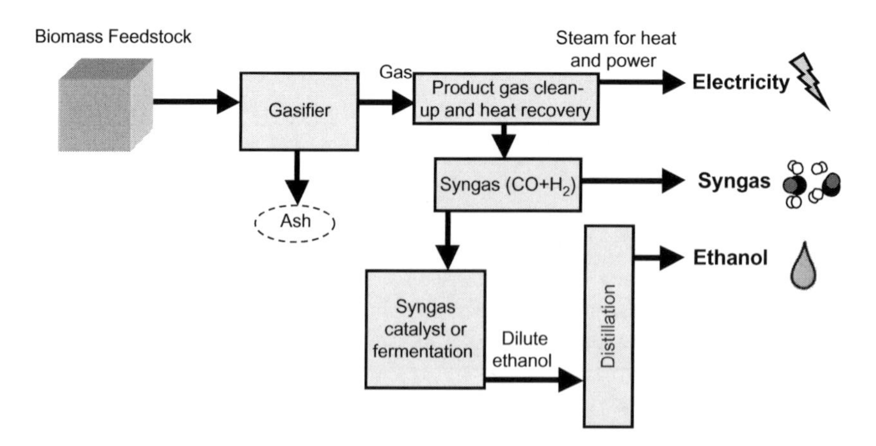


**FIGURE 1.6** Thermochemical route for production of energy, gas and ethanol.

Pyrolysis involves rapid heating in the total absence of oxygen. In liquefaction, the large molecules of solid feedstock are decomposed into liquids having smaller molecules. This occurs in the presence of a catalyst and at a still lower temperature.

Table 1.3 compares basic features of thermochemical and biochemical routes for biomass conversion. It gives that the biochemical route for ethanol production is commercially more developed than the thermochemical route, but the former requires sugar or starch for feedstock; it cannot use more plentiful lignocellulosic stuff. As a result, a larger fraction of the available biomass is not converted into ethanol. For example, in a corn plant, only the kernel is utilized for ethanol production. The stover, stalk, roots, and leaves, which constitute bulk of the corn plant, are left as wastes as being lignocellulosic. Even though the enzymatic or biochemical route is more developed, this is a batch process and takes an order of magnitude longer to complete than the thermochemical process.

In the thermochemical route (Figure 1.5), the biomass is first converted into syngas, which is then converted into ethanol through synthesis or some other means.

#### *1.2.2.1 Combustion*

Combustion represents perhaps the oldest means of utilization of biomass, given that civilization began with the discovery of fire. The burning of forest wood taught humans how to cook and how to keep themselves warm. Chemically, combustion is an exothermic reaction between oxygen and hydrocarbon in biomass. Here, the biomass is oxidized into two major

**TABLE 1.2** **Comparison of Some Major Thermochemical Conversion Processes**

| Process | Temperature (°C) | Pressure (MPa) | Catalyst | Drying |
|---|---|---|---|---|
| Liquefaction | 250–330 | 5–20 | Essential | Not required |
| Pyrolysis | 300–600 | 0.1–0.5 | Not required | Necessary |
| Combustion | 700–1400 | $\geq 0.1$ | Not required | Not essential, but may help |
| Gasification | 500–1300 | $\geq 0.1$ | Not essential | Necessary |
| Torrefaction | 200–300 | 0.1 | Not required | Necessary |

**Source**: Modified from Demirbas (2009).

**TABLE 1.3 Comparison of Biochemical and Thermochemical Routes for Biomass Conversion into Ethanol**

| | Biochemical Route (Sugar Fermentation) | Thermochemical Route |
|---|---|---|
| Feedstock | Sugarcane/starch/corn | Cellulosic stock/wood/ MSW |
| Reactor type | Batch | Continuous |
| Reaction time | Days | Minutes |
| Water usage | 3.5–170 L/L ethanol | <1 L/L ethanol |
| By-products | Distiller's dried grain | Syngas/electricity |
| Yield* | 400 L/ton | ~265–492 L/ton |
| Technology maturity | >100 plants in the United States | Pilot plant |

**Liska et al. (2009).*

stable compounds, $H_2O$ and $CO_2$. The reaction heat released is presently the largest source of human energy consumption, accounting for more than 90% of the energy from biomass.

Heat and electricity are two principal forms of energy derived from biomass. Biomass still provides heat for cooking and warmth, especially in rural areas. District or industrial heating is also produced by steam generated in biomass-fired boilers. Pellet stoves and log-fired fireplaces are a direct source of warmth in many cold-climate countries. Electricity, the foundation of all modern economic activities, may be generated from biomass combustion. The most common practice involves the generation of steam by burning biomass in a boiler and the generation of electricity through a steam turbine.

Biomass is used either as a stand-alone fuel or as a supplement to fossil fuels in a boiler. The latter option is becoming increasingly common as the fastest and least-expensive means for decreasing the emission of carbon dioxide from an existing fossil fuel plant (Basu et al., 2011). This option, called co-combustion or co-firing, is discussed in more detail in Chapter 10.

### *1.2.2.2 Pyrolysis*

Unlike combustion, pyrolysis takes place in the total absence of oxygen, except in cases where partial combustion is allowed to provide the thermal energy needed for this process. This process thermally decomposes biomass into gas, liquid, and solid by rapidly heating biomass above 300–400°C.

In pyrolysis, large hydrocarbon molecules of biomass are broken down into smaller molecules. Fast pyrolysis produces mainly liquid fuel, known as bio-oil, whereas slow pyrolysis produces some gas and solid charcoal (one of the most ancient fuels, used for heating and metal extraction before the discovery of coal). Pyrolysis is promising for conversion of waste biomass into useful liquid fuels. Unlike combustion, it is not exothermic.

#### *1.2.2.3 Torrefaction*

Torrefaction is being considered for effective utilization of biomass as a clean and convenient solid fuel. In this process, the biomass is slowly heated to 200–300°C without or little contact with oxygen. Torrefaction alters the chemical structure of biomass hydrocarbon to increase its carbon content while reducing its oxygen. Torrefaction also increases the energy density of the biomass and makes the biomass hygroscopic. These attributes thus enhance the commercial value of wood for energy production and transportation.

#### *1.2.2.4 Gasification*

Gasification converts fossil or nonfossil fuels (solid, liquid, or gaseous) into useful gases and chemicals. It requires a medium for reaction, which can be gas or supercritical water (not to be confused with ordinary water at subcritical condition). Gaseous mediums include air, oxygen, subcritical steam, or a mixture of these.

Presently, gasification of fossil fuels is more common than that of nonfossil fuels like biomass for production of synthetic gases. It essentially converts a potential fuel from one form to another. There are several major motivations for such a transformation and are as follows:

- To increase the heating value of the fuel by rejecting noncombustible components like nitrogen and water.
- To strip the fuel gas of sulfur such that it is not released into the atmosphere when the gas is burnt.
- To increase the H/C mass ratio in the fuel.
- To reduce the oxygen content of the fuel.

In general, the higher the hydrogen content of a fuel, the lower the vaporization temperature and the higher the probability of the fuel being in a gaseous state. Gasification or pyrolysis increases the relative hydrogen content (H/C ratio) in the product through one the following means:

1. *Direct*: Direct exposure to hydrogen at high pressure.
2. *Indirect*: Exposure to steam at high temperature and pressure, where hydrogen, an intermediate product, is added to the product. This process also includes steam reforming.
3. *Pyrolysis or devolatilization*: Reduction in carbon content by rejecting it via solid char or $CO_2$ gas.

Gasification of biomass also removes oxygen from the fuel to increase its energy density. For example, a typical biomass has about 40% oxygen by weight, but a fuel gas contains negligible amount of oxygen (Table 1.4). The oxygen is removed from the biomass by either dehydration (Eq. (1.1)) or decarboxylation (Eq. (1.2)) reactions. The latter reaction (Eq. (1.2)) while rejecting the oxygen through $CO_2$ also rejects carbon and thereby increasing the H/C ratio of the fuel. A positive benefit of the gasification product is that it emits less GHG when combusted:

Dehydration:

$$C_mH_nO_q \rightarrow C_mH_{n-2q} + qH_2O \quad O_2 \text{ removal through } H_2O \tag{1.1}$$

Decarboxylation:

$$C_mH_nO_q \rightarrow C_{m-q/2}\,H_n + qCO_2 \quad O_2 \text{ removal through } CO_2 \tag{1.2}$$

Hydrogen, when required in bulk for the production of ammonia, is produced from natural gas (mainly contains $CH_4$) through steam reforming, which produces syngas (a mixture of $H_2$ and CO). The CO in syngas is indirectly hydrogenated by steam to produce methanol ($CH_3OH$), an important feedstock for a large number of chemicals. These processes, however, use natural gas that is nonrenewable and is responsible for net addition of carbon

**TABLE 1.4** Carbon-to-Hydrogen (C/H) Ratio of Different Fuels

| Fuel | C/H Mass Ratio (−)[a] | Oxygen (%) | Energy Density (MJ/kg)[b] |
|---|---|---|---|
| Anthracite | ~44 | ~2.3 | ~27.6 |
| Bituminous coal | ~15 | ~7.8 | ~29 |
| Lignite | ~10 | ~11 | ~9 |
| Peat | ~10 | ~35 | ~7 |
| Crude oil | ~9 | | 42 (mineral oil) |
| Biomass/cedar | 7.6 | ~40 | ~20 |
| Gasoline | 6 | 0 | ~46.8 |
| Natural gas (~$CH_4$) | 3 | Negligible | 56 (Liquefied natural gas) |
| Syngas (CO: $H_2$ = 1:3) | 2 | Negligible | 24 |

[a]*Probstein and Hicks (2006).*
[b]*McKendry (2002).*

dioxide (a major GHG) to the atmosphere. Biomass could, on the other hand, substitute fossil hydrocarbons either as a fuel or as a chemical feedstock.

Gasification of biomass into CO and $H_2$ provides a good base for production of liquid transportation fuels, such as gasoline, and synthetic chemicals, such as methanol. It also produces methane, which can be burned directly for energy production.

#### *1.2.2.5 Liquefaction*

Liquefaction of solid biomass into liquid fuel can be done through pyrolysis, gasification, and through hydrothermal process. In the latter process, biomass is converted into an oily liquid by contacting the biomass with water at elevated temperatures (300–350°C) and high pressure (12–20 MPa) for a period of time. There are several other means including the supercritical water process (Chapter 9) for direct liquefaction of biomass. Behrendt et al. (2008) presented a review of these processes.

## 1.3 MOTIVATION FOR BIOMASS CONVERSION

Biomass conversion especially into heat and light is as ancient as human civilization. Discovery of fire from wood started the scientific development of human race that set it apart from other creatures. Its use waned due to the availability of more energy dense and convenient fossil fuels like coal and oil. However, there has been a recent surge of interest in conversion of biomass into gas or liquid. It is motivated mainly by following three factors:

1. Renewability benefits
2. Environmental benefits
3. Sociopolitical benefits.

A brief description of these benefits is given in the following sections.

### 1.3.1 Renewability Benefits

Fossil fuels like coal, oil, and gas are practical convenient sources of energy, and they meet the energy demands of society very effectively. However, there is one major problem: fossil fuel resources are finite and not renewable. Biomass, on the other hand, grows and is hence renewable. A crop cut this year could grow again next year; a tree cut today may grow up within a decade through fresh growth. Unlike fossil fuels, the biomass is not likely to be depleted with consumption. For this reason, its use is sustainable, and this feature is contributing to the growing interest in biomass use especially for energy production.

We may argue against cutting trees for energy supply because they serve as a $CO_2$ sink. This is true, but a tree stops absorbing $CO_2$ after it stops growing or dies. On the other hand, if left alone on the forest floor it can release $CO_2$ through natural degradation or in a forest fire. Furthermore, a dead tree could release more harmful $CH_4$ if it decomposes in water. The use of a tree as fuel provides carbon-neutral energy while avoiding methane gas release from decomposed deadwood. Careless use of trees for energy, however, could spell environmental disaster. But a managed utilization with fresh planting of trees following cutting, as is done by some pulp industries, could sustain its use for energy in an environment-friendly way. Energy plantation with fast-growing plants like *Switchgrass* and *Miscanthus* are being considered as fuel for new energy projects. These plants have very short growing periods that can be counted in months.

## 1.3.2 Environmental Benefits

With growing evidence of global warming, the dire need to reduce human-made GHG emissions is being recognized. Also, emission of other air pollutants, such as NO, $SO_2$, and Hg, is no longer acceptable. From elementary schools to corporate boardrooms, environment is a major issue, and it has been a major driver for biomass use for energy production. Biomass has a special appeal in this regard because, as explained below, it makes no net contribution of carbon dioxide to the atmosphere. Regulations for making biomass economically viable are in place in many countries. For example, if biomass replaces fossil fuel in a power plant, that plant could earn credits for $CO_2$ reduction equivalent to what the fossil fuel was emitting. These credits can be sold on the market for additional revenue in countries where such trades are in practice.

### *1.3.2.1 Carbon-Neutral Feature of Biomass*

When burned, biomass releases the $CO_2$ it absorbed from the atmosphere in the recent past, not millions of years ago, as is the case for fossil fuel. The net addition of $CO_2$ to the atmosphere through biomass combustion is thus considered to be zero. For this reason, biomass is considered a *carbon-neutral fuel*. One may, however, argue that $CO_2$ is released for harvesting, transporting, processing biomass, but that indirect emission is common for fossil fuels which has emissions from mining, transporting, and preparation of fossil fuel. A life cycle analysis that compares release of $CO_2$ from all direct and indirect actions shows that biomass is a clear winner over fossil fuel in this respect.

Even if one leaves aside the carbon-neutral aspect of biomass, the carbon intensity (amount of $CO_2$ released per unit energy production, g/kWhe) of biomass (35–49 g/kWhe) is much lower than that of fossil fuels like coal as the former is a low C/H ratio fuel (Weisser, D, 2007).

The $CO_2$ emission from gasification-based power plants is slightly less than that from a combustion power plant on a unit heat release basis. For example, emission from an integrated gasification combined cycle (IGCC) plant is 745 g/kWh compared to 770 g/kWh from a combustion-based subcritical pulverized coal (PC) plant (Termuehlen and Emsperger, 2003, p. 23).

Sequestration of $CO_2$ could become an important requirement for new power plants. On that note, a gasification-based power plant has an advantage over a conventional combustion-based PC power plant because $CO_2$ is more concentrated in the flue gas from an IGCC plant making it easier to sequestrate than that from a conventional PC plant where $CO_2$ is diluted with nitrogen. Table 1.5 compares the $CO_2$ emissions from different electricity-generation technologies.

Biochar produced from pyrolysis of biomass offers a new alternative to carbon capture and sequestration (CCS) (see Section 5.8).

### *1.3.2.2 Sulfur Removal*

Most virgin or fresh biomass contains little to no sulfur. Biomass-derived feedstock such as municipal solid waste (MSW) or sewage sludge contains sulfur, which requires limestone for capture. Interestingly, such derived feedstock often contains some amounts of calcium, which intrinsically aids sulfur capture.

Gasification of coal or oil has an edge over combustion in certain situations. In combustion systems, sulfur in the fuel appears as $SO_2$, which is relatively difficult to remove from the flue gas without adding an external sorbent. In a typical gasification process, 93–96% of the sulfur appears as $H_2S$ with the remaining as COS (Higman and van der Burgt, 2008, p. 351). One can easily extract sulfur from $H_2S$ by absorption. The extracted elemental sulfur in a gasification plant is a valuable by-product.

**TABLE 1.5** Comparison of Emissions and Water Use for Electricity Generation from Coal Using Two Technologies

| | PC Combustion | Gasification (IGCC) |
|---|---|---|
| $CO_2$ (kg/1000 MWh) | 0.77 | 0.68 |
| Water use (L/1000 MWh) | 4.62 | 2.84 |
| $SO_2$ emission (kg/MWh) | 0.68 | 0.045 |
| $NO_x$ emission (kg/MWh) | 0.61 | 0.082 |
| Total solids (kg/100 MWh) | 0.98 | 0.34 |

**Source**: Recompiled from graphs by Stiegel (2005).

#### *1.3.2.3 Nitrogen Removal*

A combustion system firing fossil fuels can oxidize the nitrogen in fuel and combustion air into NO, the acid rain precursor, or into $N_2O$, a GHG. Both oxides are difficult to remove. In a gasification system, on the other hand, nitrogen appears as either $N_2$ or $NH_3$, which is removed relatively easily in the syngas-cleaning stage.

Nitrous oxide emission results from the oxidation of fuel nitrogen alone. Measurement in a biomass combustion system showed a relatively low level of $N_2O$ emission (Van Loo and Koppejan, 2008, p. 295).

#### *1.3.2.4 Dust and Hazardous Gases*

Some speculate that highly toxic pollutants like dioxin and furan, which can be released in a combustion system, are not likely to form in an oxygen-starved gasifier. Particulate in the syngas is also reduced significantly by multi stage gas cleanup systems, that include primary cyclone, scrubbers, gas cooling, and acid gas-removal units. Together with these a gasification system reduces the particulate emissions by one to two orders of magnitude (Rezaiyan and Cheremisinoff, 2005).

### 1.3.3 Sociopolitical Benefits

The sociopolitical benefits of biomass use are substantial. For one, biomass is a locally grown resource. For a biomass-based power plant to be economically viable, the biomass needs to come from within a limited radius from the power plant. This means that every biomass plant can prompt the development of associated industries for biomass growing, collecting, and transporting. Some believe that a biomass fuel plant could create up to 20 times more employment locally than that by a coal- or oil-based plant (Van Loo and Koppejan, 2008, p. 1). The biomass industry thus has a positive impact on the local economy.

Another very important aspect of biomass-based energy, fuel, or chemicals is that they reduce reliance on imported fossil fuels giving a country added benefit of energy independence. The global political landscape being volatile has shown that the supply and price of fossil fuel can change dramatically within a short time, with a sharp rise in the price of feedstock. Locally grown biomass is relatively free from such uncertainties.

## 1.4 HISTORICAL BACKGROUND

The conversion of biomass into charcoal was perhaps the first large-scale application of biomass conversion process. It has been used in India, China and in the preindustrial era of Europe for extraction of iron from iron ore. Figure 5.2 shows a typical beehive oven used in early times to produce charcoal using the carbonization process. This practice continued until wood, owing to its overuse, became scarce at the beginning of the eighteenth century. Fortunately, coal was then discovered and coke was produced from coal through pyrolysis. This replaced charcoal for iron extraction.

Gasification was the next major development. Figure 1.7 shows some of the important milestones in the progression of gasification. Early developments of gasification were inspired primarily by the need for town gas for street lighting. Thomas Shirley probably performed the earliest investigation into gasification in 1659. He experimented with "carbureted hydrogen" (now called methane). The salient features of town gas from coal were demonstrated to the British Royal Society in 1733, but the scientists of the time saw no use for it. In 1798, William Murdoch used coal-gas (also known as town gas) to light the main building of the Soho Foundry, and in 1802 he presented a public display of gas lighting astonishing the local population. Friedrich Winzer of Germany patented coal-gas lighting in 1804 (www.absoluteastronomy.com/topics/coalgas).

By 1823, numerous towns and cities throughout Britain were gas-lit. At the time, the cost of gaslight was 75% less than that for oil lamps or candles, and this helped accelerate its development and deployment. By 1859, gas lighting had spread throughout Britain. It came to the United States probably in 1816, with Baltimore being the first city to use it (http://www.bge.com/aboutbge/pages/history).

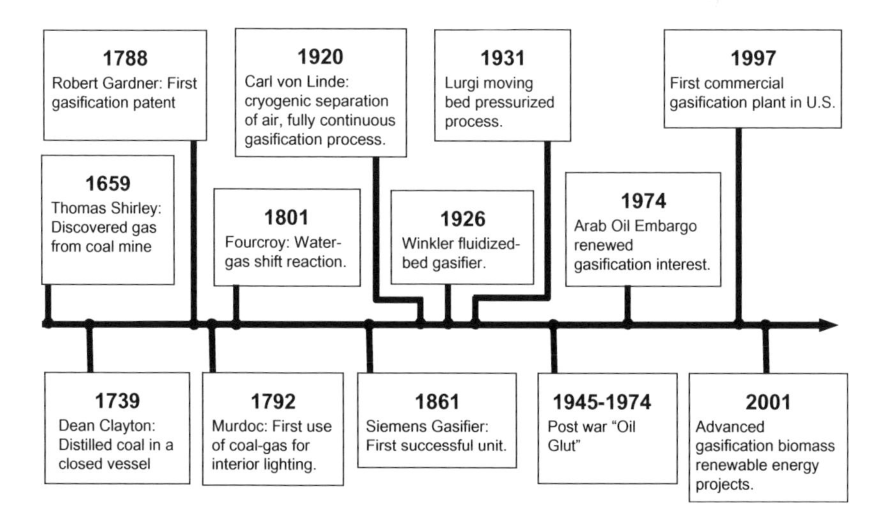


**FIGURE 1.7** Historical milestones of gasification development.

The history of gasification may be divided into four periods and are as described as follows:

*1850–1940*: During this early stage, the gas made from coal was used mainly for lighting homes and streets and for heating. Lighting helped along the Industrial Revolution by extending working hours in factories, especially on short winter days. The invention of the electric bulb *ca.* 1900 reduced the need of gas for lighting, but its use for heating and cooking continued. All major commercial gasification technologies (Winkler's fluidized-bed gasifier in 1926, Lurgi's pressurized moving-bed gasifier in 1931, and Koppers-Totzek's entrained-flow gasifier) made their debut during this period. With the discovery of natural gas, the need for gasification of coal or biomass declined.

*1940–1975*: This period saw gasification enter two fields of application as synthetic fuels: internal combustion engine and chemical synthesis into oil and other process chemicals. In the Second World War, Allied bombing of Nazi oil refineries and oil supply routes greatly diminished the crude oil supply that fueled Germany's massive war machinery. This forced Germany to synthesize oil from coal-gas using the Fischer–Tropsch (see Eq. (1.13)) and Bergius processes [$nC + (n+1)H_2 \rightarrow C_nH_{2n+2}$]. Chemicals and aviation fuels were also produced from coal.

During the Second World War, many cars and trucks in Europe operated on coal or biomass gasified in onboard gasifiers (Figure 1.4). During this period, over a million small gasifiers were built primarily for transportation. The end of the Second World War and the availability of abundant oil from the Middle East eliminated the need for gasification for transportation and chemical production.

The advent of plentiful natural gas in the 1950s dampened the development of coal or biomass gasification, but syngas production from natural gas and naphtha by steam reforming increased, especially to meet the growing demand for fertilizer.

*1975–2000*: The third phase in the history of gasification began after the Yom Kippur War, which triggered the 1973 oil embargo. On October 15, 1973, members of the Organization of Petroleum Exporting Countries (OPEC) banned oil exports to the United States and other western countries, which were at that time heavily reliant on oil from the Middle East. This shocked the western economy and gave a strong impetus to the development of alternative technologies like gasification in order to reduce dependence on imported oil.

Besides providing gas for heating, gasification found major commercial use in chemical feedstock production, which traditionally came from petroleum. The subsequent drop in oil price, however, dampened this push for gasification, but some governments, recognizing the need for a cleaner environment, gave support to large-scale development of integrated gasification combined cycle (IGCC) power plants.

*Post-2000*: Global warming and political instability in some oil-producing countries gave a fresh momentum to gasification and pyrolysis. The threat of climate change stressed the need for moving away from carbon-rich fossil fuels. Gasification came out as a natural choice for conversion of renewable carbon-neutral biomass into gas and torrefied biomass as an option for replacing coal in power plants.

The quest for energy independence and the rapid increase in crude oil prices prompted some countries to recognize the need for development of IGCC plants. The attractiveness of gasification for extraction of valuable feedstock from refinery residue was rediscovered, leading to the development of some major gasification plants in oil refineries. In fact, chemical feedstock preparation took a larger share of the gasification market than energy production.

A brief review of historical development of the pyrolysis process is given in Section 5.1.1.

## 1.5 COMMERCIAL ATTRACTION OF GASIFICATION

Gasification is a promising and important means of biomass conversion. A major attraction of gasification is that it can convert waste or low-priced fuels, as well as biomass, coal, and petcoke, into high-value chemicals like methanol. Biomass holds great appeal for industries and businesses, especially in the energy sector. For example:

1. Flue-gas cleaning downstream of a gasification plant is less expensive than that in a coal-fired plant.
2. Polygeneration is a unique feature of a gasifier plant. It can deliver steam for process, electricity for grid, and gas for synthesis, thereby providing a good product mix. Additionally, a gasifier plant produces elemental sulfur as a by-product for high-sulfur fuel.
3. For power generation, an IGCC plant can achieve a higher overall efficiency (38–41%) than a combustion-based Rankin cycle plant with a steam turbine.
4. An IGCC plant can capture and store $CO_2$ (CCS) at one-half of what it costs in a traditional PC plant (www.gasification.org). Other applications of gasification that produce transport fuel or chemicals may have even lower cost for CCS.
5. A process plant that uses natural gas as feedstock can use locally available biomass or organic waste and gasify the instead, and thereby reduce dependence on imported natural gas, which is known for exceptionally high supply and price volatility.

6. Total water consumption in a gasification-based power plant is much lower than that in a conventional power plant (Table 1.5), and such a plant can be designed to recycle its process water. For this reason, all zero-emission plants use gasification technology.
7. Gasification plants produce significantly low amounts of major air pollutants like $SO_2$, $NO_x$, and particulates. Figure 1.8 compares the emission from a coal-based IGCC plant with that from a combustion-based coal-fired steam power plant and a natural-gas-fired plant. It shows that emissions from the gasification plant are similar to those from a natural-gas-fired plant.
8. An IGCC plant produces lower $CO_2$ per MWh than that by a combustion-based steam power plant.

### 1.5.1 Comparison of Gasification and Combustion

For heat or power production, the obvious question is why a solid fuel should be gasified and then the product gas is burnt for heat, losing some part of its energy content in the process. Does it not make more sense to directly burn the fuel to produce heat? The following comparison may answer that question. The comparison is based in part on an IGCC plant and a combustion-based steam power plant, both generating electricity with coal as the fuel.

1. For a given energy throughput, the amount of flue gas obtained from gasification is less than that obtained from a direct combustion system. The lower amount of gas requires smaller equipment and hence results in lower overall costs.

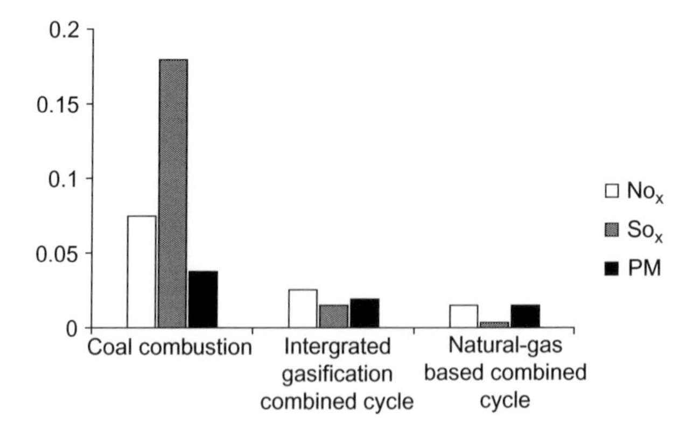


**FIGURE 1.8** Comparison emission of pollutants from power plants using coal-based steam power plant, IGCC plant and natural-gas based combined cycle plant. (Drawn in an arbitrary scale for comparison.)

2. A gasified fuel can be used in a wider range of application than solid fuel. For example, sensitive industrial processes such as glass blowing and drying cannot use dust-laden flue gas from combustion of coal or biomass, but they can use heat from the cleaner and more controllable combustion of gas produced through gasification.
3. Gas can be more easily carried and distributed than a solid fuel. Transportation of synthetic gas, or the liquid fuel produced from it, is less expensive as well as less energy intensive than transportation of solid fuel for combustion.
4. The concentration of $CO_2$ in the flue gas of a gasification-based plant is considerably higher than that of a combustion-based plant, so it is less expensive to separate and sequestrate the $CO_2$ in an IGCC plant.
5. $SO_2$ emissions are generally lower in an IGCC plant (Table 1.5). Sulfur in a gasification plant appears as $H_2S$ and COS, which can be easily converted into saleable elemental sulfur or $H_2SO_4$. In a combustion system, sulfur appears as $SO_2$, which needs a scrubber producing ash-mixed $CaSO_4$, which has less market potential.
6. Gasification produces less $NO_x$ per unit energy output than does a combustion system (Table 1.5). In gasification, nitrogen can appear as $NH_3$, which washes out with water and as such does not need a selective catalytic reducer (SCR) to meet statutory limits. A PC combustion plant, on the other hand, requires expensive SCR for this purpose.
7. The total solid waste generated in an IGCC plant is lower than that generated in a comparable combustion system (Table 1.5). Furthermore, the ash in a slagging entrained-flow gasifier appears as glassy melt, which is much easier to dispose of than the dry fly ash of a PC system.
8. For generation of electricity in a small remote location or for distributed power generation, a power pack comprising a gasifier and a compression–ignition engine is more convenient and economic than a combustion system comprising a boiler, a steam engine, and a condenser.
9. The producer gas from a gasifier can be used as a feedstock for the production of fertilizer, methanol, and gasoline. A gasification-based energy system has the option of producing value-added chemicals as a side stream. This polygeneration feature is not available in direct combustion.
10. If heat is the only product that is desired, combustion seems preferable, especially in small-scale plants. Even for a medium-capacity unit such as for district heating, central heating, and power, combustion may be more economical.
11. A gasification based system can generate power using a combined cycle (IGCC). Example 1.1 shows that such a system is more efficient than a combustion based one.

**Example 1.1**

Compare the thermodynamic efficiency of electricity generation from biomass through the following two routes:

1. Biomass is combusted in a boiler with 95% thermal efficiency (on lower heating value (LHV) basis) to generate steam, which expands in a steam turbine from 600°C to 100°C driving an electrical generator.
2. Biomass is gasified with a cold gas efficiency of 85% and the product gas is burnt to produce hot flue gas at 1200°C, which expands in a gas turbine to 600°C. Waste gas from the gas turbine enters a heat recovery steam generator that produces steam at 400°C. This steam expands to 100°C in a steam turbine.

Both turbines are connected to electricity generators. Neglect losses in the generators.

**Solution**

*For the steam power plant*:

Given:

Boiler thermal efficiency on LHV: $\eta_b = 0.95$

Inlet steam temperature: $T_1 = 600°C = 873$ K

Exhaust steam temperature: $T_2 = 100°C = 373$ K

We assume the turbine to be an ideal heat engine, operating on a Carnot cycle. So, ideal steam turbine efficiency, $\eta_{st} = 1 - ((T_2/T_1)) = 1 - (373/873) = 0.573$.

The overall efficiency of the first route, $\eta_{sc}$ is the combination of the boiler and the turbine.

$$\eta_{sc} = \eta_b \times \eta_{st} = 0.95 \times 0.573 = 0.544\% \text{ or } 54.4\%$$

*For the IGCC plant*:

Given:

Gasification efficiency: $\eta_g = 0.85$

Inlet gas temperature: $T_{g1} = 1200°C = 1473$ K

Exhaust gas temperature: $T_{g2} = 600°C = 873$ K

Ideal gas turbine efficiency, $\eta_{gt} = 1 - ((T_{g2}/T_{g1})) = 1 - (873/1473) = 0.407$.

Considering the cold gas efficiency of the gasifier efficiency

Gas turbine plant efficiency, $\eta_{gtp} = \eta_g \times \eta_{gt} = 0.85 \times 0.407 = 0.346$.

The downstream steam turbine is also a Carnot heat engine operating between 400°C or 673 K and 100°C or 373 K. So, the ideal cycle efficiency of the steam turbine in a IGCC plant is written as:

$$\eta_{st} = 1 - \left(\frac{T_2}{T_1}\right) = 1 - \frac{373}{673} = 0.446$$

Both the steam and gas turbines have been assumed to be ideal, so the ideal efficiency of the combined cycle can be calculated using the expression for combined cycle efficiency given in basic thermodynamics:

$$\eta_{igcc} = \eta_{gtp} + \eta_{st} - \eta_{st} \times \eta_{gtp} = 0.346 + 0.446 - (0.446 \times 0.346) = 0.638\% \text{ or } 63.8\%$$

Thus, the gasification-based IGCC plant has an overall efficiency of 63.8% compared to 54.4% for combustion-based Rankin cycle steam power plant.

## 1.6 BRIEF DESCRIPTION OF SOME BIOMASS CONVERSION PROCESSES

The following section presents a brief description of reactions that take place in different thermal conversion processes of biomass.

### 1.6.1 Torrefaction

It is a process of production of carbon-rich solid fuels from biomass. So, gas and liquid parts of the conversion do not form a part of the product. Torrefaction has some similarity with the process of carbonization, but there are some important differences as explained in Section 4.1.1:

$$C_nH_mO_p + \text{heat} \rightarrow \text{char} + CO + CO_2 + H_2O + \text{condensable vapors} \quad (1.3)$$

### 1.6.2 Pyrolysis

In pyrolysis, heavier hydrocarbon molecules of biomass are broken down into smaller hydrocarbon molecules, noncondensable gases like CO, $CO_2$, and solid carbon as char:

$$C_nH_mO_p + \text{heat} \rightarrow \sum_{\text{Liquid}} C_aH_bO_c + \sum_{\text{Gas}} C_xH_yO_z + \sum_{\text{Solid}} C \quad (1.4)$$

### 1.6.3 Combustion of Carbon

In subsequent discussion, we take simple carbon as the feedstock and write the chemical reaction of the process to illustrate the conversion process. The positive sign on the right side ($+Q$ kJ/kmol) of the reaction equations implies that heat is absorbed in the reaction. A negative sign ($-Q$ kJ/kmol) means that heat is released in the reaction.

When 1 kmol of carbon is burnt completely in adequate air or oxygen, it produces 394 MJ heat and carbon dioxide. This is a combustion reaction:

$$C + O_2 \rightarrow CO_2 - 393,770 \text{ kJ/kmol} \quad (1.5)$$

### 1.6.4 Gasification of Carbon

If the oxygen supply is restricted, one can gasify the carbon into carbon monoxide. The carbon then produces 72% less heat than it would have in complete combustion, but the partial gasification reaction as shown below produces a combustible gas, CO:

$$C + 1/2\, O_2 \rightarrow CO - 110,530 \text{ kJ/kmol} \quad (1.6)$$

When the gasification product, CO, is burnt subsequently in adequate oxygen, it releases the remaining 72% (283 MJ) of the heat. Thus, the CO retains only 72% of the energy of the carbon.

We can also go for complete gasification of a biomass where the energy recovery is 75–88% due to the presence of hydrogen and other hydrocarbons. The producer gas reaction is an example of gasification reaction, which produces hydrogen and carbon monoxide from carbon. This product gas mixture is also known as synthesis gas or *syngas*:

$$C + H_2O \rightarrow CO + H_2 + 131{,}000\ \text{kJ/kmol} \tag{1.7}$$

Utilization of heavy oil residues in oil refineries is an important application of gasification. Low-hydrogen hydrocarbon residues are gasified into hydrogen:

$$C_nH_m + (n/2)O_2 = nCO + (m/2)H_2 \tag{1.8}$$

This hydrogen can be used for hydrocracking of other heavy oil fractions into lighter oils.

The reaction between steam and carbon monoxide is also used for maximization of hydrogen production in the gasification process at the expense of CO:

$$CO + H_2O \rightarrow H_2 + CO_2 - 41{,}000\ \text{kJ/kmol} \tag{1.9}$$

## 1.6.5 Syngas Production

Syngas is also produced from natural gas (>80% $CH_4$) using a steam–methane-reforming reaction, instead of from solid carbonaceous fuel alone. The reforming reaction is, however, not strictly gasification but a molecular rearrangement:

$$CH_4 + H_2O\ \text{(catalyst)} \rightarrow CO + 3H_2 + 206{,}000\ \text{kJ/kmol} \tag{1.10}$$

Partial oxidation of natural gas or methane is an alternative route for production of syngas. In contrast to the reforming reaction, partial oxidation is exothermic. Partial oxidation of fuel oil also produces syngas:

$$CH_4 + 1/2\,O_2 \rightarrow CO + 2H_2 - 36{,}000\ \text{kJ/kmol} \tag{1.11}$$

## 1.6.6 Methanol Synthesis

Syngas provides the feedstock for many chemical reactions, including methanol synthesis (Eq. (1.12)). Methanol ($CH_3OH$) is a basic building block of many products, including gasoline:

$$CO + 2H_2\ \text{(catalysts)} \rightarrow CH_3OH \tag{1.12}$$

### 1.6.7 Ammonia Synthesis

Ammonia ($NH_3$) is an important feedstock for fertilizer production. It is produced from pure hydrogen and nitrogen from air:

$$3H_2 + N_2 \text{ (catalysts)} \rightarrow 2NH_3 - 92,000 \text{ kJ/kmol} \quad (1.13)$$

### 1.6.8 Fischer–Tropsch Reaction

The Fischer–Tropsch synthesis reaction can synthesize a mixture of CO and $H_2$ into a range of hydrocarbons, including diesel oil:

$$2n\, H_2 + nCO + \text{catalyst} \rightarrow (C_n\, H_{2n}) + n\, H_2O - Q \quad (1.14)$$

Here, $C_nH_{2n}$ represents a mixture of hydrocarbons ranging from methane and gasoline to wax. Its relative distribution depends on the catalyst, the temperature, and the pressure chosen for the reaction.

### 1.6.9 Methanation Reaction

Methane ($CH_4$), an important ingredient in the chemical and petrochemical industries, can come from natural gas as well as from solid hydrocarbons like biomass or coal. For the latter source, the hydrocarbon is hydrogenated to produce synthetic gas, or SNG. The overall reaction for SNG production may be expressed as:

$$C + 2H_2 \rightarrow CH_4 - 74,800 \text{ kJ/kmol} \quad (1.15)$$

More details on these reactions are given in Chapter 7.

## SYMBOLS AND NOMENCLATURE

**LHV** lower heating value of gas (kJ/mol)
$\eta$ efficiency of different components as indicated in example 1.1
$T_1$ inlet steam temperature in the steam turbine (K)
$T_2$ exhaust steam temperature in the steam turbine (K)
$T_{g1}$ inlet temperature in the gas turbine (K)
$T_{g2}$ exhaust temperature in the gas turbine (K)

Chapter 2

# Economic Issues of Biomass Energy Conversion

## 2.1 INTRODUCTION

Biomass conversion is proving to be an important option for several applications including energy and chemical production. For a project to be self-sustained, it must be economically viable and environmentally acceptable. In the short term, government subsidies, carbon tax, grants, and regional policies may help a project develop and continue for sometime, but for a technological option to be sustainable, it must be economically viable and sustainable on its own. Therefore, it is important to carry out a comprehensive analysis of a biomass conversion plant even at its conceptual stage. To do this, four important elements are to be known:

1. Availability of biomass over the projected lifespan of the plant and market for the biomass derived products.
2. Financial structure including cost of money, government subsidy if any, and loan guarantee.
3. Capital and operating costs of the plant.
4. Environmental impact and applicable regulations and the approval process.

Three potential sources of revenue for a biomass conversion plant are as follows:

1. Energy production through heat or electricity.
2. Production of chemical or metallurgical feedstock.
3. Production of solid fuel for cofiring or transport fuel as an alternative to diesel or gasoline.

Commercial viability of any of these options depends on the total cost of operation of the plant and the revenue generated from it. If the net return turns out to be negative over the lifetime of the plant, the process would not be commercially viable. It will then require some form of subsidy for sustenance of the plant.

Life-cycle assessment (LCA) is another analysis that is often considered by governments and regulating bodies in approving a new project. The LCA

**Biomass Gasification, Pyrolysis and Torrefaction.**

examines if a project is sustainable or unsustainable in the long run and in the broader context, and if it would make a net positive contribution to the health of the society. For example, ethanol production from corn is a commercially viable process under the current tax and subsidy structure in some countries like the United States. So, it would easily pass the benchmark for conventional financial analysis, but, given the life-cycle analysis of all energy input into the system and the associated greenhouse gas emissions, the sustainability of this option may be questioned especially in absence of a direct or indirect subsidy.

LCA analyzes the full range of environmental and social impact assignable to a specific biomass conversion project. It also examines the overall energy impact. A broad view of all direct and indirect inputs and the outputs of the project are taken into consideration, to help choose the least burdensome of available technological options for an energy conversion project. More details are available from other references like Jensen et al. (1997).

## 2.2 BIOMASS AVAILABILITY AND PRODUCTS

For a proper economic analysis, the input and the desired output of the biomass plant have to be known. For a biomass conversion project, the input is biomass and the output is energy, fuels, or chemicals. Unlike fossil fuel, biomass is not available in one place in a concentrated form, such that it could be collected and transported anywhere in the world where it is required. Biomass is a considerably dispersed and low-energy density fuel. Energy transportation through biomass is much more expensive than that through oil, gas, or coal. As such, biomass must be collected regionally. Thus, its local availability is critical for a biomass conversion plant.

### 2.2.1 Availability Assessment

Availability of biomass at acceptable prices over the lifespan of a plant is the very foundation of a biomass project. Being a local resource, biomass assessment and examination of availability are conducted at an early phase of a project.

It is difficult to assess the availability of biomass from an estimate of what is harvested or what is available in the forest. The entire body of biomass produced in the forest or fields is not necessarily available for energy production. Biomass can be collected from various steps of a biomass-based nonenergy process flow. For example, when wood is collected from the forest for pulp and lumber production, much residue including the tree roots are left on the forest floor. Further down the production stream, sawdust, barks, and others are also available as biomass waste. It is seen that available biomass for particular conversion option is only a fraction of what originally grew in the forest.

Primary biomass resources are categorized into two main types: virgin biomass and waste biomass. Virgin or fresh biomass is available as it grows, but secondary or waste biomass being a derived product is not available immediately. For example, an old piece of furniture in a municipality landfill site may have been in use for 20–50 years after the wood it was made of grew in the forest. Some species of aquatic biomass are also good sources of virgin biomass. These have higher net organic yields compared to most terrestrial biomass and have high growth rate. Table 2.1 gives values of potential growth rate of some aquatic biomass. Biomass being a renewable resource its growth or replacement rate is an important parameter in availability assessment.

The production of biomass depends primarily on the land availability and the climatic condition. Collection of data from the field level is important to estimate the available land areas. Analysis of historical data is needed to identify trends of land use pattern. This enables predictions for future resource potential. Biomass yields vary with type, species of the plant, agroclimatic region, rainfall, and other factors like irrigation and degree of mechanization. Information on these factors is required to assess the land resource and then to estimate the possible supply of biomass. The availability of biomass is also restricted by accessibility constrains. Cultivable land may be available, but physical difficulties in harvesting, collecting, and transporting from the point of production to the plant may render that biomass useless.

Main issues in biomass resource assessment are as follows:

- Availability of forest or suitable lands for cultivation.
- Net crop yields (for cultivated biomass).
- Energy cost for cultivation as well as the price the market would pay for the biomass.

**TABLE 2.1** Growth Rate of Some Aquatic Biomass Expressed in Dry Tons per Hectare per Year

| Biomass Species | Annual Growth Rate |
|---|---|
| *Spartina alterniflora* (in salt water) | 33 dry ton/ha year |
| Giant Cane (*Arundo donax*), bulrush (*Scirpus lacustris*)—fresh water swamp | 57–59 ton/ha year |
| Cattail (*Typha* spp.)—a wetland biomass | 25–30 dry ton/ha year |
| Water hyacinth (*Eichhornia crassipes*)—fresh water | 150 ton/ha year |
| Chlorella (algae in lake, ponds, etc.) | ~360 dry ton/ha year |

**Source**: Data compiled from Klass (1998).

#### 2.2.1.1 Energy Crop

Unlike fossil fuels, biomass is a renewable fuel. So, besides its present availability, which is equivalent to the energy reserve for fossil fuels, one must consider the annual yield, which is the amount of biomass grown in a year. For an annual species that grows every year and dies, it is the total biomass yield of that plant. For perennial species like trees that do not die every year, it is the annual growth in the plant. Table 2.1 gives the annual growth rates of some biomass species.

The other important parameter that influences the cost is the heating value of the fuel and the fraction of the total biomass available for energy production.

Energy crops can provide a biomass plant with an assured supply of feedstock over its lifetime and as such it is getting much attention as a commercial source of energy. Its use is similar to a dedicated coalmine feeding coal to a specific pithead power plant. Instead of using naturally grown trees or plants in the forest, chosen fast-growing plants are cultivated exclusively for the supply of energy. Such plants have good energy density, grow fast, and have a low maintenance cost for cultivation. It takes little water or fertilizer to grow energy crop. In many cases, they can be grown in abandoned fields. Growing these plants does not affect the cultivation of food grain. Table 2.2 presents a partial list of such energy crops.

#### 2.2.1.2 Biomass Cost

The delivered cost of biomass is an important parameter that must be known before the start of the project. It is expressed in terms of energy

**TABLE 2.2** Some Examples of Energy Crops

| Energy Crops | Annual Growth Rate (dry ton/ha year) | HHV (MJ/kg dry) |
|---|---|---|
| Miscanthus | 13–30[b] | 18.5[b] |
| SRC willow | 10–15[b] | 18.7[b] |
| Sorghum | 0.2–19[c] | |
| Switchgrass | 2.9–14[c] | 17.4[b] |
| Alfalfa | 1.6–17.4[c] | |
| Canary grass | 2.7–10.8[c] | |
| Kenaf Hybrid poplar | 10[a] | |

[a]*Drapco et al. (2008) p. 252.*
[b]*McKendry (2002), p. 45.*
[c]*Klass (1998), p. 113.*

potential, which is calculated from the delivered cost and the heating value of the fuel.

$$\text{Projected cost of biomass}\left(\frac{\$}{\text{GJ}}\right) = \frac{\text{Cost of biomass}(\$/\text{dry ton})}{\text{HHV}(\text{GJ}/\text{dry ton})} \tag{2.1}$$

$$\text{Cost of energy derived}\left(\frac{\$}{\text{GJ}}\right) = \frac{\text{Cost of biomass}(\$/\text{GJ})}{\text{Efficiency of energy conversion technology}} \tag{2.2}$$

The cost of energy derived considers only the fuel cost. So, it is lower than the actual cost of energy delivered by the plant that includes investment carrying charge and operation and maintenance expenses.

The cost of biomass could be greatly influenced by two other factors—disposal and environmental. For example, waste biomass could become an energy source with a negative price, if municipality was paid a tipping fee for picking up the waste from the producer. The other factor is environmental regulations like carbon tax or biomass utilization incentives. For marginal projects, these incentives could make a biomass economically viable.

### 2.2.2 Product Revenue from Biomass Conversion

To some extent, the economic analysis for biomass plant would depend on the end use of the product produced. The price paid or the revenue earned could be different for different products from biomass. Three main revenue sources of gasification products are as follows:

1. Energy revenue
2. Revenue from chemical production
3. Revenue from transport fuels.

#### *2.2.2.1 Energy Revenue*

This may be subdivided into (i) electricity production and (ii) thermal energy production. Electricity may be produced by operating an integrated gasification combined cycle plant (IGCC) or by operating an internal combustion engine. IGCC plants are used only for large capacity plants (>100 MWe).

In the other means of electricity generation, the produced gas, after cleaning, is fired in a reciprocating engine (e.g., diesel engine). This engine in turn drives an electric generator. This option is attractive for small capacity units and is suitable for distributed power generation.

Besides electricity, thermal energy (heat) is another commercial product of biomass. To generate heat, direct combustion is the most cost-effective method unless there are local restrictions in the use of combustion. Combined heat, power and gas production that combines several products together, is gaining popularity. Here, the plant earns revenue from the sale of electricity, heat, and gas. This type of trigeneration system is thermodynamically efficient and offers good flexibility to the operator.

Revenue from electricity and heat is often fixed by the local utility or by government regulating agencies.

#### *2.2.2.2 Revenue from Chemicals*

One of the most important uses of gasification and pyrolysis is the production of chemicals. The South African Synthetic Oil Limited (SASOL) has been operating coal gasifiers for production of oil from coal since 1950s. Here, coal is gasified to hydrogen and carbon monoxide. These gases are synthesized into liquid hydrocarbon using Fischer–Tropsch synthesis. Nowadays, many petroleum refineries are using large-scale gasifiers to gasify heavy oils into hydrogen or other gases. Of late, green chemicals are also gaining popularity. Many chemicals, traditionally produced from petroleum (e.g., resin), could be produced from pyrolysis of biomass as well. Activated charcoal, for example, is an effective reaction medium for many chemical reactions. Coke produced from biomass could be a substitute for coal-based coke.

Revenue from chemicals could be much higher than that from electricity or heat, but it fluctuates a great deal depending on the market condition.

#### *2.2.2.3 Revenue from Secondary Fuel Production*

A major commercially successful use of biomass has been the production of substitute fuels. For example, for reduction in greenhouse gases from coal-fired power plants there is a rising demand for torrefied biomass fuel. It is also being considered in iron extraction in blast furnace for the same reason. A large number of fermentation-based plants are in operation for production of corn- or sugarcane-based ethanol that is to substitute for petro-derived gasoline. Production of diesel from waste cooking oil and fat is also done on a commercial scale. Revenue from such transportation fuels varies with market price of diesel or gasoline.

Carbon credit could also be an important source of revenue in some cases. The net reduction in carbon dioxide emission from a project could bring in additional revenue to the plant through sale of carbon credits. This revenue could be added to that earned through the sale of the products of biomass gasification, pyrolysis, or torrefaction.

## 2.3 BIOMASS CONVERSION PROCESS PLANT EQUIPMENT AND COST

Depending upon the end use of the product, the process configuration of a biomass plant would change, but its basic structure will remain the same. Most of the auxiliary plant and equipment would be similar in a biomass thermal conversion plant, whether it is a combustion, gasification, pyrolysis, or torrefaction plant. Fermentation-based plants like ethanol plant would, however, have different configuration as the feedstock is cereal instead of lignocellulose biomass.

The cost of biomass conversion depends on its several major processing steps like:

1. Biomass collection
2. Preprocessing
3. Biomass conversion (torrefaction, pyrolysis, gasification, or combustion)
4. Gas cleaning/product treatment
5. Product utilization (energy or chemical production).

Once the availability of biomass is assured (see Section 2.2), the project moves to the next step of financial analysis. The cost of delivered fuel is also known from the availability analysis. The financial analysis involves assessment of capital and operation and maintenance (O&M) cost of the above components of a biomass plant. The following section briefly discusses the considerations for determination of O&M cost of the first three components of a biomass plant.

### 2.3.1 Biomass Collection System

Unlike fossil fuel, biomass is not necessarily collected from a single concentrated source. It is a relatively light and bulky material with low heat content on a volume basis. So, a biomass plant involves handling of large volumes and at times transportation over long distances. Therefore, logistical network from forest to the plant should be well planned during project development. The flow of biomass from collection point to product (solid, liquid, or gas), from production to end use should carefully identify all steps of conversion and losses. Systematic collection of data is important while ensuring that consistent units are used throughout the process chain.

Biomass collection and handling for gasification, pyrolysis, or torrefaction plants is very similar to that for a biomass steam power plant, for which a good amount of experience and cost base is available. Based on existing data, a reasonable cost estimate for new plant equipment, required for collection and feeding the biomass into the conversion unit can be arrived at.

### 2.3.2 Preprocessing

Biomass rarely comes in a form ready to be fed into the reactor. It is often mixed with metals and debris, which are to be separated (see Chapter 12) before use. The moisture content of the biomass delivered is sometimes well in excess of what the plant is designed for. In such cases, biomass needs to be dried using appropriate drying equipment, which is of course, an integral part of a torrefaction plant.

Depending upon the feedstock specification of the reactor, the biomass is to be screened, crushed, or milled. For example, a fixed bed gasifier can accept few tens of millimeter-sized particles, while an entrained flow gasifier can accept only micron-sized biomass. Thus, both capital and operating cost of biomass preparation for entrained flow gasifier would be higher than that in other types of gasifiers.

The pretreatment cost comprises the cost associated with cleaning, drying, screening, and sizing the biomass. Bridgwater (1995) gave an estimate of feed pretreatment cost in terms of electricity produced and showed that larger plants have lower specific costs. For example, for a 20 MWe plant the cost of preprocessing of feed is $600/kWe, while it is $300/kWe for a 100 MWe plant.

### 2.3.3 Gasifier Cost

The cost of gasification depends, to some extent, on the technology applied for gasification of the biomass. Technological options include:

1. Downdraft gasifier
2. Updraft gasifier
3. Side draft gasifier
4. Bubbling fluidized bed gasifier
5. Circulating fluidized bed gasifier
6. Entrained flow gasifier (atmospheric or pressurized)
7. Supercritical water gasifier.

The capital cost depends on the capacity or fuel throughput into the plant. Figure 2.1 plots the installed capital cost of a gasification plant as a function of fuel throughput into the gasifier. The total plant cost (TPC) is correlated with the fuel input for both pressurized and atmospheric pressure plants. These figures are given on the year 1994 basis. Table 2.3 presents data on three types of gasifiers in 2001 dollars. It clearly shows that entrained flow gasifier is more expensive than bubbling bed units on dry ton basis, but on heat output basis it is most economic. The comparison is not quite uniform as cost for fluidized bed gasifiers are for biomass while that for entrained flow gasifier is for coal as the feedstock.

There is a considerable difference between pressurized and atmospheric pressure gasifier plants. Even after discounting for smaller size of the gasifier reactor, high-pressure gasifiers could cost between two and four times more than atmospheric pressure plants would (Bridgwater, 1995). The feed system in a pressurized plant also adds to the cost.

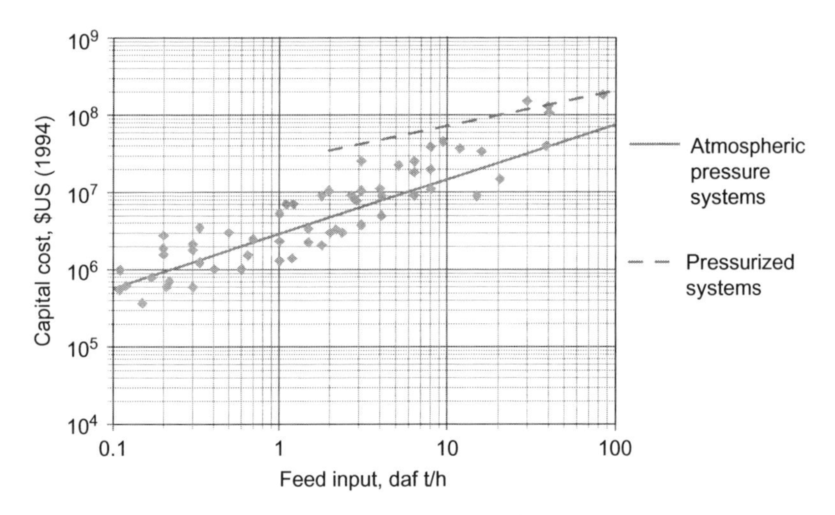


**FIGURE 2.1** TPC of gasification plant in 1994 dollars against fuel feed rate on clay ash free basis. Source: *Computed from Bridgwater (1995).*

**TABLE 2.3** Capital Cost (in 2001 dollar) of Gasifier Units Estimated by Cieferno and Ma (2002) from Several Commercial Units

| Gasifier Type | Capacity (tPD) | Specific Capital Cost ($1000/tPD) | Specific Capital Cost ($/GJ/h product gas) |
|---|---|---|---|
| Bubbling fluidized bed (biomass) | 170–960 | 13–45 | 21,600–54,900 |
| Circulating fluidized bed (biomass) | 740–910 | 24.5–28.4 | 33,000–48,000 |
| Entrained flow (coal) | 2200 | 37.3 | 1400 |

tPD—dry tons feed per day.

Bridgwater (1995) correlated the TPC of a biomass gasification plant (fuel receiving end to clean gas delivery) with fuel input ($F$) as below:

$$\text{TPC (in 1994 US dollars)} = 13(F)^{0.64} \text{ for pressurized gasifier}$$
$$= 2.9(F)^{0.7} \text{ for atmospheric pressure gasifier} \quad (2.3)$$

$F$ is fuel input in tons/h.

### 2.3.4 Torrefier Cost

Similar to the gasifier, the capital cost of a torrefaction plant also depends on the technology selected. Torrefier technologies fall under two broad categories:

1. Directly heated
2. Indirectly heated.

The cost of indirectly heated units could be a little higher than the cost of directly heated units. Uslu et al. (2008) quoted a cost in the range 4.5–11.5 million euro for a 25 MWth torrefaction plant. Peng et al. (2011) gave a comparison of capital cost of a 126,000 ton/year torrefaction unit for three technologies as listed below:

1. Moving bed torrefier: m$10.73
2. Rotary drum: m$17.58
3. Screw torrefier: m$25.45

Commercial use of torrefaction is relatively new. So, only limited data is available on its capital cost. These values are theoretical or biased on higher side by the development cost. Based on capital cost (capital cost per unit capacity) collected from some published and unpublished sources, specific torrefier cost is plotted against capacity in Figure 2.2. In spite of large variation, one could detect an economy of scale from the figure.

It may be noted that a torrefier reactor is just one component of the total torrefaction plant that needs additional units like dryer, cooler, mills, storage, biomass handling, building, and others. Such units are relatively independent of the torrefier, and could add another 50–100% to the total plant cost (TPC) depending upon the cost of the torrefier.

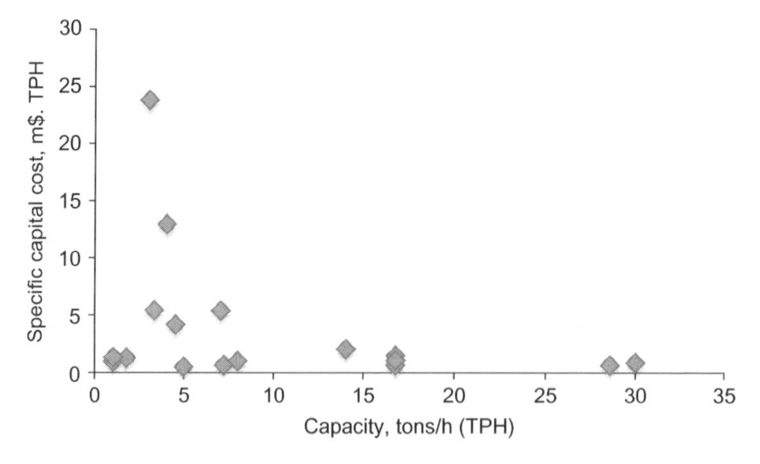


**FIGURE 2.2** Capital cost (per unit ton per hour capacity) of torrefier against capacity of unit. The scatter is due to different sources with varying conditions and technologies.

### 2.3.5 Pyrolyzer Cost

Based on a large number of data, Bridgwater et al. (2002) developed the following two empirical correlations for fast pyrolysis reactor, feeding system, and liquids recovery.

$$\text{TPC} = 40.8(1000Q)^{0.6194} \text{ in euro (reference year 2000)}$$

where $Q$ is in kilogram oven dry ton of prepared wood per hour.

Electricity may be generated of using pyrolysis oil in a multifuel diesel engine-generator set. Capital cost for this is correlated (Bridgwater et al., 2002) as below:

$$\text{TPC} = 903.1(P_{e,\text{gross}})^{0.954} \text{ in euro (reference year 2000)} \tag{2.4}$$

where $P_{e,\text{gross}}$ is gross generator output in $MW_e$.

### 2.3.6 Comparison of Capital Costs

Electricity generation from biomass has three options: gasification based combined cycle, combustion based steam cycle, and gasification based gas engine.

In an IGCC option, biomass is gasified into combustible gases, which are cleaned and burnt in the combustion chamber of a gas turbine cycle, which operates a gas turbine. The exhaust gases from the gas turbine carry substantial amount of sensible heat that is enough to generate high-pressure steam when cooled downstream. The high-pressure steam drives a steam turbine. Both turbines (gas turbine and steam turbine) drive generators to produce electricity. Such plants are generally used for capacities in excess of 100 MWe. These are expensive in absolute terms but have high energy conversion efficiency.

The next option is direct combustion of biomass in a boiler and to run a steam turbine with this steam to generate electricity based on Rankine cycle. Such steam plants are used for intermediate-sized plants.

The third option involves gasification of biomass in small gasifiers. The gas is then cleaned and used in a compression ignition engine. Such gas engine plants are generally used in small capacities especially in remote locations.

Table 2.4 gives the specific capital cost (cost per kW installed) of above three options for electricity generations from biomass. Gas engines that are used for smaller than 5 MW capacity has low absolute capital cost but exceptionally high specific capital cost. In general, a sharp decline in the capital cost is noted between 5 and 20 MWe capacities. The specific capital cost reduced at a much gentler pace between 20 and 100 MWe capacities. For that reason, the electricity generation cost reduced from 20 to 5 cents/kWh (1994 dollar and biomass cost $50/ton (dry ash free) while the IGCC plant capacity increased from 5 to 100 MWe (Bridgwater, 1995).

**TABLE 2.4** Specific Capital Cost (1994 basis) and Efficiency of Power Generation Systems Based on Biomass Gasification

| | Specific Capital Cost in 1994 ($/kWe) | | | Efficiency (%) | | |
|---|---|---|---|---|---|---|
| Plant Capacity (MWe) | IGCC | Steam | Gas Engine | IGCC | Steam | Gas Engine |
| 5 | 6500 | 6000 | 4600 | 22 | 18 | 25 |
| 10 | 4900 | 4400 | 3900 | 30 | 20 | 26 |
| 20 | 3750 | 3100 | 3000 | 37 | 22 | 28 |
| 40[a] | 2800 | 2300 | 2500[a] | 43 | 25 | 30 |
| 60[a] | 2500 | 1950 | 2200[a] | 45 | 27 | 30.5 |
| 80[a] | 2200 | 1700 | 2050[a] | 47 | 28 | 31 |
| 100[a] | 2100 | 1500 | 1950[a] | 48 | 29 | 31.5 |

[a]*Gas engines of such high capacity are rare. These are theoretical values.*
**Source**: Data compiled from Bridgwater (1995).

## 2.4 FINANCIAL ANALYSIS

Once data on all capital and operating costs are available, the financial analysis of the plant could begin. Though the analysis presented in this section is based on energy production from biomass using gasification, the analysis for chemical and fuel production would also be similar. The viability of any such project is measured in terms of following several terms:

1. Cost of electricity or product
2. Internal revenue requirement and return on investment
3. Net present value
4. Benefit–cost ratio
5. Payback period
6. Life-cycle cost.

The goal of a financial analysis is to determine one or many of the above indices to judge the economic viability of a project.

### 2.4.1 Capital Cost Adjustment for Size and Time

The capital cost for a gasifier, fuel preparation, turbine, generator, or any other major equipment discussed earlier is often derived from a database of past projects maintained by consulting companies or some major user industries. Such equipments are not necessarily of the same size as that of the planned equipment. Furthermore, cost data from those projects may not be relevant at the time when a new project is planned for. To get the projected capital cost of a specific size at a specific future date, one needs two levels of adjustment or scale-up: based on size and based on time of construction.

#### 2.4.1.1 Scale-Up with Size

The specific capital cost is defined as the total capital cost per unit capacity of the plant. In the present case of electricity generation, it is defined in terms of electrical capacity of the plant, MWe. If the specific capital cost (in \$/MWe) of a new gasifier of capacity MW is $K_g$, one may scale it up from an available database cost that gives the capital cost, $K_{g0}$ for a gasifier of capacity $MW_0$.

$$K_g = K_{g0}\left[\frac{MW_0}{MW}\right]^c \tag{2.5}$$

where $c$ is an exponent whose value is known from different capital cost estimation published (EPRI, 1991; Gerrard, 2000; Humphrey and Wellman, 1996). For example, for a coal-fired thermal power plant, $c = 0.15$ for 500–1000 MWe range and $c = 0.3$ for 300–500 MWe range.

The above equation for specific capital cost is based on per megawatt installed cost, which generally decreases with capacity increase. To find the total capital cost, the capacity of the plant is multiplied with the specific capital cost. For example, the total cost, $C_g$ for a gasifier of MW megawatt capacity is obtained by multiplying the capacity with the specific capital cost, $K_g$:

$$C_g = K_g \times MW \tag{2.6}$$

To find the total capital cost instead of specific capital cost, a different relationship is used.

$$C_g = C_{g0}\left[\frac{MW}{MW_0}\right]^n \tag{2.7}$$

where $n$ is the capacity index and MW is the capacity in appropriate units. Some representative values of the index are given in Table 2.5.

#### 2.4.1.2 Scale-Up with Time

To compute the capital cost in a certain year $C_{gA}$ based on a past cost data, $C_{gB}$, one can use the following linear relationship based on the cost index $I$.

$$C_{gA} = C_{gB}\left[\frac{I_A}{I_B}\right] \tag{2.8}$$

where $I_A$ and $I_B$ are the cost indices in year $A$ and year $B$, respectively. Such cost indices for plant and equipment are published by various organizations.

**TABLE 2.5 Capacity Index of Some Process Equipment**

| Equipment | *n* |
|---|---|
| Water–gas manufacture | 0.81 |
| Fischer–Tropsch (complete) | 0.77 |
| Centrifugal blower (1000–10000 cfm) | 0.59 |
| Reactor (300 psi) | 0.56 |
| Belt conveyor | 0.85 |
| Catalytic cracking | 0.83 |
| Oxygen plant | 0.65 |
| Centrifugal fan (1000–10,000 cfm) | 0.44 |
| Drum dryer (atm) | 0.41 |
| Screw conveyor | 0.83 |
| Packaged steam generator | 0.61 |
| $H_2S$ removal | 0.55 |
| Compressor rotary (10–400 cfm, 150 psig) | 0.69 |
| Centrifugal pump with motor | 0.33 |
| Crusher and grinder | 0.65 |

**Source**: Data compiled from Humphreys and Wellman (1996), pp. 10–18.

Some of them include Engineering News Record (ENR), Marshall and Swift (M&S), Nelson-Farrar Refinery construction index, and Chemical Engineering Plant cost index (CE).

## 2.4.2 Capital Requirement

The capital cost, CC, includes the cost of biomass energy conversion system, $C_g$, cost of fuel preparation and handling, $C_{fp}$, cost of turbine generator system, $C_{tg}$ and cost of auxiliary plants, $C_e$, which include the electrical cost, oxygen separation plant.

$$\mathrm{CC} = C_g + C_{fp} + C_{tg} + C_e \tag{2.9}$$

The total plant cost (TPC) includes the cost of plant equipment (CC), direct and indirect costs of construction ($C_{con}$), cost of project development ($C_{proj}$), most of which is office cost, and the contingency cost ($C_{cont}$).

$$\mathrm{TPC} = \mathrm{CC} + C_{con} + C_{proj} + C_{cont} \tag{2.10}$$

The contingency has two parts: process contingency $C_{\text{cont, tech}}$ and project contingency $C_{\text{cont, proj}}$. The contingency is generally taken as a percentage of the capital cost, CC.

$$C_{\text{cont}} = C_{\text{cont, tech}} + C_{\text{cont, proj}} \tag{2.11}$$

The process contingency covers the uncertainty associated with the technology, while the project contingency covers uncertainty with project execution. If the plant does not perform as designed, modifications in gasifier or associated equipment are required. The process contingency is meant to cover that expense. Gasification, pyrolysis, or torrefaction technologies are not as matured as the combustion technology. So, the process contingency of such a plant would be a higher percentage of the capital cost than it would be for a conventional combustion plant.

A large gasification plant takes several years to complete. From the beginning of the project till the date when the plant actually starts earning revenue, there is no return on the investment, which is largely borrowed. During the entire period of project development, the plant owner would have to pay interest on the borrowed capital. This amount of carrying charge is known as "allowance for funds during construction" (AFDC) and is added to the TPC to get the total plant investment required (TPI).

$$\text{TPI} = \text{TPC} + \text{AFDC} \tag{2.12}$$

The total capital requirement (TCR) must include funds required for start-up expense ($C_{\text{start}}$), working capital ($C_{\text{wc}}$), and initial catalyst and other supplies ($C_{\text{ca}}$).

$$\text{TCR} = \text{TPI} + C_{\text{start}} + C_{\text{wc}} + C_{\text{ca}} \tag{2.13}$$

## 2.4.3 Operation and Maintenance Cost

The next important aspect of the financial analysis is the O&M cost. For this, one has to know the technology and its characteristics. This will give the overall efficiency of the plant if energy generation is the product or the product mix and yield if chemical production is the goal. Table 2.4 gives a comparison of energy efficiency of three means of power (electricity) generation from biomass.

For a gasification plant, especially those, which use the product gas for fuel or chemical production, the composition of the product gas is important. This depends on the feedstock as well as on the type of gasifier reactor used. Table 2.6 presents a comparison of typical product gas composition for several types of gasifiers.

Both operating and maintenance costs are made of two major components:

1. Fixed cost
2. Variable cost.

**TABLE 2.6** **Comparison of Typical Gas Composition of Product Gas from Different Types of Gasifiers**

| Gasifier Type | $H_2$ (%) | CO (%) | $CO_2$ (%) | $CH_4$ (%) | $N_2$ (%) | HHV ($MJ/m^3$) |
|---|---|---|---|---|---|---|
| Air-blown fluid bed | 9 | 14 | 20 | 7 | 50 | 5.4 |
| Air-blown updraft | 11 | 24 | 9 | 3 | 53 | 5.5 |
| Air-blown downdraft | 17 | 21 | 13 | 1 | 48 | 5.7 |
| Oxygen-blown downdraft | 32 | 48 | 15 | 2 | 3 | 10.4 |
| Twin fluid bed | 31 | 48 | 0 | 21 | 0 | 17.4 |

**Source**: Data compiled from Bridgwater (1995).

The variable cost depends on the operation of the plant, and it does not incur if the plant is not in operation. On the other hand, the fixed operating cost is incurred even if the plant is not producing any gas or other products.

Fuel cost is the major component of the variable cost that also includes contract labor cost, maintenance cost, and cost of auxiliary supplies. The fixed operating cost includes the salary of permanent staff, supervisory staff, proportionate corporate office expenses, insurance premium, property taxes, and so on.

$$O\&M_{\text{variable}} = \text{Fuel cost} + \text{direct labor cost} + \text{maintenance labor and material} + \text{supplies cost} \quad (2.14)$$

$$O\&M_{\text{fixed}} = \text{Indirect labor} + \text{taxes} + \text{insurance} \quad (2.15)$$

The total operating expense (TOE) is therefore the sum of the above two costs:

$$\text{TOE} = O\&M_{\text{variable}} + O\&M_{\text{fixed}} \quad (2.16)$$

#### 2.4.3.1 Carrying Charge

The revenue obligations needed to support an investment or carrying charge, $C_c$ include several components:

$$C_c = R_D + R_E + T + D_B \quad (2.17)$$

Here, $R_D$ is the return on debt or the revenue required to pay for use of debt money. The return on equity ($R_E$) is the after-tax profit that is paid to the investors or shareholders for the use of equity money. $D_B$ is the book depreciation.

The depreciation may be linear over the projected lifespan or as defined by regulatory bodies. The tax $T$ is sometimes built into the return of the equity when the latter is expressed as before-tax-return on equity. These items remain constant and are independent on the financial operation of the plant.

#### 2.4.3.2 Revenue Requirement

The biomass conversion plant must sell its product at a price such that it would cover all expenses (direct or indirect) and the cost of carrying the investment. The first part (TOE) of the expense is technical in nature. One calculates it based on the technical design of the gasification plant and its performance characteristics. The second part, known as carrying charge ($C_c$), is more based on how the plant is financed. It includes all financial obligations including the expected profit or return on investment. The total revenue required from sales of all products from the plant (RR) is the sum of carrying charge and operating expenses.

$$\mathrm{RR} = C_c + \mathrm{TOE} \tag{2.18}$$

Total revenue = price of electricity (other product) × electricity (or other product) generated/year + credit earned for $CO_2$ and so on + revenue from by-product sales.

---

**Example 2.1**

A 40 MWe IGCC plant has an availability of 85% and 100% capacity factor. The total capital requirement (TCR) per kW installed including all is \$1353/kW. Debt capital is 70% of TCR. Fixed component of the yearly O&M cost is \$31/kW/year. The variable component is \$0.0031/kWh/year. Return on debt, capital is 12%, and that on equity before-tax is 16%. Book life of the plant is 30 years. Find the revenue required for the plant.

**Solution**

Let us start the calculation on a 1 kW capacity basis.

Since the return on equity is before the tax, Eq. (2.17) is modified as:

$$C_c = R_D + R_E + D_B$$

Depreciation (assuming linear) over the booked life, $D_B = 1353/30 = \$45.1/\text{year}$

Debt capital = 0.7 × \$1353 = \$947

$$R_D = (\$947) \times 0.12 = \$113.6/\text{year}$$

Equity capital = (1 − 0.7) × \$1353 = \$405.9. Return on equity is 16%. So,

$$R_E = \$405.9 \times 0.16 = \$64.9/\text{year}$$

Total carrying charge = \$45.1 + \$113.6 + \$64.9 = \$223.6/year

Availability being 85%, the actual time the plant in operation in a year = 365 × 24 × 0.85 = 7446 h/year

Total unit of electricity produced = 7446 h × 1 kW = 7446 kW/year
Variable O&M = \$0.0031/kW h × 7446 kW h = \$23.08/kW/year
Yearly fixed O&M = \$31/kW/year
Total O&M = 23.08 + 31 = \$54.08/kW/year
Total revenue requirement = \$54.08 + \$223.69 = \$277.68/kW installed/year
For 40 MW plant RR = 40,000 × 277.68 = \$107,200/year

## SYMBOLS AND NOMENCLATURE

|  |  |
|---|---|
| $F$ | fuel input to the gasification plant (tons/h) |
| $K_g$ | specific capacity of a new gasifier plant (\$/MWe) |
| MW | capacity of a new gasifier (MW) |
| $K_{g0}$ | specific capacity of an existing gasifier plant (\$/MWe) |
| $MW_0$ | capacity of an existing gasifier (MW) |
| $c$ | constant, depending on the gasifier equipment and capacity (−) |
| $C_g$ | total cost of a new gasifier (\$) |
| $C_{g0}$ | total cost of an existing gasifier (\$) |
| $n$ | capacity index (−) |
| $C_{gA}$ | capital cost in a certain year (\$) |
| $C_{gB}$ | capital cost in a past year (\$) |
| $I_A$ | cost index for year $A$ (−) |
| $I_B$ | cost index for year $B$ (−) |
| CC | capital cost of a plant (\$) |
| $C_{fp}$ | cost of fuel preparation (\$) |
| $C_{tg}$ | cost of turbine generator system (\$) |
| $C_e$ | cost of auxiliary plant (\$) |
| TPC | total cost of the plant (\$) |
| $C_{con}$ | cost of construction of the gasification plant (\$) |
| $C_{proj}$ | cost of project development (\$) |
| $C_{cont}$ | office cost and contingency (\$) |
| $C_{cont,\ tech}$ | process contingency cost (\$) |
| $C_{cont,\ proj}$ | project contingency cost (\$) |
| TPI | total project investment for a new plant (\$) |
| AFDC | interest paid on the borrowed capital (\$) |
| TCR | total capital requirement (\$) |
| $C_{start}$ | funds required for start-up expense (\$) |
| $C_{wc}$ | working capital required (\$) |
| $C_{ca}$ | funds required for the initial catalyst and other supplies (\$) |
| TOE | total operating expense in a biomass plant (\$) |
| $O\&M_{variable}$ | variable operating costs (\$) |
| $O\&M_{fixed}$ | fixed operating costs (\$) |
| $R_D$ | return on debt (\$) |
| $R_E$ | return on equity (\$) |
| $D_B$ | book depreciation (\$) |
| $T$ | income tax (\$) |
| RR | the total revenue required from sales of all products from the plant (\$) |
| $C_c$ | carrying charge (\$) |

Chapter 3

# Biomass Characteristics

## 3.1 INTRODUCTION

The characteristics of biomass greatly influence the performance of a biomass conversion system whether it is a combustor, torrefier, pyrolyzer, or gasifier. A proper understanding of the physical and the chemical properties of biomass feedstock is essential for the design of a reliable biomass conversion system. This chapter discusses some important properties of biomass that are relevant to such processes.

## 3.2 WHAT IS BIOMASS?

*Biomass* refers to any organic materials that are derived from plants or animals (Loppinet-Serani et al., 2008) that is living or was living in the recent past. A universally accepted definition is difficult to find. However, the one used by the United Nations Framework Convention on Climate Change (UNFCCC, 2005) is relevant here:

> *[A] non-fossilized and biodegradable organic material originating from plants, animals and micro-organisms. This shall also include products, by-products, residues and waste from agriculture, forestry and related industries as well as the non-fossilized and biodegradable organic fractions of industrial and municipal wastes.*

Biomass also includes gases and liquids recovered from the decomposition of nonfossilized and biodegradable organic materials. In the United States, there has been much debate on a legal definition. Appendix A gives a recent legal interpretation of renewable biomass.

As a sustainable and renewable energy resource, biomass is constantly being formed by the interaction of $CO_2$, air, water, soil, and sunlight with plants and animals. After an organism dies, microorganisms break down biomass into constituent parts like $H_2O$, $CO_2$, and its potential energy. The carbon dioxide, a biomass releases through the action of microorganisms or combustion, was absorbed by it in the recent past so, biomass combustion does not add to the total $CO_2$ inventory of the Earth. It is thus called *greenhouse gas neutral* or *GHG neutral*.

Biomass comes from botanical (plant species) or biological (animal waste or carcass) sources, or from a combination of these. It thus includes only living and recently dead biological species that can be used as fuel or in chemical production. Biomass does not include organic materials that over many millions of years have been transformed by geological processes into fossil fuels such as coal or petroleum. Common sources of biomass are:

- *Agricultural*: food grain, bagasse (crushed sugarcane), corn stalks, straw, seed hulls, nutshells, and manure from cattle, poultry, and hogs.
- *Forest*: trees, wood waste, wood or bark, sawdust (SW), timber slash, and mill scrap.
- *Municipal*: sewage sludge, refuse-derived fuel (RDF), food waste, waste paper, and yard clippings.
- *Energy Crops*: poplars, willows, switchgrass, alfalfa, prairie bluestem, corn, and soybean, canola, and other plant oils.
- *Biological*: animal waste, aquatic species, and biological waste.

### 3.2.1 Biomass Formation

Botanical biomass is formed through conversion of carbon dioxide ($CO_2$) in the atmosphere into carbohydrate by the sun's energy in the presence of chlorophyll and water. Biological species grow by consuming botanical or other biological species. Plants absorb solar energy by a process called *photosynthesis* (Figure 3.1). In the presence of sunlight of particular wavelengths, green plants break down water to obtain electrons and protons and use them to turn $CO_2$

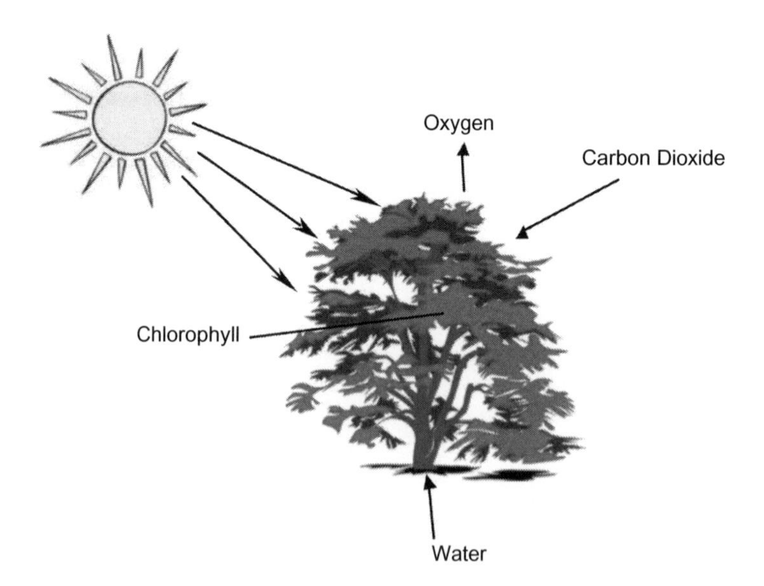


**FIGURE 3.1** Biomass grows by absorbing solar energy, carbon dioxide, and water through photosynthesis.

into glucose (represented by $CH_mO_n$), releasing $O_2$ as a waste product. The process may be represented by this equation (Hodge, 2010, p. 297):

$$\text{Living plant} + CO_2 + H_2O + \text{sunlight} \xrightarrow{\text{chlorophyll}} (CH_mO_n) + O_2 - 480\ \text{kJ/mol} \quad (3.1)$$

For every mole of $CO_2$ absorbed into carbohydrate or glucose in biomass, 1 mol of oxygen is released. This oxygen comes from water the plant takes from the ground or the atmosphere (Klass, 1998, p. 30). The chlorophyll promotes the absorption of carbon dioxide from the atmosphere, adding to the growth of the plant. Important ingredients for the growth of biomass are therefore:

- Living plant
- Visible spectrum of solar radiation
- Carbon dioxide
- Chlorophyll (serving as catalyst)
- Water.

The chemical energy stored in plants is then passed on to the animals and to the humans that take the plants as food. Animal and human wastes also constitute biomass.

### 3.2.2 Types of Biomass

Biomass comes from a variety of sources as given in Table 3.1. European committee for standardization published two standards for classification and

**TABLE 3.1** Two Major Groups of Biomass and Their Subclassification

| | | |
|---|---|---|
| A. Virgin biomass | A.1 Terrestrial biomass | i. Forest biomass<br>ii. Grasses<br>iii. Energy crops<br>iv. Cultivated crops |
| | A.2 Aquatic biomass | i. Algae<br>ii. Water plant |
| B. Waste biomass | B.1 Municipal waste | i. MSW<br>ii. Biosolids, sewage<br>iii. Landfill gas |
| | B.2 Agricultural solid waste | i. Livestock and manures<br>ii. Agricultural crop residue |
| | B.3 Forestry residues | i. Bark, leaves, floor residues |
| | B.4 Industrial wastes | i. Demolition wood, sawdust<br>ii. Waste oil/fat |

specification (EN 14961) and quality assurance (EN 15234) of biomass. Based on their origin, it classified biomass under four broad categories:

1. Woody biomass
2. Herbaceous biomass
3. Fruit biomass
4. Blend and mixtures.

Trees, bushes, and shrubs fall under woody biomass, but not the fruits or seeds that some of them bear. Herbaceous biomasses are those plants that die at the end of the growing season. These biomasses, however, include grains and cereals that grow on such plants. Fruits, though classified as a separate group, are part of woody plants. Additionally, we also have mixture or blends of biomass. Blends are intentional mixing of biomass, while mixtures are unintentional mixing of biomass.

Loosely speaking, biomass includes all plants and plant-derived materials, including livestock manures. Primary or virgin biomass comes directly from plants or animals. Waste or derived biomass comes from different biomass-derived products. Table 3.1 lists a range of biomass types grouping them as virgin or waste. Biomass may also be divided into two broad groups:

1. Virgin biomass includes wood, plants, and leaves (lignocellulose), and crops and vegetables (carbohydrates).
2. Waste biomass includes solid and liquid wastes (municipal solid waste (MSW)); sewage, animal, and human wastes; gases derived from landfilling (mainly methane); and agricultural wastes.

#### *3.2.2.1 Lignocellulosic Biomass*

A major part of biomass is lignocellulose. So this type is described in more detail. Lignocellulosic material is the nonstarch, fibrous part of plant materials. Cellulose, hemicellulose, and lignin are its three major constituents. Unlike carbohydrate or starch, lignocellulose is not easily digestible by humans. For example, we can eat rice, which is a carbohydrate, but we cannot digest the husk or the straw, which is lignocellulose. Lignocellulosic biomass is not a part of the human food chain, and therefore its use for biogas or bio-oil does not threaten the world's food supply.

A good example of lignocellulosic biomass is a woody plant—i.e., any vascular plant that has a perennial stem above the ground and is covered by a layer of thickened bark. Such biomass is primarily composed of structures of cellulose and lignin. A detailed description of wood structure is given in Section 3.3.1. Woody plants include trees, shrubs, cactus, and perennial vines. They can be of two types: (1) herbaceous and (2) nonherbaceous.

A herbaceous plant is one with leaves and stems that die annually at the end of the growing season. Wheat and rice are examples of herbaceous plants that develop hard stems with vascular bundles. Herbaceous plants do not have the thick bark that covers nonherbaceous biomass like trees.

Nonherbaceous plants are not seasonal; they live year-round with their stems above the ground. These include trees, shrubs, and vines. Nonherbaceous perennials like woody plants have stems above the ground that remain alive during the dormant season and grow shoots the next year from their above-ground parts.

The trunk and leaves of tree plants form the largest group of available biomass. These are classified as lignocellulosic, as their dominant constituents are cellulose, hemicellulose, and lignin. Table 3.2 gives a percentage of these components in some plants. Section 3.3.2 presents further discussions of the lignocellulose components.

There is a growing interest in the cultivation of plants exclusively for production of energy. These crops are called "*energy crop*," and they are lignocellulosic in nature. Such crops typically have a short growing period and high yields and require little or no fertilizer, so they provide quick return on investment. For energy production, woody crops such as *miscanthus*, willow, switchgrass, and poplar are widely utilized. These plants are densely planted. They have high-energy yield per unit of land area and require much less energy for cultivation.

#### 3.2.2.2 Crops and Vegetables

While the body of a plant or tree (e.g., trunk, branches, and leaves) is lignocellulosic, the fruit (e.g., cereal and vegetable) is a source of carbohydrate, starch, and sugar. Many plants like canola and mustard also provide fat. The fruit is digestible by humans, but the lignocellulosic body of the fruit tree is not. Some animals, however, can digest lignocellulosic biomass because of special chemicals in their stomach. The use of crops or vegetables for the production of chemicals and energy must be weighed carefully as it might affect food supplies.

**TABLE 3.2** Composition of Some Lignocellulose Wood

| Plant | Lignin (%) | Cellulose (%) | Hemicellulose (%) |
|---|---|---|---|
| Deciduous plants | 18–25 | 40–44 | 15–35 |
| Coniferous plants | 25–35 | 40–44 | 20–32 |
| Willow | 25 | 50 | 19 |
| Larch | 35 | 26 | 27 |

**Source:** Adapted from Bergman et al. (2005, p. 15).

Compared to lignocellulosic compounds, carbohydrates are easier to dissolve, so it is relatively easy to derive liquid fuels from them through fermentation or other processes. For this reason, most commercial ethanol plants use crops as feedstock.

Natural crops and vegetables are a good source of starch and sugars and, therefore can be hydrated. Some vegetables and crops (e.g., coconut, sunflower, mustard, and canola) contain fat, providing a good source of vegetable oil. Animal waste (from land and marine mammals) also provides fat that can be transformed into bio-oil. If carbohydrate is desired for the production of biogas, whole crops, such as maize, Sudan grass, millet, and white sweet clover, can be made into silage and then converted into biogas.

There are two types of crop biomass: the regularly harvested agricultural crops for food production and the energy crops for energy production.

#### 3.2.2.3 Waste Biomass

Waste biomasses are secondary biomass, as they are derived from primary biomass like trees, vegetables, meat during the different stages of their production or use. Municipal solid waste (MSW) is an important source of waste biomass, and much of it comes from renewables like food scraps, lawn clippings, leaves, and papers. Nonrenewable components of MSW like plastics, glass, and metals are not considered biomass. The combustible part of MSW is at times separated and sold as refuse derived fuel (RDF). Sewage sludge that contains human excreta, fat, grease, and food wastes is an important biomass source. Another waste is produced in sawmills during the production of lumber from wood. Table 3.3 lists the composition and heating values of some waste biomass products.

Landfills have traditionally been an important means of disposing of garbage. A designated area is filled with waste, which decomposes, producing methane gas. Modern landfilling involves careful lining of the containment

**TABLE 3.3** Typical Composition of Some Waste Biomass

| Biomass | Moisture (wt.%) | Organic Matter (dry wt.%) | Ash (dry wt.%) | HHV (MJ/dry kg) |
|---|---|---|---|---|
| Cattle manure | 20–70 | 76.5 | 23.5 | 13.4 |
| Sewage | 90–98 | 73.5 | 26.5 | 19.9 |
| RDF | 15–30 | 86.1 | 13.9 | 12.7 |
| Sawdust | 15–60 | 99.0 | 1.0 | 20.5 |

**Source:** Adapted from Klass (1998, p. 73).

cell (Figure 3.2) so that leached liquids can be collected and treated instead of leaking into groundwater. The containment cells are covered with clay or earth to avoid exposure to wind and rain.

An increasing number of municipalities are separating biodegradable wastes and are subjecting them to digestion for degradation. This avoids disposal of leachate and reduces the volume of waste. Two types of waste degradation are used: aerobic digestion and anaerobic digestion.

1. *Aerobic digestion*: This process takes place in the presence of air and so does not produce fuel gas. Here, the leachate is removed from the bottom layer of the landfill and pumped back into the landfill, where it flows over the waste repeatedly. Air added to the landfill enables microorganisms to work faster to degrade the wastes into compost, carbon dioxide, and water. Since it does not produce methane, aerobic digestion is most widely used where there is no additional need for landfill gas.
2. *Anaerobic digestion*: This process does not use air and hence produces the fuel gas methane. Here, the land-filled solids are sealed against contact with the atmospheric oxygen. The leachate is collected and pumped back into the landfill as in aerobic digestion (Figure 3.2). Additional liquids may be added to the leachate to help biodegradation of the waste. In the absence of oxygen, the waste is broken down into methane, carbon dioxide, and digestate (or solid residues). Methanogenesis bacteria like thermophiles (45–65°C), mesophiles (20–45°C), and psychophiles (0–20°C) facilitate this process (Probstein and Hicks, 2006). These biodegradation reactions are mildly exothermic. The process is represented by Eq. (3.2):

$$C_6H_{12}O_6 \text{ (representing wastes)} + \text{bacteria} = 3CO_2 + 3CH_4 + \text{digestate} \quad (3.2)$$

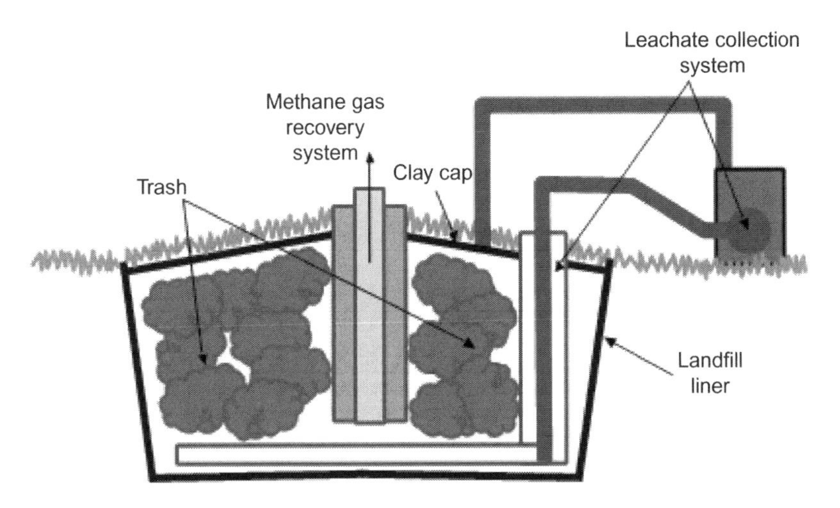


**FIGURE 3.2** Anaerobic digestion of biodegradable waste.

Methane is an important constituent of the landfill gas. It is also a powerful greenhouse gas (<21 times stronger than $CO_2$) that is often burnt in a flare or utilized in a gas engine or in similar energy applications. Anaerobic digestion is very popular in farming communities, where animal excreta are collected and stored because the gas produced can be collected in a gas holder for use in cooking and heating while the residual solid can be used as fertilizer.

## 3.3 STRUCTURE OF BIOMASS

Biomass is a complex mixture of organic materials such as carbohydrates, fats, and proteins, along with small amounts of minerals such as sodium, phosphorus, calcium, and iron. The main components of plant biomass are extractives, fiber or cell wall components, and ash (Figure 3.3).

*Extractives*: It includes substances present in vegetable or animal tissue that can be separated by successive treatment with solvents and recovered by evaporation of the solution. They include protein, oil, starch, and sugar.

*Cell wall*: It provides structural strength to the plant, allowing it to stand tall above the ground without support. A typical cell wall is made of carbohydrates and lignin. Carbohydrates are mainly cellulose or hemicellulose fibers, which impart strength to the plant structure, while the lignin holds the fibers together. These constituents vary with plant type. Some plants, such as corn, soybeans, and potatoes, also store starch (another carbohydrate polymer) and fats as sources of energy mainly in seeds and roots.

*Ash*: It is the inorganic component of the biomass.

Wood and its residues are the dominant constituents of the biomass resource base. A detailed discussion of wood-derived biomass is presented next.

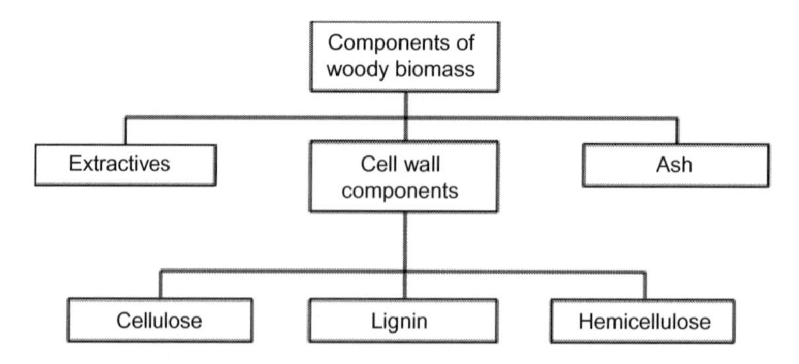


**FIGURE 3.3** Major constituents of a woody biomass.

### 3.3.1 Structure of Wood

Wood is typically made of hollow elongated and spindle-shaped cells arranged parallel to each other. Figure 3.4 is a photograph of the cross-section of a tree trunk showing the overall structure of a mature tree wood.

Bark is the outermost layer of a tree trunk or branch. It comprises an outer dead portion and an inner live portion. The inner live layer carries food from the leaves to the growing parts of the tree. It is made up of another layer known as *sapwood*, which carries sap from the roots to the leaves. Beyond this layer lies the inactive heartwood. In any cut wood, we can easily note a large number of radial marks. These radial cells (wood rays) carry food across the wood layers.

Wood cells that carry fluids are also known as fibers or *tracheids*. They are hollow and contain extractives and air. These cells vary in shape but are generally short and pointed. The length of an average tracheid is about 1000 μm for hardwood and typically 3000–8000 μm for softwood (Miller, 1999).

Tracheids are narrow. For example, the average diameter of the tracheid of softwood is 33 μm. These cells are the main conduits for the movement of sap along the length of the tree trunk. They are mostly aligned longitudinally, but there are some radial tracheids that carry sap across layers. Lateral channels, called *pith*, transport water between adjacent cells across the cell layers. Softwood (coniferous) can have cells or channels for carrying resins. A hardwood (deciduous), on the other hand, contains large numbers of pores or open vessels.

The tracheids or cells typically form an outer primary and an inner secondary wall. A layer called the *middle lamella* joins or glues together the adjacent cells. The middle lamella is predominantly made of lignin.

**FIGURE 3.4** Cross-section of a tree trunk showing outer dead bark, inner live bark, sapwood, heartwood, and wood rays. Source*: Photograph by author P. Basu.*

The secondary wall (inside the primary layer) is made up of three layers: S1, S2, and S3 (Figure 3.5). The thickest layer, S2, is made of *macrofibrils*, which consist of long cellulose molecules with embedded hemicellulose. The construction of cell walls in wood is similar to that of steel-reinforced concrete, with the cellulose fibers acting as the reinforcing steel rods and hemicellulose surrounding the cellulose microfibrils acting as the cement-concrete. The S2 layer has the highest concentration of cellulose. The highest concentration of hemicellulose is in layer S3. The distribution of these components in the cell wall is shown in Figure 3.6.

## 3.3.2 Constituents of Biomass Cells

The polymeric composition of the cell walls and other constituents of a biomass vary widely but they are essentially made of three major polymers: cellulose, hemicellulose, and lignin.

### 3.3.2.1 Cellulose

Cellulose, the most common organic compound on Earth, is the primary structural component of cell walls in biomass. Its amount varies from 90% (by weight) in cotton to 33% for most other plants. Represented by the generic formula $(C_6H_{10}O_5)_n$, cellulose is a long-chain polymer with a high degree of polymerization (<10,000) and a large molecular weight (<500,000). It has a crystalline structure of thousands of units, which are

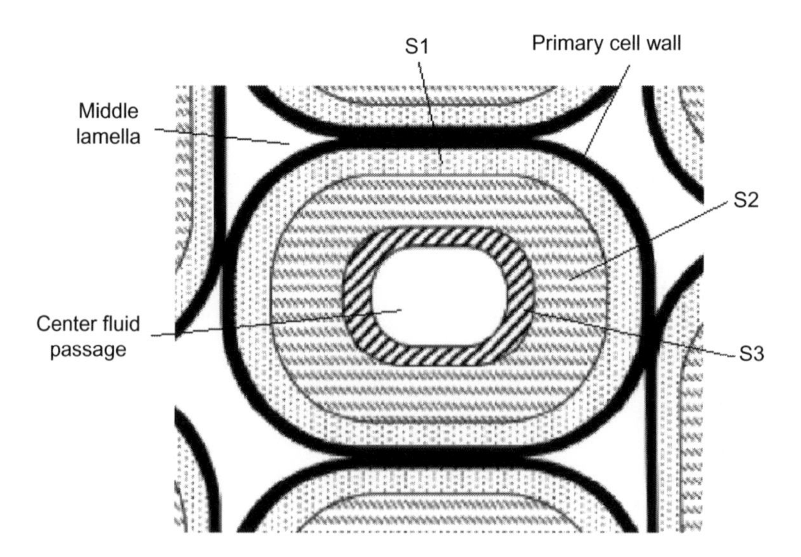


**FIGURE 3.5** Layers of a wood cell. The actual shape of the cross-section of a cell is not necessarily as shown.

made up of many glucose molecules. This structure gives it high strength, permitting it to provide the skeletal structure of most terrestrial biomass (Klass, 1998, p. 82). Cellulose is primarily composed of D-glucose, which is made of six carbons or hexose sugars (Figure 3.7).

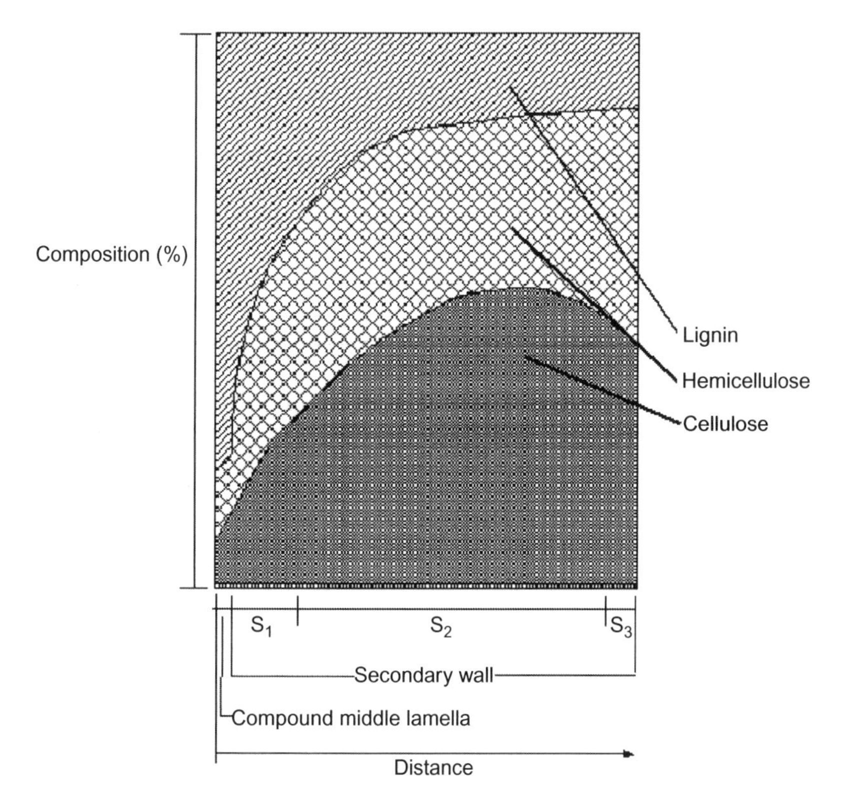


**FIGURE 3.6** Distribution of amounts of cellulose, hemicellulose, and lignin in cell walls and their layers.

**FIGURE 3.7** Molecular structure of cellulose.

Cellulose is highly insoluble and, though a carbohydrate, is not digestible by humans. It is a dominant component of wood, making up about 40–44% by dry weight. Cellulose is a major contributor of tar during gasification of biomass.

#### *3.3.2.2 Hemicellulose*

Hemicellulose is another constituent of the cell walls of a plant. While cellulose has a crystalline, strong structure that is resistant to hydrolysis, hemicellulose has a random, amorphous structure with little strength (Figure 3.8). It is a group of carbohydrates with a branched chain structure and a lower degree of polymerization (DP < 100–200) and may be represented by the generic formula $(C_5H_8O_4)_n$ (Klass, 1998, p. 84). Figure 3.8 shows the molecular arrangement of a typical hemicellulose molecule, xylan.

There is significant variation in the composition and structure of hemicellulose among different biomass. Most hemicelluloses, however, contain some simple sugar residues like D-xylose (the most common), D-glucose, D-galactose, L-ababinose, D-glucurnoic acid, and D-mannose. These typically contain 50–200 units in their branched structures.

Hemicellulose tends to yield more gases and less tar than cellulose (Milne, 2002). It is soluble in weak alkaline solutions and is easily hydrolyzed by dilute acid or base. It constitutes about 20–30% of the dry weight of most wood.

The hemicellulose content of a hardwood and a softwood sample could be comparable, but their behavior during torrefaction could be very different because of the variation in the composition of the hemicellulose of these samples. Main constituents of softwood hemicellulose are galactoglucomannans and arabino-glucuronoxylan (Table 3.4). In case of hardwood, it is glucuronoxylan, also called xylan. As one can see in Table 3.4, the glucoronoxylan (DP 200) in hardwood is in the range of 10–35%, while that in the form of arabino-glucuronoxylan (DP 100) is in the range of only 7–15%.

#### *3.3.2.3 Lignin*

Lignin is a complex highly branched polymer of phenylpropane and is an integral part of the secondary cell wall of plants. It is primarily a three-dimensional polymer of 4-propenyl phenol, 4-propenyl-2-methoxy phenol, and 4-propenyl-2.5-dimethoxyl phenol (Diebold and Bridgwater, 1997)

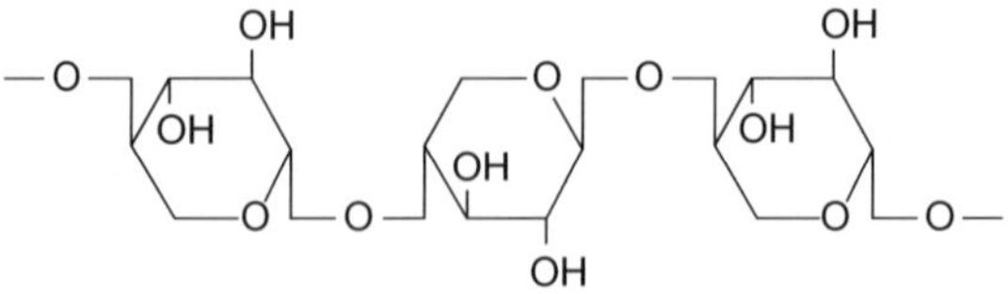

**FIGURE 3.8** Molecular structure of a typical hemicellulose, xylan.

**TABLE 3.4** Polymeric Composition of Some Wood

| | | | Hemicellulose | | | |
|---|---|---|---|---|---|---|
| Biomass Type | Cellulose | Lignin (Extractive Free Basis) | Noncellulsic | Glucomannan | Arabinogalactan | 4-0-Methylglucurono (xylan) |
| *Hardwood Average*[a] | 43–47 | 18–26 | 3 | 2–5 | 1 | 10–35 |
| Trembling aspen | 53 | 16 | 3 | 4 | 1 | 23 |
| White elm | 49 | 24 | 2 | 4 | 2 | 19 |
| Beech | 42 | 22 | 4 | 4 | 2 | 25 |
| White birch | 41 | 19 | 2 | 3 | 1 | 34 |
| Yellow birch | 40 | 21 | 3 | 7 | 1 | 28 |
| Red maple | 41 | 24 | 2 | 7 | 1 | 25 |
| *Softwood Average*[a] | 39–43 | 26–32 | 0 | 5–10[b], 10–15[c] | 2 | 7–15 |
| Balsam fir | 44 | 29 | 0 | 18 | 1 | 8 |
| Eastern white Cedar | 44 | 31 | 0 | 11 | 2 | 12 |
| Eastern hemlock | 42 | 33 | 0 | 17 | 1 | 7 |
| Jack pine | 41 | 29 | 0 | 16 | 2 | 10 |
| White spruce | 44 | 27 | 0 | 18 | 3 | 7 |
| Tamarah | 43 | 29 | 0 | 16 | 2 | 9 |

[a] *Schultz, T.P., Taylor, F.W, 1989. Wood. In: Kitani, O., Hall, C.W. (Eds.), Biomass Handbook, p. 136 (Chapter 1.2.5).*
[b] *Partially water soluble.*
[c] *Water soluble.*
**Source:** Adapted from Mullins and McKnight (1981).

(Figure 3.9). It is one of the most abundant organic polymers on Earth (exceeded only by cellulose). It is the third important constituent of the cell walls of woody biomass.

Lignin is the cementing agent for cellulose fibers holding adjacent cells together. The dominant monomeric units in the polymers are benzene rings. It is similar to the glue in a cardboard box, which is made by gluing together papers in a special fashion. The middle lamella (Figure 3.5), which is composed primarily of lignin, glues together adjacent cells or tracheids.

Lignin is highly insoluble, even in sulfuric acid (Klass, 1998, p. 84). A typical hardwood contains about 18–25% by dry weight of lignin, while softwood contains 25–35%.

## 3.4 GENERAL CLASSIFICATION OF FUELS

Classification is an important means of assessing the properties of a fuel. Fuels belonging to a particular group have similar behavior irrespective of their type or origin. Thus, when a new biomass is considered for gasification or other thermochemical conversion, we can check its classification, and then from the known properties of a biomass of that group, we can infer its conversion potential.

There are three methods of classifying and ranking fuels using their chemical constituents: atomic ratios, the ratio of lignocellulose constituents, and the ternary diagram. All hydrocarbon fuels may be classified or ranked according to their atomic ratios, but the second classification is limited to lignocellulose biomass.

### 3.4.1 Atomic Ratio

Classification based on the atomic ratio helps us to understand the heating value of a fuel, among other things. For example, the higher heating value (HHV) of a biomass correlates well with the oxygen-to-carbon (O/C) ratio, reducing from 20.5 to about 15 MJ/kg, while the O/C ratio increases from 0.86 to 1.03. When the hydrogen-to-carbon (H/C) ratio increases, the effective heating value of the fuel reduces.

The atomic ratio is based on the hydrogen, oxygen, and carbon content of the fuel. Figure 3.10 plots the atomic ratios (H/C) against (O/C) on a dry ash-free (daf) basis for all fuels, from carbon-rich anthracite to carbon-deficient

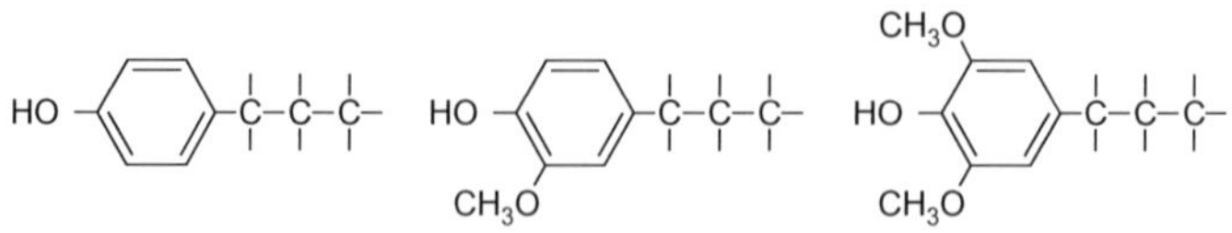

HO–C₆H₄–C–C–C– HO–C₆H₃(OCH₃)–C–C–C– HO–C₆H₂(OCH₃)₂–C–C–C–

**FIGURE 3.9** Some structural units of lignin.

woody biomass. This plot, known as *van Krevelen diagram*, shows that biomass has much higher ratios of H/C and O/C than fossil fuel has. For a large range of biomass, the H/C ratio may be expressed as a linear function of the (O/C) ratio (Jones et al., 2006).

$$\left[\frac{H}{C}\right] = 1.4125\left[\frac{O}{C}\right] + 0.5004 \tag{3.3}$$

Fresh plant biomass like leaves has very low heating values because of its high H/C and O/C ratios. The atomic ratios of a fuel decrease as its geological age increases, which means that the older the fuel, the higher its energy content. Anthracite, for example, a fossil fuel geologically formed over millions of years, has a very high heating value. The lower H/C ratio of anthracite gives it a high heating value, but the carbon intensity or the $CO_2$ emission from its combustion is high.

Among all hydrocarbon fuels, biomass is highest in oxygen content. Oxygen, unfortunately, does not make any useful contribution to heating value and makes it difficult to transform the biomass into liquid fuels. The high oxygen and hydrogen content of biomass results in high volatile and liquid yields, respectively. High oxygen consumes a part of the hydrogen in the biomass, producing less beneficial water, and thus the high H/C content does not translate into high gas yield.

### 3.4.2 Relative Proportions of Lignocellulosic Components

A biomass can also be classified on the basis of its relative proportion of cellulose, hemicellulose, and lignin. For example, we can predict the behavior

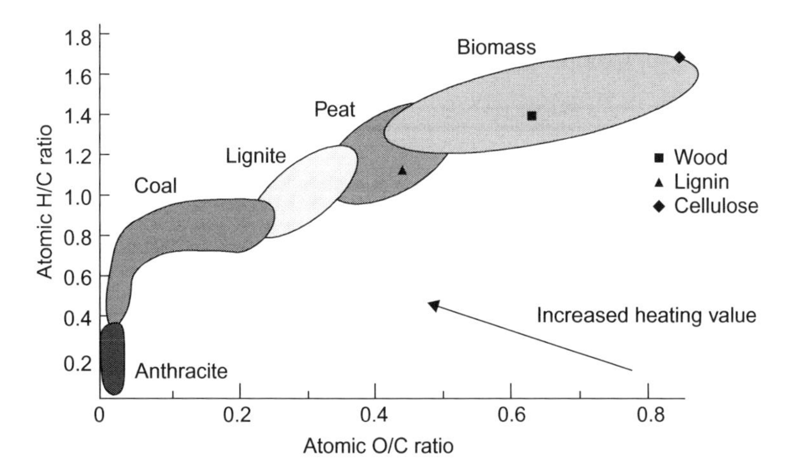


**FIGURE 3.10** Classification of solid fuels by hydrogen/carbon and oxygen/carbon ratio. Source: *Data from Jones et al. (2006).*

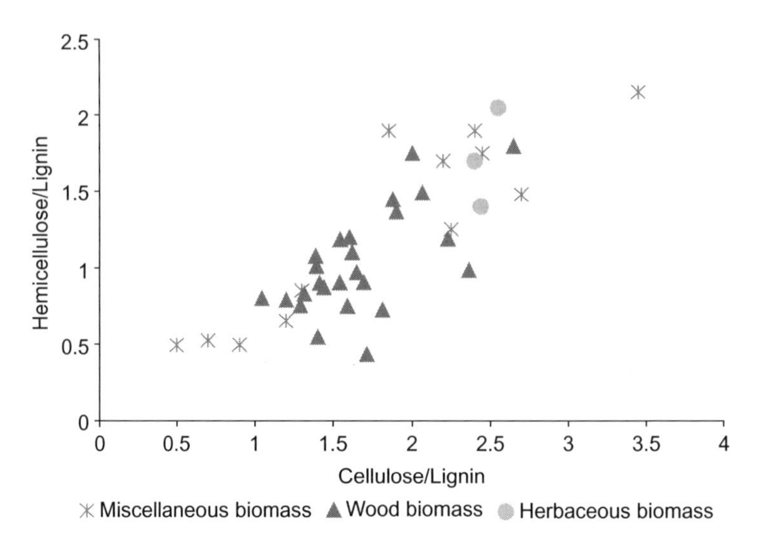


**FIGURE 3.11** Classification by constituent ratios of biomass. Source*: Data from Jones et al. (2006).*

of a biomass during pyrolysis from the knowledge of these components (Jones et al., 2006). Figure 3.11 plots the ratio of hemicellulose to lignin against the ratio of cellulose to lignin. In spite of some scatter, certain proportionality can be detected between the two. Biomass falling within these clusters behaves similarly irrespective of its type. For a typical biomass, the cellulose–lignin ratio increases from $<0.5$ to $<2.7$, while the hemicellulose–lignin ratio increases from 0.5 to 2.0.

## 3.4.3 Ternary Diagram

The ternary diagram (Figure 3.12) is not a tool for biomass classification, but it is useful for representing biomass conversion processes. The three corners of the triangle represent pure carbon, oxygen, and hydrogen—i.e., 100% concentration. Points within the triangle represent ternary mixtures of these three substances. The side opposite to a corner with a pure component (C, O, or H) represents zero concentration of that component. For example, the horizontal base in Figure 3.12 opposite to the hydrogen corner represents zero hydrogen—i.e., binary mixtures of C and O.

A biomass fuel is closer to the hydrogen and oxygen corners compared to coal. This means that biomass contains more hydrogen and more oxygen than coal contains. Lignin would generally have lower oxygen and higher carbon compared to cellulose or hemicellulose. Peat is in the biomass region but toward the carbon corner, implying that it is like a high-carbon biomass. Peat, incidentally, is the youngest fossil fuel formed from biomass.

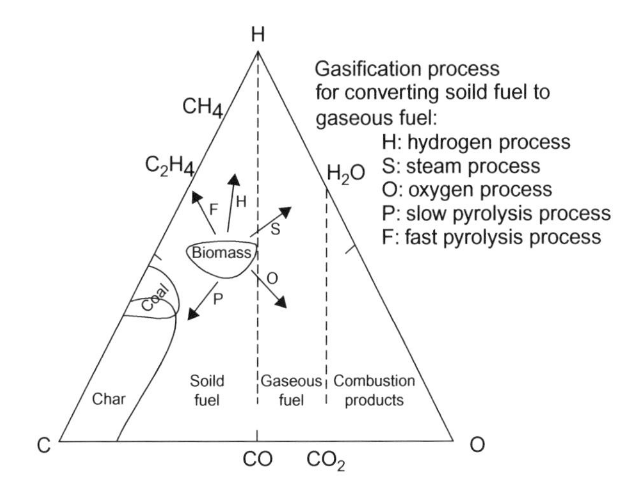


FIGURE 3.12 C—H—O ternary diagram of biomass showing the gasification process.

Coal resides further toward the carbon corner and lies close to the oxygen base in the ternary diagram, suggesting that it is very low in oxygen and much richer in carbon. Anthracite lies furthest toward the carbon corner because it has the highest carbon content. The diagram can also show the geological evolution of fossil fuels. With age, the fuel moves further away from the hydrogen and oxygen corners and closer to the carbon corner.

As mentioned earlier, the ternary diagram can depict the conversion process. For example, carbonization or slow pyrolysis moves the product toward carbon through the formation of solid char; fast pyrolysis moves it toward hydrogen and away from oxygen, which implies higher liquid product. Oxygen gasification moves the gas product toward the oxygen corner, while steam gasification takes the process away from the carbon corner. The hydrogenation process increases the hydrogen and thus moves the product toward hydrogen.

## 3.5 PROPERTIES OF BIOMASS

The following sections describe some important thermophysical properties of biomass that are relevant to gasification.

### 3.5.1 Physical Properties

Some of the physical properties of biomass affect its pyrolysis and gasification behavior. For example, permeability is an important factor in pyrolysis. High permeability allows pyrolysis gases to be trapped in the pores, increasing their residence time in the reaction zone. Thus, it increases the potential for secondary cracking to produce char. The pores in wood are generally oriented longitudinally. As a result, the thermal conductivity and diffusivity in the longitudinal direction are different from those in the lateral direction. This anisotropic behavior of wood can affect its thermochemical conversion. A densification process such as torrefaction (Chapter 4) can reduce the anisotropic behavior and therefore change the permeability of a biomass.

#### *3.5.1.1 Densities*

Density is an important design parameter for any biomass conversion system. For a granular biomass, we can define four characteristic densities: true, apparent, bulk, and biomass (growth).

##### True Density

True density is the weight per unit volume occupied by the solid constituent of biomass. Total weight is divided by actual volume of the solid content to give its true density.

$$\rho_{true} = \frac{\text{total mass of biomass}}{\text{solid volume in biomass}} \tag{3.4}$$

The cell walls constitute the major solid content of a biomass. For common wood, the density of the cell wall is typically 1530 $kg/m^3$, and it is constant for most wood cells (Desch and Dinwoodie, 1981). The measurement of true density of a biomass is as difficult as the measurement of true solid volume. It can be measured with a pycnometer or it may be estimated using ultimate analysis and the true density of its constituent elements.

##### Apparent Density

Apparent density is based on the apparent or external volume of the biomass. This includes its pore volume (or that of its cell cavities). For a regularly shaped biomass, mechanical means such as micrometers can be used to measure different sides of a particle to obtain its apparent volume. An alternative is the use of volume displacement in water. The apparent density considers the internal pores of a biomass particle but not the interstitial volume between biomass particles packed together.

$$\rho_{apparent} = \frac{\text{total mass of biomass}}{\text{apparent volume of biomass including solids and internal pores}} \tag{3.5}$$

The pore volume of a biomass expressed as a fraction of its total volume is known as its porosity, $\epsilon_p$. This is an important characteristic of the biomass.

Apparent density is most commonly used for design calculations because it is the easiest to measure, and it gives the actual volume occupied by a particle in a system. Table 3.5 gives typical apparent densities of some woods.

### Bulk Density

Bulk density is based on the overall space occupied by an amount or a group of biomass particles:

$$\rho_{bulk} = \frac{\text{total mass of biomass particles or stack}}{\text{bulk volume occupied by biomass particles or stack}} \tag{3.6}$$

Bulk volume includes interstitial volume between the particles, and as such it depends on how the biomass is packed. For example, after pouring the biomass particles into a vessel, if the vessel is tapped, the volume occupied by the particles settles to a lower value. The interstitial volume expressed as a function of the total packed volume is known as bulk porosity, $\epsilon_b$.

To determine the biomass bulk density, we can use standards like the American Society for Testing of Materials (ASTM) E-873-06. This process involves pouring the biomass into a standard-size box (305 mm × 305 mm × 305 mm) from a height of 610 mm. The box is then

**TABLE 3.5** Apparent Density of Some Wood Species

| Type | Wood Species | Apparent Density of Raw Wood, kg/m$^3$ | Shrinkage Green to Oven-Dry Volumetric, % |
|---|---|---|---|
| Softwood | Cedar, yellow | 420 | 6.4 |
| | Douglas fir | 450 | 11.9 |
| | Balsam fir | 340 | 10.7 |
| | Larch, western | 550 | 14 |
| | Pine, ponderosa | 440 | 10.5 |
| | Spruce, red | 380 | 11.7 |
| | Taramack | 480 | 11.2 |
| Hardwood | Birch yellow | 370 | 15.1 |
| | Maple, sugar | 560 | 15.7 |
| | Oak, red | 580 | 12 |

**Source:** Compiled from Mullins and McKnight (1981, p. 75.)

**TABLE 3.6 Standard Methods for Biomass Compositional Analysis**

| Biomass Constituent | Standard Methods |
|---|---|
| Carbon | ASTM E 777 for RDF |
| Hydrogen | ASTM E 777 for RDF |
| Nitrogen | ASTM E 778 for RDF |
| Oxygen | By difference |
| Ash | ASTM D 1102 for wood, E 1755 for biomass, D 3174 for coal |
| Moisture | ASTM E 871 for wood, E 949 for RDF, D 3173 for coal |
| Hemicelluloses | TAPPI T 223 for wood pulp |
| Lignin | TAPPI T222 for wood pulp, ASTM D 1106, acid insoluble in wood |
| Cellulose | TAPPI T−203 for wood pulp |

dropped from a height of 150 mm three times for settlement and refilling. The final weight of the biomass in the box divided by the box volume gives its bulk density.

The total mass of the biomass may contain the green moisture of a living plant, external moisture collected in storage, and moisture inherent in the biomass. Once the biomass is dried in a standard oven, its mass reduces. Thus, the density can be based on either green or oven-dry depending on if its weight includes surface moisture. The external moisture depends on the degree of wetness of the received biomass. To avoid this issue, we can completely saturate the biomass in deionized water, measure its maximum moisture density, and specify its bulk density accordingly.

Three of the preceding densities of biomass are related as follows:

$$\rho_{\text{apparent}} = \rho_{\text{true}}(1 - \varepsilon_{\text{p}}) \tag{3.7}$$

$$\rho_{\text{bulk}} = \rho_{\text{apparent}}(1 - \varepsilon_{\text{b}}) \tag{3.8}$$

where $\varepsilon_p$ is the void fraction (voidage) in a biomass particle and $\varepsilon_b$ is the voidage of particle packing.

## Biomass (Growth) Density

The term *biomass (growth) density* is used in bioresource industries to express how much biomass is available per unit area of land. It is defined as the total amount of above-ground living organic matter in trees expressed as oven-dry tons per unit area (e.g., tons per hectare) and includes all organic materials: leaves, twigs, branches, main bole, bark, and trees.

### 3.5.2 Thermodynamic Properties

Gasification is a thermochemical conversion process, so the thermodynamic properties of a biomass heavily influence its gasification. This section describes three important thermodynamic properties: thermal conductivity, specific heat, and heat of formation of biomass.

#### *3.5.2.1 Thermal Conductivity*

Biomass particles are subject to heat conduction along and across their fiber, which in turn influences their pyrolysis behavior. Thus, the thermal conductivity of the biomass is an important parameter in this context. It changes with density and moisture. Based on a large number of samples, MacLean (1941) developed the following correlations (from Kitani and Hall, 1989, p. 877):

$$\begin{aligned} K_{\text{eff}}(\text{W/m K}) &= \text{sp.gr}\,(0.2 + 0.004m_\text{d}) + 0.0238 \quad \text{for } m_\text{d} \text{ greater than } 40\% \\ &= \text{sp.gr}\,(0.2 + 0.0055m_\text{d}) + 0.0238 \quad \text{for } m_\text{d} \text{ less than } 40\% \end{aligned} \tag{3.9}$$

where sp.gr is the specific gravity of the fuel and $m_\text{d}$ is the moisture percentage of the biomass on a dry basis (db).

Unlike metal and other solids, biomass is highly anisotropic. The thermal conductivity along fibers of biomass is different from that across them. Conductivity also depends on the biomass' moisture content, porosity, and temperature. Some of these depend on the degree of conversion as the biomass undergoes combustion or gasification. A typical wood, for example, is made of fibers, the walls of which have channels carrying gas and moisture. Thunman and Leckner (2002) wrote the effective thermal conductivity parallel to the direction of wood fiber as a sum of contributions from fibers, moisture, and gas in it.

$$K_{\text{eff}} = G(x)K_\text{s} + F(x)\,K_\text{w} + H(x)\,[K_\text{g} + K_{\text{rad}}]\ \text{W/m K} \quad \text{for parallel to fiber} \tag{3.10}$$

where $G(x)$, $F(x)$, and $H(x)$ are functions of the cell structure and its dimensionless length; $K_\text{s}$, $K_\text{w}$, and $K_\text{g}$ are thermal conductivities of the dry solid (fiber wall), moisture, and gas, respectively; and $K_{\text{rad}}$ represents the contribution of radiation to conductivity.

These components are given by the following empirical relations, which are used to calculate the directional values of thermal conductivities (all thermal conductivities are in w/m K):

$$K_w = -0.487 + 5.887 \times 10^{-3}T - 7.39 \times 10^{-6}T^2$$

$$K_g = -7.494 \times 10^{-3} + 1.709 \times 10^{-4}T - 2.377 \times 10^{-7}T^2 + 2.202 \times 10^{-10}T^3 - 9.463 \times 10^{-14}T^4 + 1.581 \times 10^{-17}T^5$$

$$K_s = 0.52 \quad \text{in perpendicular direction} \tag{3.11}$$

$$K_{rad} = 5.33 e_{rad} \sigma T^3 d_{pore} \tag{3.12}$$

where $e_{rad}$ is the emissivity in the pores having diameter $d_{pore}$, $\sigma$ is the Stefan–Boltzmann constant, and $T$ is the temperature in K. The contribution of gas radiation in the pores, $K_{rad}$, to conductivity is important only at high temperatures.

Figure 3.13 shows the variation in the thermal conductivity of wood against its dry density. The straight line represents the thermal conductivity parallel to the fibers. The curved line gives the thermal conductivity across the fibers. The straight line is calculated from Eq. (3.10). Table C.10 that lists thermal conductivity of some wood, shows higher conductivity for hardwood, which also has higher density.

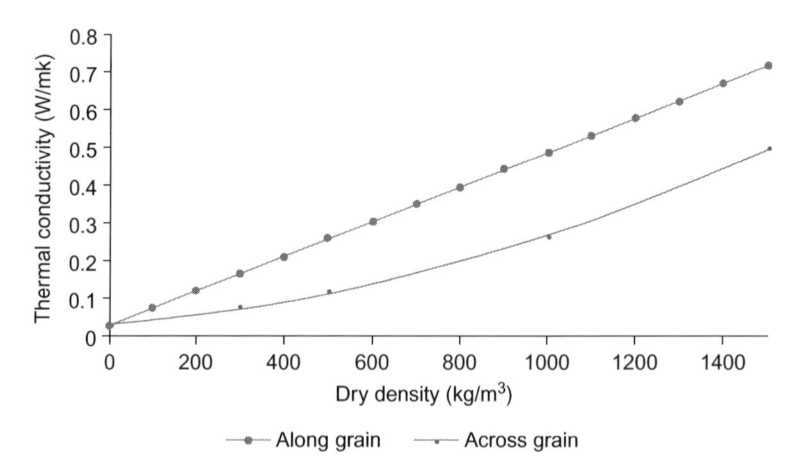


**FIGURE 3.13** The thermal conductivity of biomass along the grain (straight line) and across the grain (curved line) increases with the dry density of the biomass. The plot is for dry wood. Source: *Data from Thuman and Leckner (2002).*

### 3.5.2.2 Specific Heat

Specific heat is an important thermodynamic property of biomass often required for thermodynamic calculations. It is an indication of the heat capacity of a substance. Both moisture and temperature affect the specific heat of biomass, but density or wood species do not have much effect on the specific heat (Ragland et al., 1991). The specific heat changes much with temperature. It also depends to some extent on the type and source of biomass. Figure 3.14 shows the increase in specific heat of a softwood species with temperature. It also shows that bark of the wood has higher specific heat than its hearth wood. Char produced from this wood has interestingly much lower specific heat. Some experimental correlation of specific heat with temperature and moisture content is listed in Table 3.7.

### 3.5.2.3 Heat of Formation

Heat of formation, also known as *enthalpy of formation*, is the enthalpy change when 1 mol of compound is formed at standard state (25°C, 1 atm) from its constituting elements in their standard state. For example, hydrogen and oxygen are stable in their elemental form, so their enthalpy of formation is zero. However, an amount of energy (241.5 kJ) is released per mole when they combine to form steam.

$$H_2(\text{gas}) + 0.5O_2(\text{gas}) = H_2O\ (\text{gas}) - 241.5\ \text{kJ/mol} \tag{3.13}$$

The heat of formation of steam is thus −241.5 kJ/mol (g). This amount of energy is taken out of the system and is therefore given a negative (−) sign in the equation to indicate an exothermic reaction.

If the compound is formed through multiple steps, the heat of formation is the sum of the enthalpy change in each process step. Gases like $H_2$, $O_2$, $N_2$,

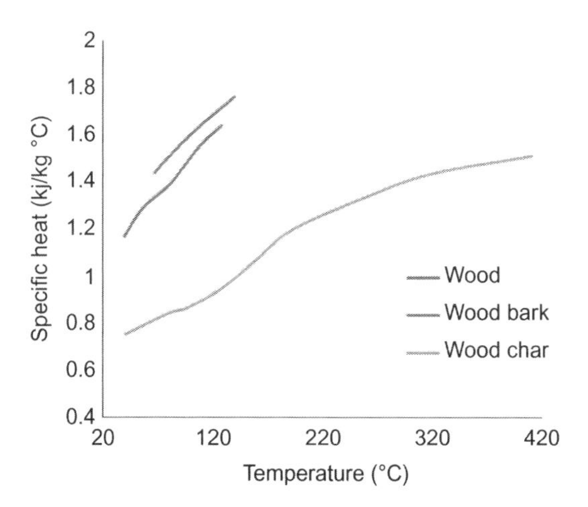


**FIGURE 3.14** Variation in specific heat with temperature for softwood, its bark, and char. Source*: Redrawn from Gupta et al. (2003).*

**TABLE 3.7** Specific Heat of Wood and Wood Char

| Reference | Fuel | Specific Heat in kJ/kg K | Validity (°C) |
|---|---|---|---|
| Ragland et al. (1991) | Dry wood | $C_{p,dry} = 0.1031 + 0.003867T$ | |
| | Wet wood | $[(C_{p,dry} + 4.19M_{dry})/(1 + M_{dry})] + A$, where $M_{dry}$ is moisture fraction on dry basis, $T$ in K, and $A = (0.02355T–1.32\ M–6.191)M_{dry}$ | |
| Ragland et al. (1991)[a] | Wood char | $1.39 + 0.00036T$ | 420–1720 |
| Gupta, et al (2003)[b] | Softwood | $0.00546T–0.524$ | 40–140 |
| | Char from softwood | $-0.0038 \times 10^{-3}T^2 + 0.00598T–0.795$ | 40–413 |
| Simpson and Tenwolde (1999)[c] | Wood | $C_{p,dry} = 0.1031 + 0.003867T$ | 7–147°C |
| | | $C_p = (C_{p,dry} + 4.19M)/(1 + 0.01M) + Ac$, where $Ac = M$ $(-0.06191 + 2.36 \times 10^{-4}T–1.33 \times 10^{-4}\ M)$ | 7–147 |
| Jenkins (1989), p. 876 | Various wood | $C_{p,dry} = 0.266 + 0.00116(T - 273)$ | 0–106°C |
| | | $C_p = C_{p,dry}(1 - M_{wet}) + 4.19M_{wet}$, where $M_{wet}$ is moisture fraction on wet basis | 0–106 |

[a]*Ragland, K.W., Aerts, D.J., Baker, A.J. (1991). Properties of wood for combustion analysis Bioresource Technol. 37, 161–168.*
[b]*Gupta, M., Yang, J., Roy, C. (2003). Specific heat and thermal conductivity of softwood bark and softwood char particles. Fuel 82, 919–927.*
[c]*Simpson, W., Tenwolde, A. (1999). Physical Properties and Moisture Relations of Wood (Chapter 3) 3–17.*

and $Cl_2$ are not compounds, and the heat of formation for them is zero. Values for the heat of formation for common compounds are given in Table 3.8.

### 3.5.2.4 Heat of Combustion (Reaction)

The *heat of reaction* (HR) is the amount of heat released or absorbed in a chemical reaction with no change in temperature. In the context of combustion reactions, HR is called *heat of combustion*, $\Delta H_{comb}$, which can be calculated from the *heat of formation* (HF) as:

$$CH_4 + 2O_2 \rightarrow 2H_2O + CO_2 \tag{3.14}$$

**TABLE 3.8 Heat of Formation of Some Important Compounds**

| Compound | $H_2O$ | $CO_2$ | CO | $CH_4$ | $O_2$ | $CaCO_3$ | $NH_3$ |
|---|---|---|---|---|---|---|---|
| Heat of formation at 25°C (kJ/mole) | −241.5 | −393.5 | −110.6 | −74.8 | 0 | −1211.8 | −82.5 |

**Source:** Data collected from Perry and Green (1997, pp. 2–186).

For example:

$$\Delta H_{comb} = 2\Delta H_{H_2O} + \Delta H_{CO_2} - \Delta H_{CH_4} - 2\Delta H_{O_2} \qquad (3.15)$$

The $\Delta H_{comb}$ for a fuel is also defined as the enthalpy change for the combustion reaction when balanced:

$$\text{Fuel} + O_2 \rightarrow H_2O + CO_2 - \text{HR} \qquad (3.16)$$

**Example 3.1**

Find the heat of formation of sawdust, the heating value of which is given as 476 kJ/mol. Assume its chemical formula to be $CH_{1.35}O_{0.617}$.

**Solution**

Using stoichiometry, the conversion reaction of SW can be written in the simplest terms as:

$$CH_{1.35}O_{0.617} + 1.029O_2 \rightarrow CO_2 + 0.675H_2O - 476\ \text{kJ/mol sawdust}$$

Similar to Eq. (3.14), we can write:

$$\text{Heat of reaction} = [HF_{CO_2} + 0.675HF_{H_2O}] - [HF_{sw} + 1.029HF_{O_2}]$$

Taking values of HF of $CO_2$, $O_2$, and $H_2O$ (g) from Table 3.8, we get:

$$HR_{sw} = [-393.5 + 0.675 \times (-241.5)] - [HF_{sw} + 1.029 \times 0] = -556.5 - HF_{sw}$$

The HR for the above combustion reaction is −476 kJ/mol. So $HF_{sw} = -556.5-(-476) = -80.5$ kJ/mol.

### *3.5.2.5 Heating Value*

The heating value of biomass is the amount of energy biomass releases when it is completely burnt in adequate oxygen. It is one of the most important properties of biomass as far as energy conversion is concerned. Compared to most fossil fuels, the heating value of biomass is low, especially on a volume basis, because its density is very low and it is high oxygen containing fuel. Section 3.6.5 discusses this in more detail.

#### 3.5.2.6 *Ignition Temperature*

Ignition temperature is an important property of any fuel because the combustion reaction of the fuel becomes self-sustaining only above this temperature. In a typical gasifier, a certain amount of combustion is necessary to provide the energy required for drying and pyrolysis and finally for the endothermic gasification reaction. In torrefier, temperature of cooled product should be lower than its ignition temperature. In this context, it is important to have some information on the ignition characteristics of the fuel.

Exothermic chemical reactions can take place even at room temperature, but the reaction rate, being an exponential function of temperature, is very slow at low temperatures. The heat loss from the fuel, on the other hand, is a linear function of temperature. At low temperatures, then, any heat released through the reaction is lost to the surroundings at a rate faster than that at which it was produced. As a result, the temperature of the fuel does not increase.

When the fuel is heated by some external means, the rate of exothermic reaction increases with a corresponding increase in the heat generation rate. Above a certain temperature, the rate of heat generation matches or exceeds the rate of heat loss. When this happens, the process becomes self-sustaining and that minimum temperature is called the *ignition temperature*.

The ignition temperature is generally lower for higher volatile matter content fuel. Because biomass particles have a higher volatile matter content than coal, they have a significantly lower ignition temperature, as Table 3.9

**TABLE 3.9** Ignition Temperatures of Some Common Fuels

| Fuel | Ignition Temperature (°C) | Volatile Matter in Fuel (dry ash-free %) | Reference |
|---|---|---|---|
| Wheat straw | 220 | 72 | Grotkjær et al. (2003) |
| Poplar wood | 235 | 75 | Grotkjær et al. (2003) |
| Eucalyptus | 285 | 64 | Grotkjær et al. (2003) |
| Ethanol | 425 | | |
| High volatile coals | 667 | 34.7 | Mühlen and Sowa (1995) |
| Medium volatile coal | 795 | 20.7 | Mühlen and Sowa (1995) |
| Anthracite | 927 | 7.3 | Mühlen and Sowa (1995) |

gives. Ignition temperature, however, is not necessarily a unique property of a fuel because it depends on several other factors like oxygen, partial pressure, particle size, rate of heating, and a particle's thermal surroundings.

## 3.6 COMPOSITION OF BIOMASS

Biomass contains a large number of complex organic compounds, moisture (*M*), and a small amount of inorganic impurities known as ash (ASH). The organic compounds comprise four principal elements: carbon (C), hydrogen (H), oxygen (O), and nitrogen (N). Biomass (e.g., MSW and animal waste) may also have small amounts of chlorine (Cl) and sulfur (S). The latter is rarely present in biomass except for secondary sources like demolition wood, which comes from torn-down buildings and structures.

Thermal design of a biomass utilization system, whether it is a gasifier or a combustor, necessarily needs the composition of the fuel as well as its energy content. In the context of thermal conversion like combustion, following two types of compositions are mostly used:

1. Ultimate or elemental composition
2. Proximate composition.

Besides these, there is also the polymeric composition of biomass, which is important for chemical conversions like torrefaction, pyrolysis, and gasification.

### 3.6.1 Ultimate Analysis

Here, the composition of the hydrocarbon fuel is expressed in terms of its basic elements except for its moisture, *M*, and inorganic constituents, ASH. A typical ultimate analysis is:

$$\mathrm{C} + \mathrm{H} + \mathrm{O} + \mathrm{N} + \mathrm{S} + \mathrm{ASH} + M = 100\% \tag{3.17}$$

Here, C, H, O, N, and S are the mass percentages of carbon, hydrogen, oxygen, nitrogen, and sulfur, respectively, in the fuel. Not all fuels contain all of these elements. For example, the vast majority of biomass may not contain any sulfur (S). The moisture or water in the fuel is expressed separately as *M*. Thus, hydrogen or oxygen in the ultimate analysis does not include the hydrogen and oxygen in the moisture, but only the hydrogen and oxygen present in the organic components of the fuel. Table 3.10 compares the ultimate analysis of several biomass materials with that of some fossil fuels.

The atomic ratios (H/C) and (O/C) plotted in Figure 3.10 is derived from the ultimate analysis of different fuels. This figure shows that biomass, cellulose in particular, has very high relative amounts of oxygen and hydrogen, which results in relatively low heating values.

The sulfur content of lignocellulosic biomass is exceptionally low, which is a major advantage in its utilization in energy conversion when $SO_2$ emission is taken into account. To reduce $SO_2$ emission from the combustion of sulfur-bearing fuels, such as fuel-oil, coal, and petcoke, one can use limestone. Theoretically, for every mole of sulfur captured, only 1 mol of limestone

**TABLE 3.10 Comparison of Ultimate Analysis (dry basis) of Some Biomass and its Comparison with Other Fossil Fuels**

| | C (%) | H (%) | N (%) | S (%) | O (%) | Ash (%) | HHV (kJ/kg) | Source |
|---|---|---|---|---|---|---|---|---|
| Maple | 50.6 | 6.0 | 0.3 | 0 | 41.7 | 1.4 | 19,958 | Tillman (1978) |
| Douglas fir | 52.3 | 6.3 | 9.1 | 0 | 40.5 | 0.8 | 21,051 | Tillman (1978) |
| Douglas fir (bark) | 56.2 | 5.9 | 0 | 0 | 36.7 | 1.2 | 22,098 | Tillman (1978) |
| Redwood | 53.5 | 5.9 | 0.1 | 0 | 40.3 | 0.2 | 21,028 | Tillman (1978) |
| Redwood waste | 53.4 | 6.0 | 39.9 | 0.1 | 0.1 | 0.6 | 21,314 | |
| Sewage sludge | 29.2 | 3.8 | 4.1 | 0.7 | 19.9 | 42.1 | 16,000 | |
| Straw-rice | 39.2 | 5.1 | 0.6 | 0.1 | 35.8 | 19.2 | 15,213 | Tillman (1978) |
| Husk-rice | 38.5 | 5.7 | 0.5 | 0 | 39.8 | 15.5 | 15,376 | Tillman (1978) |
| SW | 47.2 | 6.5 | 0 | 0 | 45.4 | 1.0 | 20,502 | Wen et al. (1974) |
| Paper | 43.4 | 5.8 | 0.3 | 0.2 | 44.3 | 6.0 | 17,613 | Bowerman (1969) |
| MSW | 47.6 | 6.0 | 1.2 | 0.3 | 32.9 | 12.0 | 19,879 | Sanner et al. (1970) |
| Animal waste | 42.7 | 5.5 | 2.4 | 0.3 | 31.3 | 17.8 | 17,167 | Tillman (1978) |
| Peat | 54.5 | 5.1 | 1.65 | 0.45 | 33.09 | 5.2 | 21,230 | |
| Lignite | 62.5 | 4.38 | 0.94 | 1.41 | 17.2 | 13.4 | 24,451 | Bituminous Coal Research (1974) |
| PRB coal | 65.8 | 4.88 | 0.86 | 1.0 | 16.2 | 11.2 | 26,436 | Probstein and Hicks (2006), p. 14 |
| Anthracite | 83.7 | 1.9 | 0.9 | 0.7 | 10.5 | 2.3 | 27,656 | Basu et al. (2000), p. 25 |
| Petcoke | 82 | 0.5 | 0.7 | 0.8 | 10.0 | 6.0 | 28,377 | Basu et al. (2000), p. 25 |

**Source:** Reed (2002).Fuel

($CaCO_3$) is required which will release 1 mol of additional carbon dioxide during the production of CaO from $CaCO_3$. In reality, much more 2–4 mol of limestone is required due to imperfect sulfation of the calcined limestone. Thus, for capture of sulfur dioxide, considerable amount of additional carbon dioxide is generated.

Biomass, in addition to being $CO_2$ neutral, results in additional reduction in $CO_2$ emission because of the absence of sulfur capture-related $CO_2$ emission as needed for many fossil fuels as described above.

### 3.6.2 Proximate Analysis

Proximate analysis gives the composition of the biomass in terms of gross components such as moisture (*M*), volatile matter (VM), ash (ASH), and fixed carbon (FC). It is a relatively simple and inexpensive process. Table 3.11 compares the proximate analysis of corn cob and rice husk measured in two different techniques.

#### *3.6.2.1 Volatile Matter*

The volatile matter of a fuel is the condensable and noncondensable vapor released when the fuel is heated. Its amount depends on the rate of heating and the temperature to which it is heated. Options for its measurement are discussed in Chapter 13.

#### *3.6.2.2 Ash*

Ash is the inorganic solid residue left after the fuel is completely burned. Its primary ingredients are silica, aluminum, iron, and calcium; small amounts of magnesium, titanium, sodium, and potassium may also be present.

Strictly speaking, this ash does not represent the original inorganic mineral matter in the fuel, as some of the ash constituents can undergo oxidation during burning. For exact analysis, correction may be needed. The ash

**TABLE 3.11** Comparison of Proximate Analysis of Biomass Measured by Two Methods

| Fuel | FC (% dry) | VM (% dry) | ASH (% dry) | Technique Used |
|---|---|---|---|---|
| Corncobs | 18.5 | 80.1 | 1.4 | ASTM |
| | 16.2 | 80.2 | 30.6 | TG |
| Husk-rice | 16.7 | 65.5 | 17.9 | ASTM |
| | 19.9 | 60.6 | 19.5 | TG |

**Source:** Compiled from Klass (1998, p. 239).

content of biomass is generally very small but may play a significant role in biomass utilization especially if it contains alkali metals such as potassium or halides such as chlorine. Straw, grasses, and demolition wood are particularly susceptible to this problem. These components can lead to serious agglomeration, fouling, and corrosion in boilers or gasifiers (Mettanant et al., 2009).

The ash obtained from biomass conversion does not necessarily come entirely from the biomass itself. For collection, biomass is often scraped off the forest floor and then undergoes multiple handlings, during which it can pick up a considerable amount of dirt, rock, and other impurities. In many plants, these impurities constitute the major inorganic component of the biomass feedstock.

### 3.6.2.3 Moisture

High moisture is a major characteristic of biomass. The root of a plant biomass absorbs moisture from the ground and pushes it into the sapwood. The moisture travels to the leaves through the capillary passages. Photosynthesis reactions in the leaves use some of it, and the rest is released to the atmosphere through transpiration. For this reason, there is more moisture in the leaves than in the tree trunk.

The total moisture content of some biomass can be as high as 90% (db), as seen in Table 3.12. Moisture drains much of the deliverable energy from a gasification plant, as the energy used in evaporation is not recovered. This important input parameter for design must be known for assessment of the cost of transportation or energy penalty in drying the biomass. The moisture in biomass can remain in two forms: (1) free or external and (2) inherent or equilibrium.

Free moisture is that which is above the equilibrium moisture content. It generally resides outside the cell walls. Inherent moisture, on the other hand, is absorbed within the cell walls. When the walls are completely saturated, the biomass is said to have reached the fiber saturation point or equilibrium moisture. Equilibrium moisture is a strong function of the relative humidity

**TABLE 3.12** Typical Moisture Content of Some Biomass

| | Corn Stalks | Wheat Straw | Rice Straw | Rice Husk | Dairy Cattle Manure | Wood Bark | Sawdust | Food Waste | RDF Pellets | Water Hyacinth |
|---|---|---|---|---|---|---|---|---|---|---|
| Moisture (wet basis) | 40–60 | 8–20 | 50–80 | 7–10 | 88 | 30–60 | 25–55 | 70 | 25–35 | 95.3 |

**Source:** Compiled from Kitani and Hall (1989, p. 863).Biomass

and weak function of air temperature. For example, the equilibrium moisture of wood increases from 3% to 27% when the relative humidity increases from 10% to 80% (Kitani and Hall, 1989, p. 864). The moisture content of some biomass fuels is given in Table 3.12.

### Basis of Expressing Moisture

Biomass moisture is often expressed on a db. For example, if $W_{wet}$ kg of wet biomass becomes $W_{dry}$ after drying, its db ($M_{dry}$) is expressed as:

$$M_{dry} = \frac{W_{wet} - W_{dry}}{W_{dry}} \tag{3.18}$$

This can give a moisture percentage greater than 100% for very wet biomass, which might be confusing. For that reason, the basis of moisture should always be specified.

The wet-basis moisture is:

$$M_{wet} = \frac{W_{wet} - W_{dry}}{W_{wet}} \tag{3.19}$$

The wet basis ($M_{wet}$) and the db ($M_{dry}$) are related as:

$$M_{dry} = \frac{M_{wet}}{1 - M_{wet}} \tag{3.20}$$

### Fixed Carbon

Fixed carbon (FC) in a fuel is determined from the following equation, where $M$, VM, and ASH stand for moisture, volatile matter, and ash, respectively.

$$\mathrm{FC} = 1 - M - \mathrm{VM} - \mathrm{ASH} \tag{3.21}$$

This represents the solid carbon in the biomass that remains in the char in the pyrolysis process after devolatilization. With coal, FC includes elemental carbon in the original fuel plus any carbonaceous residue formed while heating, in the determination of VM (standard D-3175).

Biomass carbon comes from photosynthetic fixation of $CO_2$ and thus all of it is organic. During the determination of VM, a part of the organic carbon is transformed into a carbonaceous material called pyrolytic carbon. Since FC depends on the amount of VM, it is not determined directly. VM also varies with the rate of heating. In a real sense, then, fixed carbon is not a fixed quantity, but its value, measured under standard conditions, gives a useful evaluation parameter of the fuel. For gasification analysis, FC is an important parameter because in most gasifiers the conversion of fixed carbon into gases determines the rate of gasification and its yield. This conversion reaction, being the slowest, is used to determine the size of the gasifier.

#### Char

Char, though a carbon residue of pyrolysis or devolatilization, is not a pure carbon; it is not the fixed carbon of the biomass either. Known as *pyrolytic char*, it contains some volatiles and ash in addition to fixed carbon. Biomass char is very reactive. It is highly porous and does not cake.

### 3.6.3 Thermogravimetric Analysis

Because of the time and expense involved in proximate analysis by ASTM D-3172, Klass (1998) proposed an alternative method using thermogravimetry (TG) or differential thermogravimetry. It is discussed in Chapter 13.

### 3.6.4 Bases of Expressing Biomass Composition

The composition of a fuel is often expressed on different bases depending on the situation. The following four bases of analysis are commonly used:

1. As-received
2. Air-dry
3. Total dry
4. Dry and ash-free.

A comparison of these bases is shown in Figure 3.15.

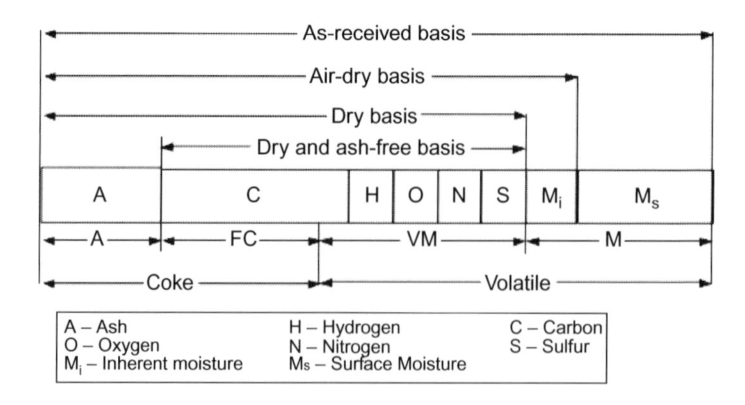


**FIGURE 3.15** Basis of expressing fuel composition.

### 3.6.4.1 *As-Received Basis*

When using the as-received (ar) basis, the results of ultimate and proximate analyses may be written as follows:

$$\text{Ultimate analysis:} \quad C + H + O + N + S + ASH + M = 100\% \tag{3.22}$$

$$\text{Proximate analysis:} \quad VM + FC + M + ASH = 100\% \tag{3.23}$$

where VM, FC, $M$, and ASH represent the weight percentages of volatile matter, fixed carbon, moisture, and ash, respectively, measured by proximate analysis, and C, H, O, N, and S represent the weight percentages of carbon, hydrogen, oxygen, nitrogen, and sulfur, respectively, as measured by ultimate analysis. The ash and moisture content of the fuel is the same in both analyses. "As received" can be converted into other bases.

### 3.6.4.2 *Air-Dry Basis*

When the fuel is dried in air, its surface moisture is removed while its inherent moisture is retained. So, to express the constituent on an air-dry (ad) basis, the amount is divided by the total mass less the surface moisture. For example, the carbon percentage on the "ad" basis is calculated as:

$$C_{ad} = \frac{100\ C}{100 - M_a} \tag{3.24}$$

where $M_a$ is the mass of surface moisture removed from 100 kg of moist fuel after drying in air. Other constituents of the fuel can be expressed similarly.

### 3.6.4.3 *Total Dry Basis*

Fuel composition on the "ad" basis is a practical parameter and is easy to measure, but to express it on a totally moisture-free (td) basis, we must make allowance for surface as well as inherent moisture, $M_i$. This gives the carbon percentage on a total db, $C_{td}$:

$$C_{td} = \frac{100\ C}{100 - M_i} \tag{3.25}$$

where $M$ is the total moisture (surface + inherent) in the fuel: $M = M_a + M_i$.

### 3.6.4.4 *Dry Ash-Free Basis*

Ash is another component that at times is eliminated along with moisture. This gives the fuel composition on a daf basis. Following the aforementioned examples, the carbon percentage on a "daf" basis, $C_{daf}$, can be found.

$$C_{daf} = \frac{100C}{100 - M - ASH} \tag{3.26}$$

where $(100 - M - ASH)$ is the mass of biomass without moisture and ash.

The percentage of all constituents on any basis totals 100. For example:

$$C_{daf} + H_{daf} + O_{daf} + N_{daf} + S_{daf} = 100\% \quad (3.27)$$

### 3.6.5 Heating Value of Fuel

The heating value of biomass is relatively low, especially on a volume basis, because its density is very low.

#### *3.6.5.1 Higher Heating Value (HHV)*

It is defined as the amount of heat released by the unit mass or volume of fuel (initially at 25°C) once it is combusted and the products have returned to a temperature of 25°C. It includes the latent heat of vaporization of water. HHV is also called gross calorific value. In North America, the thermal efficiency of a system is usually expressed in terms of HHV, so it is important to know the HHV of the design fuel. Table 3.10 gives HHV of some biomass.

#### *3.6.5.2 Lower Heating Value (LHV)*

The temperature of the exhaust flue gas of a boiler is generally in the range 120–180°C. The products of combustion are rarely cooled to the initial temperature of the fuel, which is generally below the condensation temperature of steam. So, the water vapor in the flue gas does not condense, and therefore the latent heat of vaporization of this component is not recovered. The effective heat available for use in the boiler is a lower amount, which is less than the chemical energy stored in the fuel.

The lower heating value (LHV), also known as the net calorific value, is defined as the amount of heat released by fully combusting a specified quantity less the heat of vaporization of the water in the combustion product.

The relationship between HHV and LHV is given by:

$$\mathrm{LHV} = \mathrm{HHV} - h_g\left(\frac{9\mathrm{H}}{100} - \frac{M}{100}\right) \quad (3.28)$$

where LHV, HHV, $H$, and $M$ are lower heating value, higher heating value, hydrogen percentage, and moisture percentage, respectively, on an "ar" basis. Here, $h_g$ is the latent heat of steam in the same units as HHV. The latent heat of vaporization when the reference temperature is 100°C is 2260 kJ/kg.

Many European countries define the efficiency of a thermal system in terms of LHV. Thus, an efficiency expressed in this way appears higher than that expressed in HHV (as is the norm in many countries, including the United States and Canada), unless the basis is specified.

#### 3.6.5.3 Bases for Expressing Heating Values

Similar to fuel composition, heating value (HHV or LHV) may also be expressed in any of the following bases:

- "ar" basis
- db, also known as moisture-free basis (mf)
- daf, also known as moisture ash-free basis

If $M_f$ kg of fuel contains $Q$ kJ of heat, $M_w$ kg of moisture, and $M_{ash}$ kg of ash, HHV can be written in different bases as follows:

$$\mathrm{HHV_{ar}} = \frac{Q}{M_f}\ \mathrm{kJ/kg}$$

$$\mathrm{HHV_{db}} = \frac{Q}{(M_f - M_w)}\ \mathrm{kJ/kg}$$

$$\mathrm{HHV_{daf}} = \frac{Q}{(M_f - M_w - M_{ash})}\ \mathrm{kJ/kg} \tag{3.29}$$

#### 3.6.5.4 Estimation of Biomass Heating Values

Experimental methods are the most reliable means of determining the heating value of biomass. If these are not possible, empirical correlations like the Dulong–Berthelot equation, originally developed for coal with modified coefficients for biomass, may be used. Channiwala and Parikh (2002) developed the following unified correlation for HHV based on 15 existing correlations and 50 fuels, including biomass, liquid, gas, and coal.

$$\mathrm{HHV} = 349.1\mathrm{C} + 1178.3\mathrm{H} + 100.5\mathrm{S} - 103.4\mathrm{O} - 15.1\mathrm{N} - 21.1\mathrm{ASH}\ \mathrm{kJ/kg} \tag{3.30}$$

where C, H, S, O, N, and ASH are percentages of carbon, hydrogen, sulfur, oxygen, nitrogen, and ash as determined by ultimate analysis on a dry basis. This correlation is valid within the range:

$0 < \mathrm{C} < 92\%$; $0.43 < \mathrm{H} < 25\%$
$0 < \mathrm{O} < 50$; $0 < \mathrm{N} < 5.6\%$
$0 < \mathrm{ASH} < 71\%$; $4745 < \mathrm{HHV} < 55{,}345$ kJ/kg

Ultimate analysis is necessary with this correlation, but it is expensive and time consuming. Zhu and Venderbosch (2005) developed an empirical method to estimate HHV without ultimate analysis. This empirical relationship between the stoichiometric ratio (SR) and the HHV is based on data for 28 fuels that include biomass, coal, liquid, and gases. The relation is useful for preliminary design:

$$\mathrm{HHV} = 3220 \times \mathrm{SR}\ \mathrm{kJ/kg} \tag{3.31}$$

where the SR is the theoretical mass of the air required to burn 1 kg fuel.

### 3.6.6 Stoichiometric Calculations for Complete Combustion

Noting that dry air contains 23.16% oxygen, 76.8% nitrogen, and 0.04% inert gases by weight, the dry air required for complete combustion of a unit weight of dry hydrocarbon, $M_{da}$, is given by:

$$M_{da} = \left[0.1153\text{C} + 0.3434\left(\text{H} - \frac{\text{O}}{8}\right) + 0.043\text{S}\right] \text{ kg/kg of dry fuel} \quad (3.32)$$

where C, H, O, and S are the percentages of carbon, hydrogen, oxygen, and sulfur, respectively, on a dry basis.

The actual air including excess air EAC and moisture, $X_m$, in air is $M_{wa} = (1 + \text{EAC})M_{da}(1 + X_m)$.

#### 3.6.6.1 Amount of Product Gas of Complete Combustion

The total weight of the flue gas, $W_c$, produced through combustion of 1 kg of biomass may be found from stoichiometry as (Basu, 2006, p. 448):

$$W_c = M_{wa} - 0.2315\, M_{da} + 3.66\text{C} + 9\text{H} + \text{N} + \text{O} + 2.5\text{S} \quad (3.33)$$

#### 3.6.6.2 Composition of the Product of Combustion

**a.** *Carbon dioxide*

Carbon dioxide produced from fixed carbon in coal = 3.66 C kg/kg fuel.

**b.** *Water vapor*

Water in the flue gas comes from the combustion of hydrogen in the coal and the moisture present in the combustion air, coal, and limestone.

$$\text{Water vapor in the flue gas} = [9\text{H} + \text{EAC}.\, M_{da}X_m + M_f + L_qX_{ml}] \text{ kg/kg fuel} \quad (3.34)$$

**c.** *Nitrogen*

Nitrogen in the flue gas comes from the coal as well as from the combustion air.

$$\text{Nitrogen from the air and fuel} = [\text{N} + 0.768\ \text{EAC}.\, M_{da}] \text{ kg/kg fuel} \quad (3.35)$$

**d.** *Oxygen*

The oxygen in the flue gas comes from oxygen in the coal, excess oxygen in the combustion air, and the oxygen left in the flue gas for incomplete capture of sulfur. For each mole of unconverted sulfur, ½ mol of oxygen is saved. Thus:

$$\text{Oxygen in the flue gas} = [\text{O} + 0.2315M_{da}(\text{EAC} - 1)] \text{ kg/kg fuel} \quad (3.36)$$

**e.** *Sulfur dioxide*

The $SO_2$ present in the flue gas is given below:

$$\text{Sulfur dioxide in flue gas} = 2\text{S kg/kg fuel} \quad (3.37)$$

### 3.6.7 Composition of the Product Gas of Gasification

The product gas of gasification is generally a mixture of several gases including moisture or steam. Its composition may be expressed in any of the following ways:

- Mass fraction $m_i$
- Mole fraction $n_i$
- Volume fraction $V_i$
- Partial pressure $P_i$

It may also be expressed on a dry or a wet basis. The wet basis is the composition gas expressed on the basis of total mass of the gas mixture including any moisture in it. The dry basis is the composition with the moisture entirely removed.

The following example illustrates the relationship between different ways of expressing the product gas composition.

---

**Example 3.2**

The gasification of a biomass yields $M$ kg/s product gas, with the production of its individual constituents as follows:

Hydrogen: $M_H$, kg/s
Carbon monoxide: $M_{CO}$, kg/s
Carbon dioxide: $M_{CO_2}$, kg/s
Methane: $M_{CH_4}$ kg/s
Other hydrocarbon (e.g., $C_3H_8$): $M_{HC}$, kg/s
Nitrogen: $M_N$, kg/s
Moisture: kg/s

Find the composition of the product gas in mass fraction, mole fraction, and other fractions.

**Solution**

Since the total gas production rate, $M$, is:

$$M = M_H + M_{CO} + M_{CO_2} + M_{CH_4} + M_H + M_N + M_{H_2O} \text{ kg/s} \quad \text{(i)}$$

the mass fraction of each species is found by dividing the individual production rate by the total. For example, the mass fraction of hydrogen is $m_H = M_H/M$.

The mole of an individual species is found by dividing its mass by its molecular weight:

$$\text{Moles of hydrogen } n_H = \text{mass/molecular weight of } H_2 = m_H/2 \quad \text{(ii)}$$

The total number of moles of all gases is found by adding the moles of $i$ species of gases, $n = \sum(n_i)$ moles. So the mole fraction of hydrogen is $x_H = n_H/n$. Similarly for any gas, the mole fraction is:

$$X_i = n_i/n \quad \text{(iii)}$$

where the subscript refers to the $i$th species.

The volume fraction of a gas can be found by noting that the volume that 1 kmol of any gas occupies at normal temperature and pressure (NTP) (at 0°C and 1 atm) is 22.4 $m^3$. So, taking the example of hydrogen, the volume of 1 kmol of hydrogen in the gas mixture is 22.4 $nm^3$ at NTP.

The total volume of the gas mixture is $V$ = summation of volumes of all constituting gases in the mixture $= \sum([\text{number of moles } (n_i) \times 22.4])/\text{nm}^3 = 22.4n$. The volume fraction of hydrogen in the mixture is volume of hydrogen/total volume of the mixture:

$$V_H = \frac{22.4 n_H}{22.4 \sum n_i} = \frac{n_h}{n} = x_H \quad \text{(iv)}$$

Thus, we note that:

Volume fraction = mole fraction

The partial pressure of a gas is the pressure it exerts if it occupies the entire mixture volume $V$. Ideal gas law gives the partial pressure of a gas component, $i$, as

$$P_i = \frac{n_i RT}{V} P_a$$

The total pressure, $P$, of the gas mixture containing total moles, $n$, is

$$P = n\frac{RT}{V}$$

So we can write:

$$x_i = \frac{n_i}{n} = \frac{p_i}{p} = \frac{v_i}{V} \quad \text{(v)}$$

Partial pressure as fraction of total pressure = mole fraction = volume fraction.

The partial pressure of hydrogen is $P_H = x_H P$.

The molecular weight of the mixture gas, $MW_m$, is known from the mass fraction and the molecular weight of individual gas species

$$MW_m = \sum[x_i MW_i] \quad \text{(vi)}$$

where $MW_i$ is the molecular weight of gas component $i$ with mole fraction $x_i$.

## SYMBOLS AND NOMENCLATURE

| | |
|---|---|
| **ASH** | weight percentage of ash (%) |
| **C** | weight percentage of carbon (%) |
| $C_p$ | specific heat of biomass (J/g K) |
| $C_{p\theta}$ | specific heat of biomass at temperature $\theta$°C (J/g C) |
| $C_w$ | specific heat of water (J/g K) |
| $d_{pore}$ | pore diameter (m) |
| $e_{rad}$ | emissivity in the pores (−) |
| **FC** | weight percentage of fixed carbon (%) |

| | |
|---|---|
| **$G(x)$, $F(x)$, $H(x)$** | functions of the cell structure and its dimensionless length of the biomass in Eq. (3.10) (–) |
| **HR** | heat of combustion or heat of reaction (kJ/mol) |
| **HF** | heat of formation (kJ/mol) |
| **H** | weight percentage of hydrogen (%) |
| **HHV** | high heating value of fuel (kJ/kg) |
| $h_g$ | latent heat of vaporization (kJ/kg) |
| $K_{eff}$ | effective thermal conductivity of biomass (W/m K) |
| $K_s$ | thermal conductivity of the solid in dry wood (W/m K) |
| $K_w$ | thermal conductivity of the moisture in dry wood (W/m K) |
| $K_g$ | thermal conductivity of the gas in dry wood (W/m K) |
| $K_{rad}$ | radiative contribution to the conductivity of wood (W/m K) |
| **LHV** | low heating value of fuel (kJ/kg) |
| $M_{wet}$ | biomass moisture expressed in wet basis (–) |
| $M_{dry}$ | biomass moisture expressed in dry basis (–) |
| $m_d$ | moisture percentage (by weight, %) |
| $M$ | weight percentage of moisture (%) |
| $M_a$ | mass of surface moisture in biomass (kg) |
| $M_i$ | mass of inherent moisture in biomass (kg) |
| $M_f$ | mass of fuel (kg) |
| $M_w$ | mass of moisture in the fuel (kg) |
| $M_{ash}$ | mass of ash in the fuel (kg) |
| $m_i$ | mass fraction of the *i*th gas (–) |
| **MW** | molecular weight of gas mixture (–) |
| $n$ | number of moles (–) |
| $n_i$ | mole fraction of the *i*th gas (–) |
| **N** | weight percentage of nitrogen (%) |
| **O** | weight percentage of oxygen (%) |
| $P_i$ | partial pressure of the *i*th gas (Pa) |
| $P$ | total pressure of the gas (Pa) |
| $Q$ | heat content of fuel (kJ) |
| $R$ | universal gas constant (8.314 J/mol K) |
| **sp.gr** | specific gravity (–) |
| $S$ | weight percentage of sulfur (%) |
| $T$ | temperature (K) |
| $W_{wet}$ | weight of wet biomass (kg) |
| $W_{dry}$ | weight of dry biomass (kg) |
| **VM** | weight percentage of volatile matter (%) |
| $V$ | volume of gas ($m^3$) |
| $V_i$ | volume fraction of the *i*th gas (–) |
| $\Delta H_{comb}$ | heat of combustion or reaction, kJ/mol |
| $\rho_{true}$ | true density of biomass ($kg/m^3$) |
| $\rho_{apparent}$ | apparent density of biomass ($kg/m^3$) |
| $\rho_{bulk}$ | bulk density of biomass ($kg/m^3$) |
| $\varepsilon$ | void fraction (–) |

$\in_p$ porosity of biomass (−)
$\sigma$ Steven–Boltzmann's constant ($5.67 \times 10^{-8}$ W/m$^2$ K$^4$)
$\theta$ temperature (°C)

## Subscripts

**ad, ar, db** subscripts representing air dry, as-received basis, and dry basis
**daf** dry ash-free basis
**td** total-dry basis
***i*** *i*th component
**m** mixture

Chapter 4

# Torrefaction

## 4.1 INTRODUCTION

Biomass can provide a full range of convenient feedstock for energy, metallurgical and chemical industries. This feedstock can be in the form of solid, liquid, or gases. Production of solid fuels from biomass using carbonization has been practiced for many thousands of years. It provided early people with charcoal, the first convenient solid fuel as well as a feedstock for iron extraction at a later date. The art of torrefaction (French word for "roasting") has been used in a host of industries for tea and coffee making, but only in recent time, it has caught the attention of power industries for the production of a coal substitute from biomass. Torrefaction is often called a pretreatment process as it prepares biomass for further use instead of direct use in its raw form. Torrefied biomass finds use in fields such as:

- Cofiring biomass with coal in large coal-fired power plant boilers
- Use as fuel in decentralized or residential heating system
- Use as a convenient fuel for gasification
- Potential feedstock for chemical industries
- Substitute for coke in blast furnace for reduction in carbon foot print.

This chapter discusses the production of solid fuels from biomass, its principle, technologies available, and design considerations.

## 4.2 WHAT IS TORREFACTION?

Though no generally accepted definition of torrefaction is available at the moment, by examining various features of the process and attributes of the product, one may describe torrefaction as:

*a thermochemical process in an inert or limited oxygen environment where biomass is slowly heated to within a specified temperature range and retained there for a stipulated time such that it results in near complete degradation of its hemicellulose content while maximizing mass and energy yield of solid product.*

Typical temperature range for this process is between 200°C and 300°C (Bergman et al., 2005). Though other ranges (Table 4.1) have been suggested, none exceeds the maximum temperature of 300°C. Torrefaction above this

**Biomass Gasification, Pyrolysis and Torrefaction.**

**TABLE 4.1** Torrefaction Temperature Range as Suggested by Different Researchers

| Researchers | Temperature Range (°C) |
|---|---|
| Arias et al. (2008) | 220–300 |
| Chen and Kuo (2010), Prins (2005), Zwart et al. (2006) | 225–300 |
| Pimchuai et al. (2010), Prins et al. (2006) | 230–300 |
| Bergman et al. (2005), Tumuluru et al. (2011a), Rouset et al. (2011), Sadaka and Negi (2009) | 200–300 |

temperature would result in extensive devolatilization and carbonization of the polymers both of which are undesirable for torrefaction (Bergman et al., 2005). Also, the loss of lignin in biomass is very high above 300°C. This loss could make it difficult to form pellets from torrefied products. Furthermore, fast thermal cracking of cellulose causing tar formation starts at temperature 300–320°C (Prins et al., 2006). These reasons fix the upper limit of torrefaction temperature as 300°C.

Another important aspect of torrefaction's definition is oxygen concentration in the reactor. Studies (Basu et al., 2013; Uemura et al., 2011) on the effect of oxygen concentration on torrefaction suggest that it is not essential to have oxygen-free environment for torrefaction. Presence of a modest amount of oxygen can be tolerated and may even have a beneficial effect on the torrefaction.

A major motivation of torrefaction is to make the biomass lose its fibrous nature such that it is easily grindable, while it is still possible to form into pellets without binders. Such requirements limit the torrefaction temperature range to 200–300°C.

Slow heating rate is an important characteristic of torrefaction. Unlike in pyrolysis, the heating rate in torrefaction must be sufficiently slow to allow maximization of solid yield of the process. Typically the heating rate of torrefaction is less than 50°C/min (Bergman et al., 2005). A higher heating rate would increase liquid yield at the expense of solid products as is done for pyrolysis.

The thermal decomposition of biomass occurs via a series of chemical reactions coupled with heat and mass transfer. Within the temperature range of 100–260°C, hemicellulose is chemically most active, but its major degradation starts above 200°C. Cellulose degrades at still higher temperature (>275°C), but its major degradation occurs within a narrow temperature band of 270–350°C (Chen et al., 2011). Lignin degrades gradually over the temperature range of 250–500°C, though it starts softening in the temperature range of 80–90°C (Cielkosz, 2011).

### 4.2.1 Pyrolysis, Carbonization, and Torrefaction

The torrefaction process is sometimes confused with related processes like carbonization, mild pyrolysis, roasting, and wood cooking, but the motivation and process conditions of these processes are not necessarily the same. A major objective of torrefaction is to increase the energy density of the biomass by increasing its carbon content while decreasing its oxygen and hydrogen content. This objective is similar to that of carbonization that produces charcoal but with an important difference that the latter does not retain maximum amount of energy of the biomass, and thereby gives low energy yield. Difference between torrefaction, pyrolysis, and carbonization is delineated further in this section and in Table 4.2.

Pyrolysis, carbonization, and torrefaction are all parts of the thermal decomposition process of biomass. Table 4.2 illustrates the changes that occur when a piece of wood or any biomass is heated in an inert atmosphere. Though the thermal degradation processes are listed as separate sequential processes they could overlap each other to some extent during the heating process. Each process has its own motivation. For example, the major motivation of pyrolysis is the production of liquid stuff, while that of carbonization is solid.

The term "pyrolysis" means thermal decomposition or chemical change brought about by heat (see Chapter 5). This dictionary definition could cover torrefaction, carbonization, and pyrolysis processes, but generally we use the term "pyrolysis" in a more restrictive sense for the thermal process for production of liquid extracts from biomass.

Carbonization is perhaps the oldest biomass conversion process that came to the service of humankind. It requires relatively high temperature and is a slow and long process (Table 4.2). For centuries, people have been using carbonization to produce charcoal from biomass. Charcoal has been used for thousands of years for many applications including heating, production of gunpowder, and metal extraction. Even now, charcoal has important commercial use in a number of applications including the following:

**a.** Fuel in domestic oven or barbeque. Fuel for steam generation or cement production.
**b.** Reducing agent in metallurgical industries.
**c.** Filter medium for water filter.
**d.** Pollutant capture and reaction site in chemical industries.

Traditional carbonization process uses beehive retort (Figure 5.2) where wood is piled inside a mud covered pit to restrict air entry, and it is ignited at the base. A part of the combustion heat provides the energy needed for carbonization. Such plants suffer from a high level of smoke production. Modern plants are relatively smoke free and typically operate at about 900°C. The amount of charcoal produced per unit weight of raw wood is low, and it depends on the peak temperature of carbonization.

**TABLE 4.2 Changes Taking Place in Biomass as It Is Heated in Inert Atmosphere (based on www.FAO.org/documents/x.5) and the Thermochemical Process Taking Place in It**

| Temperature Range of Heating (°C) | Process That Occurs | Heating Rate | Process | Solid Product |
|---|---|---|---|---|
| 20–110 | The wood is preheated and it approaches 100°C, moisture starts evaporating | Low/fast | Drying | Bone dry wood |
| 110–200 | Further preheating removes traces of moisture and slight decomposition starts | Low/fast | Postdrying preheating | Preheated dry wood |
| 200–270 | Wood decomposes releasing volatile (e.g., acetic acid, methanol, CO, and $CO_2$) that escape | Low | Torrefaction | Mildly torrefied wood |
| 270–300 | Exothermic decomposition starts releasing condensable and noncondensable vapors | Low | Torrefaction | Severely torrefied wood |
| 300–400 | Wood structure continues to break down. Tar release starts to predominate | Low | Low temperature carbonization | Low fixed carbon charcoal |
| | | High | Pyrolysis | Liquid |
| 400–500 | Residual tar from charcoal is released | Low | Carbonization | High fixed carbon charcoal |
| | | High | Pyrolysis | Liquid |
| >500 | Carbonization is complete | | High temperature carbonization | Tar-free charcoal |
| | | | Pyrolysis | Liquid, higher gas yield |

www.fao.org/docrep/x5555e/x5555e03.htm; FAO document repository. Industrial charcoal making, Wood carbonization and the product it makes (Chapter 2).

#### *4.2.1.1 Difference Between Carbonization, Pyrolysis, and Torrefaction*

As explained earlier, the most important difference between pyrolysis, carbonization, and torrefaction lies in their product motivation. For example, the primary motivation of pyrolysis is to maximize its liquid production while minimizing the char yield. The objective of carbonization, on the other hand, is to maximize fixed carbon and minimize hydrocarbon content of the solid product, while that of torrefaction is to maximize energy and mass yields with reduction in oxygen to carbon (O/C) and hydrogen to carbon (H/C) ratios.

Carbonization is similar to torrefaction in many respects, but there are some important differences between them. For example, carbonization drives away most of the volatiles, but torrefaction retains most of it, driving away only the early volatilized low energy dense compounds and chemically bound moistures. Therefore, during torrefaction the carbonization reactions that remove the volatiles should be avoided (Tumuluru et al., 2011). Table 4.3 compares the properties of products of torrefaction and carbonization. Here, we see a comparison of the properties of carbonized and torrefied wood with those of raw wood with coal as a reference.

**TABLE 4.3 Comparison of Typical Properties of Raw Wood, Torrefied Wood, Charcoal, and Coal**

| | Typical Wood[a] | Torrefied Wood | Charcoal (Carbonization) | Coal (Bituminous)[b] |
|---|---|---|---|---|
| Temperature (°C) | | 200–300 | >300 | |
| Moisture (%)(wb) | 30–60 | 1–5 | 1–5 | 3–20 |
| Volatile (%db) | 70–80 | 55–65 | 10–12 | 28–45 |
| FC (%db) | 15–25 | 28–35 | 85–87 | 45–60 |
| Mass yield | | ~80% | ~30% | |
| Energy density (db) (MJ/kg) | ~18[a] | 20–24 | 30–32 | 24–33 |
| Volumetric energy density (GJ/m$^3$) | ~5.8[a] | 6.0–10.0 | 18.5–19.8 | 30–40[c] |
| Apparent density (kg/m$^3$) | 350–680[e] | 300–500[d] | 600–640 | 1100–1350 |
| Hydrophobicity | Hygroscopic | Hydrophobic | Hydrophobic | Hydrophobic |

[a]*Bergman, P.C.A., 2005. Combined torrefaction and pelletisation. ECN Report: ECN-C-05-073.*
[b]*Steam 41 edition, p. 9–6, 9–10, 12–7*
[c]*Based on 1225 kg/m$^3$.*
[d]*Mullins and McNight, p. 75.*
[e]*Jenkins 1989, p. 866.*
wb- wet basis; db- dry basis.
**Source:** Adapted from Kleinschmidt (2010).

Both carbonization and torrefaction require relatively slow rates of heating, while pyrolysis relies on fast pace of heating to maximize the liquid yield.

Carbonization takes place at higher temperatures with a certain level of oxygen that allows sufficient combustion to supply the heat for the process. The torrefaction process on the other hand tries to avoid oxygen as well as combustion.

Torrefaction is a thermal decomposition that takes place at low temperature and within a narrow temperature range of 200–300°C, while carbonization is a high temperature (>300–600°C) destructive distillation process. Carbonization produces more energy dense fuel than torrefaction, but it has a much lower energy yield.

The following sections provide further details of these processes. Since Chapter 5 presents a comprehensive discussion on pyrolysis, that process is not described here. Torrefaction is discussed in detail in this chapter with a brief introduction to carbonization.

## 4.3 CARBONIZATION

Solid products of carbonization of biomass include fuel charcoal, activated charcoal, biocoke, and biochar. All of these are produced in processes similar to that of torrefaction, that is, slow heating in absence or low oxygen. The major difference lies in the process temperature. While torrefaction is carried out at a low and narrow temperature range of 200–300°C, other processes are carried out at much higher temperature. Such high temperature product of wood has the generic name charcoal. Within the charcoal class there are several divisions like fuel charcoal, activated charcoal, biocoke, and biochar. These have some fine differences primarily for application considerations. Because of its stable pore structure with high surface area, charcoal is a good reducing and adsorbent agent. It finds use in the following industries:

1. Fuel charcoal for energy, barbeque, and so on
2. Manufacture of carbon disulfide, sodium cyanide, and carbides
3. Smelting and sintering of iron ores, case hardening of steel, and purification in smelting of nonferrous metals
4. Water purification, gas purification, solvent recover, and waste water treatment
5. Carbon sequestration and soil remediation.

The carbonization product may be divided into several types from its application consideration. A brief description of these products is given in the following sections.

### 4.3.1 Charcoal Fuel

Charcoal is one of the earliest fuels used by human race. Presently it is used as a smokeless fuel in many countries and a feedstock for barbeque fuel.

Figure 5.2 shows a sketch of an oven typically used for the production of charcoal. Wood is stacked on the ground and a clay covering is built over this leaving a small opening at the bottom. This helps reduce oxygen supply to the wood. The small opening provides just the amount of oxygen to burn some wood to provide heat for carbonization. Since the oven is closed and well insulated, whatever heat is generated is retained inside the oven and that helps slow down the thermal degradation of the wood into charcoal. The temperature inside the carbonizer could be as high as 800°C.

Modern industrial processes for charcoal making employ internal heating (Missouri kiln), external heating (VMR retort), or heating by gas recirculation (the Degussa process) (Antal and Gronli, 2003). A review of technologies for production of charcoal is given in FAO (2008). Fuel charcoal has high fixed carbon content and a modest amount of volatile matter (Table 4.3).

### 4.3.2 Activated Charcoal

Activated charcoal is a valuable product used in a host of chemical and environmental industries. Its large pore surface area gives it an exceptionally high adsorption capacity. As a result, this type of charcoal fetches a considerably higher price from the market than by normal fuel charcoal.

Activated charcoal is produced by removing the tarry products from conventional fuel charcoal. This makes the pores in charcoal more accessible for adsorption. The activation process increases the pore surface area by orders of magnitude.

There are several methods for making activated charcoal, but the basic process is essentially the same. It involves heating ground charcoal to about 800°C in an atmosphere of superheated steam. The charcoal thus avoids contact with oxygen while distilling away the tar that was blocking the fine structures of the charcoal. Steam carries away the tarry residues. After this the solid product is poured into a sealed container and allowed to cool.

### 4.3.3 Biocoke

This type of charcoal is produced specifically for metal extraction as a substitute for conventional coke that is produced from coking coal. When heated with metallic ores with oxides or sulfides, carbon in biocoke combines with oxygen, and sulfur allowing easy metal extraction. It is acknowledged to be a better reductant than coke (FAO, 1983). Biocoke has been used for extraction of iron from iron ore during the very early days of metallurgical industries. Biocoke needs certain specific properties for its use in blast furnace. It must have adequate compressive strength to withstand the pressure of heavy burden of solids in the blast furnace. Additionally, it needs to have good fracture resistance to maintain constant permeability of the furnace charge to the air blast (FAO, 1985, p. 8). A major motivation of using biocoke in blast furnace is to replace coal and thereby reduce net $CO_2$ emission from the iron and steel industries. Pulverized biocoke may be injected directly into blast furnace. Alternately, torrefied biomass or charcoal could be mixed with coking coal and fed into a coke oven where formed coke is produced and charged into the blast furnace from the top.

### 4.3.4 Biochar

Biochar is a charcoal product of pyrolysis. Here, carbonization takes place at relatively high temperatures. Biochar is known for its carbon sequestration potential and soil remediation properties. Vegetation or forest residues are often burnt down in some parts of the world for making room for cultivation and to provide biochar to the soil that improves the fertility and other properties of the soil. An important aspect of this otherwise inefficient process is that at least a part of the total carbon in biomass that would have been released to the atmosphere as a greenhouse gas is now retained as stable solid char in the soil. The higher the carbonization, the better is the property of biochar though carbon retention as solid is less. It is discussed in some further details in Section 5.8.

## 4.4 TORREFACTION PROCESS

Torrefaction is a thermal process accomplished by contact with a heating medium or heat carrier. A simple illustration of the torrefaction process is shown in Figure 4.1 that shows how the mass and energy content of biomass changes as it is converted into a torrefied product. Section 4.5.2 gives more details. The heating medium here is represented by a flame but it could be a

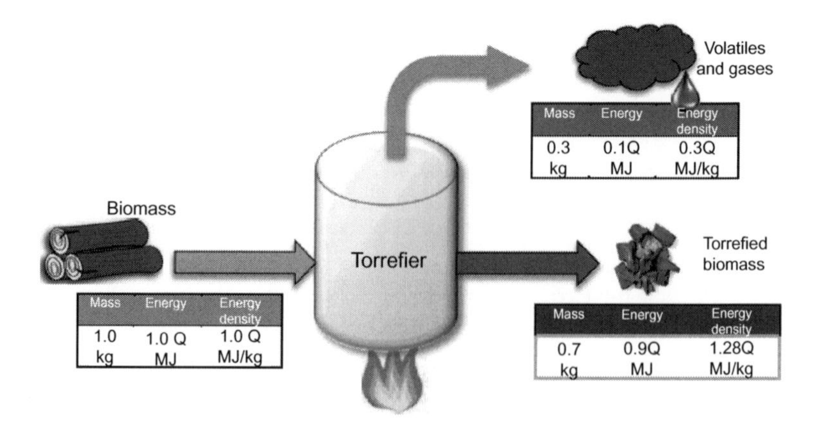


**FIGURE 4.1** Mass and energy changes of a feed undergoing torrefaction.

hot substance, dry or wet. In wet torrefaction, the biomass is subjected to heating in hot compressed water (Yan et al., 2009). Dry torrefaction involves heating either by a hot inert gas or by indirect heating. The dry process is the accepted method for commercial torrefaction.

Figure 4.2 describes the sequential process of torrefaction with the help of photographs of a fresh twig cut from a maple tree and taken through different stages of the process. Mass loss of the wood in each stage is also shown on the photograph. Figure 4.2A shows a photograph of the twig after its bark is peeled off exposing its wet surface of the branch. Thereafter, the twig is left in an air-drying oven at 70°C. After about an hour, its surface moisture evaporates, and the surface no longer appears wet (Figure 4.2B). The mass loss of the wood at this stage is about 20%.

After that the oven temperature is raised to 110°C for drying when the inherent moisture within the pores of the wood escapes, and then it is bone dried at 140°C for an hour. The wood appears a little reddish (Figure 4.2C), and the mass loss nearly doubles because nearly all of the intrinsic moisture in the biomass is released at this stage. Changes that occurred up to this stage are primarily physical as very little chemical decomposition took place.

Thereafter, the piece is heated in an inert medium at 200°C for 1 h and the chemical decomposition starts (Figure 4.2D), but the extent of decomposition being small at this temperature, the piece lost mass by only a meager amount of 4%. The oven temperature is then raised to 250°C when the piece is baked for another hour. Now one notes evidence of major reaction by the color change to dark brown and 11% additional loss in mass. At this stage of

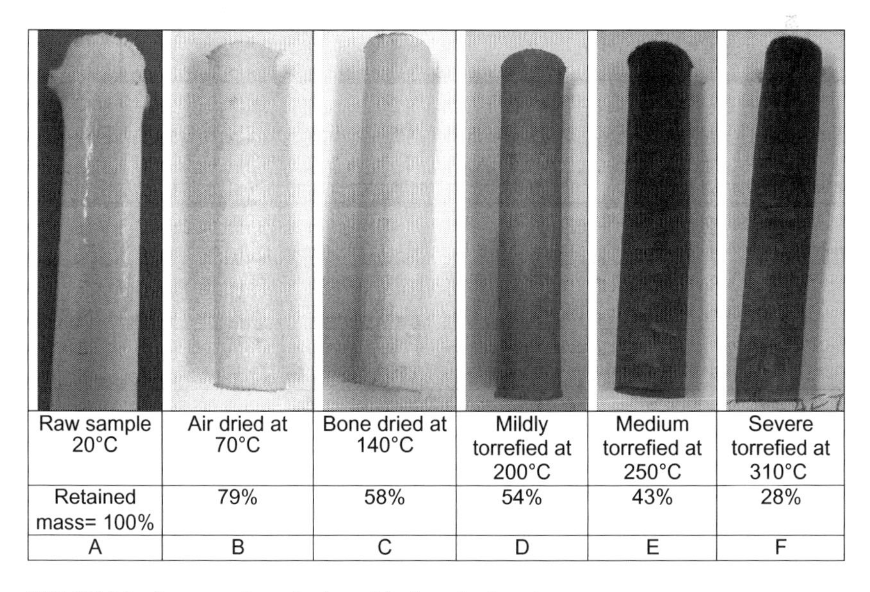


FIGURE 4.2 Progress of torrefaction of the branch of maple tree.

torrefaction, the piece retains only 43% of the original mass but loses only 13% of its energy content.

The wood sample is then heated to a further high temperature of 310°C for 1 h when the sample wood becomes much darker and brittle. This is a severe form of torrefaction when it loses another 11% of the original mass due to loss of moisture, volatiles, and gases. The energy density increases but much of the energy is lost in the process. The mass-based energy density increases at the expense of decreased energy yield.

### 4.4.1 Heating Stages

Different stages of heating as illustrated in Figure 4.2 are examined in more detail in the following sections. Figure 4.3 shows schematically the historical changes (in ideal condition) in mass, temperature, and the energy consumption of a biomass piece during torrefaction. Temperature of the heating medium (furnace) was kept at the design torrefaction temperature, while biomass was heated and its temperature change was recorded. From this diagram, one notes the following stages of thermal treatment.

#### *4.4.1.1 Predrying*

This is the first step in the process. When the biomass is heated from room to the drying temperature (~100°C), its temperature rises steadily receiving sensible heat from the reactor or the heating medium. For predrying heating, the energy required, $Q_{ph}$, is $M_f C_{pw}(100 - T_0)$. Accounting for heat loss, one can write the heat required, $Q_{pd}$, as:

$$Q_{pd} = \frac{M_f C_{pw}(100 - T_0)}{h_{upd}} \tag{4.1}$$

where, $C_{pw}$ is the specific heat of wet or as-received biomass, $M_f$ is mass of raw biomass, and $T_0$ is feed temperature. A heat utilization efficiency factor, $h_{u,pd}$ of the system is used here to account for some heat that may be lost from the drier. Energy required, $Q_{pd}$, is generally a relatively small fraction of the total heat requirement, $Q_{total}$.

#### *4.4.1.2 Drying*

Drying on the other hand is the most energy-intensive step of torrefaction. It is especially so for high-moisture biomass because the moisture in biomass is evaporated during this stage. There is very little change in biomass temperature as evaporation takes place at constant temperature till all the surface moisture or free water is driven off. The temperature starts to climb after the critical moisture is reached when the rate of evaporation starts to decrease. This stage makes the biomass bone dry. In most cases, one notes a sharp increase in the total energy demand in this stage (Figure 4.3).

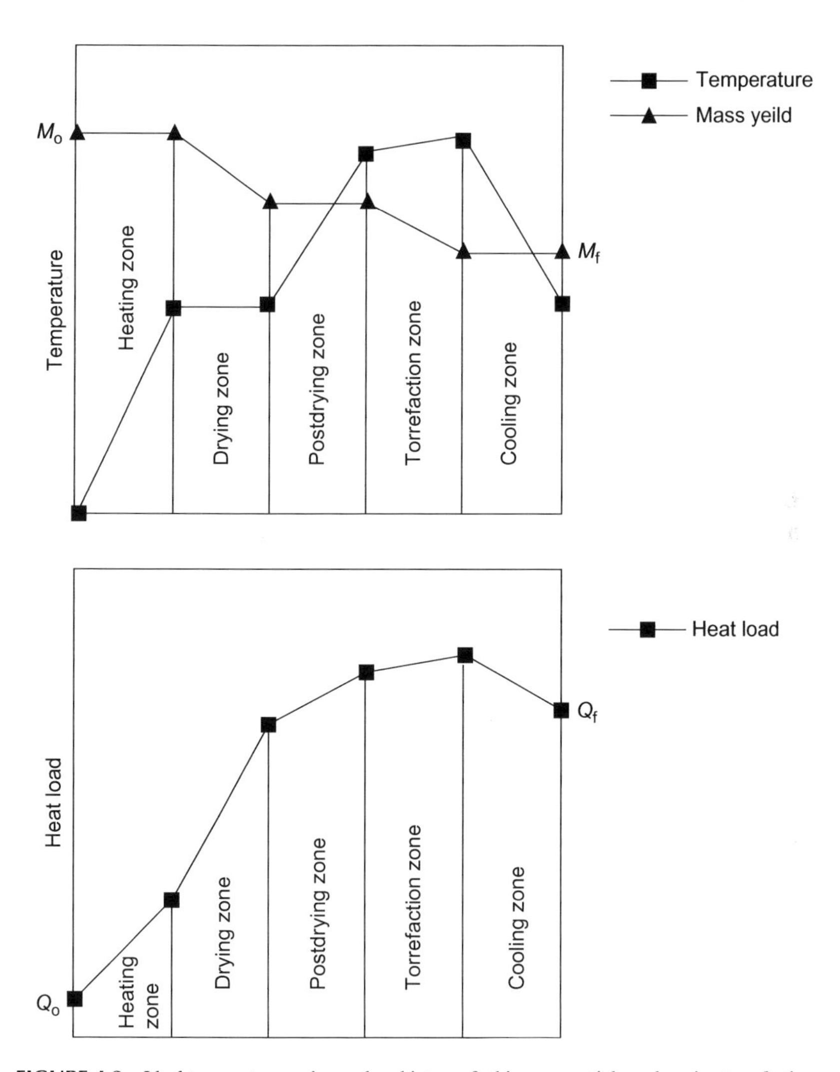


**FIGURE 4.3** Ideal temperature and mass loss history of a biomass particle undergoing torrefaction.

The heat load or energy required for this stage, $Q_d$, is $L\ M_fM$. By taking into account, the heat utilization efficiency of the dryer section, $h_{u.d}$, the load may be written as:

$$Q_d = \frac{LM_fM}{h_{ud}} \tag{4.2}$$

where $L$ is the latent heat of vaporization (~2260 kJ/kg) of water at 100°C. The moisture fraction of the biomass as-received ($M_f$) is $M$.

### 4.4.1.3 Postdrying Heating

After the biomass is dried, it needs to be heated further to the designed torrefaction temperature, $T_t$. This temperature is generally in excess of 200°C because very little decomposition of the biomass takes place below this temperature.

During this stage, all physically bound moisture along with some light organic compounds escape from the biomass (Bergman et al., 2005). One can see in Figure 4.3 that the energy demand of this stage is relatively small because it provides only sensible heat to the dried biomass.

$$\text{Energy required,}\quad Q_{\text{pdh}} = \frac{M_f(1-M)C_{\text{pd}}(T_t - 100)}{h_{u,\text{pdh}}} \tag{4.3}$$

where $C_{pd}$ is the specific heat of dry biomass and $h_{u,pdh}$ is the heat utilization efficiency of this section.

### 4.4.1.4 Torrefaction

Torrefaction stage is key to the whole process as the bulk of depolymerization of the biomass takes place in this stage. A certain amount of time is needed to allow the desired degree of depolymerization of the biomass to occur. The degree of torrefaction depends on the reaction temperature as well as on the time the biomass is subjected to torrefaction. This time is also called reactor *residence time* or *torrefaction time*.

The torrefaction time should be measured from the instant the biomass reaches the temperature for the onset of torrefaction (200°C) because the degradation of biomass below this temperature is negligible. The torrefaction process is mildly exothermic (Prins, 2005) over the temperature range of 250–300°C. So, except for heat loss, the torrefaction stage should require very little energy (Figure 4.3), but in practice it could require some heat to make up for the unavoidable heat loss from the torrefaction section of the reactor.

$$Q_{\text{tor}} = H_{\text{loss}} + M_f(1-M)X_t \tag{4.4}$$

Here, $X_t$ is a parameter (kJ/kg product) that determines the amount of heat absorbed during torrefaction. It is positive for endothermic and negative for exothermic torrefaction reactions. The amount of heat loss $H_{loss}$ to the ambience from the torrefaction section is a function of reactor design.

### 4.4.1.5 Cooling

Biomass leaves the torrefier at the torrefaction temperature, which is the highest temperature in the system. This being generally above the ignition temperature of most torrefied biomass (Table 3.8), unless cooled down sufficiently the product could catch fire on contact with air. Additionally, handling of such a hot product is unsafe and dangerous. So, the torrefied product must be cooled down from the torrefaction temperature ($T_t$), to acceptable final temperature ($T_p$) of the product for further processing or storage. By extracting the energy, $Q_{cool}$, in the form of either hot air or vaporized liquid, the product is cooled.

$$Q_{cool} = M_f(1 - M)\,MY_{db}\,C_{pt}\,(T_t - T_p) \tag{4.5}$$

where $C_{pt}$ is the specific heat of torrefied biomass and $MY_{db}$ is the mass yield.

The extracted energy, $Q_{cool}$, may be partially recovered in the form of hot air or vaporized liquid like steam, which could be gainfully utilized in providing a part of the energy required for drying or preheating the biomass.

### 4.4.2 Mechanism of Torrefaction

The thermochemical changes in biomass during torrefaction may be divided into five regimes following Bergman et al. (2005a):

1. *Regime A (50–120°C)*: This is a *nonreactive drying* regime where there is a loss in physical moisture in biomass but no change in its chemical composition. The biomass shrinks but may regain its structure if rewetted (Tumuluru et al., 2011). Upper temperature is higher for cellulose.
2. *Regime B (120–150°C)*: This regime is separated out only in case of lignin that undergoes softening, which make it serve as a binder.
3. *Regime C (150–200°C)*: This is called "*reactive drying*" regime that results in structural deformity of the biomass that cannot be regained upon wetting. This stage initiates breakage of hydrogen and carbon bonds and depolymerization of hemicellulose. This produces shortened polymers that condense within solid structures (Bergman et al., 2005a).
4. *Regime D (200–250°C)*: This regime along with regime (E) constitutes torrefaction zone for hemicellulose. This regime is characterized by limited devolatilization and carbonization of solids structure formed in regime (C). It results in the breakdown of most inter- and intramolecular hydrogen, C–C and C–O bonds forming condensable liquids and non-condensable gases (Tulumuru et al., 2011).
5. *Regime E (250–300°C)*: This is the higher part of torrefaction process. Extensive decomposition of hemicellulose into volatiles and solid products takes place. Lignin and cellulose, however, undergo only a limited amount of devolatilization and carbonization. Biomass cell structure is completely destroyed in this regime making it brittle and nonfibrous.

Major devolatilization and carbonization of the biomass polymers take place in a different temperature range. Some qualitative values taken from Prins (2005, p. 89) are given below.

Hemicellulose: 225–300°C
Cellulose: 305–375°C
Lignin: 250–500°C

The basic polymeric constituents of biomass, namely hemicellulose, cellulose, and lignin, are believed to react independently, and as such they do not show the synergetic effect (Chen and Kuo, 2011). Thus, mass loss of individual components can be simply added to get the total mass loss during torrefaction as shown in Figure 4.5B.

Major attractions of torrefaction pretreatment stem from the degradation of the hemicellulose content of the biomass. So, torrefaction is characterized primarily by the degradation of hemicellulose. Dehydration and decarboxylation are the main reactions in this degradation that produce both condensable and noncondensable products. The torrefaction process produces solid, liquid, and gaseous products as shown in Figure 4.4. The solid component is made primarily of char along with items like some sugar and polymeric structures and ash (Bergman et al., 2005a). The noncondensable gases comprise CO, $CO_2$, and small amounts of $CH_4$. Condensed liquid contains water from thermal decomposition, lipids such as terpenes and waxes, and organics such as alcohols and furans.

Torrefaction products comprise carbon water, carbon dioxide, carbon monoxide, acetic acid, methanol, and formic acid. The formation of $CO_2$ is due to decarboxylation. The acetic acid comes from the decomposition of acetyl pendant group in cellulose. Carbon monoxide comes mainly from the reaction between $CO_2$ and steam with porous char surface of the biomass (White and Dietenberger, 2001).

Though torrefaction is characterized mainly by the degradation of hemicellulose, other polymers, cellulose, and lignin also degrade to some extent that depends on the temperature (Figure 4.5A). The mass loss due to torrefaction at a given temperature is the sum of degradation of each of the three polymers (Chen et al., 2011) and moisture if any. Figure 4.5B shows a simple qualitative diagram of decomposition of the polymers. Three figures here give mass losses of hemicellulose, cellulose, and lignin when subjected to torrefaction at different temperatures. It plots the mass as percentage of the original mass of the biomass (dry ash free (daf)). By drawing a horizontal line at the given temperature and by adding the intercept, one can get the projected mass loss at that temperature. This is plotted on the extreme right graph in Figure 4.5B as the same function of temperature.

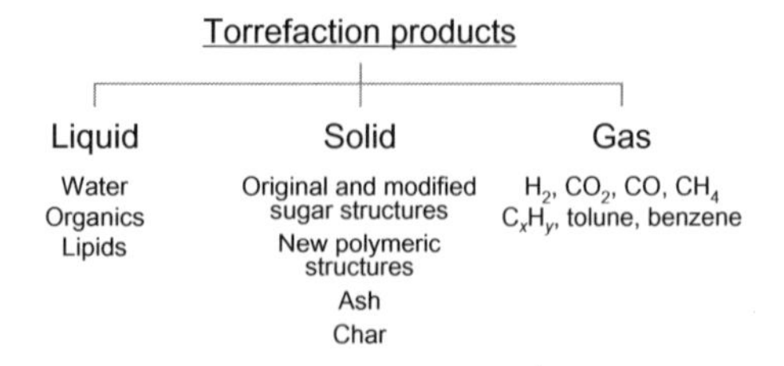


**FIGURE 4.4** Products of torrefaction of biomass.

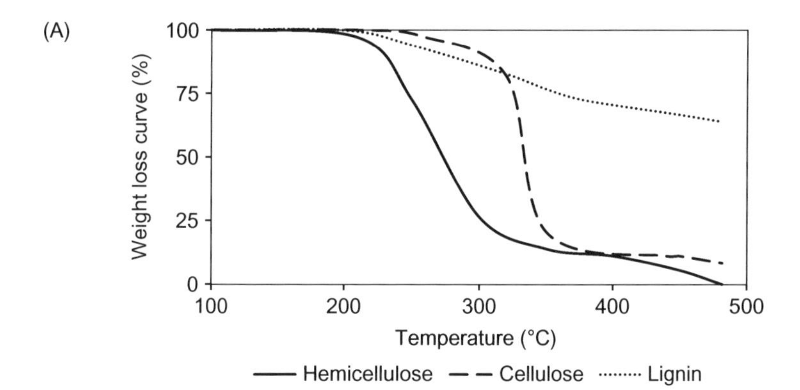


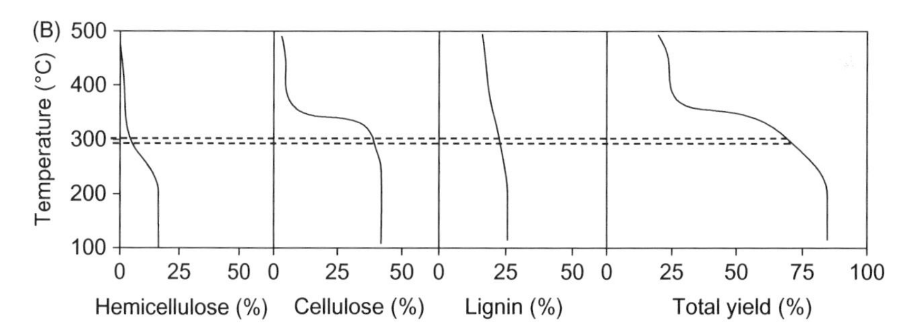


**FIGURE 4.5** (A) A comparison of degradation of lignin, cellulose, and hemicellulose in inert atmosphere. Lignin represents acid lignin. (B) A qualitative diagram of mass loss of torrefaction of different polymeric (initial composition of yellow poplar: hemicellulose = 16.6%, cellulose = 42.2%, lignin = 25.6%). Source*: (A) Drawn from the experiment of Shafidazeh and McGinnis (1971) with cottonwood in a TGA.*

### 4.4.3 Effect of Design Parameters on Torrefaction

The following section discusses how some feed and operating parameters influence the torrefaction process.

#### *4.4.3.1 Temperature*

Torrefaction temperature has the greatest influence on torrefaction as the degree of thermal degradation of biomass depends primarily on the temperature. Figure 4.6A illustrates this effect showing how the mass yield decreases with increasing temperature. Figure 4.6B shows that energy yield also decreases with increasing temperature. Higher temperature gives lower mass and energy yields but higher energy density. The fraction of fixed carbon in a sample increases while that of hydrogen and oxygen decreases as the torrefaction temperature increases (Bridgeman et al., 2008). Cielkosz and Wallace (2011) observed that mass yield variation is related to the temperature, $T_t$, and residence time, $t$, by an exponential function of $(t/T_t)$.

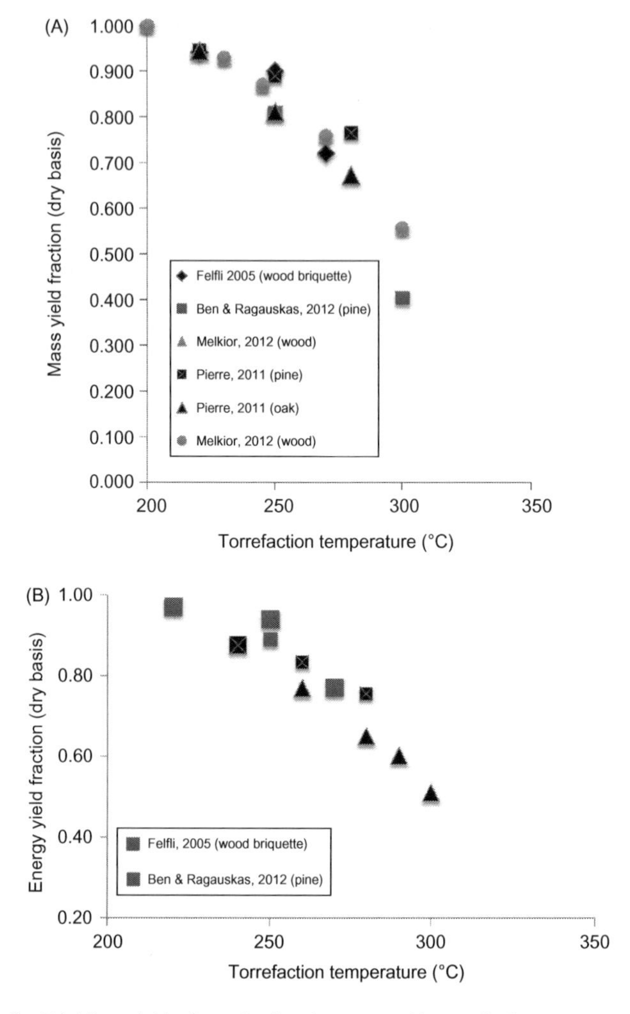


**FIGURE 4.6** (A) Mass yield of torrefaction decreases with torrefaction temperature. (B) The energy yield of torrefaction reduces with increasing torrefaction temperature. Source*: Data taken from several sources.*

## Core Temperature Rise

The torrefaction process in a coarse biomass particle (millimeter or centimeter size) takes place mainly within its interior. Temperature inside the particle is thus more important than one on its surface for the decomposition. Torrefaction process is mildly endothermic below 270°C, but it is mildly exothermic above 280°C possibly due to exothermic breakdown of sugars at higher temperatures (Cielkosz and Wallace, 2011). The magnitude of heat of reaction is however small (Yan et al., 2010). In any case, during torrefaction heat is transported from a heat source first to the biomass particle's outer surface by convection.

Thereafter, the heat is transferred to its interior by conduction and pore convection. Thus, one would expect a negative temperature gradient between the torrefaction reactor, particle surface, and its interior (core).

For finer particles with low Biot number, the temperature difference between the particle surface and its core is small. In case of large particles, however, the Biot number being larger, one could expect a finite temperature difference between the biomass core and its outer surface.

Figure 4.7 shows simultaneous measurements of temperatures in the core and outside of a large biomass particle as it goes through torrefaction. Here, we observe that after the particle enters the reactor, its core temperature is much below the reactor or outer surface temperature, but the former starts rising steadily receiving heat from the reactor. The core temperature interestingly rises above the reactor temperature suggesting that the torrefaction reaction has become net exothermic.

After reaching a peak, the temperature starts declining but asymptotically remains slightly above the reactor temperature. This suggests that the overall reaction in the core remains slightly exothermic. The peak temperature reached at the biomass depends on the heat and mass transfer to the biomass interior and as such it is influenced by the size, shape, and temperature. The core temperature is of major importance as the torrefaction reaction depends on the core temperature rather than on the reactor temperature. For that

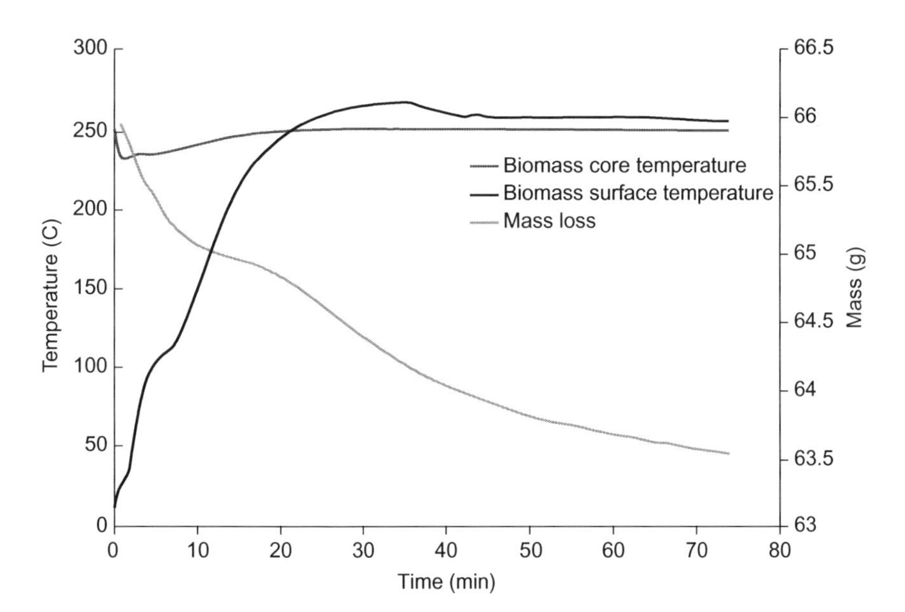


**FIGURE 4.7** Historical changes in temperature at the core of a 22 mm diameter poplar wood cylinder along with the corresponding furnace temperature measured just outside the wood and its mass loss.

reason, two particles of different sizes could have different mass or energy yields in otherwise identical torrefaction condition.

### 4.4.3.2 Residence Time

The residence time of biomass in the torrefier also influences the thermal degradation of biomass. Slow heating rate is one of the distinguishing characteristics of torrefaction that makes it different from the rapidly heated pyrolysis process. It is typically less than 50°C/min (Bergman et al., 2005a). For this reason, the residence time of biomass in a torrefaction reactor is much longer in tens of minutes. Longer residence time gives lower mass yield and higher energy density. Figure 4.8 illustrates the effect of residence time on the mass yield as well as on the energy yield of torrefaction. Both yields reduce with residence time. The influence of residence time on the torrefaction product is, however, not as dominant as that of the torrefaction temperature. The effect of residence time on mass loss diminishes after about 1 h (Stelt et al., 2011).

### 4.4.3.3 Biomass Type

The biomass type is another important parameter that could influence torrefaction. As hemicellulose degrades most within the torrefaction temperature range, one would expect a higher mass loss in a biomass with high hemicellulose content. However, it is interesting that a hardwood and softwood with similar hemicellulose content when torrefied under identical conditions could show very different mass yields (Prins et al., 2006). Torrefaction of hardwood gives lower mass yield than that of softwood because xylan, the active content of hemicellulose of hardwood (deciduous), constitutes 80–90%, while in softwood (coniferous), it constitutes only 15–30% (Sudo et al., 1989).

The xylan or 4-*O*-methylglucuronoxylan content of the hemicellulose is most reactive within the torrefaction temperature range, and it degrades faster than

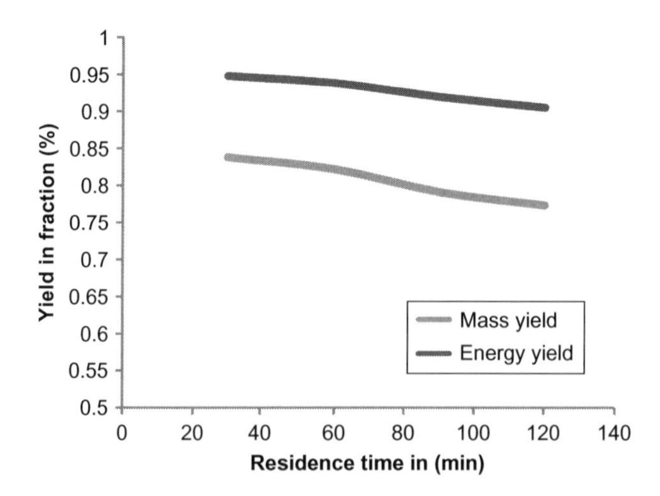


**FIGURE 4.8** Effect of residence time on the mass and energy yield of torrefaction (poplar wood: 25 mm diameter, 76 mm long, 250 °C).

any other solid components of the biomass (Basu et al., 2013a–c, 2012; Prins, 2005). For this reason, the mass loss is greatly influenced by the xylan content in the biomass rather than the hemicellulose content alone. Table 4.4 shows values of xylan in some hardwood and softwood.

During torrefaction, hardwood releases mainly acetic acid and water during torrefaction, while softwood releases mostly formic acid. Since hardwood experience higher mass loss on torrefaction without much effect on the energy loss, it would have a higher gain in energy density compared to that in softwood (Prins, 2005; Prins et al., 2006).

#### *4.4.3.4 Feed Size*

The size of biomass particles or pieces is another parameter that could affect the torrefaction yield. This effect may not be prominent for fine size of particles but could be measurable for large sizes.

The mass yield showed a modest increase with increase in volume mean diameter (Basu et al., 2013a) or the length of a piece of constant diameter. An opposite result (mass yield reduced) was found when the diameter was increased keeping the length constant (Basu et al., 2013a). These are direct result of heat transfer to biomass interior and the temperature dependent reaction within it. It is discussed further below.

Torrefaction involves convective heat transfer from the reactor to the biomass surface, conduction of the heat into the biomass interior, and finally the reaction within it. Relative magnitude of these three rates decides which parameter might influence the overall torrefaction process. Biot number, which is the ratio of heat convection to the outer surface and conduction of the same into the interior of the particle, is given as:

$$Bi = \frac{hV}{\lambda S} \sim \frac{hr_p}{\lambda} \tag{4.6}$$

where $V$ is the particle volume, $S$ the external surface area, $h$ the convective heat transfer coefficient on biomass particle, $r_p$ is characteristic particle size taken here as the radius of particle, and $\lambda$ is thermal conductivity of biomass particle.

Pyrolysis number, $Py$, is another parameter that could influence the torrefaction process. It relates the external surface heat transfer rate to the torrefaction reaction rate (Pyle and Zaror, 1984).

$$Py = \frac{h}{K\rho C_p r_p} \tag{4.7}$$

where, $K$ is reaction rate of torrefaction, $s^{-1}$, $\rho$ is density of particle, and $C_p$ is specific heat of particle.

If the Biot number is sufficiently small, as is the case for fine particles, the internal thermal resistance is negligible, and if the pyrolysis number is very high, the reaction will be rate controlled. Larger particles on the other hand would have higher $Bi$ and lower $Py$. So, the torrefaction would be

**TABLE 4.4** Composition of Wood Extractive on Free Basis

| | | | | | Breakdown of Hemicellulose | | | |
|---|---|---|---|---|---|---|---|---|
| Type | Wood | Cellulose | Lignin | Hemicellulose | Noncellulosic Glucan (%) | Glucomanan (%) | Arbinogalacta (%) | Xylan (%) |
| Hardwood | Beech | 42 | 22 | 36 | 11 | 11 | 6 | 69 |
| | White birch | 41 | 19 | 40 | 5 | 8 | 3 | 85 |
| | White elm | 49 | 24 | 27 | 7 | 15 | 7 | 70 |
| | Tremblin Aspen | 53 | 16 | 31 | 10 | 13 | 3 | 74 |
| | Average value | 45 | 21 | 34 | 9 | 15 | 3 | 74 |
| Softwood | Balsam fir | 44 | 29 | 27 | 0 | 67 | 4 | 30 |
| | White spruce | 44 | 27 | 29 | 0 | 59 | 7 | 34 |
| | Eastern White cedar | 44 | 31 | 25 | 0 | 44 | 8 | 48 |
| | Jack pine | 41 | 29 | 30 | 0 | 53 | 7 | 40 |
| | Average value | 43 | 29 | 28 | 0 | 57 | 7 | 32 |

**Source:** Adapted from Mullins and McKnight (1981), p. 98.

controlled by heat conduction into it and that may lead to a higher temperature in the biomass core.

## 4.5 DEGREE OF TORREFACTION

The torrefaction process is similar to that of roasting of coffee beans. The higher the roasting temperature the darker is the color of the bean. The price and taste of roasted coffee depends to some degree on the roasting temperature. Table 4.5, for example, presents a range of roasting temperature with the corresponding name of the coffee brand produced from it. One can thus see how the degree of roasting varies with the torrefaction temperature. So, for such a type of process, a gradation is indeed necessary for the quantification of the process of torrefaction of biomass. A tentative gradation or severity of the process (not the product) may be offered as below (Chen and Kuo, 2011):

1. Light torrefaction: Occurs at a temperature of 200–240°C or at 230°C when only hemicellulose is degraded leaving lignin and cellulose unaffected.
2. Medium torrefaction: Occurs at a temperature of 240–260°C or at about 250°C, when cellulose is mildly affected.
3. Severe torrefaction: Occurs at a temperature of 260–300°C or at 275°C characterized by depolymerization of lignin, cellulose, as well as hemicellulose.

**TABLE 4.5** Degree of Coffee Roasting Classified by Roasting Temperature

| Name of the Coffee Product | Roasting Temperature (°C) |
|---|---|
| Green unroasted coffee | 75 |
| Arabic coffee | 165 |
| Cinnamon roast | 195 |
| England roast | 205 |
| American roast | 210 |
| City roast | 220 |
| Full city | 225 |
| Vienna roast | 230 |
| French roast | 240 |
| Italian roast | 245 |
| Spanish roast | 250 |
| Immanent fire | 497 |

**Source:** From wikipedia.org/wiki/Coffee_roasting.

Within each temperature range, the rate and degree of torrefaction change from low to high.

Severe torrefaction results in the greatest mass and energy loss, but it gives highest energy density in the torrefied biomass. Light torrefaction on the other hand retains maximum amount of mass and energy of the dry biomass but attains the lowest energy density. The choice of torrefaction regime would depend on the specific technocommercial need of the torrefaction plant.

The above definition of the degree of torrefaction does not take into account the residence time of torrefaction, which also affects torrefaction (Section 4.3.3). For example, a biomass torrefied for 1 min at a given temperature will have a substantially lower degree of torrefaction than if it had been torrefied for 100 min at that temperature. So, the degree of torrefaction should be a combined product of temperature, residence time, and other influencing factor like oxygen concentration.

As the quality of roasted coffee is identified by its color and or taste, the quality of torrefied biomass for energetic use could also be expressed in terms of the following three attributes:

1. Mass yield
2. Energy density
3. Energy yield.

These are described in some detail in the following sections.

### 4.5.1 Mass Yield

Mass yield gives a measure of the solid yield of the torrefaction process. It defines what fraction of the original mass of biomass would remain in the torrefied product. Torrefaction concerns the change in the hydrocarbon content of the biomass. So, a simple definition based on a ratio of product mass and the original (feed) mass may not give a true picture of the process.

Drying is a physical change while torrefaction primarily concerns changes in the organic component of the biomass. A typical biomass contains physically bound water, inorganic materials (ash)[1], and organic substances. As ash and water do not carry any part of the chemical energy of the biomass their removal is of little consequence as far as energy content of the product is concerned. The drying of biomass increases its energy density but that does not bring about any chemical change in the biomass. Torrefaction on the other hand takes the process to a further height where chemical changes make the biomass both physically and chemically more attractive. Thus, a definition of mass yield on dry

[1]Strictly speaking the inorganic materials in biomass are not necessarily equal to its ash content. For example, Na and K in ash exist as inorganic form as a salt in biomass. On the contrary, P and S in ash originate from DNA and proteins, respectively.

basis (db) is more scientific, and it distinguishes torrefaction from drying or deashing processes of a biomass.

So, logically mass yield, MY, should be defined as the fraction of the original organic component of biomass that is converted into solid char. Thus, mass yield should be defined on a "dry ash free" basis.

Mass yield of torrefied biomass on dry ash free basis, $MY_{daf}$, is

$$MY_{daf} = \frac{\text{mass of torrefied biomass on daf basis}}{\text{mass of original biomass on daf basis}} \tag{4.8}$$

Although, the above definition is most accurate, its effectiveness in design calculations is somewhat restricted. It does not give the actual solid mass in the torrefied product; it gives only the residue of the organic component. When one calculates the sensible heat or the mass of solid handled, the total solid amount that includes the ash is needed. This difference could be substantial for organic feedstock such as rice husk and sewage sludge where inorganic content is high. For this reason, an alternative definition of mass yield on "db" or dry basis alone can be used.

Mass yield on dry basis, $MY_{db}$:

$$MY_{db} = \frac{\text{mass of torrefied biomass on dry basis}}{\text{mass of original biomass on dry basis}} \tag{4.9}$$

Mass yield is not generally expressed on "as-received" (ar) basis. However, for quick but rough design calculations, one may use it to determine the overall material flow in and out of the system. For such design purpose, one may use a more practical definition on as-received basis as below:

Mass yield on "as-received" basis, $MY_{ar}$:

$$MY_{ar} = \frac{\text{total mass of torrefied biomass}}{\text{mass of wet biomass as-received}} \tag{4.10}$$

Relationship between these three definitions of mass yield may be derived as below:

$$MY_{ar} = (1 - M)\, MY_{db} \tag{4.11}$$

$$MY_{daf} = \frac{MY_{db} - ASH_{db}}{1 - ASH_{db}} \tag{4.12}$$

or

$$MY_{db} = MY_{daf}(1 - ASH_{db}) + ASH_{db}$$

where $M$ and ASH are fractions of moisture and ash, respectively on as-received basis.

The ash fraction on dry basis can be related to that on as-received basis as:

$$ASH_{db} = \frac{ASH}{1 - M}$$

**Example 4.1**

A biomass company plans to build a commercial torrefaction plant in British Columbia, Canada, to utilize the beetle-infested pine forest. This waste product contains 35% moisture ($M$) on "as-received" basis. The composition of the feed on "dry basis" is as below:

Proximate analysis (db):

Volatiles: 80.71%, fixed carbon: 16.16%, ash: 3.13%.

Ultimate analysis (db):

C: 47.99%, H: 6.25%, O: 40.73%, N: 1.31%, S: 0.58%, ASH: 3.13%.

Pilot plant tests suggested an optimum torrefaction temperature and residence time for the biomass as 280°C and 20 min, respectively, such that 20% of the dry biomass is converted into volatiles carrying 5% of the total thermal energy.

**Calculate**

1. The lower and higher heating value (HHV) of the biomass feed on (a) wet basis, (b) dry basis, and (c) dry ash free basis.
2. Mass yield on dry basis and on dry ash free basis.

**Solution**

1. *Heating Value*

The composition is given on dry basis. So, calculate HHV on dry basis $HHV_{f,db}$, first using the correlation Eq. (3.32):

**a.**

$$\begin{aligned} HHV_{f,db} &= 349.1C + 1178.3H + 100.5S - 103.4O - 15.1N - 21.1\,A\ (\text{MJ/kg}) \\ &= 349.1 \times 47.99 + 1178.3 \times 6.25 + 100.5 \times 0.58 - 103.4 \times 40.73 \\ &\quad - 15.1 \times 1.31 - 21.1 \times 3.13 \\ &= 19{,}788\ \text{kJ/kg} \sim 19.8\ \text{MJ/kg} \end{aligned}$$

We can calculate lower heating value (LHV) from HHV by using Eq. (3.30).

$$\text{LHV} = \text{HHV} - 2241.7\left(\frac{9\text{H}}{100} - \frac{M}{100}\right)\ \text{kJ/kg}$$

As it is on dry basis $M = 0$

$$\begin{aligned} \text{LHV}_{f,db} &= 19{,}878 - 2241.7\left(\frac{9 \times 6.25}{100} - \frac{0}{100}\right) \\ &= 18{,}504\ \text{kJ/kg} \\ &= 18.50\ \text{MJ/kg} \end{aligned}$$

**b.** Heating values on wet or "as-received" basis is found from Eq. (3.31).

$$\text{HHV}_{db} = \left(\frac{\text{HHV}_{ar}}{1 - M}\right)$$

So,

$$\begin{aligned}HHV_{f,ar} &= HHV_{f,db}(1-M)\\ &= 19.88\times(1-0.35)\\ &= 12.92\ \text{MJ/kg}\end{aligned}$$

Using Eq. (3.30) again noting that $H$ was given on "dry" basis:

$$H_{ar} = 6.25\times(1-0.35) = 4.06$$

$$\begin{aligned}LHV_{f,ar} &= 12{,}922 - 2241.7\left(\frac{9\times4.06}{100} - \frac{35}{100}\right)\\ &= 11{,}317\ \text{kJ/kg}\\ &= 11.31\ \text{MJ/kg}\end{aligned}$$

**c.** Similarly, values of HHV and LHV on dry ash free basis are found using Eq. (3.31), which gives:

$$HHV_{daf} = \frac{Q}{(M_f - M_w - M_{ash})} = \frac{Q}{M_f}\frac{1}{(1-M-\text{ASH})} = HHV_{daf}\frac{1}{(1-M-\text{ASH})}$$

$$HHV_{f,daf} = HHV_{db}\frac{1}{(1-\text{ASH})} = \frac{19.88}{(1-0.0313)} = 20.52\ \text{MJ/kg}$$

$$HHV_{db} = HHV_{ar}\frac{1}{(1-M)}$$

$$LHV_{f,daf} = \frac{LHV_{f,db}}{1-\text{ASH}_{db}}$$

$$= 18.51 - 0.0313 = 19.09\ \text{MJ/kg}$$

where $M$ and ASH are moisture and ash fraction in raw biomass, respectively.

The biomass contains 3.13% ash on dry basis.

$$\begin{aligned}\text{So ash percentage on ''as-received or wet basis''(ASH)} &= \text{ASH}_{db}\times(1-M)\\ &= 3.13\times(1-0.35)\\ &= 2.03\%\end{aligned}$$

Using Eq. (3.31), we can find ASH of the feed on "daf" basis from that on "as-received" basis as follows;

$$\text{ASH}_{daf} = \frac{\text{ASH}_{ar}}{(1-M-\text{ASH})} = \frac{3.13}{(1-035-0.00203)} = 4.83\%$$

**2.** *Mass yield*

The wood loses 20% of its dry mass during torrefaction. So, the mass yield on "db" ($MY_{db}$) = 1 − dry mass loss = 100 × (1 − 0.2) = 80%

From Eq. (4.11):

$$\begin{aligned}\text{Mass yield wet basis } (MY_{ar}) &= MY_{db}\times(1-M)\\ &= 80\times(1-0.35)\\ &= 52\%\end{aligned}$$

From Eq. (4.12):
Mass yield on dry and ash free basis:

$$MY_{daf} = \frac{MY_{db} - ASH_d}{(1 - ASH_d)}$$

$$MY_{daf} = \frac{80 - 3.13}{(1 - 3.13)} = 79.35\%$$

## 4.5.2 Energy Density

Energy density is another important parameter of the product of torrefaction. It gives the amount of energy released when unit mass of the torrefied product is burnt and its product is cooled. Energy density is also associated with terms like specific energy, calorific value, and heating value. Most applications use energy density on mass basis such as kJ/kg, Btu/lb, kCal/kg. Energy density as defined here is equivalent to the HHV described in Section 3.6.5.

$$\begin{aligned}\text{Energy density} &= \text{amount of energy released when unit mass of the torrefied} \\ &\quad \text{biomass is fully combusted} \\ &= \text{higher heating value (HHV)}\end{aligned} \tag{4.13}$$

Energy density may also be defined on volume basis, where it gives the amount of useful thermal energy stored in unit volume of a substance. Here, energy density is expressed as kJ/m$^3$, kCal/m$^3$, or BTU/ft$^3$. The volume-based definition is used only in special cases like shipment of fuels. For example, for ocean freight, the rate is generally based on volume basis subjected to a maximum weight.

Biomass contains appreciable amount of moisture which evaporates during combustion. So, energy density may also be expressed as lower heating value (LHV), which is lower than the HHV as the former does not consider the heat used in evaporating moisture. The heat of vaporization water being high (~2260 kJ/kg), the difference between the HHV and net or LHV could be appreciable for a wet, raw, or "as-received" biomass (see Section 3.6.5). Therefore, one should pay particular attention to how the energy density or the heating value is expressed as.

Energy density can be expressed on "as-received" basis, "dry basis", or on "dry ash free" basis. Equation (3.31) gives relations between these definitions of energy density or heating value.

Figure 4.1 illustrates how the energy density of a sample biomass could increase after torrefaction. Here, 100-unit mass of biomass with 100-unit energy is torrefied losing certain amount (viz. 30 unit) of mass. The solid product of torrefaction generally retains a higher fraction of the energy (viz. 90 unit) of the biomass because the lost mass comprises mostly of water, carbon dioxide, and energy-lean gases. So, the energy density of the solid mass will increase from [100/100] or 1.0 to [90/(100−30)] or 1.28 in that unit. The gaseous mass or the volatile part of the biomass (30 unit) carries the residual energy (100−90)/30 or 0.33 energy density. Thus, we see from Figure 4.1 how the torrefaction process increases the energy density of the biomass through this pretreatment.

### 4.5.3 Energy Yield

Energy yield gives the fraction of the original energy in the biomass retained after torrefaction. After torrefaction, energy-rich components remain in the biomass, but some energy-lean components are lost. This leads to some loss in the overall energy content of the biomass, though there is an increase in the energy density as illustrated in Figure 4.1. Energy yield defines this retention, and as such it is of great practical importance especially where the biomass is used for energy conversion.

Torrefaction makes biomass use convenient especially in energy systems like a boiler but at the expense of some energy loss. Energy yield gives quantitative value of this loss and is defined as:

$$\text{Energy yield (EY)} = \frac{\text{energy in torrefied product}}{\text{energy in raw biomass}} \tag{4.14}$$

Energy yield may be written in terms of heating values of the biomass before and after torrefaction:

$$\text{EY} = \frac{\text{mass of product} \times \text{heating value of product}}{\text{mass of biomass feed} \times \text{heating value of feed}}$$

By expressing the heating value on dry ash free basis ($HHV_{daf}$), one can relate it to the mass yield, $MY_{daf}$, as:

$$EY_{daf} = \left.\frac{\text{product mass}}{\text{feed mass}}\right|_{daf} \times \left.\frac{HHV_{product}}{HHV_{feed}}\right|_{daf} = MY_{daf} \times \left.\frac{HHV_{product}}{HHV_{feed}}\right|_{daf} \tag{4.15}$$

Unlike mass yield, energy yield does not depend on how the product or feed is expressed as:

---

**Example 4.2**

Using data from Example 4.1, calculate the following:

**a.** Energy yield on "dry" and "dry ash free" basis.

**b.** HHV of torrefied biomass on "dry" and on "dry ash free" basis.

**Solution**

**a.** *Energy yield*:

Example 4.1 states that volatiles carried 5% of the total energy. So, the amount of energy that remains in the solid is

$$\text{Energy yield (EY)} = 1 - \text{energy lost} = 100 \times (1 - 0.05) = 95\%$$

As energy yield is independent of whether the mass is expressed on dry or dry ash free basis. We can write

$$EY_{ar} = EY_{daf} = EY_{db} = 95\%$$

**b.** *HHV of torrefied biomass*:

Equation (4.15) relates the HHV of torrefied product with that of the raw feed as:

$$EY_{daf} = MY_{daf} \times \left. \frac{HHV_{product}}{HHV_{feed}} \right|_{daf}$$

$$HHV_{t,daf} = HHV_{f,daf} \times \frac{EY_{daf}}{MY_{daf}}$$

Example 4.1 gives $HHV_{f,daf} = 20.52$ MJ/kg and $MY_{daf} = 79.35\%$.

$$HHV_{t,daf} = 20.52 \times \frac{0.95}{0.7935} = 25.36 \text{ MJ/kg}$$

The ash of the biomass is not lost during torrefaction, though the overall mass of the biomass is reduced. The absolute amount of ash in the feed does not change while the overall mass reduces. That is, the ash percentage changes after torrefaction. Torrefaction does not change the absolute amount of ash in the biomass though the overall mass reduces.

Example 4.1 gives mass yield and ash in feed on "as-received" basis as 52% and 2.03% respectively. So, the ash content of torrefied mass, $ASH_t$ is

$$ASH_t = 2.03/0.52 = 0.039 = 3.9\%$$

From Eq. (3.31) we can get:

$$HHV_{t,db} = HHV_{t,daf} \frac{(1 - M - ASH)}{(1 - M)} = 25.36 \times \frac{(1 - 0 - 0.039)}{(1 - 0)} = 24.37 \text{ MJ/kg}$$

## 4.6 PHYSICAL PROPERTIES OF TORREFIED BIOMASS

Torrefaction brings about several important changes in the physical properties of biomass. Some of them are much relevant for cofiring of biomass with coal (Chapter 10). This section discusses the modification of several physical properties of biomass through torrefaction pretreatment.

### 4.6.1 Density and Volume

We have seen earlier that the mass of biomass reduces during torrefaction. This change often brings about some reduction in volume as well. As a result, the density change due to torrefaction could not be defined by the mass yield alone. There is thus a distinct effect of torrefaction on the density of biomass, which is a reduction with torrefaction (Table 4.6).

The information on the effect of torrefaction on density is important for the design of a torrefaction plant as well as for a detail analysis of the process. For a good understanding of the effect of torrefaction on biomass density, it is worth recalling three types of densities explained in Section 3.5.1.

1. True density of particle (based on solid or cell wall volume alone)
2. Apparent density of particle (based on biomass solid and internal pore volume)
3. Bulk density of particles packed in an enclosure or piled on a surface (based on biomass solid, pore volume, and the void between particles in the packing).

The density of the cell walls that is the "*true density*" for most lignocellulose biomass is typically of the order of 1400 kg/m$^3$ (Jenkins, 1989). After torrefaction, there is only a marginal (<5%) reduction in the true density (Phanphanich and Mani, 2011), but the reduction in the apparent density is noticeable (Table 4.6). The bulk density of biomass in packing also reduces with torrefaction temperature.

Exploratory work carried out on coarse pieces of poplar wood showed (Basu et al., 2013b) that it is apparent density under torrefied condition decreases with severity and temperature of torrefaction. The rate of this decline, however, reduces at higher temperatures. The external volume of the torrefied biomass also reduces with increasing torrefaction temperature but to a lesser degree than done by its mass. This causes the apparent density to reduce with torrefaction temperature.

### 4.6.2 Grindability

Raw biomass is highly fibrous in nature, and its surface fibers often lock in with each other like in Velcro. This greatly increases interparticle friction.

**TABLE 4.6** Change of Densities with Torrefaction Temperature for Several Types and Sizes of Wood

| Temperature (°C) | 25 | 200 | 220 | 225 | 240 | 250 | 260 | 275 | 280 | 300 |
|---|---|---|---|---|---|---|---|---|---|---|
| Bulk density[a] (kg/m$^3$) | 381 | | | 342 | | 332 | | 376 | | 400[b] |
| Apparent density[c] (kg/m$^3$) | 500 | 489 | 445 | | 444 | | 395 | | 340 | 297 |
| True density[a] (kg/m$^3$) | 1400 | | | 1410 | | 1400 | | 1370 | | 1340 |

[a]*Pine wood chips 20.94–70.59 mm long, 1.88–4.94 mm thick, and 15.08–39.70 mm wide (Phanphanich and Mani, 2011).*
[b]*This reduced density could be a result of an experimental error.*
[c]*Poplar wood 25.4 mm diameter × 32 mm (Basu et al., 2012).*

These along with the plastic behavior of biomass cause handling difficult especially its pneumatic transportation through pipes. The soft and plastic nature of biomass makes it also difficult to grind and pulverize biomass to fine sizes.

Co-combustion of coal with biomass requires biomass to be ground to sizes comparable to those of coal (~75 μm), and then conveyed pneumatically through pipes (see Chapter 10). Because of its soft, nonbrittle characteristics, considerably more energy is required to grind untreated biomass to required fineness. For example, to grind a ton of coal to a fineness ($d_{50} \sim 500\ \mu m$) 7–36 kW h of grinding energy would be required, while 130–170 kW h of energy is needed to grind the same amount of raw poplar wood to that fineness (Esteban and Carrasco, 2006). There is thus nearly an order of magnitude increase in energy consumption when a coal pulverizer is used for biomass grinding. Additionally, torrefaction also influences the final particle size distribution.

Torrefaction results in complete breakdown of the cell structures of biomass making its particle brittle, smooth, and less fibrous. By making biomass particles more brittle, smoother, and less fibrous torrefaction addresses above problems to a great extent. An absence of fibrous exterior, sharp ends of the biomass particles after torrefaction (Phanphanich et al., 2011) reduces the friction created by the interlocking of these fibers during handling a pneumatic transportation.

#### *4.6.2.1 Effect of Torrefaction Parameters on Grinding*

All torrefaction parameters like temperature, residence time, and original particle size play a role in the reduction in the energy required for grinding to a given fine fraction. Torrefaction parameters influence the grinding of torrefied product in the following order (Joshi, 1979):

$$\text{Temperature} > \text{residence time} > \text{original particle size} \tag{4.16}$$

Thus, torrefaction temperature is the most influential parameter for grinding. The higher the torrefaction temperature, the lower the energy required for grinding or for a given energy input a greater amount of finer particles are obtained after grinding. After torrefaction, the particles are not only smaller but their size distribution is also more uniform. The grinding energy requirement for specified level of grinding decreases with torrefaction temperature. For example, Phanphanich et al. (2011) noted that the specific energy consumption reduced from about 237 KW h/t for raw biomass to about 24 kW h/t for that torrefied at 280°C.

### 4.6.3 Hydrophobicity of Torrefied Biomass

Biomass is hygroscopic in nature. So, it absorbs moisture even when it is stored after drying. Thus, extended storage of biomass is very expensive in terms of the energy spent in evaporating the moisture during combustion or gasification. Coal on the other hand is hydrophobic, that is, it does not absorb moisture or is less hygroscopic. So, this penalty for extended storage is absent for coal. The extent of water repellant property is described by its hydrophobicity.

Presence of moisture in fuel is undesirable for several reasons:

1. Moisture not only reduces the heating value of the fuel but also it greatly increases the stack loss in a combustion system. For example, 1 kg evaporated moisture at 150°C carries away 2698 kJ of moisture while the same mass of dry flue gas will carry only 160 kJ at that temperature.
2. Moisture increases the potential for fungus development in biomass when stored.
3. Moisture increases the cost of transportation and handling and feed preparation without making any useful contribution to the fuel's use.

Thus, the lower the moisture in the fuel, the better is its end use. Torrefaction can address these problems in the following ways:

1. Drying in pretorrefaction stage reduces the moisture of raw biomass from 10–50% to about 1–5%.
2. After torrefaction, biomass becomes largely hydrophobic, or resistant to water, and thus it absorbs very little moisture.
3. The hydrophobic character of torrefied biomass allows its extended storage without biological degradation (Tumuluru et al., 2011).

#### *4.6.3.1 Why Biomass Becomes Hydrophobic after Torrefaction?*

In biomass, the moisture absorption capacity of its hemicellulose constituent is highest. The capacity of cellulose and lignin follows that (Li et al., 2012). Since torrefaction involves near-complete breakdown of hemicellulose, the process makes biomass hydrophobic. Raw biomass readily absorbs moisture due to the presence of its hydroxyl (–OH) groups that form hydrogen bonds to retain additional water. The torrefaction process destroys the OH groups and thereby reduces its capacity to absorb water (Pastorova et al., 1993). Additionally, due to the chemical rearrangement during torrefaction, nonpolar unsaturated structures are formed in biomass after torrefaction. The nonpolar character of condensed tar on the solid also prevents condensation of water vapor inside the pores. Felfli et al. (2005) attributed the hydrophobicity of torrefied biomass to tar condensation inside the pores that obstruct the passage of moist air through the solid, which then avoids the condensation of water vapor.

Table 4.7 compares the hygroscopic character of raw biomass with that of torrefied biomass. After immersing in water for 2 h, the water uptake of the torrefied biomass is nearly two orders of magnitude lower than that of

**TABLE 4.7** Rise in Moisture Content After Submerging in Water for 2 h

| | Moisture % on Dry Ash Free Basis | |
|---|---|---|
| Condition of the Feed | Sawdust | Water Hyacinth |
| 25°C, Raw biomass before torrefaction | 150.3 | 197.5 |
| After torrefaction at 250°C | 7.8 | 17.7 |
| After torrefaction at 270°C | 3.3 | 14.9 |
| After torrefaction at 300°C | 2.1 | 8.8 |

**Source:** From Pimchua et al. (2010).

raw biomass. It further shows that the higher the torrefaction temperature, the lower is its water absorption ability. Additionally, it also depends on the type of biomass.

A more severe torrefaction (higher temperature and or longer residence time) could make torrefied products more hydrophobic (Verhoeff et al., 2011), but some researchers (Medic et al., 2012) noted that improvement in hydrophobicity above 250°C is not significant (Yan et al., 2009).

Wet torrefaction, though done in water, interestingly makes the product more hydrophobic (Yan et al., 2009) than dry torrefaction.

### 4.6.4 Explosion Potential of Torrefied Dust

Dust explosion is a major problem in handling and conveying fine dusts especially of easily ignitable materials. Torrefaction makes biomass brittle and could result in more dust during handling. Additionally, due to its high reactivity and low moisture content, torrefied biomass could more easily ignite than coal, which in turn increases the explosion potential of the torrefied biomass within mills or conveying pipes. Chapter 10 discusses this aspect further.

In addition to the explosion potential, torrefied biomass also carries a risk of fire because of its low ignition temperature. Some biomass plants have experienced this.

### 4.6.5 Densification or Pelletization

Biomass is an energy-lean fuel. This makes its transportation more expensive in terms of megajoule energy transported. So, to improve its energy density, biomass is often compressed into denser pellets or briquettes.

Table 4.8 compares energy density and several related characteristics of wood, torrefied wood, pelletized green wood, and pelletized torrefied wood. Data for a typical subbituminous coal is also given here for reference.

Pelletization increases the bulk density of green or untreated biomass because of its higher apparent density and regular shapes. The densification of biomass through pelletization can be made more effective if the wood is torrefied and then pelletized. In a typical process, the biomass is torrefied, cooled, ground to required size and then subjected to densification under pressure and slight heating.

Densification through pelletization increases the mass energy density of wood from 10.5 to 20.7 MJ/kg and volume energy density from 5.8 to 16.6 GJ/$m^3$. Thus, further densification of wood through pelletization and torrefaction could make transportation and handling of wood competitive with that of coal.

Torrefaction could also remove a major limitation of raw wood pellets. When stored for long periods of time, the pellet absorbs much moisture reducing the strength of the pellet, even to the extent of crumbling when wet.

Densification of raw biomass through pelletization requires it to be ground typically to 3.2–6.4 mm size (Mani et al., 2006). After that the ground biomass is compressed into pellets to increase its density. Some form of external binding agent needs to be used to give the pellet a good binding strength. Alternatively, the natural binder of biomass, lignin, may also be

**TABLE 4.8** Comparison of Mass and Energy Densities

| | Fresh Wood | Pellet of Wood[a] | Torrefied Wood | Pellet of Torrefied Wood[a] | Subbituminous Coal[a] |
|---|---|---|---|---|---|
| Moisture content (%) | 35 | 8.5 | 3 | 3 | 31 |
| Mass energy density (LHV, as-received) (MJ/kg) | 10.5 | 15.9 | 19.9 | 20.7 | 20.4 HHV |
| Volume energy density (GJ/$m^3$) | 5.8 | 9.2 | 4.6 | 16.6 | 26.6 |
| Bulk density (kg/$m^3$) | 550 | 575 | 230 | 800 | 850 |

[a]*Average values taken form Bergman et al. (2005b). Volume energy density calculated based on apparent density of 1300 kg/$m^3$.*
#Rank III, Group 2, Steam, p. 9–7.

utilized, but the temperature of the biomass must be increased to the range of 50–150°C so that the lignin within biomass softens (Gilbert et al., 2009). When the heated biomass particles are compressed, they form good physical bonds between them. Thereafter, when cooled the lignin hardens holding the compressed particles together providing it with a good mechanical strength without an external binding agent.

Higher pressures produce pellets with higher bulk densities and improved tensile strengths but slightly elevated temperatures (~70°C) could have a greater beneficial effect on the quality of pellets due to the softening of the lignin in biomass. So, the higher the lignin contents, the higher the pelletization quality (Gilbert et al., 2009).

During torrefaction (200–300°C), the hemicellulose content of biomass largely degrades while only a small part of its lignin breaks down. Thus, after torrefaction the amount of lignin as a percentage of the total biomass should ideally increase improving its binding property. Some experimental data, however, suggest pelletization of torrefied wood to be harder due to its brittle nature.

Additionally, torrefaction opens more lignin-active sites by breaking down the hemicellulose matrix and forming fatty unsaturated structures, which creates better binding. So, for torrefied wood, one could use lower pressure and lower temperature for densification of biomass. Pelletization of torrefied biomass, therefore, needs less energy than that by pelletization of raw biomass pelletization (Tumuluru et al., 2011).

One drawback of torrefied pellets could be that due to the loss of hemicellulose, pellets can be more brittle and less strong (Gilbert et al., 2009). To avoid this shortcoming, one could carry out torrefaction and pelletization simultaneously.

**Example 4.3**

The wood of Example 4.1 (*M:* 35%; $HHV_{ar}$: 12.92 MJ/kg; apparent density: 300 kg/m$^3$) is being considered for either pelletization or torrefaction followed by pelletization. The pelletized raw wood is expected to have a moisture content of 7% and an apparent density of 650 kg/m$^3$.

The raw wood, when torrefied, suffers 39% reduction in density but is free from moisture and its HHV increases to 24.59 MJ/kg. Neglect any change in its energy content due to pelletization and assume a 20% reduction in volume due to torrefaction.

Taking necessary values from Example 4.1:

**a.** Compare the volumetric energy density between raw wood and pelletized wood.

**b.** Compute the increase in energy density if the wood is torrefied and the pellets made from that.

**Solution**

**a.** *Volume energy density:*

Raw wood:

The volumetric density of energy of raw wood, $EV_f$, that contains 35% moisture would be

$$EV_f = 12.92\frac{MJ}{kg} \times 300\frac{kg}{m^3} = 3876\ MJ/m^3$$

Pelletized wood:

$$\text{Density of pelletized raw wood} = 650\ kg/m^3$$

Since the energy content did not change during pelletization, the volume energy density of the pellet is

$$EV_f = 12.92\frac{MJ}{kg} \times 650\frac{kg}{m^3} = 8398\ MJ/m^3$$

So, we note a large increase in volumetric energy density through pelletization.

**b.** *Energy density:*

Moisture content of wood pellets is 7%.

Pelletization affects moisture content but not the energy density on dry basis. It does not affect the energy content. So, the mass energy density of pelletized wood is the same as the feed wood as-received:

$$HHV_{p,d} = HHV_{f,d}$$

$$\frac{HHV_{p,ar}}{(1-0.07)} = HHV_{f,d} = \frac{12.92}{(1-0.35)} = 19.87\ MJ/kg$$

$$HHV_{p,ar} = 19.87 \times 0.93 = 18.48\ MJ/kg$$

But through pelletization, biomass mass density, $\rho_p$, has increased to 650 kg/m$^3$. So, the volume density of energy of pelletized wood, $EV_p$, will be:

$$EV_p = HHV_p \times \rho_p = HHV_r(wt) \times 650 = 18.48 \times 650\ MJ/m^3 = 12.01\ GJ/m^3$$

Torrefied wood:

$$\text{Density of the torrefied wood reduced by 39\%} = (1-0.39) \times 300 = 183.5\ kg/m^3$$

Hence, volumetric energy density of torrefied product, $EV_t$, is calculated as:

$$EV_t = 24.59 \times 183.5 = 4512\ MJ/m^3$$

Torrefied pellet:

When the wood is torrefied, its mass density of energy increases to 24.59 MJ/kg. Since the problem neglects any change in mass density between torrefied pellet and raw wood pellet, we take this value to be 650 kg/m$^3$. So, the volumetric density of energy of torrefied pellet, $EV_{p,t}$, would be

$$EV_{p,t} = HHV_t \times \rho_p = 24.59 \times 650\ MJ/m^3 = 15.98\ GJ/m^3$$

| | **Raw Wood** | **Torrefied Wood** | **Wood Pellet** | **Torrefied Pellet** |
|---|---|---|---|---|
| Volume energy density, MJ/m$^3$ | 3876 | 4512 | 8398 | 15,983 |
| Mass energy density (as-received basis), MJ/kg | 12.92 | 24.59 | 19.28 | 24.59 |

**c.** *Increase in energy density*:

The volumetric energy density of raw wood was based on initial value = (15,983−3876)/3876 = 3.12.

There is thus an increase of 312% of volumetric energy density of biomass after torrefaction.

The increase in volumetric energy density due to pelletization alone = (12,530 − 3876)/3876 = 2.23 or 223%.

## 4.7 TORREFACTION TECHNOLOGIES

A typical torrefaction plant would include several units like biomass handling, preparation, dryer, torrefier, and product cooler. Among these, the torrefaction reactor is most important. Presently, vendors are offering many designs of torrefier. This section presents a broad overview of those designs of torrefier by classifying them under specific groups. Table 4.9 presents a

**TABLE 4.9 Torrefaction Reactor Technologies Supplied by Different Developers**

| Torrefier Technology | Technology Supplier |
|---|---|
| Rotary drum reactor | CDS(UK), Torr-coal (NL), BIO3D(FR), EBES AG(AT), 4Energy Invest (BE), BioEndev/EPTS (SWE), Atmosclear S.A. (CH) |
| Screw conveyor | BTG (NL), Biolake (NL),FoxCoal (NL), Agri-tech (US) |
| Multiple hearth | CMI-NESA (BE), Wyssmont (US) |
| Entrained | Topell (NL), Airex (Canada) |
| Fluidized | Topell (NL) |
| Mircowave | Rorowave (UK) |

**Source:** Updated from Kleinschmidt (www.kema.com).

list of torrefaction technologies currently available in the market. Figure 4.9 shows the schematic of some of the available technologies.

### 4.7.1 Classification of Torrefaction Reactors

A wide range of torrefier or torrefaction reactors is in use or in development. Such designs often evolve from other biomass processing units like dryer, pyrolyzer, and carbonizer. These reactors may, however, be divided into some specific generic groups based on two aspects of the torrefaction process: heat transfer and solid contacting.

#### *4.7.1.1 Classification on Mode of Heating*

Heating is an important part of the torrefaction process. A medium carries heat and transfers it to the biomass particles. The transfer of heat to biomass particles could take place through one of the following means:

- Gas–particle convection
- Wall–particle conduction
- Electromagnetic heating of biomass
- Particle–particle heat transfer
- Liquid–particle heat transfer.

Based on the mode of heating, torrefaction reactors may be grouped into two basic types:

1. Directly heated type
2. Indirectly heated type.

Some reactors may, however, have a combination of these basic modes of heating. The following sections describe some of the common types of directly and indirectly heated reactors.

#### Directly Heated Reactors

In directly heated reactors, biomass is heated directly by a heat-carrying medium, and the heat is exchanged through direct contact between the biomass and the heat carrier. The heat carrier could be either a hot gas without oxygen or one with limited amount of oxygen. It could also be hot nonreactive solids or hot fluid-like pressurized water, steam, or waste oil.

***Convective Reactor (Moving/Fixed/Entrained Bed)*** This is the most common type of reactor used for torrefaction. Here, the heat carrier is a hot gas percolating through or flowing past biomass particles that are either stationary or moving (Figure 4.9B). The hot gas may be completely inert (Energy research centre of the Netherlands (ECN) moving bed) or with a small amount (2–3%) of oxygen (Thermya moving bed) (Ryall, 2012). If the biomass particles remain stationary with respect to reactor wall, it is called *fixed bed*. Such beds are used when biomass is loaded into multiple containers, which are then slowly drawn through a long hot but nonoxidizing tunnel furnace.

If particles move with respect to reactor wall, it is called *moving bed*. The wall of the reactor can be horizontal, vertical, or inclined. The particles may be moved, by gravity, by force of a mechanical device like rotating disk (Figure 4.9C) or vibrating motion of belt (Figure 4.9D). Particles that flow through the reactor are unidirectional without backmixing. The heat transfer occurs primarily through gas–solid convection that depends on the relative velocity between the biomass and the hot fluid.

Entrained flow type reactors carry finely ground biomass in hot inert or low oxygen gas. Because of high heat transfer coefficient between fine (sawdust like) biomass and high-velocity gas carrying them, the particles are heated to torrefaction temperature quickly. As such, it requires much shorter residence time in the reactor. Rapid heating reduces the solid yield increasing the liquid yield.

Some directly heated convective-type torrefiers use a rotating drum where the biomass is heated directly by hot gas passing through the tumbling drum (Figure 4.9E). In this case, the drum simply serves as a mixing device while heat transfer takes place through gas–particle convection.

***Fluidized Bed*** In this type of torrefier, hot inert gas is blown through a bed of granular heat-carrier solids or appropriately sized biomass particles (Li et al., 2012) in a way that the solids behave like a fluid. These heat-carrier particles being in vigorously mixed and agitated state can easily heat up any fresh biomass particle dropped into it (Basu, 2006). The biomass particles thus undergo torrefaction in a well-mixed state with uniform temperature distribution. The system, therefore, ensures a product quality more uniform than that is available from moving or fixed bed reactors. Separation of heat-carrier solids from torrefied biomass and entrainment of fine biomass particles are some of the limitations of this technology.

The dominant mode of heat transfer in a fluidized bed is particle-to-particle heat transfer. The "torbed" technology works on this principle where biomass particles are fluidized above a grate of inclined slots and is subjected to cross flow of gas (Figure 4.9H). This type can provide very uniform quality of the torrefaction product.

***Hydrothermal Reactor*** Here, the biomass is subjected to pressurized heating in water and thus obviates the need for drying (Yan et al., 2009). It is especially suitable for high-moisture or wet biomass. The process could bring about a slight improvement in the hydrophobicity of the biomass (Medic et al., 2012). The dominant mode of heat transfer in a hydrothermal reactor is that between hot water (steam) and biomass. While this process has several potential advantages, the energy required for pressurization and movement of a large volume of biomass across a pressure barrier could be an issue. One could, however, avoid this by adopting batch process as used for fermentation reactors. Multiple digester-type reactors could be used to produce a torrefied product that is dewatered using some conventional means.

Another form of hydrothermal reactor could use hot liquid (waste liquid or superheated steam) as the heat carrier. If the liquid is inexpensive and has some heating value like waste oil, it could be a viable heating medium that does not have to be removed from the product. On the other hand, superheated steam may condense making the product less attractive.

### Indirectly Heated Reactors

In indirectly heated torrefiers, the heat-carrying medium does not contact the biomass directly. Heat is transported across a wall or through electromagnetic radiation. So, here it is relatively easy to avoid contact with oxygen and therefore avoid undesired combustion during torrefaction. Such reactors have two major advantages: one that the heating fluid and medium can be anything hotter and the other is that the volatiles released during torrefaction are not diluted by the heating medium passing through it. So, the gaseous product of torrefaction can be combusted separately to supplement the thermal load of the reactor.

Since the heat is conducted slowly from the biomass layer in contact with the hot reactor wall to the core of biomass pack, one could expect a temperature gradient resulting in nonuniform heating of the biomass inventory in the reactor. A microwave that heats by electromagnetic irradiation may also result in nonuniform heating of the biomass particles (Basu et al., 2012).

***Rotating Drum*** Such torrefiers use a indirectly heated rotating drum that tumbles the biomass in an environment of inert gaseous medium (Figure 4.9E). The biomass is heated by hot drum walls or by hot internals in the drum. The heat transfer from the wall to the biomass particles is the primary controlling factor and not the heat transfer from gas to particle. Biomass is generally dried separately.

There is another version of such indirectly heated torrefier. Here, the biomass is contained in thermally conducting containers, which are carried through a hot furnace in an inert ambience at a slow speed. Heat conducted into the biomass through the heated wall slowly torrefies the biomass. Because of the relatively low heat exchange coefficient, residence time for this type is large and could be in the order of hours.

***Screw or Stationary Shaft*** Here, the torrefaction reactor (circular or rectangular cross section) is stationary, and it could be vertical, horizontal, or inclined (Figure 4.9F). The reactor is generally heated from outside to avoid contact with hot gases, though some may have holes for the products of torrefaction to escape. Heat is conducted to the biomass by means of conduction across reactor walls. A rotating screw churns and moves the biomass through the reactor to enhance heat transfer between the wall and the bulk of the biomass and at the same time to move the biomass along its length (Figure 4.9F).

***Microwave*** Microwave irradiation involves electromagnetic wave in the range of 300 MHz to 300 GHz. Typical microwave ovens or microwave reactors work at a frequency of 2.45 GHz. The microwave irradiation produces efficient internal heating by direct coupling of microwave energy with the molecules of biomass. The electric component of electromagnetic microwave radiation causes heating by two main mechanisms: dipolar polarization and ionic conduction. The heating depends on the ability of the materials being heated to absorb microwaves and convert it into heat (Figure 4.9A). Metals, for example, reflect microwave, while biomass absorbs it.

The microwave reactor (Figure 4.9A) is different from other indirectly heated reactors, where biomass particles are heated externally, that is, heat from the reactor wall arrives at the surface of biomass particles, and then it is conducted into the interior of the biomass. Contrary to this, biomass particles in a microwave reactor are heated from within. Microwave heating biomass may not be very efficient because biomass is a poor thermal conductor. In a microwave reactor, the heating is internal; every part of the biomass in the path of microwave radiation are heated simultaneously. Limited data available (Basu et al., 2012) show that microwave torrefaction creates an extremely fast rate of heating of the biomass interior, a matter of seconds. So, it does not allow heat to be conducted adequately to its exterior, and it causes a large temperature gradient in wood sizes of 25 mm or larger. Some investigators (Ren et al., 2012) found more encouraging results like 67–90% energy yield with 79–88% overall energy recovery.

### 4.7.1.2 Classification on Mode of Gas–Solid Mixing

In chemical processing plants, reactors are often classified by their gas–solid contacting modes. As such, this classification better helps understand the mixing that is vital for the reactor. In the list of reactors shown in Table 4.10, one can identify the following four modes of contact.

1. Plug flow (gas percolates through static solids; gas and solid both move unidirectional)
2. Partial back-mixed (e.g., fluidized bed, where gas is unidirectional but solids are back-mixed)
3. Tumbling (solids tumbles or moves around in a drum or cylindrical tunnel)
4. Entrained (solids are pneumatically transported by gas).

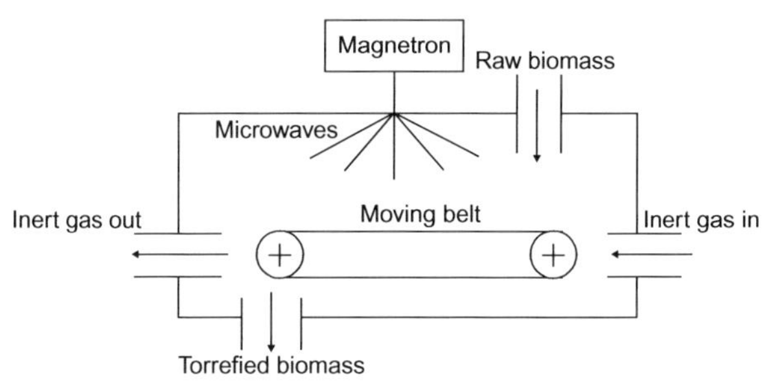


A. Microwave reactor

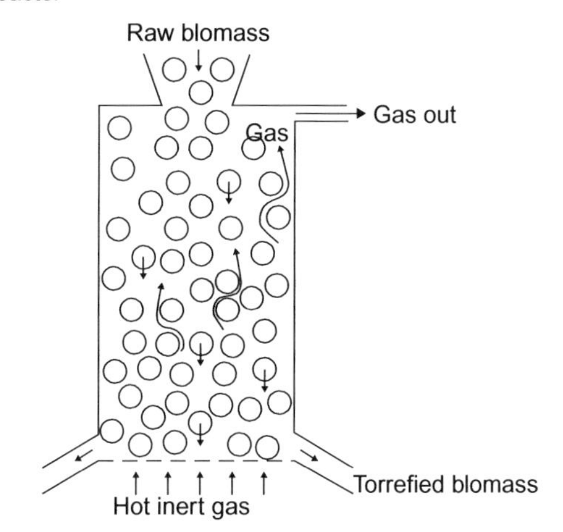


B. Moving bed reactor

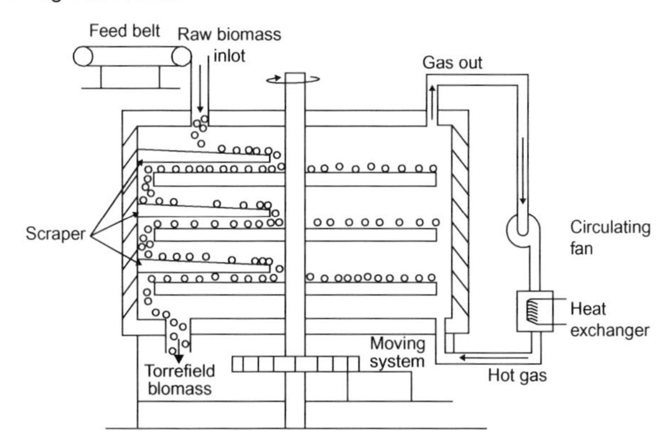


C. Multiple hearth furnace

**FIGURE 4.9** Schematic of some torrefaction technologies.

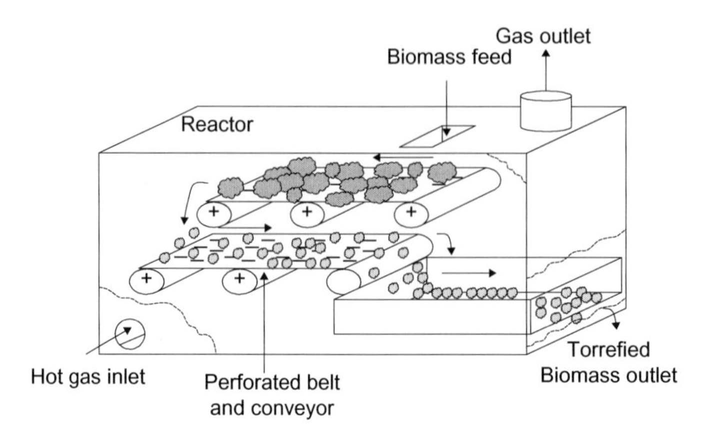


D. Oscillating belt reactor

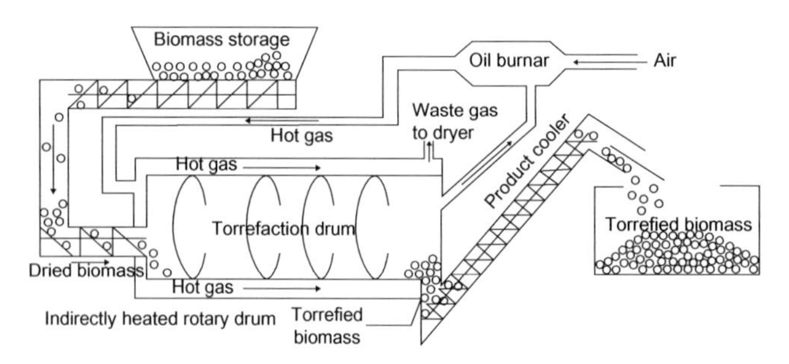


E. Rotary drum reactor

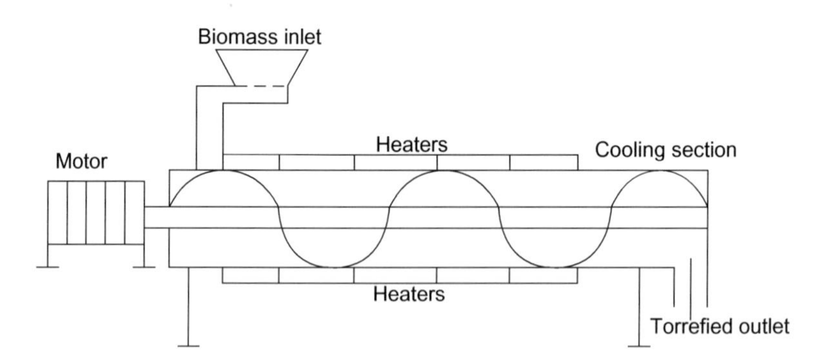


F. Screw conveyor reactor

**FIGURE 4.9** (Continued)

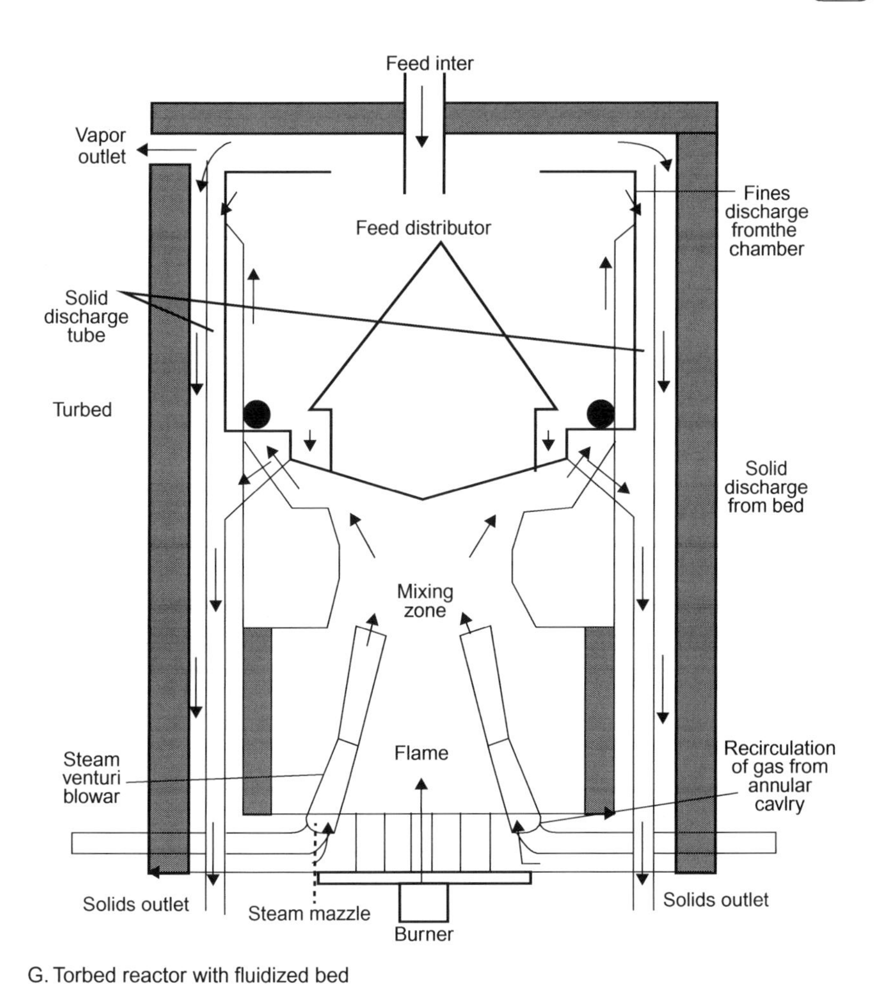


G. Torbed reactor with fluidized bed

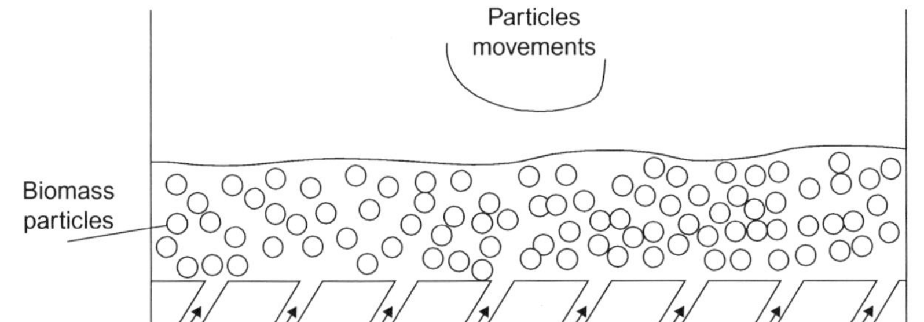


H. Fluidized bed section

**FIGURE 4.9** (Continued)

**TABLE 4.10** Reactor Classification by Heat Transfer Mode and Fluid–Solid Contacting Mode

| No. | | Mode of Heat Transfer | Gas–Solid Motion |
|---|---|---|---|
| 1 | Convective bed reactor (fixed, moving, entrained) | Gas–particle convection | Plug flow, entrained flow |
| 2 | Rotating drum reactor | Wall–particle conduction | Solids tumbling or moving around drum |
| 3 | Fluidized-bed reactor | Particle–particle convection | Back-mixed solids, plug-flow gas |
| 4 | Microwave reactor | Electromagnetic heating of water molecules in biomass | Plug-flow solids |
| 5 | Hydrothermal reactor | Water–particle heat transfer | Fixed bed for batch reactor |

Table 4.9 lists some major technology suppliers with specific types of torrefaction plants they are supplying. The table also lists the type of reactor technology they can be classified into. The hydrothermal reactor, though a potential reactor, is not being offered by anyone at the moment.

In the fluidized-bed reactor of "torbed flow technology," (Figure 4.9G) gas at high velocity flows through angled stationary blades supporting the biomass (Figure 4.9H) at temperatures up to 280°C, which gives a reactor residence time less than 5 min (Ontario Power Generation, 2010). In both belt conveyor and multiple hearth technologies, biomass moves on surfaces at a defined rate while the heating medium (hot flue gas, hot nitrogen, or superheated steam) flows over them providing heat to the biomass by convection. The heating is therefore mixed convective type.

## 4.8 DESIGN METHODS

This section presents a simplified method for the design of a torrefier.

### 4.8.1 Design of Torrefaction Plant

The first step in the design is the choice of reactor type. Design of the rest of the plant will to a great extent depend on this choice. A typical torrefaction plant compromises biomass handling and pretreatment (like chipping), grinding of the biomass if fine particles are needed, its drying, and finally torrefaction.

Energy required in each step is to be estimated for the overall design of the plant. The following is an order of magnitude estimate of energy requirement for these stages:

- Chipping of wood: 180–2360 kJ/kg wood (Cielkosz and Wallace, 2011).
- Grinding: 270–450 kJ/kg of feedstock (Cielkosz and Wallace, 2011).
- Drying of raw wood: 3000–9000 kJ/kg water removed (Cielkosz and Wallace, 2011).
- Torrefaction of dried wood: 130–350 kJ/kg torrefied wood (estimated).

#### *4.8.1.1 Choice of Reactor Type*

Choice of a reactor depends on several factors or considerations. For example, a choice made from capital cost consideration may not give the best operating cost or highest yield, while one with the highest yield may not suit the available feedstock or may require high capital investment. Applying proper weightage to a selection criterion, one could make a final selection.

#### *4.8.1.2 Design Approach*

The following section discusses an approach to the design of a continuous-type torrefaction plant. The whole process of torrefaction can be divided into five stages (Figure 4.3). For the sake of convenience, we combine them into three functional units:

1. Drying of raw feed in dryer
2. Torrefaction of dried feed in torrefier
3. Cooling of torrefied product in cooler

As shown in Figure 4.3, the energy demand for each of the above functional units or zones of a torrefaction plant is different. Hence, a reactor needs to provide the right amount of heat to the specific zone. Torrefaction is a slow conversion process. Hence, average-sized feed would need a relatively long residence time in the torrefaction reactor. This has to be provided by an adequate volume of the torrefaction zone. The product leaves the torrefier at the torrefaction temperature, which happens to be the highest temperature of the biomass. So, the hot product needs to be cooled before it is taken to storage. This is accomplished in the cooler or cooling zone of an integrated system.

For a desired set of product properties like HHV, grindability, hydrophobicity, and the type of torrefier reactor, the following two important parameters must be known in advance:

1. Torrefaction temperature
2. Torrefier residence time

Such information can be gathered from a bench/pilot scale test, published research, or experience from similar plants. To illustrate the process, we take

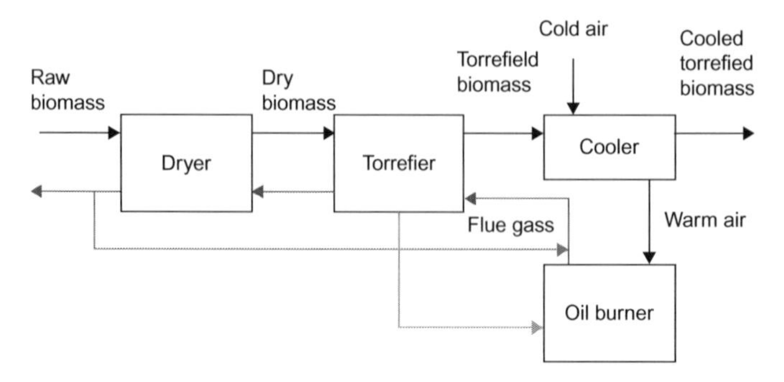


**FIGURE 4.10** Schematic of a torrefaction system.

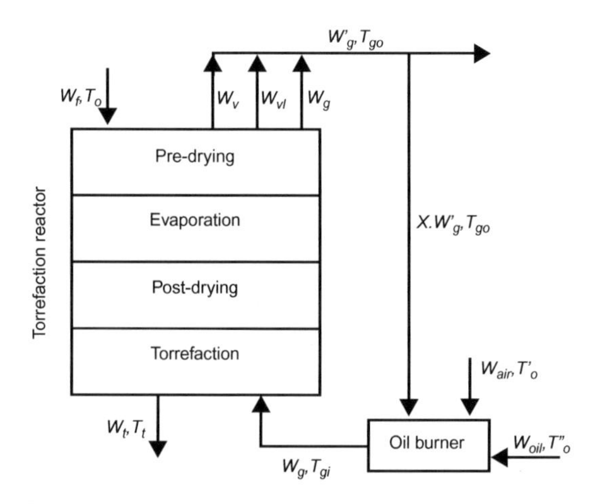


**FIGURE 4.11** A directly heated integrated torrefaction unit.

an example of a simple torrefaction plant where an oil burner provides the energy required for the process. For further simplification, we assume that the volatiles released during torrefaction are not utilized to reduce the oil consumption.

## Design Input

Figure 4.10 shows a schematic of a generic torrefaction unit, while Figure 4.11 shows an integrated single shaft moving bed reactor that is directly heated by hot oxygen-free flue gas. The heating medium, hot flue gas, moves up through the biomass while heating it through gas–particle convection. Fresh biomass drops from the top of the vertical reactor and descends slowly through the reactor while undergoing different phases of the process (Figure 4.11). Biomass and the heating medium are thus in counter-current mode.

*Input Parameter:*

Capacity of the torrefier = $W_t$ kg/s of torrefied wood including ash and moisture
Desired mass yield = $MY_{daf}$
Moisture fraction of raw biomass = $M$
Ash fraction of raw biomass = ASH
Corresponding values of the torrefied product are $M'$ and ASH$'$.

## Mass and Energy Balance

For the given capacity, $W_t$, of the unit, we calculate the required feed rate of raw biomass, $W_f$, entering the drier, and the flow-rate of dried biomass $W_d$ entering the torrefier section.

From Eq. (4.1), the flow-rate of the torrefied product on dry ash free basis is

$$W_{t,daf} = W_t(1 - M' - ASH') = MY_{daf} \times W_{daf}\ kg/s$$

where $W_{daf} = W_f\,(1 - M - ASH)$

Combining the above two, we get the feed rate of raw biomass as:

$$W_f = \frac{W_t(1 - M' - ASH')}{MY_{daf}(1 - M - ASH)}\ kg/s \tag{4.17}$$

Moisture in the biomass is reduced through torrefaction, but that is not the case for ash. Its absolute amount remains unchanged after torrefaction. Neglecting any loss of ash between the units, we get:

$$W_t ASH' = W_f ASH \tag{4.18}$$

For further analysis, we consider a single integrated torrefaction system as shown in Figure 4.11 where each zone requires specific amounts of heat.

---

**Example 4.4**

Design a moving bed torrefier to produce 1 ton/h (daf) of torrefied biomass from raw biomass containing 30% moisture but negligible amount of ash. Torrefaction at 280°C yields 70% mass (daf). Biomass and air enter the unit at ambient temperature of 20°C. Hot gas leaves torrefier at 105°C.

**Solution**

Desired output on daf basis:

$$W_t = 1\,TPH = 1000/3600 = 0.277\ kg/s$$

Moisture in raw feed = 30%

Using Eq. (4.17), we calculate the feed rate of raw biomass (neglecting any loss):

$$W_f = ((W_t/MY_{daf}\,(1 - M - A))) = [0.277/(0.7 \times (1 - 0.3))] = 0.566\ kg/s$$

Moisture in biomass: $W_v = 0.566 \times 0.3 = 0.17$ kg/s

Flow-rate of dry biomass: $W_d = (1-0.3) \times 0.566 = 0.396$ kg/s

$$\begin{aligned} W_{vl} &= W_f - W_t - M\,W_f \\ &= 0.566 - 0.278 - 0.566 \times 0.3 = 0.118 \text{ kg/s} \end{aligned}$$

---

***Dryer*** The feed is heated to the torrefaction temperature, $T_t$, in three stages:

**i.** $Q_{ph}$ is needed for preheating feed from room temperature, $T_0$, to the drying temperature, taken as 100°C.

$$Q_{ph} = W_f C_{pw}(100 - T_0)$$

**ii.** $Q_{dr}$ is required for complete evaporation of moisture in biomass. For this reason, there is hardly any change in the temperature in this zone. The heat duty of this stage is generally highest especially for a high-moisture feed:

$$Q_{dr} = W_f ML$$

Here, $L$ denotes the heat of vaporization at 100°C. So, we take this as 2260 kJ/kg.

**iii.** $Q_{pd}$ is needed for postdrying heating of the dry feed to the torrefaction temperature, $T_t$.

$$Q_{pd} = W_d\, C_{pd}(T_t - 100)$$

Total energy required or heat load, $Q_d$, for drying biomass to the torrefaction temperature, $T_t$, is sum of above three.

$$\begin{aligned} Q_d &= Q_{ph} + Q_{dr} + Q_{pd} \\ &= W_f C_{pw}(100 - T_0) + W_f ML + W_d C_{pd}(T_t - 100) \text{ kW} \end{aligned} \tag{4.19}$$

where $C_{pw}$ and $C_{pd}$ are specific heats of raw biomass and dry biomass, respectively.

Unlike in Section 4.5.1, we consider an overall heat loss fraction $X_d$ to calculate the actual heat load $Q'_d$ of the dryer as below:

$$Q'_d = \frac{Q_d}{(1 - X_d)} \tag{4.20}$$

Though torrefaction starts from its onset temperature, which is about 200°C, for convenience the entire heating, from 100°C to the torrefaction temperature, $T_t$, is included in the postdrying zone.

***Torrefier*** One can see from Figure 4.3 that the heat load or the energy required for torrefaction is very low. It is because the overall torrefaction reaction is either mildly exothermic or endothermic depending upon the torrefaction temperature. So, neglecting this heat, we take heat required in the torrefier $Q'_T$ as the loss from the reactor $Q_{tL}$.

$$Q'_T = Q_{tL} \quad (4.21)$$

The heat loss from the torrefier is a function of the size and level of insulation of the torrefaction reactor. It is not necessarily a fraction of the input energy.

So, the total heat load of the torrefier and dryer, $Q_{total}$, is sum of the above two:

$$Q_{total} = Q'_d + Q'_T \quad (4.22)$$

***Cooler*** The cooling section (Figure 4.10) cools the torrefied product from the torrefaction temperature to a safe temperature $T_c$ ($\sim < 50°C$) that is close to that of the atmospheric temperature. Heat extracted from the torrefied product $Q_c$ is therefore:

$$Q_c = W_t\, C_d(T_t - T_c) \quad (4.23)$$

This heat $Q_c$ can be utilized through suitable arrangements to preheat the burner air, and thereby reduce the oil consumption in the burner. Assuming a heat loss fraction $X_c$ in the cooler, the preheat temperature $T'_0$ of the burner air $W_{air}$ can be found from the following:

$$W_{air}\, C_{air}\, (T'_0 - T_0) = (1 - X_c)Q_c \quad (4.24)$$

***Burner*** Burner provides energy for the process. A major challenge in a directly heated system is to avoid oxygen in the flue gas. Biomass ignition temperature being exceptionally low, it could ignite even at temperatures as low as 200°C. Thus, a low excess air burner is to be used for such systems. Even then there could be air infiltration in a negative draft system raising the oxygen in the heating medium. Thus, special care is needed in the design of a burner system in a directly heated system like described here.

The total energy required $Q_{total}$ for the system is provided by the enthalpy of flue gas from the burner. This energy may be supplemented by burning along with the oil the volatiles released, $W_{vl}$, in the torrefier (Figure 4.10). The heating value of the volatile, $LHV_{vl}$, is however relatively low.

The temperature of the oil flame generally exceeds 1000°C, which is rather high, and could set fire to the torrefied biomass on first contact. So, it is necessary to reduce the temperature of the heating medium, $W_g$, to a lower value, $T_{gi}$, by diluting the burner flue gas with a part, $x$, of the relatively cold product gas, $W'_g$, leaving the torrefier. So, a mixture, $W_g$ comprising the burner gas and the recycled torrefier gas, $(X\, W'_g)$, enters the torrefier at temperature $T_{gi}$, which should be no more than the ignition temperature of the torrefied product ($\sim 300°C$) to avoid any risk of fire.

In a directly heated torrefier (Figure 4.11), the mixture flue gas would typically first enter the torrefier that may absorb a small amount of heat.

Thereafter, the hot gas would enter the drier in a directly heating system. So, the temperature of the gas leaving the drier ($T_{g0}$) should preferably be above 100°C to avoid condensation.

The flue gas, $W'_g$, leaves the drier carrying with it moisture and product gases from the torrefaction. So, the total flow-rate of product gas:

$$W'_g = W_f + W_g - W_t \tag{4.25}$$

The burner burns an amount of oil, $W_{oil}$, in combustion air, $W_{air}$, to produce a product gas that is mixed with recycled flue gas ($x\ W'_g$) to produce the hot gas, $W_g$, to be used as the heating medium. The combustion air could be preheated in the cooling section of the torrefier to a preheat temperature of $T'_0$. The mass balance around the burner system (Figure 4A.2) may be written as:

$$W_g = x\ W'_g + W_{oil} + W_{air} \tag{4.26}$$

An energy balance of the torrefier can give the amount of diluted flue gas $W_g$ as:

$$W_g = \frac{A}{C_g(T_{gi} - T_{g0})}\ \text{kg/}s \tag{4.27}$$

where, $A = (W_v C_v + W_{vl} C_g) T_{g0} + W_t C_d T_t - W_f C_b T_0 + W_v L + Losses$

$C_v$ is the specific heat of steam and $C_g$ is the average specific heat of flue gas between 300°C and $T_{g0}$.

An energy balance of the burner could give the amount of oil that must be consumed to provide necessary energy for torrefaction (see Appendix).

$$W_{oil} = \frac{1}{K-P} \times \frac{W_g}{W'_g}\left[\frac{C_g T_{gi}}{C_g T_{g0} + \text{VL}'_{fr}\text{LHV}_{v1}} - 1\right] \tag{4.28}$$

where,

$$K = \frac{(\alpha(\text{A/F})C_a T'_0 + \text{LHV}\eta + C_{oil}T''_0)}{W'_g C_g T_{g0} + W_{vl}\text{LHV}_{vl}\eta} = \frac{(\alpha(A/F)+1)}{W'_g} \quad \text{and} \quad \text{VL}'_{fr} = \frac{W_{vl}}{W'_g}$$

where, $\eta$ is the efficiency of the burner, $T'_0$ is the temperature of the preheated air entering the burner, and $\text{VL}_{fr}$ is the fraction of volatiles in the product gas of torrefaction. (The Appendix at the end of Chapter 4 gives the derivations of Eqs. (4.27) and (4.28).)

## Unit Sizing

The above calculations give the heat and mass balance of the system. Now, we find the sizes of different functional sections of the torrefier that would permit the required heat and mass transfer for the system to take place.

We take a simple case of a moving bed torrefier characterized by a uniform downward flow of solids and upward flow of gas through a vertical uniform cross-section shaft of the unit (Figure 4.11). Hot gas, $W_g$, enters the bottom of the torrefaction section at temperature, $T_{gi}$, and this along with other gaseous products leaves the top of the drying section at a temperature $T_{g0}$. Fresh biomass, $W_f$, enters the top of the drier at $T_0$, and hot torrefied biomass, $W_t$, leaves the bottom of the torrefier at $T_t$.

If the reactor volume is $V_r$, the total particle surface area, $S_r$, available for heat transfer, and the reaction can be calculated for known values of the surface to volume ratio $S_p$ of biomass particles and $\varepsilon_p$, the voidage in this section.

$$S_r = (1 - \varepsilon_p)V_r S_p \tag{4.29}$$

The average value of surface to volume ratio of solid particles is the ratio of external surface area and volume of the particles, $S_p = (A_p/V_p)$, which is a function of the particle shape.

***Predrying Section*** We assume that hot gas enters its bottom at temperature, $T_{g_3}$ and leaves from the top at $T_{g0}$. The biomass enters from the top at $T_0$ and leaves the bottom at 100°C. Gas to biomass particle heat transfer takes place across the log mean temperature difference, LMTD, which is given as:

$$\text{LMTD} = \frac{[(T_{g_3} - 100) - (T_{g_0} - T_0)]}{\ln[((T_{g_3} - 100)/(T_{g_0} - T_0))]} \tag{4.30}$$

If the particle–gas overall heat transfer coefficient is $U_t$, the heat transferred to biomass, $Q_{ph}$, in the predrying section is

$$Q_{ph} = U_t S_{r,ph} \text{LMTD}_{ph} \tag{4.31}$$

where the subscript "ph" refers to the values in the preheating section.

The preheat section's volume, $V_{r,ph}$, should be such that it accommodates at least an amount of biomass that would give the required gas–particle surface area of $S_{r,ph}$. Using the relationship (Eq. (4.29)) between particle surface area and volume occupied by it in this section, we can find the minimum volume of this section of the reactor, $V_{r,ph}$, as:

$$V_{r,ph} = \frac{Q_{ph}}{U_t \text{LMTD}_{ph}} \times \frac{1}{S_p(1 - \epsilon_p)} \tag{4.32}$$

The minimum height of this section, $L_{ph}$, with cross-sectional area, $A_r$, is given as:

$$L_{ph} = \frac{V_{r,ph}}{A_r} \tag{4.33}$$

Besides the volume, there are two other stipulations for such a moving bed system.

**a.** Solid moves under gravity with an acceptable space velocity $U_s$, that will provide the required residence time in the drying sections.
**b.** Upward gas velocity is not too high to blow fine particles away.

1. Data for solid space velocity in a torrefier is not available at the moment but a value of 0.5–5.0 m/h is used in an updraft cool gasifier (see Section 8.8.1.1). For a chosen value of space velocity and the designed feed rate of biomass, $W_f$, one calculates the minimum cross-sectional area, $A_r$, of the reactor as:

$$A_r = \frac{W_f}{U_s \rho_{bulk}} \tag{4.34}$$

where $\rho_{bulk}$ is the bulk density of solids. It is to be noted that the calculation is based on $W_f$, instead of $W_t$ to allow easier flow of solids.

2. The superficial velocity of the gas, $U_g$, needs to be chosen to avoid fluidization or entrainment of fine solids. It is calculated from the volume flow-rate of gas products $W'_g$ through this section:

$$U_g = \frac{W'_g}{\rho_g A_r} \tag{4.35}$$

where $\rho_g$ is the density of the gas at average temperature of the section which, for the preheating section, is $(T_{g3} + T_{g0})/2$.
The gas–solid relative space velocity is

$$U_{rel} = U_g + U_s \tag{4.36}$$

The value of $U_{rel}$ should not exceed the entrainment velocity of fine biomass particles or minimum fluidization velocity of average particles.

***Drying and Postdrying Sections*** A similar method is used for the design of these two successive sections. The log mean temperature difference is calculated from the temperatures shown in Figure 4.12.

***Torrefier Section*** The torrefaction zone is designed in the same way but with an additional stipulation that it must provide the specified solid residence time for the biomass feed and required yield.

The volume, $V_{tor}$, of the torrefaction zone for the given residence time of and production rate, $W_t$, is:

$$V_{tor} = \frac{\tau W_t}{\rho_{bulk}} \tag{4.37}$$

The height of the torrefaction zone is therefore calculated from $V_{tor}$:

$$L_{tor} = \frac{V_{tor}}{A_r} \tag{4.38}$$

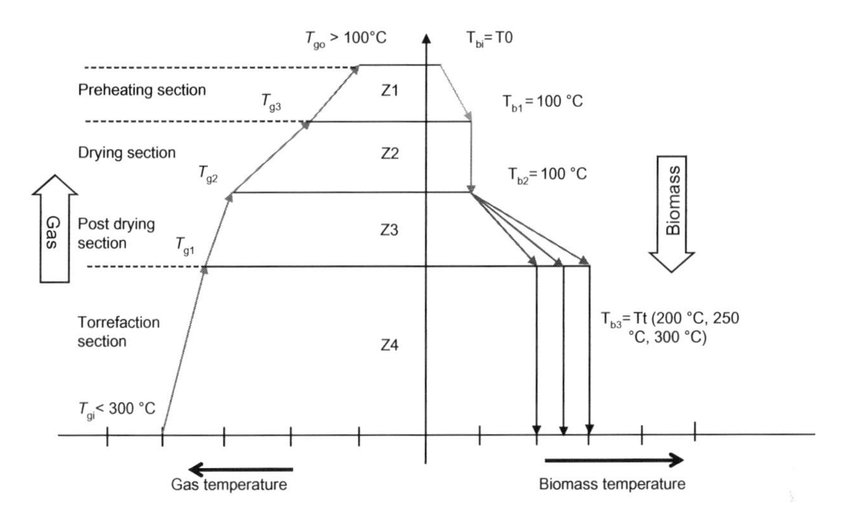


**FIGURE 4.12** A qualitative diagram for gas and solid temperature distribution along the height of a vertical torrefaction system. right hand side shows temperature of biomass while left hand side shows that for gas percolating through the system.

The space velocity of solid in the torrefier section is generally very low to allow the required torrefaction time in this zone.

---

**Example 4.5**

In Example 4.4, hot gas from an oil burner is diluted by recycled flue gas to reduce its temperature to 300°C, and it is then fed into the bottom of the torrefier. Heating value of oil is 45.5 MJ/kg and the burner operates at 20% excess air with an efficiency of 95%. Take latent heat of vaporization as 2260 kJ/kg.

Neglecting all heat losses, find the amount of oil consumption and what fraction of flue gas needs to be recycled through the burner.

Given:

Specific heat of flue gas, $C_g = 1.13$ kJ/kg C
Specific heat of steam, $C_v = 1.89$ kJ/kg C
Specific heat of air, $C_{air} = 1.006$ kJ/kg C
Specific heat of raw biomass, $C_b = 1.46$ kJ/kg C
Specific heat of dry or torrefied biomass, $C_d = 0.269$ kJ/kg C
Specific heat of oil, $C_{oil} = 1.7$ kJ/kg C
Stoichiometric air–oil ratio = 14.6
LHV of volatiles = 1286 kJ/kg

**Solution**

Neglecting losses, we calculate the following.

Energy required for raising 0.566 kg (Example 4.4) of raw biomass to 100°C in preheater is calculated using Eq. (4.19):

$$Q_{ph} = 0.566 \times 1.46 \times (100 - 20) = 66.1 \text{ kW}$$

Energy required for evaporation of the 30% moisture (Example 4.4) in biomass in dryer section is calculated using Eq. (4.19):

$$Q_{dry} = (0.566 \times 0.3) \times 2260 = 383.8 \text{ kW}$$

Energy required for heating 0.396 kg dried biomass (Example 4.4) to 280°C is calculated using Eq. (4.19):

$$Q_{pd} = 0.396 \times 0.269 \times (280 - 100) = 19.2 \text{ kW}$$

Total load = 66.1 + 414.6 + 19.2 = 469.0 kW
So, $Q_{total} = 469$ kW

***Burner Design*** Given:

$T_{gi} = 300°C$, $T_t = 280°C$, $T_{g0} = 105°C$, $T_0 = 20°C$
Theoretical (A/F) ratio = 14.6
Burner efficiency, $\eta = 0.95$; $LHV_{oil} = 45{,}500$ kJ/kg

The amount of flue gas, $W_g$, fed into the torrefier is calculated using Eq. (A.5) of Appendix:

$$W_g = \frac{A}{C_g(T_{gi} - T_{g0})}$$

$$A = (W_v C_v + W_{vl} C_g) T_{g0} + W_t C_d T_t - W_f C_b T_0 + W_v L$$

We use the data given below for calculation:

$$T_{gi} = 300°C, \quad T_t = 280°C, \quad T_{g0} = 105°C, \quad L = 2260 \text{ kJ/kg}$$

Values calculated in Example 4.4 are as follows:

$W_t = 0.277$ kg/s
$W_v = 0.170$ kg/s
$W_{vl} = 0.119$ kg/s
$W_f = 0.566$ kg/s
$A = (0.17 \times 1.89 + 0.119 \times 1.13) \times 105 + 0.277 \times 0.269 \times 280 - 0.567 \times 1.46 \times 20 + 0.17 \times 2260 = 436.4$ kg/s

$$W_g = \frac{A}{C_g(T_{gi} - T_{g0})} = \frac{436.4}{1.13(300 - 105)} = 1.98 \text{ kg}/s$$

$$W'_g = W_f + W_g - W_t = 0.566 + 1.98 - 0.277 = 2.27 \text{ kg}/s$$

To find oil consumption, we use Eq. (A.7) of Appendix:

$$W_{oil} = \frac{1}{K - P} \times \frac{W_g}{W'_g} \left[ \frac{C_g T_{gi}}{C_g T_{g0} + VL'_{fr} LHV_{vl} \eta} - 1 \right]$$

$W_g = 1.99$ kg/s
$W_v = 0.17$ kg/s
$W_{vl} = 0.119$ kg/s
$W'_g = 2.27\ \mathrm{kg}/s$

Excess air coefficient, $\alpha = 1 + 0.2 = 1.2$.

Fraction of generated volatile, $\mathrm{VL}'_{fr} = \dfrac{W_{vl}}{W'_g} = \dfrac{0.119}{2.28} = 0.052$

The parameters $K$ and $P$ are calculated as:

$$K = \frac{[((A/F)\alpha)C_{air}T_0 + \mathrm{LHV}\eta + C_{oil}T_0]}{W'_g C_g T_{g0} + W_{vl}\mathrm{LHV}_{vl}\eta}$$

$$= \frac{(14.6 \times 1.2 \times 1.006 \times 20 + 45500 \times 0.95 + 1.7 \times 20)}{(2.27 \times 1.13 \times 105 + 0.119 \times 1286 \times 0.95)} = 105.1$$

$$P = \frac{[((A/F)\alpha) + 1]}{W'_g} = \frac{(14.6 \times 1.2 + 1)}{2.27} = 8.16$$

$$W_{oil} = \frac{1}{(104.9 - 8.13)}\left(\frac{1.98}{2.27}\right)\left[\frac{1.13 \times 400}{1.13 \times 105 + 0.079 \times 1286 \times 0.95} - 1\right] = 0.0077\ \mathrm{kg}/s$$

Equation (A.4) gives the fraction of flue gas recirculated back into the torrefier:

$$X = \frac{(W_g - W_{oil}.(((A/F)\alpha) + 1)}{W'_g} = \frac{[1.99 - 0.0077(14.6 \times 1.2 + 1)]}{2.27} = 0.81$$

Percentage of gas recirculation = 81%

## SYMBOLS AND NOMENCLATURE

**ASH** ash fraction in raw or as-received biomass (–)
$A_r$ reactor cross-sectional area ($m^2$)
$Bi$ Biot number ($hV/\lambda S$)
$C_{pd}$ specific heat of dried biomass (kJ/kg C)
$C_{pw}$ specific heat of wet or as-received biomass (kJ/kg C)
$C_{pg}$ mean specific heat of flue gas (kJ/kg C)
**EV** energy density on volume basis (kJ/$m^3$)
**EY** energy yield (–)
**HHV** higher heating value on mass basis (kJ/kg)
$h$ biomass particle to heating medium (gas) heat transfer coefficient (kJ/kg C)
$h_u$ heat utilization efficiency (–)
$K$ kinetic rate of torrefaction ($s^{-1}$)
$L$ latent heat of vaporization of water (kJ/kg)
**LHV** lower heating value on mass basis (kJ/kg)
**MY** mass yield (–)
$M$ moisture fraction in raw or as-received biomass (–)

$M_f$ mass of feedstock (kg)
$M_w$ mass of water in feedstock (kg)
$M_{ash}$ mass of ash in feedstock (kg)
$M_g$ mass of diluted hot gas per unit mass of oil burnt (kg)
$M_{oil}$ mass flow rate of oil burnt (kg/s)
$Py$ Pyrolysis number (–)
$Q$ energy content (kJ)
$Q_d$ theoretical heat load of dryer (kW)
$Q'_d$ actual heat load of dryer (kW)
$Q_t$ theoretical heat load of torrefier (kW)
$Q'_t$ actual heat load of torrefier (kW)
$Q_{tl}$ heat loss from torrefier (kW)
$r_p$ radius of biomass particle (m)
$S$ external surface area of a biomass particle ($m^2$)
$T_{gi}$ temperature of gas at inlet of torrefier plant (°C)
$T_0$ ambient temperature of gas, (°C)
$T_{g0}$ temperature of gas at exit of torrefier plant (°C)
$T_t$ torrefaction temperature (°C)
$U_g$ space velocity of gas (m/s)
$U_s$ space velocity of solids (m/s)
$V$ volume of biomass particle ($m^3$)
$W_d$ dry biomass feed rate (kg/s)
$W_f$ feed rate of wet or as-received biomass (kg/s)
$W_g$ flow rate of diluted hot gas entering the torrefier (kg/s)
$W_t$ production rate of torrefied biomass (kg/s)
$X_d$ fractional heat loss from the drier (–)

## Subscripts

**ar** as-received basis
**bulk** bulk
**cooler** cooling section of torrefaction plant
**d** dryer
**daf** dry ash free basis
**db** dry basis
**feed** feedstock or raw/wet/as-received biomass
**pd** predrying section
**pdh** postdrying heater
**t** torrefied product
**tor** torrefier
**product** value in product

## Greek symbol

$\rho$ density ($kg/m^3$)
$\lambda$ thermal conductivity of biomass (kJ/C m)
$\eta$ combustion efficiency of burner (–)

# APPENDIX
# MASS AND ENERGY BALANCE OF TORREFIER

## Assumptions

- Torrefaction is mildly exothermic.
- Torrefaction occurs at constant atmospheric pressure.
- Negligible moisture in air.
- A/F = Air−fuel ratio for stoichiometric combustion = 14.6 for diesel oil.
- $T_{g0}$ is chosen higher than 100°C to avoid condensation in drier (assume 105°C).
- $T_{gi}$ is chosen less than 300°C to avoid combustion of torrefied wood (assume 300°C).
- Heating values of volatile is estimated to be 1.286 MJ/kg on dry basis.

## MASS BALANCE

### Torrefier

From Figure 4A.1:

$$W_t + W'_g = W_f + W_g$$

On rearranging, one gets:

$$W'_g = W_f + W_g - W_t \tag{A.1}$$

$$W'_g = W_v + W_{vl} + W_g$$

where $W_{vl}$ is the volatile in fluid product leaving torrefier.

$$\text{Moisture in fuel} \quad W_v = MW_f \tag{A.2}$$

where $\alpha$ is the excess air coefficient used in oil burner.

Substituting $W'_g$ from Eq. (A.1), we get:

$$W_f + W_g - W_t = W_v + W_{vl} + W_g$$

Substituting $W_v$ from Eq. (A.2):

$$W_{vl} = W_f - W_t - M\, W_f \tag{A.3}$$

### Oil Burner

From Figure 4A.2:

$$W_g = W_{air} + W_{oil} + X\, W'_g$$

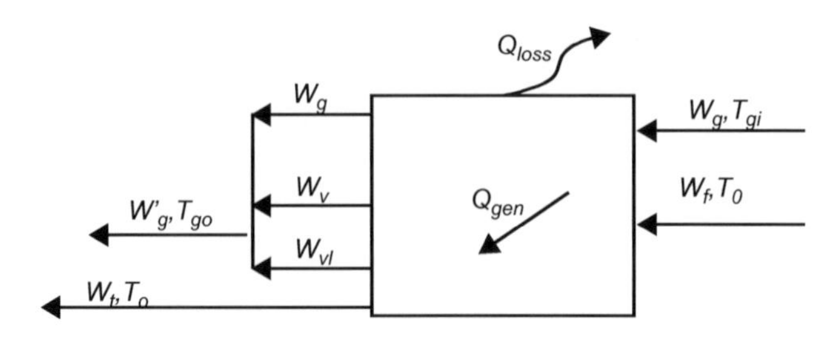


**FIGURE 4A.1** Control volume of torrefier.

Air required for burning oil with an excess air coefficient, $\alpha$ is

$$W_{\mathrm{air}} = \alpha \frac{\mathrm{A}}{\mathrm{F}} W_{\mathrm{oil}}$$

From here, the fraction of gas recirculation, $X$, is calculated as:

$$X = \frac{[W_{\mathrm{g}} - W_{\mathrm{oil}}.(\alpha(A/F) + 1)]}{W'_{\mathrm{g}}} \tag{A.4}$$

## ENERGY BALANCE

### Torrefier: Control Volume of Torrefaction Zone (Figure 4.11)

From the mass balance around the torrefier (Figure 4A.1), we get:

Enthalpy in + energy generated by mild exothermic process = enthalpy out + energy loss through reactor + latent heat carried away by the moisture content

$$W_{\mathrm{g}} C_{\mathrm{g}} T_{\mathrm{gi}} + W_{\mathrm{f}} C_{\mathrm{b}} T_{\mathrm{o}} + Q_{\mathrm{gen}} = (W_{\mathrm{g}} C_{\mathrm{g}} + W_{\mathrm{v}} C_{\mathrm{v}} + W_{\mathrm{vl}} C_{\mathrm{g}}) T_{\mathrm{g0}} + W_{\mathrm{t}} C_{\mathrm{d}} T_{\mathrm{t}} + Q_{\mathrm{loss}} + W_{\mathrm{v}} L$$

where $L$ is latent heat of vaporization of moisture.

Assuming that heat losses and heat generated by mild exothermic reaction are negligible, the above can be simplified as:

$$W_{\mathrm{g}} C_{\mathrm{g}} (T_{\mathrm{gi}} - T_{\mathrm{g0}}) + W_{\mathrm{f}} C_{\mathrm{b}} T_0 - (W_{\mathrm{v}} C_{\mathrm{v}} + W_{\mathrm{vl}} C_{\mathrm{g}}) T_{\mathrm{g0}} - W_{\mathrm{t}} C_{\mathrm{d}} T_{\mathrm{t}} - W_{\mathrm{v}} L = 0$$

So, the mass flow of heating medium, $W_{\mathrm{g}}$, is

$$W_{\mathrm{g}} = \frac{(W_{\mathrm{v}} C_{\mathrm{v}} + W_{\mathrm{vl}} C_{\mathrm{g}}) T_{\mathrm{g0}} + W_{\mathrm{t}} C_{\mathrm{d}} T_{\mathrm{t}} - W_{\mathrm{f}} C_{\mathrm{b}} T_0 + W_{\mathrm{v}} L}{C_{\mathrm{g}} (T_{\mathrm{gi}} - T_{\mathrm{g0}})}$$

or

$$W_{\mathrm{g}} = \frac{A}{C_{\mathrm{g}} (T_{\mathrm{gi}} - T_{\mathrm{g0}})} \tag{A.5}$$

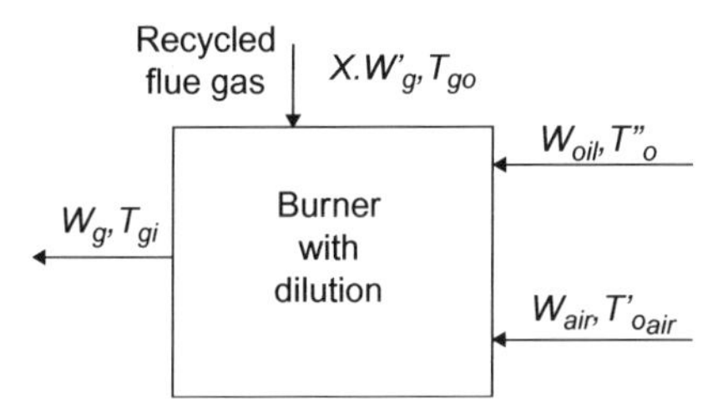


**FIGURE 4A.2** Control volume of burner with gas mixing chamber.

where,

$$A = (W_v C_v + W_{vl} C_g) T_{g0} + W_t C_d T_t - W_f C_b T_0 + W_v L$$

## Oil Burner: Control Volume of Oil Burner

We assume that $X$ fraction of the flue gas leaving the torrefier is fed into the burner along with fresh air $W_{air}$. Since the recycled gas ($X\,W'_g$) contains some unburnt volatiles ($XW_{vl}$), we assume this to supplement oil in the burner. We assume fuel oil is preheated to temperature $T''_0$.

$$\begin{aligned} Q_{vl} &= \text{heat energy released from the combustion of volatile gases} \\ &= XW_{vl}\text{LHV}_{vl}\eta \end{aligned}$$

where $\eta$ is combustion efficiency.

From Figure 4A.2, we write the energy balance as:

$$W_g C_g T_{gi} = XW'_g C_g T_{g0} + XW_{vl}\text{LHV}_{vl}\eta + W_{air} C_a T'_0 + W_{oil}\text{LHV}\eta + W_{oil} C_{oil} T''_0$$

So, the mass fraction, $X$, of torrefier product gas recycled is

$$X = \frac{W_g C_g T_{gi} \alpha (A/F) C_a T'_0 - W_{oil}\text{LHV}\eta - W_{oil} C_{oil} T''_0}{W'_g C_g T_{g0} + W_{vl}\text{LHV}_{vl}\eta} \tag{A.6}$$

Substituting values from Eqs. (A.4) and (A.6), we have

$$\frac{[W_g - W_{oil}(\alpha(A/F)+1)]}{W'_g} = \frac{W_g C_g T_{gi} - W_{oil}\alpha(A/F) C_a . T'_0 - W_{oil}\text{LHV}\eta - W_{oil} C_{oil} T''_0}{W'_g C_g T_{g0} + W_{vl}\text{LHV}_{vl}\eta}$$

Amount of oil consumed in the burner is calculated from here as:

$$W_{oil} = \frac{1}{K-P} \times \frac{W_g}{W'_g}\left[\frac{C_g T_{gi}}{C_g T_{g0} + \text{VL}'_{fr}\text{LHV}_{vl}\eta} - 1\right] \tag{A.7}$$

where,

$$K = \frac{(\alpha(A/F) C_a . T'_0 + \text{LHV}\eta + C_{oil} T''_0)}{W'_g C_g T_{g0} + W_{vl}\text{LHV}_{vl}\eta}; \quad P = \frac{(\alpha A/F + 1)}{W'_g} \quad \text{and} \quad \text{VL}'_{fr} = \frac{W_{vl}}{W'_g}$$

Chapter 5

# Pyrolysis

## 5.1 INTRODUCTION

Pyrolysis is a thermochemical decomposition of biomass into a range of useful products, either in the total absence of oxidizing agents or with a limited supply that does not permit gasification to an appreciable extent. It also forms several initial reaction steps of gasification. During pyrolysis, large complex hydrocarbon molecules of biomass break down into relatively smaller and simpler molecules of gas, liquid, and char (Figure 5.1).

Pyrolysis has similarity to or overlaps with processes like cracking, devolatilization, carbonization, torrefaction, dry distillation, destructive distillation, and thermolysis, but it has no similarity with the gasification process, which involves chemical reactions with an external agent known as *gasification medium*. Pyrolysis of biomass is typically carried out in a temperature range of 300–650°C compared to 800–1000°C for gasification and 200–300°C for torrefaction. A discussion on the difference between these processes is given in Section 4.2.1 and Table 5.2.

Biochar is the solid product of biomass pyrolysis. This provides an alternative to complete burning of agricultural product or forest residues releasing the carbon to the atmosphere. Biochar could retain a part of that carbon in stable solid form in soil for hundreds of years. Owing to its growing importance for carbon sequestration, many institutions are taking a closer look at this. A discussion on this is presented in Section 5.8.

This chapter explains the basics of pyrolysis. A brief discussion of the design implications of the two is also presented.

### 5.1.1 Historical Background

Charcoal from wood via pyrolysis was essential for extraction of iron from iron ore in the preindustrial era. Figure 5.2 shows a typical beehive oven used in early times to produce charcoal from biomass using a slow pyrolysis process. This practice continued until wood supplies nearly ran out and coal, produced inexpensively from underground mines, replaced charcoal for iron production.

The modern petrochemical industry owes a great deal to the invention of a process for production of kerosene using pyrolysis. In the mid-1840s,

**Biomass Gasification, Pyrolysis and Torrefaction.**

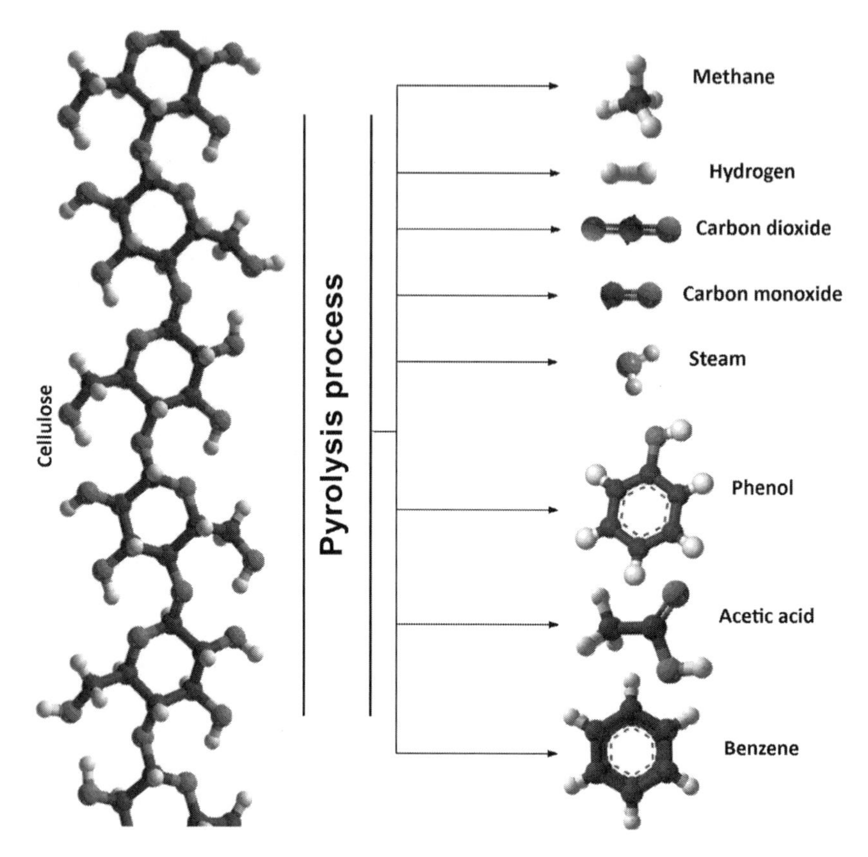


**FIGURE 5.1** Decomposition of large hydrocarbon molecules into smaller ones during pyrolysis.

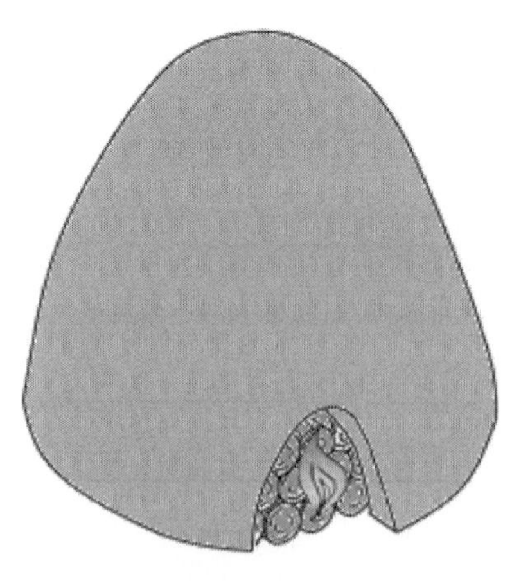

**FIGURE 5.2** Beehive oven for charcoal production through slow pyrolysis of wood.

**FIGURE 5.3** Abraham Gesner, inventor of kerosene and his kerosene lamp.

Abraham Gesner, a physician practicing in Halifax, Canada (Figure 5.3), began searching for a cleaner-burning mineral oil to replace the sooty oil from whales, the primary fuel used during those times on the eastern seaboard of the United States and in Atlantic Canada. By carefully distilling a few lumps of coal at 427°C, purifying the product through treatment with sulfuric acid and lime, and then redistilling it, Gesner obtained several ounces of a clear liquid (Gesner, 1861). When this liquid was burned in an oil lamp similar to the one shown in Figure 5.3, it produced a clear bright light that was much superior to the smoky light produced by the burning of whale oil. Dr. Gesner called his fuel *kerosene*—from the Greek words for wax and oil. Later, in the 1850s, when crude oil began to flow in Pennsylvania and Ontario, Gesner extracted petro-based kerosene from that.

The invention of kerosene, the first transportable liquid fuel, brought about a revolution in lighting that touched and is still the case in even the remotest parts of the world. It also had a major positive impact on the ecology. For example, in 1846, more than 730 ships hunted whales to meet the huge demand for whale oil. In just a few years after the invention of kerosene, the hunt was reduced to only a few ships, saving whales from possible extinction.

## 5.2 PYROLYSIS

Pyrolysis involves rapid heating of biomass or other feed in the absence of air or oxygen at a maximum temperature, known as the *pyrolysis temperature*, and holding it there for a specified time to produce noncondensable gases,

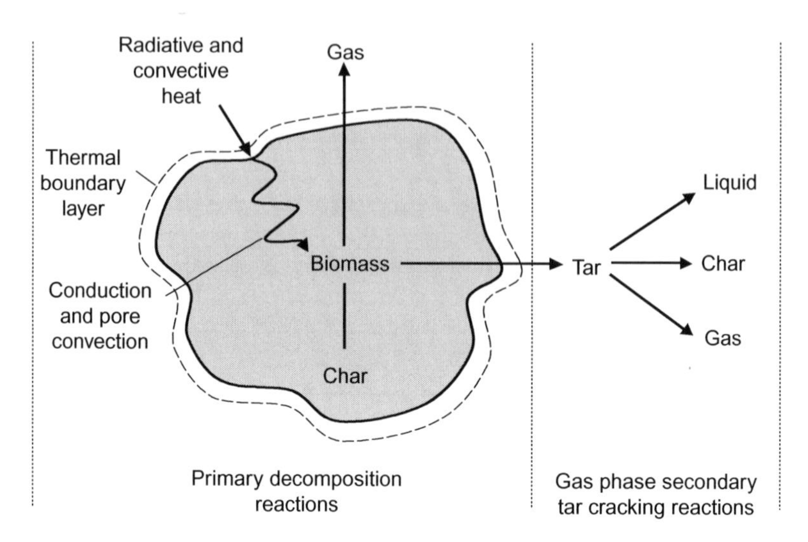


**FIGURE 5.4** Pyrolysis process in a biomass particle.

solid char, and liquid product. The liquid product is of primary interest in pyrolysis. The nature of its product depends on several factors, including pyrolysis temperature and heating rate.

The initial product of pyrolysis is made of condensable gases and solid char. The condensable gas may break down further into noncondensable gases ($CO$, $CO_2$, $H_2$, and $CH_4$), liquid, and char (Figure 5.4). This decomposition occurs partly through gas-phase homogeneous reactions and partly through gas–solid-phase heterogeneous thermal reactions. In gas-phase reactions, the condensable vapor is cracked into smaller molecules of noncondensable permanent gases such as $CO$ and $CO_2$.

The pyrolysis process may be represented by a generic reaction such as:

$$C_nH_mO_p(\text{biomass}) \xrightarrow{\text{heat}} \sum_{\text{liquid}} C_xH_yO_z + \sum_{\text{gas}} C_aH_bO_c + H_2O + C\ (\text{char}) \tag{5.1}$$

Pyrolysis is an essential prestep in a gasifier. This step is relatively fast, especially in reactors with rapid mixing.

Figure 5.5 shows the process by means of a schematic of a fluidized bed pyrolysis plant. Biomass is fed into a pyrolysis chamber containing hot solids (fluidized bed) that heat the biomass to the pyrolysis temperature, at which its decomposition starts. The condensable and noncondensable vapors released from the biomass leave the chamber, while the solid char produced remains partly in the chamber and partly in the gas. The gas is separated from the char and cooled downstream of the reactor. The condensable vapor condenses as bio-oil or pyrolysis oil. The noncondensable gases leave the

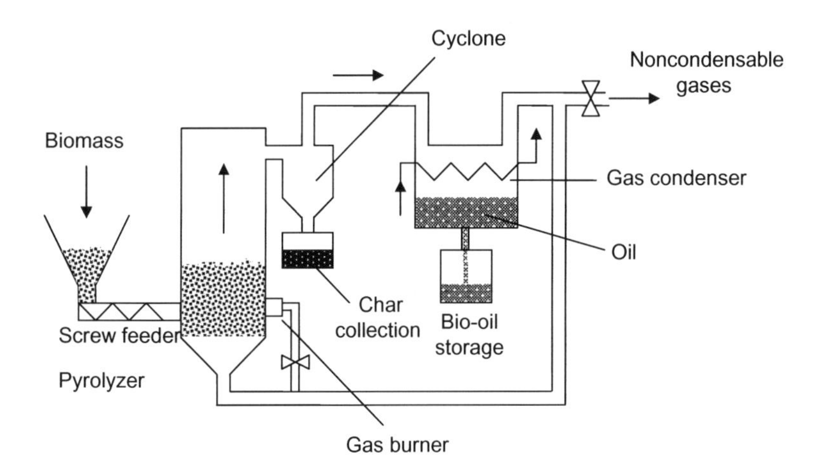


**FIGURE 5.5** Simplified layout of a pyrolysis plant.

chamber as product gas, which is the product of interest. Similarly, the solid char may be collected as a commercial product or burned in a separate chamber to produce heat that is necessary for pyrolysis. As this gas is free from oxygen, part of it may be recycled into the pyrolysis chamber as a heat carrier or fluidizing medium. There are, of course, variations of the process, which will be discussed later.

## 5.2.1 Pyrolysis Products

As mentioned earlier, pyrolysis involves a breakdown of large complex molecules into several smaller molecules. Its product is classified into three principal types:

1. Liquid (tars, heavier hydrocarbons, and water)
2. Solid (mostly char or carbon)
3. Gas (e.g., $CO_2$, $H_2O$, $CO$, $C_2H_2$, $C_2H_4$, $C_2H_6$, $C_6H_6$).

The relative amounts of these products depend on several factors including the heating rate and the final temperature reached by the biomass.

The pyrolysis product should not be confused with the "volatile matter" of a fuel as determined by its *proximate analysis*. In proximate analysis, the liquid and gas yields are often lumped together as "volatile matter," and the char yield as "fixed carbon." Since the relative fraction of the pyrolysis yields depends on many operating factors, determination of the volatile matter of a fuel requires the use of standard conditions as specified in test codes such as ASTM D-3172 and D-3175. The procedure laid out in D-3175, for example, involves heating a specified sample of the fuel in a furnace at 950°C for 7 min to measure its volatile matter.

#### 5.2.1.1 Liquid

The liquid yield, known as tar, bio-oil, or biocrude, is a black tarry fluid containing up to 20% water. It consists mainly of homologous phenolic compounds. Bio-oil is a mixture of complex hydrocarbons with large amounts of oxygen and water. While the parent biomass has a lower heating value (LHV) in the range of 19.5–21 MJ/kg dry basis, its liquid yield has a lower LHV in the range of 13–18 MJ/kg wet basis (Diebold et al., 1997).

Rapid and simultaneous depolymerization and fragmentation of the cellulose, hemicellulose, and lignin components of biomass produce bio-oil. In a typical operation, the biomass is subjected to a rapid increase in temperature followed by an immediate quenching to "freeze" the intermediate pyrolysis products. Rapid quenching is important, as it prevents further degradation, cleavage, or reaction with other molecules (see Section 5.4.2 for more details).

Bio-oil is a microemulsion, in which the continuous phase is an aqueous solution of the products of cellulose and hemicellulose decomposition, and small molecules from lignin decomposition. The discontinuous phase is largely composed of pyrolytic lignin macromolecules (Piskorz et al., 1988). Bio-oil typically contains molecular fragments of cellulose, hemicellulose, and lignin polymers that escaped the pyrolysis environment (Diebold and Bridgwater, 1997). The molecular weight of the condensed bio-oil may exceed 500 Daltons (Diebold and Bridgwater, 1997, p. 10). Compounds found in bio-oil fall into the following five broad categories (Piskorz et al., 1988):

- Hydroxyaldehydes
- Hydroxyketones
- Sugars and dehydrosugars
- Carboxylic acids
- Phenolic compounds.

#### 5.2.1.2 Solid

Biochar is the solid yield of pyrolysis. It is primarily carbon (~85%), but it can also contain some oxygen and hydrogen. Unlike fossil fuels, biomass contains very little inorganic ash. The LHV of biomass char is about 32 MJ/kg (Diebold and Bridgwater, 1997), which is substantially higher than that of the parent biomass or its liquid product. It is characterized by large pore surface area.

#### 5.2.1.3 Gas

Primary decomposition of biomass produces both condensable gases (vapor) and noncondensable gases (*primary gas*). The vapors, which are made of heavier molecules, condense upon cooling, adding to the liquid yield of pyrolysis. The noncondensable gas mixture contains lower-molecular-weight gases like carbon dioxide, carbon monoxide, methane, ethane, and ethylene.

**TABLE 5.1** Comparison of Heating Values of Some Fuels

| Fuel | Petcoke | Bituminous Coal | Sawdust | Bio-oil | Pyrolysis Gas |
|---|---|---|---|---|---|
| Units | MJ/kg | MJ/kg | MJ/kg dry | MJ/kg | $MJ/Nm^3$ |
| Heating value | ~29.8 | ~26.4 | ~20.5 | 13–18 | 11–20 |

These do not condense on cooling. Additional noncondensable gases are produced through secondary cracking of the vapor (see Section 5.4.2) at higher temperature (Figure 5.6) these are called *secondary gases*. The final noncondensable gas product is thus a mixture of both primary and secondary gases. The LHV of primary gases is typically 11 $MJ/Nm^3$, but that of pyrolysis gases formed after severe secondary cracking of the vapor is much higher: 20 $MJ/Nm^3$ (Diebold and Bridgwater, 1997). Table 5.1 compares the heating values of pyrolysis gas with those of bio-oil, raw biomass, and two fossil fuels.

## 5.2.2 Types of Pyrolysis

Based on the heating rate, pyrolysis may be broadly classified as slow and fast. It is considered slow if the time, $t_{heating}$, required to heat the fuel to the pyrolysis temperature is much longer than the characteristic pyrolysis reaction time, $t_r$, and vice versa. That is:

- Slow pyrolysis: $t_{heating} \gg t_r$
- Fast pyrolysis: $t_{heating} \ll t_r$.

By assuming a simple linear heating rate ($T_{pyr}/t_{heating}$, K/s), these criteria may be expressed in terms of heating rate as well. Here, $T_{pyr}$ is the pyrolysis temperature.

There are a few other variants depending on the medium and pressure at which the pyrolysis is carried out. Given specific operating conditions, each process has its characteristic products and applications. Slow and fast pyrolysis are based on the heating rate while hydropyrolysis is based on the environment or medium in which the pyrolysis is carried out.

Slow and fast pyrolysis are carried out generally in the absence of a medium. Two other types are conducted in a specific nonoxidizing medium: hydrous pyrolysis (in $H_2O$) and hydropyrolysis (in $H_2$). These types are used mainly for the production of chemicals.

In slow pyrolysis, the residence time of vapor in the pyrolysis zone (vapor residence time) is on the order of minutes or longer. This process is not used for traditional pyrolysis, where production of liquid is the main goal. Slow pyrolysis is used primarily for char production and is broken

**TABLE 5.2 Characteristics of Some Thermal Decomposition Processes**

| Pyrolysis Process | Residence Time | Heating Rate | Final Temperature (°C) | Products |
|---|---|---|---|---|
| Torrefaction | 10–60 min | Very small | 280 | Torrefied biomass |
| Carbonization | Days | Very low | >400 | Charcoal |
| Fast | <2 s | Very high | ~500 | Bio-oil |
| Flash | <1 s | High | <650 | Bio-oil, chemicals, gas |
| Ultrarapid | <0.5 s | Very high | ~1000 | Chemicals, gas |
| Vacuum | 2–30 s | Medium | 400 | Bio-oil |
| Hydropyrolysis | <10 s | High | <500 | Bio-oil |
| Methanopyrolysis | <10 s | High | >700 | Chemicals |

down into two types: carbonization and torrefaction. Torrefaction takes place in a very low and narrow temperature (200–300°C), while carbonization takes place at much higher and broad temperature.

In fast pyrolysis, the vapor residence time is on the order of seconds or milliseconds. This type of pyrolysis, used primarily for the production of bio-oil and gas, is of two main types: flash and ultrarapid.

Table 5.2 compares the characteristics of different thermal decomposition processes and shows carbonization as the slowest and ultrarapid as the fastest. Carbonization produces mainly charcoal; fast pyrolysis processes target production of liquid or gas.

### *5.2.2.1 Slow Pyrolysis*

Carbonization is a slow pyrolysis process, in which the production of charcoal or char is the primary goal. It is the oldest form of pyrolysis, which is in use for thousands of years. The biomass is heated slowly in the absence of oxygen to a relatively low temperature (~400°C) over an extended period of time, which in ancient times ran for several days to maximize the char formation. Figure 5.2 is a sketch of a typical beehive oven in which large logs were stacked and covered by a clay wall. It allows a certain amount of oxygen for partial combustion of wood. A small fire at the bottom provided the required heat for carbonization. The fire essentially stayed in the well-insulated closed chamber. Carbonization allows adequate time for the condensable vapor to be converted into char and noncondensable gases.

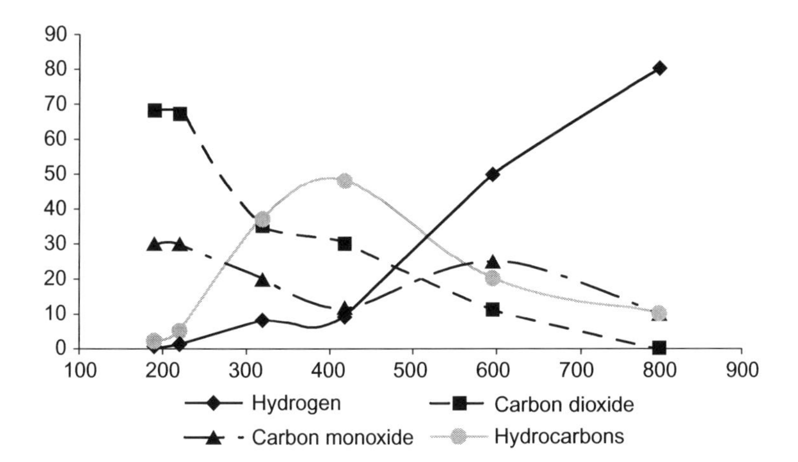


**FIGURE 5.6** Release of gases during dry distillation of wood. Source: *Drawn based on the data of Nikitin (1966).*

Conventional pyrolysis involves all three types of pyrolysis product (gas, liquid, and char). As such, it heats the biomass at a moderate rate to a moderate temperature (~600°C). The product residence time is on the order of minutes.

### *5.2.2.2 Fast Pyrolysis*

The primary goal of fast pyrolysis is to maximize the production of liquid or bio-oil. The biomass is heated so rapidly that it reaches the peak (pyrolysis) temperature before it decomposes. The heating rate can be as high as 1000–10,000°C/s, but the peak temperature should be below 650°C if bio-oil is the product of interest. However, the peak temperature can be up to 1000°C if the production of gas is of primary interest. Fluidized beds similar to the one shown in Figures 5.5 and 5.7A,B may be used for fast pyrolysis.

Four important features of the fast pyrolysis process that help increase the liquid yield are (i) very high heating rate, (ii) reaction temperature within the range of 425–600°C, (iii) short residence time (<3 s) of vapor in the reactor, and (iv) rapid quenching of the product gas.

### *5.2.2.3 Flash Pyrolysis*

In flash pyrolysis, biomass is heated rapidly in the absence of oxygen to a relatively modest temperature range of 450–600°C. The product, containing condensable and noncondensable gas, leaves the pyrolyzer within a short residence time of 30–1500 ms (Bridgwater, 1999). Upon cooling, the condensable vapor is then condensed into a liquid fuel known as "bio-oil." Such an operation increases the liquid yield while reducing the char production. A typical yield of bio-oil in flash pyrolysis is 70–75% of the total pyrolysis product.

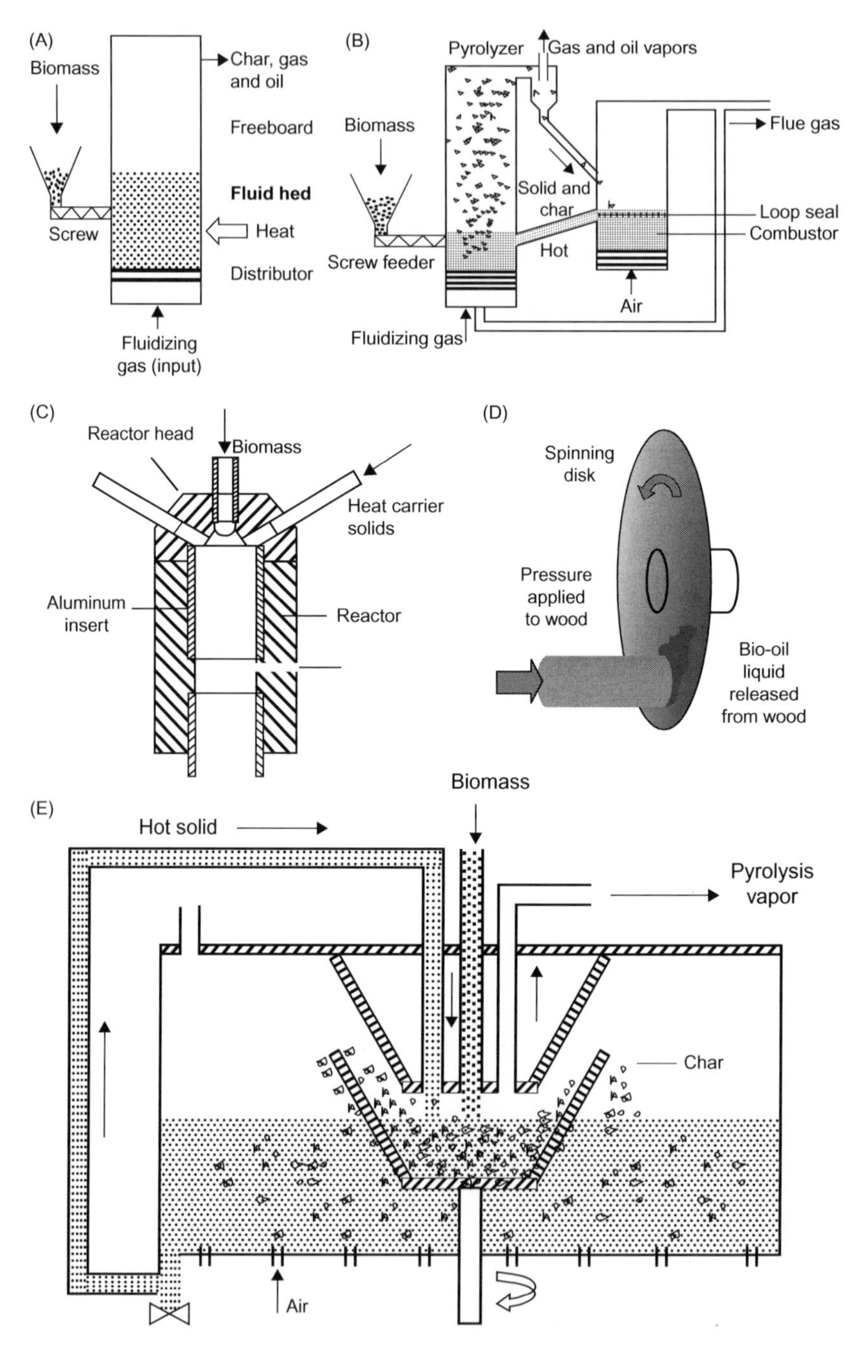


**FIGURE 5.7** (A) Bubbling fluidized bed pyrolyzer; (B) Circulating fluidized bed pyrolyzer; (C) Ultrarapid pyrolysis; (D) Ablative pyrolysis; (E) Rotating cone pyrolyzer; and (F) Vacuum pyrolysis of biomass.

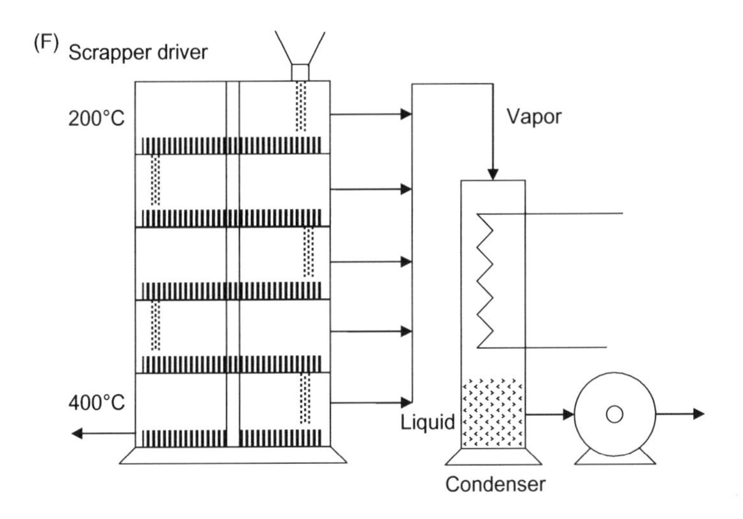


**FIGURE 5.7** (Continued)

### 5.2.2.4 Ultrarapid Pyrolysis

Ultrarapid pyrolysis involves extremely fast heating of biomass. In one method, heat-carrier solids impact on biomass steam, resulting in a very high (Figure 5.7C) heating rate. A rapid quenching of the primary product follows the pyrolysis, occurring in its reactor. A gas–solid separator separates the hot heat-carrier solids from the noncondensable gases and primary product vapors and returns them to the mixer. They are then heated in a separate combustor. Then a nonoxidizing gas transports the hot solids to the mixer as shown in Figure 5.7C. A precisely controlled short uniform residence time is an important feature of ultrarapid pyrolysis. To maximize the product yield of gas, the pyrolysis temperature is around 1000°C for gas and around 650°C for liquid.

### 5.2.2.5 Pyrolysis in the Presence of a Medium

Normal pyrolysis is carried out in the absence of a medium such as air, but a special type is conducted in a medium such as water or hydrogen.

Hydropyrolysis is one such type where this thermal decomposition of biomass takes place in an atmosphere of high-pressure hydrogen. Hydropyrolysis can increase the volatile yield and the proportion of lower-molar-mass hydrocarbons (Rocha et al., 1997). This process is different from the hydrogasification of char. Its higher volatile yield is attributed to hydrogenation of free-radical fragments sufficient to stabilize them before they repolymerize and form char (Probstein and Hicks, 2006, p. 99).

Hydrous pyrolysis is the thermal cracking of the biomass in high-temperature water. It could convert, for example, turkey offal into light hydrocarbon for production of fuel, fertilizer, or chemicals. In a two-stage process, the first stage takes place in water at 200–300°C under pressure; in the second stage, the produced hydrocarbon is cracked into lighter hydrocarbon at a temperature of around 500°C (Appel et al., 2004). High oxygen content is an important shortcoming of bio-oil. Hydropyrolysis can produce bio-oil with reduced oxygen.

## 5.3 PYROLYSIS PRODUCT YIELD

The product of pyrolysis depends on the design of the pyrolyzer, the physical and chemical characteristics of the biomass, and important operating parameters such as:

- Heating rate
- Final temperature (pyrolysis temperature)
- Residence time in the reaction zone.

Besides these, the tar and the yields of other products depend on (i) pressure, (ii) ambient gas composition, and (iii) presence of mineral catalysts (Shafizadeh, 1984).

By changing the final temperature and the heating rate, it is possible to change the relative yields of the solid, liquid, and gaseous products of pyrolysis. Rapid heating yields higher volatiles and more reactive char than those produced by a slower heating process; slower heating rate and longer residence time result in secondary char produced from a reaction between the primary char and the volatiles.

### 5.3.1 Effect of Biomass Composition

The composition of the biomass, especially its hydrogen-to-carbon (H/C) ratio, has an important bearing on the pyrolysis yield. Each of the three major constituents of a lignocellulosic biomass has its preferred temperature range of decomposition. Analysis of data from thermogravimetric apparatus differential thermogravimetry on some selected biomass suggests the following temperature ranges for initiation of pyrolysis (Kumar and Pratt, 1996):

- Hemicellulose: 150–350°C
- Cellulose: 275–350°C
- Lignin: 250–500°C.

The individual constituents undergo pyrolysis differently, making varying contributions to yields. For example, cellulose and hemicellulose are the main sources of volatiles in lignocellulose biomass. Of these, cellulose is a primary source of condensable vapor. Hemicellulose, on the other hand, yields more noncondensable gases and less tar than released by cellulose (Reed, 2002, p. II-109). Owing to its aromatic content, lignin degrades slowly, making a major contribution to the char yield.

Cellulose decomposes over a narrow temperature range of 300–400°C (Figure 4.5A). In the absence of any catalyst, pure cellulose pyrolyzes predominantly to a monomer, levoglucosan (Diebold and Bridgwater, 1997). Above 500°C, the levoglucosan vaporizes, with negligible char formation, thus contributing mainly to gas and oil yields. Hemicelluloses are the least-stable components of wood, perhaps because of their lack of crystallinity (Reed, 2002, p. II-102). It decomposes within 200–300°C (Figure 4.5A).

Unlike cellulose, lignin decomposes over a broader temperature range of 280–500°C, with the maximum release rate occurring at 350–450°C (Kudo and Yoshida, 1957). Lignin pyrolysis produces more aromatics and char than produced by cellulose (Soltes and Elder, 1981). It yields about 40% of its weight as char under a slow heating rate at 400°C (Klass, 1998). Lignin makes some contribution to the liquid yield (~35%), which contains aqueous components and tar. It yields phenols via cleavage of ether and carbon–carbon linkages (Mohan et al., 2006). The gaseous product of lignin pyrolysis is only about 10% of its original weight.

## 5.3.2 Effect of Pyrolysis Temperature

During pyrolysis, a fuel particle is heated at a defined rate from the ambient to a maximum temperature, known as the *pyrolysis temperature*. The fuel is held there until completion of the process. The pyrolysis temperature affects both composition and yield of the product. Figure 5.6 is an example of how, during the pyrolysis of a biomass, the release of various product gases changes with different temperatures. We can see that the release rates vary widely for different gaseous constituents.

The amount of char produced also depends on the pyrolysis temperature. Low temperatures result in greater amount of char; high temperatures result in less. Figure 5.8 shows how the amount of solid char produced from the pyrolysis of a grape bagasse decreases with increasing temperature, but the heating value of the char increases with temperature. This happens because the fixed carbon, which has a higher heating value, in the char increases while the volatile content of the char decreases. The amount of noncondensable gas ($CO_2$, CO, $H_2$, $CH_4$) increases with temperature.

## 5.3.3 Effect of Heating Rate

The rate of heating of the biomass particles has an important influence on the yield and composition of the product. Rapid heating to a moderate temperature (400–600°C) yields higher condensable volatiles and hence more liquid, while slower heating to that temperature produces more char. For example, Debdoubi et al. (2006) observed that during pyrolysis of Esparto, when the heating rate increased from 50°C/min to 250°C/min, the liquid yield increased from 45% to 68.5% at a pyrolysis temperature of 550°C.

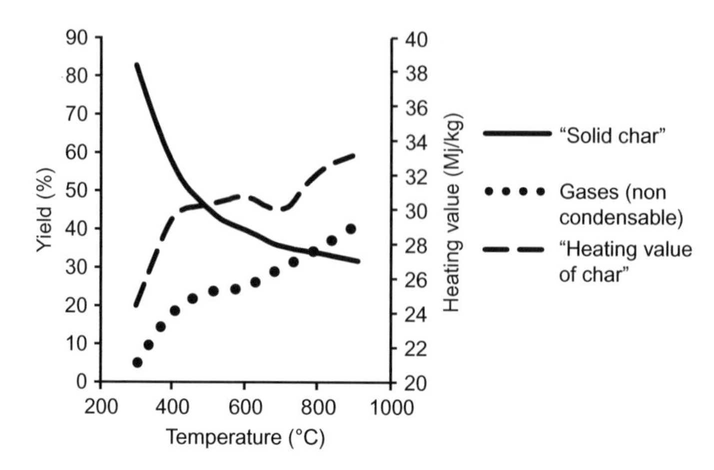


**FIGURE 5.8** Char yield from pyrolysis decreases with temperature while gas yield increases. Heating value of the solid char produced increases with temperature. Source*: Data for grape bagasse (0.63–1.0 mm and HHV–17.2 MJ/kg dry) replotted from Encinar et al. (1996).*

The heating rate alone, however, does not define the product. The residence time of the product in the reactor is also important. During slow heating, a slow or gradual removal of volatiles from the reactor permits a secondary reaction to occur between char particles and volatiles, leading to a secondary char formation.

The operating parameters of a pyrolyzer are adjusted to meet the requirement of the final product of interest. Tentative design norms for heating in a pyrolyzer include the following:

**a.** To maximize char production, use a slow heating rate (<0.01–2.0°C/s), a low final temperature, and a long gas residence time.

**b.** To maximize liquid yield, use a high heating rate, a moderate final temperature (450–600°C), and a short gas residence time.

**c.** To maximize gas production, use a moderate to slow heating rate, a high final temperature (700–900°C), and a long gas residence time.

Production of charcoal through carbonization uses step (a). Fast pyrolysis uses step (b) to maximize liquid yield. Step (c) is used when gas production is to be maximized.

### 5.3.4 Effect of Particle Size

The composition, size, shape, and physical structure of the biomass could exert some influence on the pyrolysis product through their effect on heating rate. Finer biomass particles offer less resistance to the escape of condensable gases, which therefore escape relatively easily to the surroundings before undergoing secondary cracking. This results in a higher liquid yield.

Larger particles, on the other hand, facilitate secondary cracking due to the higher resistance they offer to the escape of the primary pyrolysis product. For this reason, older methods of charcoal production used stacks of large-size wood pieces in a sealed chamber (Figure 5.2).

## 5.4 PYROLYSIS KINETICS

A study of pyrolysis kinetics provides important information for the engineering design of a pyrolyzer or a gasifier. It also helps explain how different processes in a pyrolyzer affect product yields and composition. Three major processes that influence the pyrolysis rate are chemical kinetics, heat transfer, and mass transfer. This section describes the physical and chemical aspects that govern the process.

### 5.4.1 Physical Aspects

From a thermal standpoint, we may divide the pyrolysis process into four stages. Although divided by temperature, the boundaries between them are not sharp; there is always some overlap:

1. *Drying (~100°C).* During the initial phase of biomass heating at low temperature, the free moisture and some loosely bound water is released. The free moisture evaporates, and the heat is then conducted into the biomass interior (Figure 5.4). If the humidity is high, the bound water aids the melting of the lignitic fraction, which solidifies on subsequent cooling. This phenomenon is used in *steam bending* of wood, which is a popular practice for shaping it for furniture (Diebold and Bridgwater, 1997).
2. *Initial stage (100–300°C).* In this stage, exothermic dehydration of the biomass takes place with the release of water and low-molecular-weight gases like CO and $CO_2$. Torrefaction takes place in this stage.
3. *Intermediate stage (>200°C).* This is *primary pyrolysis*, and it takes place in the temperature range of 200–600°C. Most of the vapor or precursor to bio-oil is produced at this stage. Large molecules of biomass particles decompose into char (*primary char*), condensable gases (vapors and precursors of the liquid yield), and noncondensable gases.
4. *Final stage (~300–900°C).* The final stage of pyrolysis involves secondary cracking of volatiles into char and noncondensable gases. If they reside in the biomass long enough, relatively large-molecular-weight condensable gases can crack, yielding additional char (called *secondary char*) and gases. This stage typically occurs above 300°C (Reed, 2002, p. III-6). The condensable gases, if removed quickly from the reaction site, condense outside in the downstream reactor as tar or bio-oil. It is apparent from Figure 5.6 that a higher pyrolysis temperature favors production of hydrogen, which increases quickly above 600°C. An additional contribution of the shift reaction (Eq. (7.16)) further increases the hydrogen yield above 900°C.

Temperature has a major influence on the product of pyrolysis. The carbon dioxide yield is high at lower temperatures and decreases at higher temperatures. The release of hydrocarbon gases peaks at around 450°C and then starts decreasing above 500°C, boosting the generation of hydrogen.

Hot char particles can catalyze the primary cracking of the vapor released within the biomass particle and the secondary cracking occurring outside the particle but inside the reactor. To avoid cracking of condensable gases and thereby increasing the liquid yield, rapid removal of the condensable vapor is very important. The shorter the residence time of the condensable gas in the reactor, the less the secondary cracking and hence the higher the liquid yield.

Some overlap of the stages in the pyrolysis process is natural. For example, owing to its low thermal conductivity (0.1−0.05 W/m K), a large log of wood may be burning outside while the interior may still be in the drying stage, and water may be squeezed out from the ends. During a forest fire, this phenomenon is often observed. The observed intense flame comes primarily from the combustion of the pyrolysis products released from the wood interior rather than from the burning of the exterior surface.

### 5.4.2 Chemical Aspects

As mentioned earlier, a typical biomass has three main polymeric components: (i) cellulose, (ii) hemicellulose, and (iii) lignin. These constituents have different rates of degradation and preferred temperature ranges of decomposition.

#### *5.4.2.1 Cellulose*

Decomposition of cellulose is a complex multistage process. A large number of models have been proposed to explain it. The Broido−Shafizadeh model (Bradbury et al., 1979) is the best known and can be applied, at least qualitatively, to most biomass (Bridgwater et al., 2001).

Figure 5.9 is a schematic of the Broido−Shafizadeh model, according to which the pyrolysis process involves an intermediate prereaction (I) followed by two competing first-order reactions:

- *Reaction II*: dehydration (dominates at low temperature and slow heating rates)
- *Reaction III*: depolymerization (dominates at fast heating rates).

Reaction II involves dehydration, decarboxylation, and carbonization through a sequence of steps to produce char and noncondensable gases like water vapor, carbon dioxide, and carbon monoxide. It is favored at low temperatures, of less than 300°C (Soltes and Elder, 1981, p. 82) and slow heating rates (Reed, 2002, p. II-113).

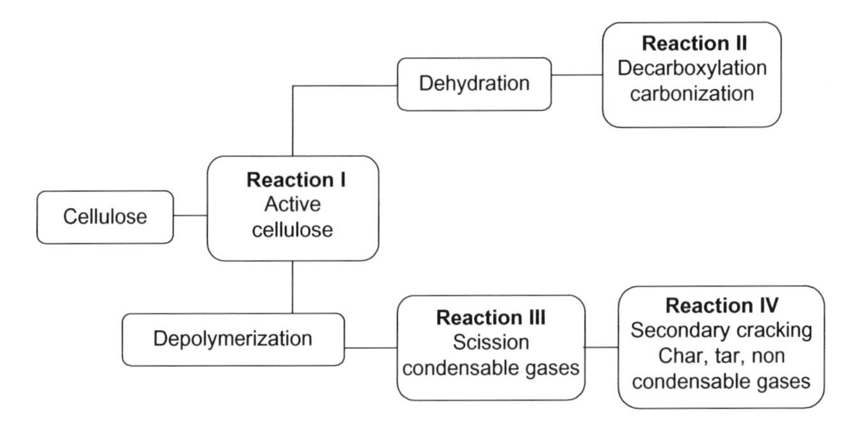


**FIGURE 5.9** Modified "Broido–Shafizadeh" model of cellulose, which can be reasonably applied to the whole biomass.

Reaction III involves depolymerization and scission, forming vapors including tar and condensable gases. Levoglucosan is an important intermediate product in this path (Klass, 1998, p. 228), which is favored under faster heating rates (Reed, 2002, p. II-113) and higher temperatures of over 300°C (Soltes and Elder, 1981, p. 82).

The condensable vapor, if permitted to escape the reactor quickly, can condense as bio-oil or tar. On the other hand, if it is held in contact with biomass within the reactor, it can undergo secondary reactions (reaction IV), cracking the vapor into secondary char, tar, and gases (Figure 5.9). Reactions II and III are preceded by reaction I, which forms a very short-lived intermediate product called *active cellulose* that is liquid at the reaction temperature but solid at room temperature (Boutin and Lédé, 2001; Bradbury et al., 1979; Bridgwater et al., 2001).

There is speculation on the existence of reaction I, as this unstable species is not detected in the final product in most pyrolysis processes. It is, however, apparent in *ablative pyrolysis*, where wood is dragged over a hot metal surface (Figure 5.7D) to produce the feeling of smooth lubrication due to the presence of the intermediate liquid product "active cellulose."

The Broido–Shafizadeh model, though developed for one biomass component (cellulose), can be applied to the pyrolysis of an entire biomass such as wood. Depolymerization (reaction III) (Figure 5.9) has activation energies higher than those of dehydration (reaction II) (Bridgwater et al., 2001). Thus, a lower temperature and a longer residence time favor this reaction, producing primarily char, water, and carbon dioxide. On the other hand, owing to its higher activation energy, reaction III is favored at higher temperatures, fast heating rate, and longer residence times, yielding mainly gas. Moderate temperature and short vapor residence time avoid secondary cracking, producing mainly condensable vapor—the precursor of bio-oil, which is of

**TABLE 5.3 Rate Constants for Pyrolysis of Cellulose According to Broido–Shafizadeh Model**

| Reaction $(dm/dt) = A_i(V_i^3 - V_i)e^{-E_i/RT}$ | $A_i$ ($s^{-1}$) | $E_i$ (kJ/mol) |
|---|---|---|
| I—First degradation (active cellulose), Bradbury et al. (1979) | $2.8 \times 10^{19}$ | 243 |
| II—Dehydration (char + gas), Bradbury et al. (1979) | $1.31 \times 10^{10}$ | 153 |
| III—Depolymerization (tars), Bradbury et al. (1979) | $3.16 \times 10^{14}$ | 198 |
| IV—Secondary cracking (gas, char), Uden et al. (1988) | $4.28 \times 10^{6}$ | 107.5 |

great commercial importance. For cellulose pyrolysis, Table 5.3 gives some suggested reaction rate constants for reactions I, II, III, and IV.

If a log of wood is heated very slowly, it shows glowing ignition, because reaction II predominates under this condition, producing mostly char, which ignites in contact with air without a yellow flame. If the wood is heated faster, it burns with a yellow flame, because at a higher heating rate, reaction III predominates, producing more vapors or tar, both of which burn in air with a bright yellow flame.

#### *5.4.2.2 Hemicellulose*

Hemicellulose produces not only more gas and less tar but also less char in comparison to cellulose. However, it produces the same amount of aqueous product of pyroligneous acid (Soltes and Elder, 1981, p. 84). Hemicellulose undergoes rapid thermal decomposition (Demirbas, 2000), which starts at a temperature lower than that for cellulose or lignin. It contains more combined moisture than lignin, and its softening point is lower as well. The exothermic peak of hemicellulose appears at a temperature lower than that for lignin (Demirbas, 2000). In slow pyrolysis of wood, hemicellulose pyrolysis begins at 130–194°C, with most of the decomposition occurring above 180°C (Mohan et al., 2006, p. 126).

#### *5.4.2.3 Lignin*

Pyrolysis of lignin typically produces about 55% char (Soltes and Elder, 1981), 15% tar, 20% aqueous components (pyroligneous acid), and about 12% gases. It is more difficult to dehydrate lignin than cellulose or hemicellulose (Mohan, 2006, p. 127). The tar produced from it contains a mixture of phenolic compounds, one of which, phenol, is an important raw material of green resin (a resin produced from biomass). The aqueous portion comprises methanol, acetic acid, acetone, and water. The decomposition of lignin in wood can begin at 280°C, continuing to 450–500°C and can reach a peak rate at 350–450°C (Kudo and Yoshida, 1957).

### 5.4.3 Kinetic Models of Pyrolysis

To optimize the process parameters and maximize desired yields, knowledge of the kinetics of pyrolysis is important. However, it is very difficult to obtain reliable data of kinetic rate constants that can be used for a wide range of biomass and for different heating rates. This is even more difficult for fast pyrolysis as it is a nonequilibrium and non-steady-state process. For engineering design purposes, a "black-box" approach can be useful, at least for the first approximation. The following discussion presents a qualitative understanding of the process based on data from relatively slow heating rates.

Kinetic models of the pyrolysis of lignocellulosic fuels like biomass may be broadly classified into three types (Blasi, 1993):

1. *One-stage global single reactions.* The pyrolysis is modeled by a one-step reaction using experimentally measured weight-loss rates.
2. *One-stage, multiple reactions.* Several parallel reactions are used to describe the degradation of biomass into char and several gases. A one-stage simplified kinetic model is used for these parallel reactions. It is useful for determination of product distribution.
3. *Two-stage semiglobal reactions.* This model includes both primary and secondary reactions, occurring in series.

#### 5.4.3.1 One-Stage Global Single-Reaction Model

This reaction model is based on a single overall reaction:

$$\text{Biomass} \rightarrow \text{volatile} + \text{char}$$

It neglects presence of ash and assumes moisture remains in volatile. The rate of pyrolysis depends on the unpyrolyzed mass of the biomass. Thus, the decomposition rate of mass, $m_b$, in the primary pyrolysis process may be written as:

$$\frac{dm_b}{dt} = -k(m_b - m_c) \tag{5.2}$$

Here, $m_c$ is the mass of char remaining after complete conversion (kg), $k$ is the first-order reaction rate constant ($\sigma^{-1}$), and $t$ is the time (s).

The fractional change, $X$, in the mass of the biomass may be written in nondimensional form as:

$$X = \frac{(m_0 - m_b)}{(m_0 - m_c)} \tag{5.3}$$

where $m_0$ is the initial mass of the biomass (kg).

Substituting fractional conversion for the mass of biomass in Eq. (5.2),

$$\frac{dX}{dt} = -k(1 - X) \tag{5.4}$$

**TABLE 5.4 Kinetic Rate Constants for One-Step Single-Reaction Global Model**

| Fuel | Temperature (K) | $E$ (kJ/mol) | $A$ ($s^{-1}$) | References |
|---|---|---|---|---|
| Cellulose | 520–1270 | 166.4 | $3.9 \times 10^{11}$ | Lewellen et al. (1977) |
| Hemicellulose | 520–1270 | 123.7 | $1.45 \times 10^{9}$ | Min (1977) |
| Lignin | 520–1270 | 141.3 | $1.2 \times 10^{8}$ | Min (1977) |
| Wood | 321–720 | 125.4 | $1.0 \times 10^{8}$ | Nolan et al. (1973) |
| Almond shell | 730–880 | 95–121 | $1.8 \times 10^{6}$ | Font et al. (1990) |
| Beech sawdust | 450–700 | 84 ($T$ > 600K) | $2.3 \times 10^{4}$ | Barooah and Long (1976) |

Solving this equation we get:

$$X = 1 - A\ \exp(-kt) \tag{5.5}$$

where $A$ is the preexponential coefficient, $k = E/RT$ $E$ is the activation energy (J/mol), $R$ is the gas constant (J/mol K), and $T$ is the temperature (K).

Owing to the difficulties in extracting data from dynamic thermogravimetric analysis, reliable data on the preexponential factor, $A$, and the activation energy, $E$, are not easily available for fast pyrolysis (Reed, 2002, p. II-103). However, for slow heating, we can obtain some reasonable values. If the effect of secondary cracking and the heat-transfer limitation can be restricted, the weight-loss rate of pure cellulose during pyrolysis can be represented by an irreversible, one-stage global first-order equation.

For the one-step global reaction model, Table 5.4 lists values of the activation energy $E$ and the preexponential factor $A$, for the pyrolysis of various biomass types at a relatively slow heating rate.

Other models are not discussed here, but details are available in several publications, including Blasi (1993).

## 5.5 HEAT TRANSFER IN A PYROLYZER

The preceding discussions assume that the heat or mass transport rate is too high to offer any resistance to the overall rate of pyrolysis. This is true at a temperature of 300–400°C (Thurner and Mann, 1981), but at higher temperatures heat and mass transport influence the overall rate and so cannot be neglected. This section deals with heat transport during pyrolysis.

During pyrolysis, heat is transported to the particle's outer surface by radiation and convection. Thereafter, it is transferred to the interior of the particle by conduction and pore convection (Figure 5.4). The following modes of heat transfer are involved in this process (Babu and Chaurasia, 2004b).

- Conduction inside the particle
- Convection inside the particle pores
- Convection and radiation from the particle surface.

In a commercial pyrolyzer or gasifier, the system heats up a heat-transfer medium first; that, in turn, transfers the heat to the biomass. The heat-transfer medium can be one or a combination of the following:

- Reactor wall (for vacuum reactor)
- Gas (for entrained-bed or entrained-flow reactor)
- Heat-carrier solids (for fluidized bed).

Bubbling fluidized beds use mostly solid–solid heat transfer. Circulating fluidized beds (CFB) and transport reactors make use of gas–solid heat transfer in addition to solid–solid heat transfer.

Since heat transfer to the interior of the biomass particle is mostly by thermal conduction, the low thermal conductivity of biomass (~0.1 W/m K) is a major deterrent to the rapid heating of its interior. For this reason, even when the heating rate of the particle's exterior is as fast as 10,000°C/s, the interior can be heated at a considerably slower rate for a coarse particle. Because of the associated slow heating of the interior, the secondary reactions within the particles become increasingly important as the particle size increases, and as a result the liquid yield reduces (Scott and Piskorz, 1984). For example, Shen et al. (2009) noted that oil yield decreased with particle size within the range of 0.3–1.5 mm, but no effect was noted when the size was increased to 3.5 mm. Experimental results (Seebauer et al., 1997), however, do not show much effect of particle size on the biomass.

### 5.5.1 Mass Transfer Effect

Mass transfer can influence the pyrolysis product. For example, a sweep of gas over the fuel quickly removes the products from the pyrolysis environment. Thus, secondary reactions such as thermal cracking, repolymerization, and recondensation are minimized (Sensoz and Angin, 2008).

### 5.5.2 Is Pyrolysis Autothermal?

An important question for designers is whether a pyrolyzer can meet its own energy needs or is dependent on external energy. The short and tentative answer is that a pyrolyzer as a whole is not energy self-sufficient. The reaction heat is inadequate to meet all energy demands, which include heat required to raise the feed and any inert heat-transfer media to the reaction temperature, heat consumed by endothermic reactions, and heat losses from the reactor. In most cases, it is necessary to burn the noncondensable gases and the char produced to provide the heat required. If that is not adequate, other heat sources are necessary to supply the energy required for pyrolysis. The following section discusses the heat requirement of reactions taking place in a pyrolyzer.

The dehydration (reaction II) process is exothermic, while depolymerization (reaction III) and secondary cracking (reaction IV) are endothermic (Bridgwater et al., 2001). Among reactions between intermediate products of pyrolysis, some are exothermic and some are endothermic. In general, pyrolysis of hemicellulose and lignin is exothermic. Cellulose pyrolysis is endothermic at lower temperatures (<400–450°C), and it becomes exothermic at higher temperatures owing to the following exothermic reactions (Klass, 1998):

$$CO + 3H_2 \rightarrow CH_4 + H_2O - 226 \text{ kJ/gmol} \tag{5.6}$$

$$CO + 2H_2 \rightarrow CH_3OH - 105 \text{ kJ/gmol} \tag{5.7}$$

$$0.17C_6H_{10}O_5 \rightarrow C + 0.85H_2O - 80 \text{ kJ/gmol} \tag{5.8}$$

$$CO + H_2O \rightarrow CO_2 + H_2 - 42 \text{ kJ/gmol} \tag{5.9}$$

(All equations refer to a temperature of 1000 K, and $C_6H_{10}O_5$ represents the cellulose monomer.)

For this reason a properly designed system initially requires external heat only until the required temperature is reached.

Char production from cellulose (Eq. (5.8)) is slightly exothermic. However, at a higher temperature, when sufficient hydrogen is produced by reaction (Eq. (5.9)), other exothermic reactions (Eqs. (5.6 and 5.7)) can proceed. At low temperatures and short residence times of volatiles, only endothermic primary reactions are active (heat of reaction −225 kJ/kg), while at high temperatures exothermic secondary reactions (heat of reaction 20 kJ/kg) are active (Blasi, 1993).

In conclusion, for design purposes, one may neglect the heat of reaction for the pyrolysis process, but it is necessary to calculate the energy required for vaporization of products and for heating feedstock gases to the pyrolysis temperature (Boukis et al., 2007).

## 5.6 PYROLYZER TYPES

Pyrolyzers have been used since ancient times to produce charcoal (Figure 5.2). Early pyrolyzers operated in batch mode using a very slow rate of heating and for long periods of reaction to maximize the production of char. If the objective of pyrolysis was to produce the maximum amount of liquid or gas, then the rate of heating, the peak pyrolysis temperature, and the duration of pyrolysis

**TABLE 5.5** Effect of Operating Variables on the Pyrolysis Yield

| To Maximize Yield of | Maximum Temperature | Heating Rate | Gas Residence Time |
|---|---|---|---|
| Char | Low | Slow | Long |
| Liquid | Low (~500°C[a]) | High | Short |
| Gas | High | Low | Long |

[a]*Bridgwater (1999).*
**Source:** Table compiled from Demirbas (2001).

had to be chosen accordingly. These choices also decided what kind of reactor was to be used. Table 5.5 lists the choice of heating rate, temperature, and gas residence time for maximization of the yield.

Modern pyrolyzers are more concerned with gas and liquid products, and require a continuous process. A number of different types of pyrolysis reactor have been developed. Based on the gas–solid contacting mode, they can be broadly classified as fixed bed, fluidized bed, and entrained bed, and then further subdivided depending on design configuration. The following are some of the major pyrolyzer designs in use:

- Fixed or moving bed
- Bubbling fluidized bed
- Circulating fluidized bed (CFB)
- Ultrarapid reactor
- Rotating cone
- Ablative reactor
- Vacuum reactor.

Except for the moving bed, other pyrolyzer types are shown in Figure 5.7.

### 5.6.1 Fixed-Bed Pyrolyzer

Fixed-bed pyrolysis, operating in batch mode, is the oldest pyrolyzer type. Heat for the thermal decomposition of biomass is supplied either from an external source or by allowing limited combustion as in a beehive oven (Figure 5.2). The product may flow out of the pyrolyzer because of volume expansion while the char remains in the reactor. In some designs, a sweep gas is used for effective removal of the product gas from the reactor. This gas is necessarily inert and oxygen free. The main product of this type is char owing to the relatively slow heating rate and the long residence time of the product in the pyrolysis zone.

### 5.6.2 Bubbling-Bed Pyrolyzer

Figure 5.9A shows a bubbling fluidized-bed pyrolyzer. Crushed biomass (2–6 mm) is fed into a bubbling bed of hot sand or other solids. The bed is fluidized by an inert gas such as recycled flue gas. Intense mixing of inert bed solids (sand is commonly used) offers good and uniform temperature control. It also provides high heat transfer to biomass solids. The residence time of the solids is considerably higher than that of the gas in the pyrolyzer.

The required heat for pyrolysis may be provided either by burning a part of the product gas in the bed, as shown in Figure 5.5, or by burning the solid char in a separate chamber and transferring that heat to the bed solids (Figure 5.7B). The pyrolysis product would typically contain about 70–75% liquid on dry wood feed. As shown in the figure, the char in the bed solids acts as a vapor-cracking catalyst, so its separation through elutriation or otherwise is important if the secondary cracking is to be avoided to maximize the liquid product. The entrained char particles are separated from the product gas using single- or multistage cyclones. A positive feature of a bubbling fluidized-bed pyrolyzer is that it is relatively easy to scale up.

### 5.6.3 CFB Pyrolyzer

A CFB pyrolyzer, shown in Figure 5.7B, works on the same principle as the bubbling fluidized bed except that the bed is highly expanded and solids continuously recycle around an external loop comprising a cyclone and loop seal (Basu, 2006, p. 35). The riser of the CFB operates in a special hydrodynamic regime known as *fast bed*. It provides good temperature control and uniform mixing around the entire height of the unit. The superficial gas velocity in a CFB is considerably higher than that in a bubbling bed. High velocity combined with excellent mixing allows a CFB to have large throughputs of biomass. Here, gas and solids move up the reactor with some degree of internal refluxing. As a result, the residence time of average biomass particles is longer than that of the gas, but the difference is not as high as it is in a bubbling bed. A major advantage of this system is that char entrained from the reactor is easily separated and burnt in an external fluidized bed. The combustion heat is transferred to the inert bed solids that are recycled to the reactor by means of a loop seal.

Rapid thermal pyrolysis (RTP), a commercial process developed by Ensyn of Canada probably originated from the ultrarapid fluidized-bed pyrolyzer developed at the University of Western Ontario in Canada. RTP uses a riser reactor. Here, biomass is introduced into a vessel and rapidly heated to 500°C by a tornado of upflowing hot sand; it is then cooled within seconds. The heating rate is on the order of 1000°C/s, and the reactor residence time is from a few hundredths of a millisecond to a maximum of 5 s, which gives a liquid yield as high as 83% for wood (Hulet et al., 2005).

### 5.6.4 Ultrarapid Pyrolyzer

High heating rate and short residence time in the pyrolysis zone are two key requirements of high liquid yield. The ultrarapid pyrolyzer, shown in Figure 5.7C, developed by the University of Western Ontario provides extremely short mixing (10–20 ms), reactor residence (70–200 ms), and quench (~20 ms) times. Because the reactor temperature is also low (~650°C), one can achieve a liquid yield as high as 90% (Hulet et al., 2005). The inert gas nitrogen is heated at 100°C above the reactor temperature and injected at very high velocity into the reactor to bombard a stream of biomass injected in the reactor. The reactor can also use a heat-carrier solid like sand that is heated externally and bombarded on a biomass stream through multiple jets. Such a high-velocity impact in the reactor results in an exceptionally high heating rate. The biomass is thus heated to the pyrolysis temperature in a few milliseconds. The pyrolysis product leaves the reactor from the bottom and is immediately cooled to suppress a secondary reaction or cracking of the oil vapor. This process is therefore able to maximize the liquid yield during pyrolysis.

### 5.6.5 Ablative Pyrolyzer

This process, shown in Figure 5.7D, involves creation of high pressure between a biomass particle and a hot reactor wall. This allows uninhibited heat transfer from the wall to the biomass, causing the liquid product to melt out of the biomass the way frozen butter melts when pressed against a hot pan. The biomass sliding against the wall leaves behind a liquid film that evaporates and leaves the pyrolysis zone, which is the interface between biomass and wall. As a result of high heat transfer and short gas residence time, a liquid yield as high as 80% is reported (Diebold and Power, 1988). The pressure between biomass and wall is created either by mechanical means or by centrifugal force. In a mechanical system, a large piece of biomass is pressed against a rotating hot plate.

### 5.6.6 Rotating-Cone Pyrolyzer

In this process, biomass particles are fed into the bottom of a rotating cone (360–960 rev/min) together with an excess of heat-carrier solid particles (Figure 5.7E). Centrifugal force pushes the particles against the hot wall; the particles are transported spirally upward along the wall. Owing to its excellent mixing, the biomass undergoes rapid heating (5000 K/s) and is pyrolyzed within the small annular volume. The product gas containing bio-oil vapor leaves through another tube, while the solid char and sand spill over the upper rim of the rotating cone into a fluidized bed surrounding it, as shown in Figure 5.7E. The char burns in the fluidized bed, and this combustion helps heat the cone as well as the solids that are recycled to it to supply heat for pyrolysis. Special features of this reactor include very short solids residence time (0.5 s) and a small gas-phase residence time (0.3 s). These typically provide a liquid yield of 60–70% on dry feed (Hulet et al., 2005). The absence of a carrier gas is another advantage of this process. The complex geometry of the system may raise some scale-up issues.

### 5.6.7 Vacuum Pyrolyzer

A vacuum pyrolyzer, as shown in Figure 5.7F, comprises a number of stacked heated circular plates. The top plate is at about 200°C while the bottom one is at about 400°C. Biomass fed to the top plate drops into successive lower plates by means of scrapers. The biomass undergoes drying and pyrolysis while moving over the plates. No carrier gas is required in this pyrolyzer. Only char is left when the biomass reaches the lowest plate. Though the heating rate of the biomass is relatively slow, the residence time of the vapor in the pyrolysis zone is short. As a result, the liquid yield in this process is relatively modest, about 35–50% on dry feed, with a high char yield. This pyrolyzer design is complex, especially given the fouling potential of the vacuum pump.

## 5.7 PYROLYZER DESIGN CONSIDERATIONS

This section discusses design considerations in the production of liquid fuel and charcoal through pyrolysis.

### 5.7.1 Production of Liquid Through Pyrolysis

Pyrolysis is one of several means of production of liquid fuel from biomass. The maximum yield of organic liquid (pyrolytic oil or bio-oil) from thermal decomposition may be increased to as high as 70% (dry weight) if the biomass is rapidly heated to an intermediate temperature and if a short residence time in the pyrolysis zone is allowed to reduce secondary reactions. Table 5.2 gives the effect of heating rate, pyrolysis temperature, and residence time on the pyrolysis product. These findings may be summarized as follows:

- A slower heating rate, a lower temperature, and a longer residence time maximize the yield of solid char.

- A higher heating rate, a higher temperature, and a shorter residence time maximize the gas yield.
- A higher heating rate, an intermediate temperature, and a shorter residence time maximize the liquid yield.

There is an optimum pyrolysis temperature for maximum liquid yield. The yield is highest at 500°C and drops sharply above and below this temperature (Boukis et al., 2007). The residence time is generally in the range of 0.1–2.0 s. These values depend on several factors, including the type of biomass (Klass, 1998). We can use a kinetic model for a reasonable assessment of the yield. The one proposed by Liden et al. (1988) may be used for predicting pyrolysis liquid yields over a wide range of conditions.

Heat transfer is a major consideration in the design of a pyrolyzer. The heat balance for a typical pyrolyzer may be written as:

$$\begin{aligned}&[\text{Heat released by char combustion}] + [\text{Heat in incoming stream}]\\&= [\text{Heat required for pyrolysis}] + [\text{Heat loss}]\end{aligned} \tag{5.10}$$

Assessing heat loss accurately is difficult before the unit is designed. So, for preliminary assessment, we can take this to be 10% of the heat in the incoming stream (Boukis et al., 2007, p. 1377).

Fast, or flash, pyrolysis is especially suitable for pyrolytic liquefaction of biomass. The product is a mixture of several hydrocarbons, which allows production of fuel and chemicals through appropriate refining methods. The heating value of the liquid produced is a little lower or in the same range (13–18 MJ/kg) as that of the parent biomass. The pyrolytic liquid contains several water-soluble sugars and polysaccharide-derivative compounds and water-insoluble pyrolytic lignin.

Pyrolytic liquid contains a much higher amount of oxygen (~50%) than does most fuel oil. It is also heavier (specific gravity ~1.3) and more viscous. Unlike fuel oil, pyrolytic oil increases in viscosity with time because of polymerization. This oil is not self-igniting like fuel oil, and as such it cannot be blended with diesel for operating a diesel engine.

Pyrolytic oil is, however, a good source of some useful chemicals, like natural food flavoring, that can be extracted, leaving the remaining product for burning. Alternately, we can subject the pyrolytic oil to hydrocracking to produce gasoline and diesel.

## 5.8 BIOCHAR

Charcoal, also known as biochar, is a preferred product of slow pyrolysis at a moderate temperature. Thermodynamic equilibrium calculation shows that the char yield of most biomass may not exceed 35%. Table 5.6 gives the theoretical equilibrium yield of different products of biomass at different

**TABLE 5.6 Thermodynamic Equilibrium Concentration of Pyrolysis of Cellulose at Different Temperatures**

| | Temperature (°C) | | | | |
|---|---|---|---|---|---|
| **Products** | **200** | **300** | **400** | **500** | **600** |
| C | 32 | 28 | 27 | 27 | 25.2 |
| $H_2O$ | 36.5 | 32.5 | 9.5 | 27 | 22.5 |
| $CH_4$ | 8.5 | 10 | 10.5 | 10 | 9 |
| $CO_2$ | 23.9 | 28 | 32 | 35 | 36 |
| CO | 0 | 0 | 0.1 | 1.2 | 4.5 |

**Source:** Derived from Antal (2003).

temperatures. Actual yield, however, could be much different. Assuming that cellulose represents biomass, the stoichiometric equation for production of charcoal (Antal, 2003) may be written as:

$$C_6H_{10}O_5 \rightarrow 3.74C + 2.65H_2O + 1.17CO_2 + 1.08CH_4 \quad (5.11)$$

Charcoal production from biomass requires slow heating for a long duration but at a relatively low temperature of around 400°C. An example of severe pyrolysis or carbonization is seen in the coke oven in an iron and steel plant, which pyrolyzes (carbonizes) coking coal to produce hard coke used for iron extraction. Coke oven is an indirectly heated fixed-bed pyrolyzer that operates at a temperature exceeding 1000°C and for a long period of time to maximize gas and solid coke production.

Biochar has a special appeal in greenhouse gas reduction as its production can greatly increase the amount of carbon retained in ground in stable form similar to that is done for carbon sequestration. The carbon in agricultural residues and forest residues when left on the ground is released over the time to the atmosphere as $CO_2$ or $CH_4$. On the other hand, if biomass is converted into biochar, as much as 50% of the carbon contained in the biomass could stay in the soil as a stable biochar residue. In most shifting cultivation systems around the world, the natural vegetation is burned after slashing. Between 38% and 84% of the biomass carbon is released to the atmosphere during such burn (Lehman et al., 2006). This is a very inefficient way of producing biochar. Pyrolysis provides the best means of production of biochar. It is the solid residue of pyrolysis, which makes it a by-product of this process.

### 5.8.1 Potential Benefits of Biochar

Biochar has a number of benefits as listed below (www.biochar.ca, www.biocharfarms.org):

1. Sequesters carbon and thereby minimize climate change
2. Carbon negative emission
3. Displaces carbon positive fossil fuels
4. Reduces nutrient losses in soils
5. Reduces fertilizer use
6. Enhances marginal soil productivity
7. Increases sustainable food production
8. Improves water retention, aeration, and tilth
9. Higher cation exchange capacity (CEC)
10. Improves water quality by reducing contaminated runoff and nutrient loss
11. Soil remediation
12. Reversal of desertification on massive scales and can work in tandem with reforestation
13. A better alternative to slash-and-burn of agricultural residues
14. Decreases nitrous oxide and methane emissions from solids
15. Net primary production
16. Generates carbon offsets and increased on-farm profitability for the company.

As mentioned earlier, biochar is produced through pyrolysis as the solid by-product. The quality of biochar is defined by its following characteristics:

- The BET or internal surface area
- pH of the char
- CEC of biochar
- Carbon recovery in char.

Above characteristics of biochar depend on how biochar is produced. It is thus influenced by the following three processes and feed parameters:

1. Temperature
2. Type of biomass
3. Residence time.

Pyrolysis temperature is the most important parameter influencing the properties of biochar. Figure 5.10 shows that there is a sharp increase in surface area between temperature 450°C and 550°C. The CEC also increases during the period. The pH increases steadily within the temperature range of 200–800°C. The carbon yield or carbon recovery decreases with temperature

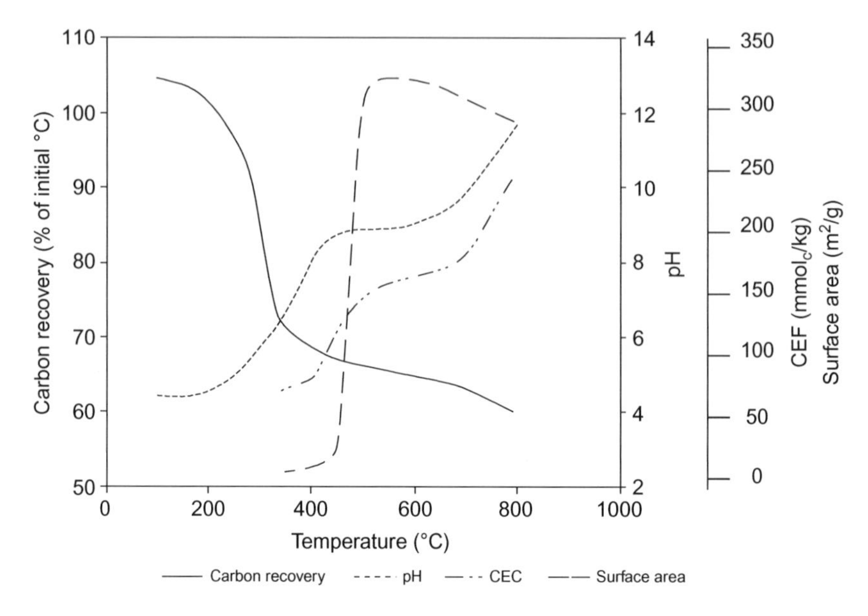


**FIGURE 5.10** A qualitative diagram showing changes in properties of biochar with temperature of its production. Source*: Drawn after Lehman et al. (2007). Higher than 100% carbon recovery below 200°C is likely to be an experimental error of the authors.*

rise to 60% of the initial carbon, but it is less influenced above 600°C. These suggests that pyrolysis at higher temperature is desirable from a biochar production standpoint.

## SYMBOLS AND NOMENCLATURE

| | |
|---|---|
| $A$ | preexponential factor ($s^{-1}$) |
| $E$ | activation energy (J/mol) |
| $k$ | reaction rate ($s^{-1}$) |
| $m_b$ | mass of biomass at time $t$ (kg) |
| $m_c$ | mass of char residue (kg) |
| $m_0$ | initial mass of biomass (kg) |
| $R$ | universal gas constant (J/mol K) |
| $T$ | temperature (K) |
| $T_{pyr}$ | pyrolysis temperature (K) |
| $t$ | time (s) |
| $t_{heating}$ | heating time (s) |
| $t_r$ | reaction time (s) |
| $X$ | fractional change in mass of biomass |

Chapter 6

# Tar Production and Destruction

## 6.1 INTRODUCTION

Tar is a major nuisance in both gasification and pyrolysis. It is a thick, black, highly viscous liquid that condenses in the low-temperature zones of a gasifier, clogging the gas passage and leading to system disruptions. Tar is highly undesirable, as it can create many problems including:

- Condensation and subsequent plugging of downstream equipment.
- Formation of tar aerosols.
- Polymerization into more complex structures.

Nevertheless, tar is an unavoidable by-product of the thermal conversion process. This chapter discusses what tar is, how it is formed, and how to influence its formation such that plants and equipment can live with this "necessary evil" while minimizing its detrimental effects.

## 6.2 TAR

Tar is a complex mixture of condensable hydrocarbons, including, among others, oxygen-containing, 1- to 5-ring aromatic, and complex polyaromatic hydrocarbons (Devi et al., 2003). Neeft et al. (2003) defined tar as "all organic contaminants with a molecular weight larger than 78, which is the molecular weight of benzene." The International Energy Agency (IEA) Bioenergy Agreement, the US Department of Energy (DOE), and the DGXVII of the European Commission agreed to identify as tar all components of product gas having a molecular weight higher than that of benzene (Knoef, 2005, p. 278).

A common perception about tar is that it is a product of gasification and pyrolysis that can potentially condense in colder downstream sections of the unit. While this is a fairly good description, a more specific and scientific definition may be needed for technical, scientific, and legal work. Presently, there is no universally accepted definition of tar. As many as 30 definitions are available in the literature (Knoef, 2005, p. 279). Of these, one of the

**Biomass Gasification, Pyrolysis and Torrefaction.**

IEA's gasification task force appears most appropriate (Milne et al., 1998). It is as follows:

> The organics, produced under thermal or partial-oxidation regimes (gasification) of any organic material, are called "tar" and are generally assumed to be largely aromatic.

### 6.2.1 Acceptable Limits for Tar

Tar remains vaporized until the gas carrying it is cooled, when it either condenses on cool surfaces or remains in fine aerosol drops (<1 μm). This makes the product gas unsuitable for use in gas engines, which have a low tolerance for tar. Thus, there is a need for tar reduction in product gas when the gas is to be used in an engine. This can be done through appropriate design of the gasifier and the right choice of operating conditions, including reactor temperature and heating rate. Even these adjustments may not reduce tars in the gas to the required level, necessitating further downstream cleanup.

Standard gas cleaning involves filtration and/or scrubbing, which not only removes tar but also strips the gas of particulate matters and cools it to room temperature. These practices clean the gas adequately, making it acceptable to most gas engines. However, they result in a great reduction in overall efficiency in the production of electricity or mechanical power using a gas engine. Furthermore, gas cleaning greatly adds to the capital investment of the plant.

Biomass gasification is at times used for distributed power generation in remote locations in small- to medium-capacity plants. For such plants, the addition of a scrubber or a filtration system significantly increases the overall plant costs. This limitation makes biomass-based distributed power-generation projects highly sensitive to the cost of tar cleanup.

The presence of tar in the product gas from gasification can potentially decide the usefulness of the gas. The following are the major applications of the product gas:

**a.** Direct-combustion systems
**b.** Internal-combustion engines
**c.** Syngas production.

Table 6.1 presents data on the tolerance levels of tar and particulate contents for several applications of gas.

**a.** In applications where the raw gas is burnt directly without cooling, there is no need for cleaning. Such systems have little restriction on the amount of tar and particulates as long as the gas travels freely to the burner and as long as the burner design does not impose any restrictions of its own.

**TABLE 6.1 Upper Limits of Biomass Gas Tar and Particulates**

| Application | Particulate (g/Nm$^3$) | Tar (g/Nm$^3$) |
|---|---|---|
| Direct combustion | No limit specified | No limit specified |
| Syngas production | 0.02 | 0.1 |
| Gas turbine | 0.1–120 | 0.05–5 |
| IC engine | 30 | 50–100 |
| Pipeline transport | | 50–500 for compressor |
| Fuel cells | | <1.0 |

**Source**: Data compiled from Milne et al. (1998).

However, the flue gas produced after combustion must be clean enough to meet local emission requirements.

Cofiring of gasified biomass in fossil-fuel-fired boilers is an example of direct firing. Industrial units like ovens, furnaces, and kilns are also good examples of direct firing. In such applications, it is not necessary to cool the gas after production. The gas is fired directly in a burner while it is still hot, in the temperature range of 600–900°C. Thus, there is little chance of tar condensation. However, the pipeline between the gasifier exit and the burner inlet should be such that the gas does not cool down below the dewpoint of tar. If that happens, tar deposition might clog the pipes, leading to hazardous conditions.

**b.** Internal-combustion engines, such as diesel or Otto engines, are favorite applications of gasified biomass, especially for distributed power generation. In such applications, the gas must be cooled, and as such there is a good chance of condensation of the tar in the engine or in fuel-injection systems. Furthermore, the piston-cylinder system of an internal-combustion engine is not designed to handle solids, which imposes tighter limits on the tar as well as on the particulate level in the gas. Particulate and tar concentrations in the product gas should therefore be below the tolerable limits, which are 30 mg/Nm$^3$ for particulates and 100 mg/Nm$^3$ for tar (Milne et al., 1998, p. 41). The gas turbine, another user of biomass gas, imposes even more stringent restrictions on the cleanliness of the gas because its blades are more sensitive to deposits from the hot gas passing through them after combustion. Here, the tar concentration should be between 0.5 and 5.0 mg/Nm$^3$ (Milne et al., 1998, p. 39).

**c.** The limits for particulates and tar in syngas applications are even more stringent, as tar poisons the catalyst. For these applications, Graham and Bain (1993) suggested an upper limit as low as 0.02 mg/Nm$^3$ for particulates and

0.1 mg/$Nm^3$ for tar. Interest in fuel cells is rising, especially for the direct production of electricity from hydrogen through gasification. The limiting level of tar in the gas fed into a fuel cell that produces electricity directly from fuel gas is specific to the organic constituents of the gas.

#### *6.2.1.1 Level of Tar Production*

The amount of tar in product gas depends on the gasification temperature as well as on the gasifier design. Typical average tar levels in gases from downdraft and updraft biomass gasifiers are 1 and 50 g/$Nm^3$, respectively (Table 6.2). Average tar levels in product gas from bubbling and circulating fluidized-bed gasifiers are about 10 g/$Nm^3$. For a given gasifier, the amount of the tar yield (percentage of dry mass of biomass) reduces with temperature. Actual amount of tar yield depends on a number of factors like gasifier-type temperature. Figure 6.1 shows the range of tar yield at different temperatures.

### 6.2.2 Tar Formation

Tar is produced primarily through depolymerization during the pyrolysis stage of gasification. Biomass (or other feed), when fed into a gasifier, first undergoes pyrolysis that can begin at a relatively low temperature of 200°C and complete at 500°C. In this temperature range, the cellulose, hemicellulose, and lignin components of biomass break down into *primary tar*, which is also known as *wood oil* or *wood syrup*. This contains oxygenates and primary organic condensable molecules called *primary tar* (Milne et al., 1998, p. 13). Char is also produced at this stage. Above 500°C, the primary tar components start reforming into smaller, lighter noncondensable gases and into a series of heavier molecules called *secondary tar* (Figure 6.2). The noncondensable gases include $CO_2$, CO, and $H_2O$. At still higher temperatures, primary tar products are destroyed and tertiary products are produced.

**TABLE 6.2** Typical Levels of Tar in Biomass Gasifier by Type

| Gasifier Type | Average Tar Concentration in Product Gas (g/$Nm^3$) | References |
|---|---|---|
| Downdraft | 0.01–6 | Hasler (1999) |
| Circulating fluidized bed | 1–30 | Han and Kim (2008) |
| Bubbling fluidized bed | 1–23 | Han and Kim (2008) |
| Updraft | 10–150 | Milne and Evans (1998), p. 15 |
| Entrained flow | Negligible | |

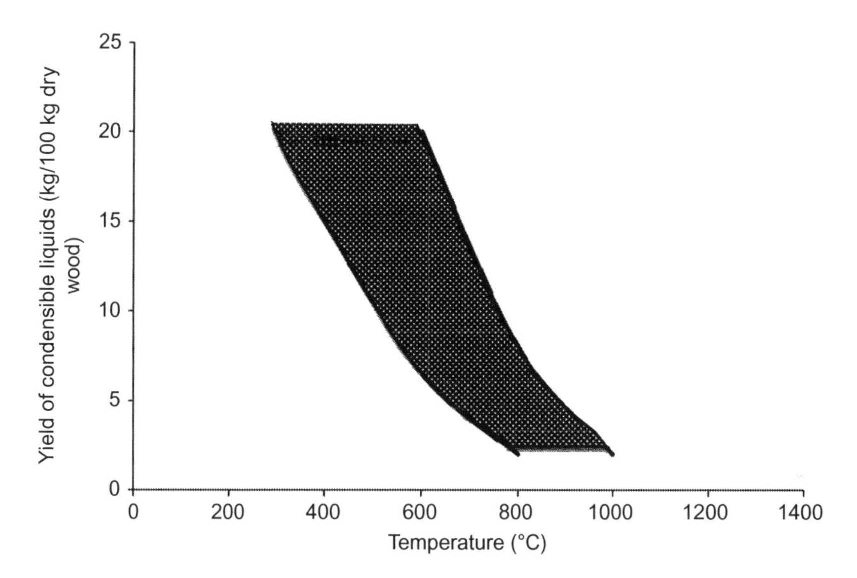


**FIGURE 6.1** Effect of maximum gasification temperature on the range of tar yield. Source: *From Baker et al. (1988).*

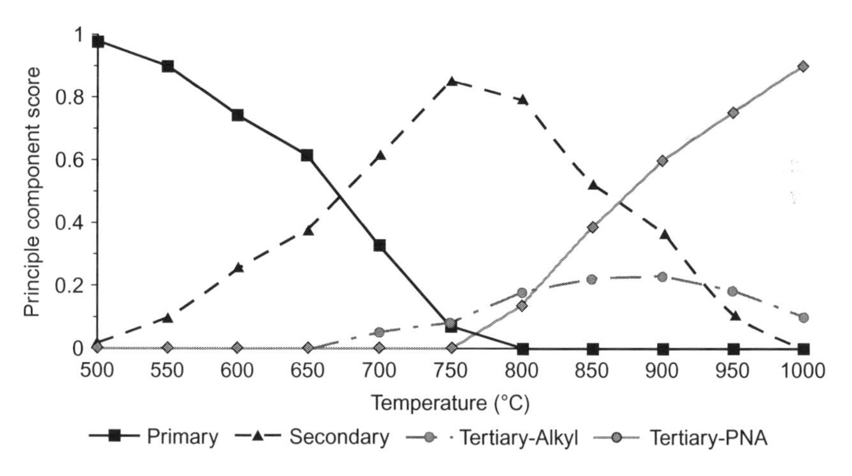


**FIGURE 6.2** Variation of primary, secondary and tertiary tar products with temperature for 0.3 s residence time. Source: *From Evans and Milne (1997), p. 807.*

### 6.2.3 Tar Composition

As we can see in Table 6.3, tar is a mixture of various hydrocarbons. It may also contain oxygen-containing compounds, derivatives of phenol, guaiacol, veratrol, syringol, free fatty acids, and esters of fatty acids (Razvigorova et al., 1994). The yield and composition of tar depend on the reaction

**TABLE 6.3** Typical Composition of Tar

| Component | Weight (%) |
|---|---|
| Benzene | 37.9 |
| Toluene | 14.3 |
| Other 1-ring aromatic hydrocarbons | 13.9 |
| Naphthalene | 9.6 |
| Other 2-ring aromatic hydrocarbons | 7.8 |
| 3-ring aromatic hydrocarbons | 3.6 |
| 4-ring aromatic hydrocarbons | 0.8 |
| Phenolic compounds | 4.6 |
| Heterocyclic compounds | 6.5 |
| Others | 1.0 |

**Source**: Adapted from Milne et al. (1998).

temperature, the type of reactor, and the feedstock. Table 6.3 shows that benzene is the largest component of a typical tar.

Tar may be classified into four major product groups: primary, secondary, alkyl tertiary, and condensed tertiary (Evans and Milne, 1997). Short descriptions of these follow.

#### *6.2.3.1 Primary Tar*

Primary tar is produced during primary pyrolysis. It comprises oxygenated, primary organic, condensable molecules. Primary products come directly from the breakdown of the cellulose, hemicellulose, and lignin components of biomass. Milne and Evans (1998) listed a large number of compounds of acids, sugars, alcohols, ketones, aldehydes, phenols, guaiacols, syringols, furans, and mixed oxygenates in this group.

#### *6.2.3.2 Secondary Tar*

As the gasifier's temperature rises above 500°C, primary tar begins to rearrange, forming more noncondensable gases and some heavier molecules called *secondary tar*, of which phenols and olefins are important constituents. As such, one notes a rise in secondary product at the expense of primary product (Figure 6.2).

#### 6.2.3.3 Tertiary Tar

The *alkyl tertiary product* includes methyl derivatives of aromatics, such as methyl acenaphthylene, methylnaphthalene, toluene, and indene (Evans and Milne, 1997). These are formed at higher temperature.

*Condensed tertiary aromatics* make up a polynuclear aromatic hydrocarbon (PAH) series without substituents (atoms or a group of atoms substituted for hydrogen in the parent chain of hydrocarbon). This series contains benzene, naphthalene, acenaphthylene, anthracene/phenanthrene, and pyrene.

The secondary and tertiary tar products come from the primary tar. The primary products are destroyed before the tertiary products appear (Milne et al., 1998).

Figure 6.2 shows that above 500°C with increasing temperature, the secondary tar increases at the expense of the primary tar. Once the primary tar is nearly destroyed, tertiary tar starts appearing with increasing temperature. At this stage, the secondary tar begins to decrease. Thus, high temperatures destroy the primary tar but not the tertiary tar products.

## 6.3 TAR REDUCTION

The tar in coal gasification comprises benzene, toluene, xylene, and coal tar, all of which have good commercial value and can be put to good use. Tar from biomass, on the other hand, is mostly oxygenated and has little commercial use. Thus, it is a major headache in gasifiers, and a major roadblock in the commercialization of biomass gasification. Research over the years has improved the situation greatly, but the problem has not completely disappeared. Tar removal remains an important part of the development and design of biomass gasifiers.

Several options are available for tar reduction. These may be divided into two broad groups (Figure 6.3):

1. Postgasification (or secondary) reduction, which strips the product gas of the tar already produced.
2. In situ (or primary) tar reduction, which avoids tar formation.

In situ reduction is carried out by various means so that the generation of tar inside the gasifier is less, thereby eliminating the need for any removal to occur downstream. As this process is carried out in the gasifier, it influences the product gas quality. Postgasification reduction, on the other hand, does not interfere with the process in the reactor, and therefore the quality of the product gas is unaffected (Figure 6.3A).

At times, it may not be possible to remove the tar to the desired degree while retaining the quality of the product gas. In such cases, a combination of in situ and postgasification reduction can prove very effective. The tar

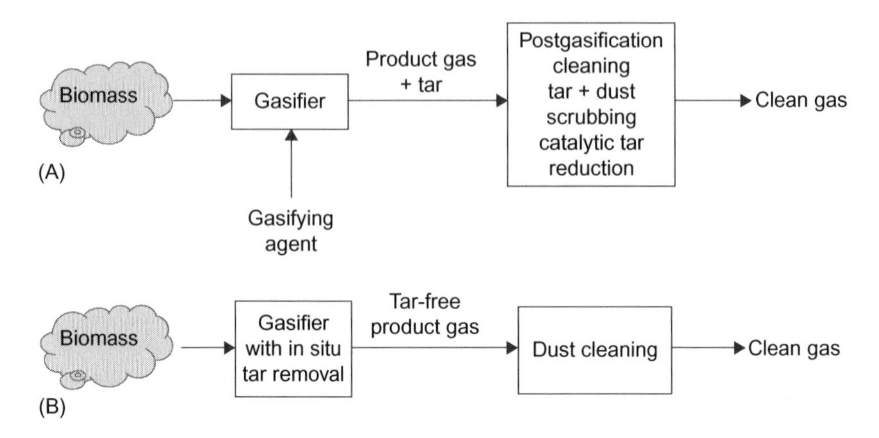


**FIGURE 6.3** Schematic of two major means (in situ (B) and postgasification (A)) of tar reduction.

removed is separated after the product gas leaves the gasifier (Figure 6.3B). Details of these two approaches are given in the following sections.

### 6.3.1 In Situ Tar Primary Reduction

In this approach, the operating conditions in the gasifier are adjusted such that tar formation is reduced. Alternately, the tar produced is converted into other products before it leaves the gasifier. Reduction is achieved by one of the following means:

- Modification of the operating conditions of the gasifier.
- Addition of catalysts or alternative bed materials in the fluidized bed.
- Modification of the gasifier design.

#### *6.3.1.1 Reduction Reactions*

Biomass type also influences the tar product. The appropriate choice of one or a combination of these factors can greatly reduce the amount of tar in the product gas leaving the gasifier. Reforming, thermal cracking, and steam cracking are the three major reactions responsible for tar destruction (Delgado et al., 1996). They convert tar into an array of smaller and lighter hydrocarbons as shown here:

$$\text{Tar} \Rightarrow \begin{bmatrix} \text{reforming} \\ \text{thermal cracking} \\ \text{steam cracking} \end{bmatrix} \Rightarrow [CO_2 + CO + H_2 + CH_4 + \cdots + \text{coke}]$$

**i.** *Tar reforming.* We can write the reforming reaction as in Eq. (6.1) by representing tar as $C_nH_x$. The reaction takes place in steam gasification, where steam cracks the tar, producing simpler and lighter molecules like $H_2$ and CO.

$$C_nH_x + nH_2O \rightarrow (n + x/2)H_2 + nCO \quad (6.1)$$

**ii.** *Dry tar reforming.* The dry reforming reaction takes place when $CO_2$ is the gasifying medium instead of steam. Here tar is broken down into $H_2$ and CO (Eq. (6.2)). Dry reforming is more effective than steam reforming especially when dolomite is used as the catalyst (Sutton et al., 2001).

$$C_nH_x + nCO_2 \rightarrow (x/2)H_2 + 2nCO \quad (6.2)$$

**iii.** *Thermal cracking.* Thermal cracking can reduce tar, but it is not as attractive as reforming because it requires high (>1100°C) temperature and produces soot (Dayton, 2002). Because this temperature is higher than the gas exit temperature for most biomass gasifiers, external heating or internal heat generation with the addition of oxygen may be necessary. Both options have major energy penalties.

**iv.** *Steam cracking.* In steam cracking, the tar is diluted with steam and is briefly heated in a furnace in the absence of oxygen. The saturated hydrocarbons are broken down into smaller hydrocarbons.

The following sections elaborate the operating conditions used in in situ reduction of tar.

### 6.3.1.2 Operating Conditions

Operating parameters that influence tar formation and conversion include reactor temperature, reactor pressure, gasification medium, equivalence ratio, and residence time.

#### Temperature

Reactor operating temperature influences both the quantity and the composition of tar. The quantity in general decreases with an increase in reaction temperature, as does the amount of unconverted char. Thus, high-temperature operation is desirable on both counts. The production of oxygen-containing compounds like phenol, cresol, and benzofuran reduces with temperature, especially below 800°C. With increasing temperature, the amount of 1- and 2-ring aromatics with substituents decreases but that of 3- and 4-ring aromatics increases. Aromatic compounds without substituents (e.g., naphthalene and benzene) are favored at high temperatures. The naphthalene and benzene content of the gas increases with temperature (Devi et al., 2003). High temperature also reduces the ammonia content of the gas and improves the char conversion but has a negative effect of reducing the product gas' useful heating value.

An increase in the freeboard temperature in a fluidized-bed gasifier can also reduce the tar in the product gas. A reduction in tar was obtained by

Narváez et al. (1996) by injecting secondary air into the freeboard. This may be due to increased combustion in the freeboard. Raising the temperature through secondary air injection in the freeboard may have a negative impact on heating value.

### Reactor Pressure

With increasing pressure, the amount of tar decreases, but the fraction of PAH increases (Knight, 2000).

### Residence Time

Residence time has a nominal effect on tar yield in a fluidized-bed gasifier. Kinoshita et al. (1994) noted that with increasing gas residence time (bed height/superficial gas velocity), the yield of oxygenated compounds and 1- and 2-ring compounds (benzene and naphthalene excepted) decreased, but the yield of 3- and 4-ring compounds increased.

### Gasification Medium

Four mediums—air, steam, carbon dioxide, and steam–oxygen mixture that are typically used for gasification—may have different effects on tar formation and conversion. The ratio of fuel to gasification medium is an important parameter that influences the product of gasification, including tar. This parameter is expressed differently for different mediums. For example, for air gasification, the parameter is the *equivalence ratio* (ER); for steam gasification, it is the *steam-to-biomass ratio* (S/B); and for steam–oxygen gasification, it is the *gasifying ratio* (Table 6.4). An example of the range of tar production for three gasification mediums for typical values of their characteristic parameters is given in Table 6.5.

In general, the yield of tar in steam gasification is greater than that in steam–oxygen gasification. Of these, air gasification is the lowest tar producer (Gil et al., 1999). The tar yield in a system depends on the amount of gasifying medium per unit biomass gasified.

*Gasification in air*: Both yield and concentration of tar in the product gas decreases with an increase in the ER. Higher ER (see Section 8.6.2 for a definition) allows greater amounts of oxygen to react with the volatiles in the *flaming pyrolysis* zone (Figure 6.4). Above an ER of 0.27, phenols are nearly all converted and less tar is formed (Kinoshita et al., 1994). This decrease is greater at higher temperatures. At a higher ER, the fraction of PAH, benzene, naphthalene, and other 3- and 4-ring aromatics increases in the product gas. While higher ER reduces the tar, it reduces the quality of the gas as well. The heating value of the gas is reduced because of nitrogen dilution from air.

*Gasification in steam*: When steam reacts with biomass to produce $H_2$ (Eq. (6.3)), the tar-reforming reaction reduces the tar.

$$C_nH_x + nH_2O \rightarrow (n + x/2)H_2 + nCO \tag{6.3}$$

**TABLE 6.4 Gasification Mediums and Characteristic Parameters**

| Medium | Parameter |
|---|---|
| Air | ER = ratio of air used to stoichiometric air |
| Steam | Steam-to-biomass (S/B) ratio |
| Carbon dioxide | $CO_2$-to-biomass ratio |
| Steam and oxygen | Gasifying ratio (GR): (steam + $O_2$)-to-biomass ratio |

**TABLE 6.5 Effect of Gasification Medium on Characteristics of Tar Production**

| Medium | Operating Condition | Tar Yield (g/ $Nm^3$) | LHV (MJ/ $Nm^3$ dry) | Tar Yield (g/kg $BM_{daf}$) |
|---|---|---|---|---|
| Steam | S/B = 0.9 | 30–80 | 12.7–13.3 | 70 |
| Steam and oxygen | GR = 0.9, $H_2O/O_2 = 3$ | 4–30 | 12.5–13.0 | 8–40 |
| Air | ER = 0.3, H/C = 2.2 | 2–20 | 4.5–6.5 | 6–30 |

**Source**: Data compiled from Gil et al. (1999).

A large reduction in tar yield was seen over an S/B ratio range of 0.5–2.5 (Herguido et al., 1992). Further reduction is possible in the presence of catalyst, which encourages the tar-reforming reaction (García et al., 1999).

*Gasification in a steam–oxygen mixture*: The addition of oxygen with steam further improves tar reduction. Additionally, it provides the heat needed to make the gasification reaction autothermal. When one uses oxygen along with steam, the mass ratio of (steam + oxygen) to biomass, known as the *gasification ratio* (GR), is used to characterize this reaction. The tar yield reduces with an increase in the gasifying ratio. For example, an 85% reduction in tar is obtained when the GR is increased from 0.7 to 1.2 (Aznar et al., 1997). Light tars are produced at a low GR.

*Gasification in carbon dioxide*: The tar may be reformed on the catalyst surface in a carbon dioxide medium. Such a reaction is called *dry reforming* and is shown in Eq. (6.4) (Sutton et al., 2001).

$$C_nH_x + nCO_2 \rightarrow 2nCO + (x/2)H_2 \tag{6.4}$$

The effect of gasifying agents on tar reduction or tar yield is compared in Table 6.5 (Gil et al., 1999).

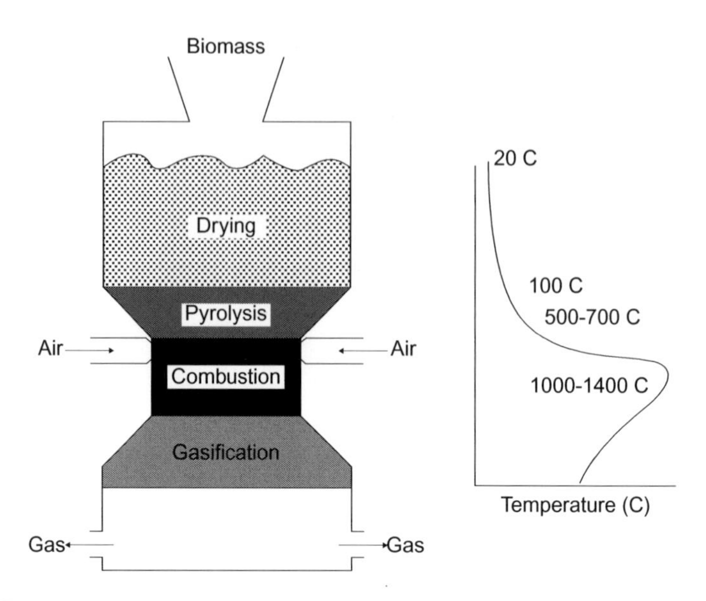


**FIGURE 6.4** Tar generation is downdraft gasifier. The tar produced passes through the highest temperature zone in a downdraft gasifier, and as such it is easily cracked.

### *6.3.1.3 Tar Reduction by Catalysts in Fluidized-Bed Gasifiers*

Catalysts accelerate the two main chemical reactions of tar reduction. In a steam-reforming reaction, we have

$$C_nH_x + nH_2O \xrightarrow{\text{Catalyst}} (n + x/2)H_2 + nCO \tag{6.5}$$

In a dry-reforming reaction, we have

$$C_nH_x + nCO_2 \xrightarrow{\text{Catalyst}} (x/2)H_2 + 2nCO \tag{6.6}$$

Catalysts can facilitate tar reduction reactions either in the primary reactor (gasifier) or downstream in a secondary reactor. Three main types of catalysts used are dolomite, alkali metal, and nickel. Olivine, and char have also found successful use as catalysts for tar reduction. Effects of these catalysts are detailed below.

#### Dolomite

Dolomite ($MgCO_3$, $CaCO_3$) is relatively inexpensive and is readily available. It is more active if calcined and used downstream in the postgasification secondary reactor at above 800°C (Sutton et al., 2001). The reforming reaction of tar on a dolomite surface occurs at a higher rate with $CO_2$ (Eq. (6.6)) than with steam (Eq. (6.5)). Under proper conditions, it can entirely convert the tar but cannot convert methane if that is to be avoided for syngas production.

Carbon deposition deactivates dolomite, which, being less expensive, may be discarded.

### Olivine

Olivine is a magnesium–iron silicate mineral (Mg, $Fe_2$) $SiO_4$ that comes in sizes (100–400 μm) and density ranges (2500–2900 $kg/m^3$) similar to those of sand. Thus, it is conveniently used with sand in a fluidized-bed gasifier. The catalytic activity of olivine is comparable to that of calcined dolomite. When using olivine, Mastellone and Arena (2008) noted a complete destruction of tar from a fluidized-bed gasifier for plastic wastes, while Rapagnà et al. (2000) obtained a 90% reduction in a biomass-fed unit.

### Alkali

Alkali metal catalysts are premixed with biomass before they are fed into the gasifier. Some of them are more effective than others. For example, the order of effectiveness of some alkali catalysts can be shown as follows:

$$K_2CO_3 > Na_2CO_3 > (Na_3H(CO_3)_2 \times 2H_2O) > Na_2B_4O_7 \times 10H_2O \quad (6.7)$$

Unlike dolomite, alkali catalysts can reduce methane in the product gas, but it is difficult to recover them after use. Furthermore, alkali cannot be used as a secondary catalyst. Its use in a fluidized bed makes the unit prone to agglomeration (Mettanant et al., 2009).

### Nickel

Many commercial nickel catalysts are available in the market for reduction of tar as well as methane in the product gas. They contain various amounts of nickel. For example, catalyst R-67-7H of Haldor Topsøe has 12–14% Ni on an $Mg/Al_2O_3$ support (Sutton et al., 2001). Nickel catalysts are highly effective and work best in the secondary reactor. Use of dolomite or alkali as the primary catalyst and nickel as the secondary catalyst has been successfully demonstrated for tar and methane reduction. Catalyst activity is influenced by temperature, space–time, particle size, and composition of the gas atmosphere. The optimum operating temperature for a nickel catalyst in a downstream fluidized bed is 780°C (Sutton et al., 2001). Steam-reforming nickel catalysts for heavy hydrocarbons are effective for reduction of tar while nickel catalysts for light hydrocarbons are effective for methane reduction. Deactivation due to carbon deposition and particle growth is a problem for nickel-reforming catalysts.

### Char

Char, a carbonaceous product of pyrolysis, also catalyzes tar reforming when used in the secondary reactor. Chembukulam et al. (1981) obtained a nearly total reduction in tar with this. As it is a major gasification element, char is not easily available in a gasifier's downstream. Design modification is needed to incorporate char as a catalyst.

#### 6.3.1.4 Gasifier Design

The design of the gasifier can be a major influence on the amount of tar in the product gas. For example, an entrained flow gasifier can reduce the tar content to less than 0.1 g/$Nm^3$, while in an updraft gasifier, the tar can well exceed 100 g/$Nm^3$. To understand how gasifier design might influence tar production, we will examine the tar production process.

As we see in Figure 6.2, primary tar is produced at fairly low temperatures (200–500°C). It is a mixture of condensable hydrocarbons that undergoes molecular rearrangement (reforming) at higher temperatures (700–900°C), producing some noncondensable gases and secondary tar. Tar is produced at an early stage when biomass (or another fuel) undergoes pyrolysis following drying. Char is produced further downstream in the process and is often the final solid residue left over from gasification. The gasifier design determines where pyrolysis takes place, how the tar reacts with oxidants, and the temperature of the reactions. This in turn determines the net tar production in the gasifier.

Updraft, downdraft, fluidized bed, and entrained bed are the four major types of gasifier with their distinct mode of tar formation. Table 6.2 compares their tar production, and a brief discussion of formation of tar in these reactors follows here.

##### Updraft Gasifier

Biomass is fed from the top and a gasifying medium (air) is fed from the bottom. In this countercurrent reactor, the product gas leaves from the top while solids leave from the bottom. Figure 6.5 illustrates the motion of biomass, gas, and tar. The temperature is highest close to the grate, where oxygen meets with char and burns the char. The hot gas travels up, providing heat to the endothermic gasification reactions and meets pyrolyzing biomass at a low temperature (200–500°C). Primary tar is produced in this temperature range (Figure 6.5). This tar travels upward through cooler regions and therefore has no opportunity for conversion into gases and secondary tar. For this reason, updraft gasifiers generate the highest amount of tar—typically 10–20% by weight of the feed.

##### Downdraft Gasifier

Figure 6.4 shows the tar production in a downdraft gasifier. It is a cocurrent reactor where both gas and feed travel downward. The temperature is highest in the downstream combustion zone. The tar is produced just after drying in a zone close to the feed point where the temperature is relatively low (200–500°C). The oxygen in the air, along with the tar, travels downward to

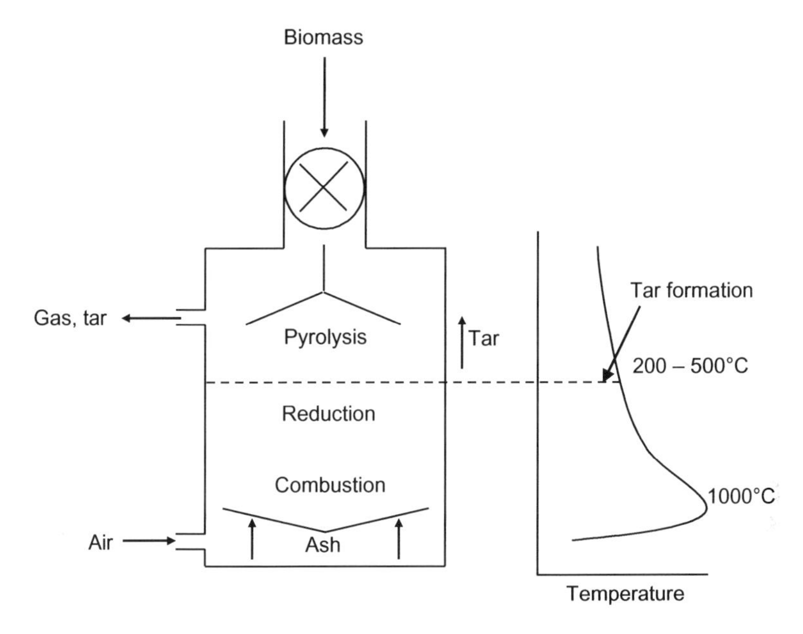


**FIGURE 6.5** Tar production in updraft gasifier. Here, the tar passes through only the low temperature (200–500°C) zone. So it does not get any opportunity to crack.

the hotter zone. Owing to the availability of oxygen and high temperature, the tar readily burns in a flame, raising the gas temperature to 1000–1400°C. The flame occurs in the interstices between feed particles, which remain at 500–700°C (Milne et al., 1998, p. 14). This phenomenon is called *flaming pyrolysis*. While passing through the highest temperature zone, the pyrolysis product, tar, contacts oxygen and as such it has the greatest opportunity to be converted into noncondensable gases. For this reason, a downdraft gasifier has the lowest tar production (<1 g/Nm$^3$).

## Fluidized-Bed Gasifier

In a typical fluidized bed (bubbling or circulating), the gasification medium enters from the bottom, but the fuel is fed from the side or the top. In either case, the fuel is immediately mixed throughout the bed owing to its exceptionally high degree of mixing (Figure 6.6). Thus, the gasification medium entering the grid comes into immediate contact with fresh biomass particles undergoing pyrolysis as well as with spent char particles from the biomass, which has been in the bed for some time. When air or oxygen is present in the gasification medium, the oxygen on contact with pyrolyzing feed burns the tar released, while its contact with the spent char particles causes the char to burn.

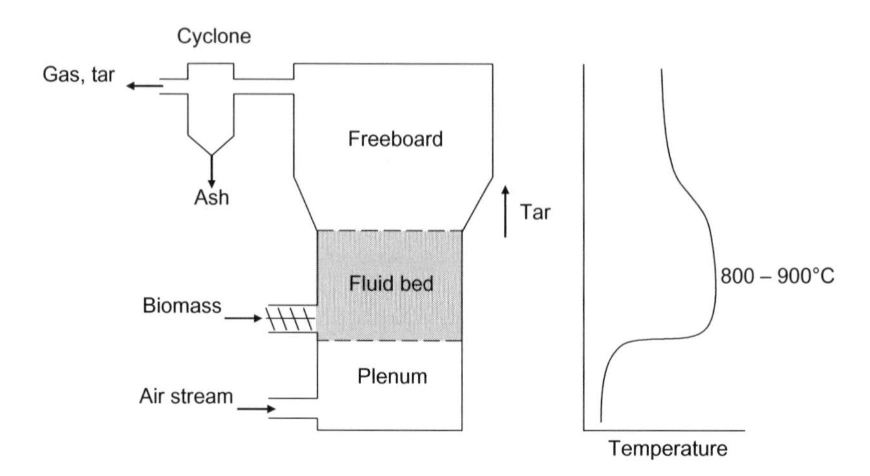


**FIGURE 6.6** Bubbling fluidized-bed gasifier. Here tar is not produced at any specific location. As such it passes through average temperature zone of the fluidized-bed gasifier.

Though the solids are back-mixed, the gases flow upward in plug-flow mode. This means that further up in the bed neither older char particles nor fresh pyrolyzing biomass particles come in contact with the oxygen. Any tar released moves up in the bed and leaves along with the product gas. For this reason, tar generation in a fluidized-bed gasifier is between the two extremes represented by updraft and downdraft gasifiers, averaging about 10 mg/Nm$^3$.

### Entrained-Flow Gasifier

Tar production is negligible, as whatever is released passes through a very-high-temperature (>1000°C) combustion zone and is therefore nearly all converted into gases.

### *6.3.1.5 Design Modifications for Tar Removal*

Modification of a reactor design for tar removal involves the following:

- Secondary air injection
- Separation of the pyrolysis zone from the char gasification zone
- Passage of pyrolysis products through the char.

We saw earlier that char is effective in aiding tar decomposition. A moving-bed two-stage gasifier that uses the first stage for pyrolysis and the second stage for conversion of tar in a bed of char succeeds in reducing the tar by 40 times (Bui et al., 1994). Air addition in the second stage increases the temperature and thereby reduces the tar (Knoef, 2005, p. 170).

A large commercial unit (70 MW fuel power) uses this concept, where biomass dries and pyrolyzes in a horizontal moving bed, heated by waste heat from a diesel engine. The tar concentration of the product gas is about 50 g/Nm$^3$. This gas passes through the neck of a vertical chamber, where injection of preheated gas raises the temperature above 1100°C, reducing the tar amounts to 0.5 g/Nm$^3$. It then passes through a fixed bed of char or carbon being gasified. Tar in the gas leaving the gasifier is very low (<0.025 g/Nm$^3$). It is further cleaned to 0.005 g/Nm$^3$ in a bag filter (Knoef, 2005, p. 159).

Another design involves twin fluidized beds. Biomass fed into the first bed is pyrolyzed. The pyrolysis product can be burnt to provide heat. The char then travels to a parallel fast fluidized-bed combustor that burns part of it. A commercial unit (8 MW fuel power) operates on this principle, where gas leaving the gasifier contains 1.5–4.5 g/Nm$^3$ tar. A fabric filter that separates dust and some tar reduces its concentration to 0.75 g/Nm$^3$, which is finally reduced to 0.010–0.04 g/Nm$^3$ in a scrubber.

### 6.3.2 Postgasification—Secondary Reduction of Tar

As indicated earlier, the level of cleaning needed for the product gas depends greatly on its end use. For example, combustion in an engine or a gas turbine needs a substantially cleaner product gas than that required by a boiler. Most commercial plants use particulate filters or scrubbers to attain the required level of cleanliness. A substantial amount of tar can be removed from the gas in a postgasification cleanup section. It can be either catalytically converted into useful gases like hydrogen or simply captured and scrubbed away. The two basic postgasification methods are physical removal and cracking (catalytic or thermal).

#### *6.3.2.1 Physical Tar Removal*

Physical cleaning is similar to the removal of dust particles from a gas. It requires the tar to be condensed before separation. Tar removal by this means could typically vary between 20% and 97% (Han and Kim, 2008). Table 6.6 shows some typical values of extent of tar removal by different methods.

The energy content of the tar is often lost in this process such that it remains as mist or drops on suspended particles in gas. Physical tar removal can be accomplished by cyclones, barrier filters, wet electrostatic precipitators (ESP), wet scrubbers, or alkali salts. The choice depends on the following:

- Inlet concentration of particulate and tar
- Inlet particle size distribution
- Particulate tolerance of the downstream application of the gas.

**TABLE 6.6** Range of Tar Removal by Different Physical Separation Methods

| Physical Methods | Tar Removal (%) | References |
|---|---|---|
| Sand bed filter | 50–97 | Hasler (1999) |
| Venturi scrubber | 50–90 | Han and Kim (2008) |
| Rotational particle separator | 30–70 | Han and Kim (2008) |
| Wash tower | 10–25 | Han and Kim (2008) |
| Wet ESP | 50–70 | Paasen (2004) |
| Fabric filter | 0–50 | Han and Kim (2008) |
| Catalytic tar cracker | >95 | Hasler (1999) |

The size distribution of the inlet particulates is difficult to measure, especially for finer particulates, but its measurement is important in choosing the right collection devices. For example, submicron ($<1$ μm) particulates need a wet ESPs, but this device is significantly more expensive than others. A fabric filter may work for fines, but it may fail if there is any chance of condensation.

## Cyclones

Cyclones are not very effective for tar removal because of the tar's stickiness and because cyclones cannot remove small ($<1$ μm) tar droplets (Knoef, 2005, p. 196). It is however effective in removing particulates from the product gas.

## Barrier Filters

Barrier filters present a physical barrier in the path of tar and particulates while allowing the clean gas to pass through. One of their special features is that they allow coating of their surface with appropriate catalytic agents to facilitate tar cracking. These filters are of two types: candle and fabric.

*Candle filters* are porous, ceramic, or metallic. The porosity of the material is chosen such that the finest particles do not pass through. Particles failing to pass through the filter barrier deposit on the wall (Figure 6.7), forming a porous layer of solids called a "filter cake." Gas passes through the porous layer as well as through the filter. One major problem with the filter cake is that as it grows in thickness, the pressure drop across the filter increases. Thus, provision is made for its occasional removal. A popular means of removal is pressure pulse in opposite directions.

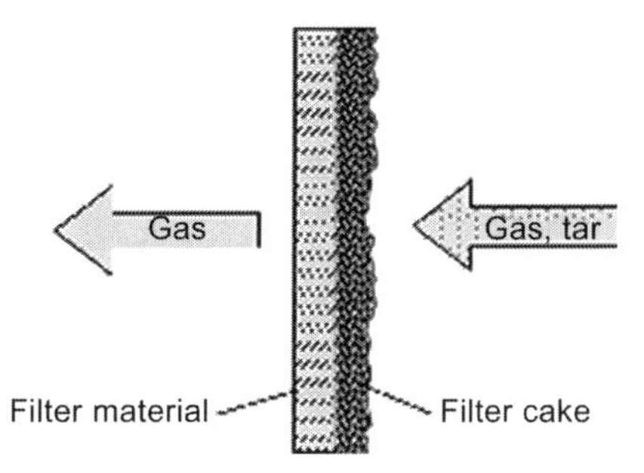


**FIGURE 6.7** Mechanism of separation of dust in a barrier filter.

Besides their high-pressure drop, barrier filters also suffer from the problem that if a filter is broken or cracked, dust and tar-laden gas preferentially flow through that passage, adversely affecting downstream equipment. The condensation of tar on the filter elements can block the filter, and this is a major concern. Ceramic filters can be designed to operate in temperatures as high as 800–900°C.

*Fabric filters* are made of woven fabric as opposed to porous materials as in candle filters. Unlike candle filters, they can operate only in lower temperatures (<350°C). Here, the filter cake is removed by either back-flushing as with a candle filter or shaking. Condensation of tar on the fabric is a problem here if the gas is cooled excessively.

One could use a fabric filter with a precoat, which is removed along with the dust cake formed on the filter. Such precoat can effectively remove undesired substances from the product gas.

## Wet Electrostatic Precipitators

Wet ESPs are used in some gasification plants. The gas is passed through a strong electric field with electrodes. High voltage charges the solid and liquid particles. As the flue gas passes through a chamber containing anode plates or rods with a potential of 30–75 kV, the particles in the flue gas pick up the charge and are collected downstream by positively charged cathode collector plates. Grounded plates or walls also attract the charged particles and are often used for design simplicity. Although collection efficiency does not decrease as particles build up on the plates, periodic mechanical wrapping is required to clean the plates to prevent the impediment of the gas flow or the short-circuiting of the electrodes through the built-up ash.

The collected solid particles are cleaned by mechanical means, but a liquid like tar needs cleaning by a thin film of water. Wet ESPs have very high (>90%) collection efficiency over the entire range of particle size down to about 0.5 μm, and they have very low pressure drop (few inches water gauge). Sparking due to high voltage is a concern with an ESP, especially when it is used to clean highly combustible syngas. Thus, the savings from lower fan power due to low pressure drop is offset by a higher safety cost. Additionally, the capital cost for ESP is 3–4 times higher than that for a wet scrubber.

### Wet Scrubbers

Here, water or an appropriate scrubbing liquid is sprayed on the gas. Solid particles and tar droplets collide with the drops, forming larger droplets because of coalescence. Such larger droplets are easily separated from the gas by a demister like cyclone. The gas needs to be cooled until it is below 100°C before cleaning. The tar-laden scrubbing liquid may be fed back into the gasifier or its combustion section. Alternatively, stripping the tar away may regenerate the scrubbing liquid.

Some commercial methods, such as the OLGA and TARWTC technologies, use proprietary oil as the scrubbing liquid. The tar-laden liquid is then reinjected into the gasifier for further conversion (Knoef, 2005, p. 196). Figure 6.8 shows a schematic of the OLGA process.

Wet scrubbers have a high (>90%) collection efficiency, but the efficiency drops sharply below 1μm-sized particles. They consume much fan power owing to the large (~50 in water gauge) pressure drop across the scrubber. While their operating cost is high, their capital cost is much less than that for ESPs.

A system with a tar removal scrubber produces cleaned gas with a lower outlet temperature and higher energy content, but it contains tars that are

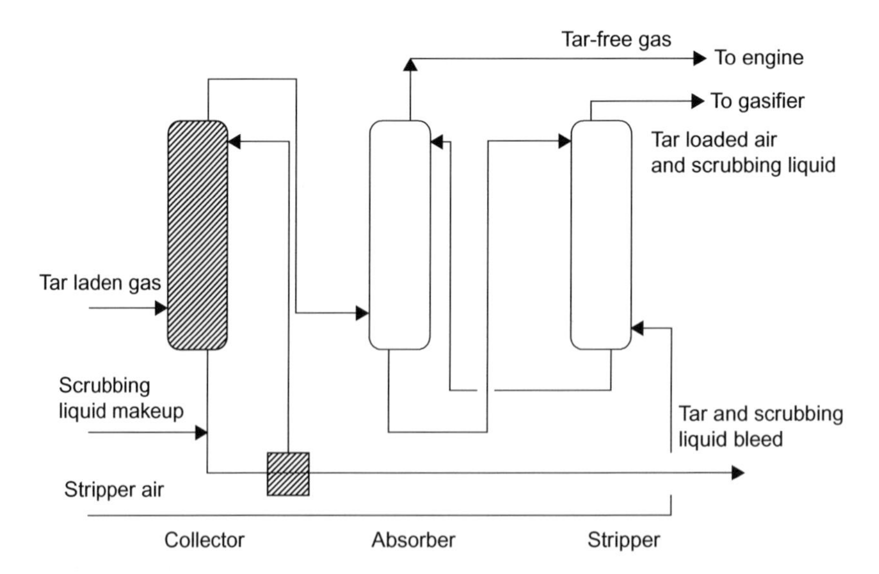


**FIGURE 6.8** Schematic of the OLGA process. Source: *Redrawn from Han and Kim (2008).*

more difficult to remove. The main challenge of tar removal is the formation of "tar balls," which are long-chained hydrocarbons that have a tendency to agglomerate and stick together, fouling equipment in the initial stages of tar condensing, and collecting.

The tar-laden stripper gas, if fed into the gasifier, lowers its dew point well below that of water. This allows condensation of the tar, while flue gas containing tar vapor can be recycled back to the combustion section of the gasifier for combustion.

### Alkali Remover

Compared to fossil fuels, biomass is rich in alkali salts that typically vaporize at high gasifier temperatures but condense downstream below 600°C. Because condensation of alkali salts causes serious corrosion problems, efforts are made to strip the gas of alkali. If the gas can be cooled to below 600°C, the alkali will condense onto fine solid particles ($<5\ \mu m$) that can be captured in a cyclone, ESPs, or filters. Some applications do not permit cooling of the gas. In such cases, the hot gas may be passed through a bed of active bauxite maintained at 650–725°C.

### Disposal of Collected Tar

Tar removal processes produce liquid wastes with higher concentration of organic compound, which increase the complexity of water treatment. Wastewater contaminants include dissolved organics, inorganic acids, $NH_3$, and metals. Collected tars are classified as hazardous waste, especially if they are formed at high temperatures (Stevens, 2001). Several technologies are available for treatment of these contaminants before their final disposal. Some examples include extraction with organic solvent, distillation, adsorption on activated carbon, wet oxidation, oxidation with hydrogen peroxide ($H_2O_2$), oxidation with ozone ($O_3$), incineration, and biological treatment.

## *6.3.2.2 Cracking*

Postgasification cracking could break large molecules of tar into smaller molecules of permanent gases such as $H_2$ or CO. The energy content of the tar is thus mostly recovered through the smaller molecules formed. Unlike in physical cleaning, the tar need not be condensed for cracking. This process involves heating the tar to a high temperature (~1200°C) or exposing it to catalysts at lower temperatures (~800°C). There are two major types of cracking: thermal and catalytic.

1. Thermal cracking without a catalyst is possible at a high temperature (~1200°C). The temperature requirement depends on the constituents of the tar. For example, oxygenated tars may crack at around 900°C (Stevens, 2001). Oxygen or air may be added to allow partial combustion

of the tar to raise its temperature, which is favorable for thermal cracking. Thermal decomposition of biomass tars in electric arc plasma is another option. This is a relatively simple process, but it produces gas with a lower energy content.

2. Catalytic cracking is commercially used in many plants for the removal of tar and other undesired elements from product gas. It generally involves passing the dirty gas over catalysts. The main chemical reactions taking place in a catalytic reactor are represented by Eq. (6.5) in the presence of steam (steam reforming) and Eq. (6.6) in the presence of $CO_2$ (dry reforming). The main reactions for tar conversion are endothermic in nature. So, a certain amount of combustion reaction is allowed in the reactor by adding air.

Nonmetallic catalysts include less-expensive disposable catalysts: dolomite, zeolite, calcite, and so on. They can be used as bed materials in a fluidized bed through which tar-laden gas is passed at a temperature of 750–900°C. Attrition and deactivation of the catalyst are a problem (Lammars et al., 1997). A proprietary nonmetallic catalyst, D34, has been used with success in a fluidized bed at 800°C followed by a wet scrubber (Knoef, 2005, p. 153).

Metallic catalysts include Ni, Ni/Mo, Ni/Co/Mo, NiO, Pt, and Ru on supports like silica–alumina and zeolite (Aznar et al., 1997). Some of them are used in the petrochemical industry and are readily available. A blend of Ni/Co/Mo converts $NH_3$ along with tars. Catalysts deactivate during tar cracking and so need reactivation. Typically, the catalysts are placed in a fixed or fluidized bed. Tar-laden gas is passed through at a temperature of 800–900°C.

Dolomite (calcined) and olivine sand are very effective in in situ reduction in tar cracking. This type of catalytic cracking takes place in the typical temperature of fluidized bed. A good improvement in the gas yield and tar reduction is noted when catalytic bed materials were used.

Chapter 7

# Gasification Theory

## 7.1 INTRODUCTION

The design and operation of a gasifier require an understanding of the gasification process, its configuration, size, feedstock, and operating parameters influence on the performance of the plant. A good comprehension of the basic reactions is fundamental to the planning, design, operation, troubleshooting, and process improvement of a gasification plant, as is learning the alphabet to read a book. This chapter introduces the basics of the gasification process through a discussion of the reactions involved and the kinetics of the reactions with specific reference to biomass. It also explains how this knowledge can be used to develop a mathematical model of the gasification process.

## 7.2 GASIFICATION REACTIONS AND STEPS

Gasification is the conversion of solid or liquid feedstock into useful and convenient gaseous fuel or chemical feedstock that can be burned to release energy or used for production of value-added chemicals.

Gasification and combustion are two closely related thermochemical processes, but there is an important difference between them. Gasification packs energy into chemical bonds in the product gas; combustion breaks those bonds to release the energy. The gasification process adds hydrogen to and strips carbon away from the hydrocarbon feedstock to produce gases with a higher hydrogen-to-carbon (H/C) ratio, while combustion oxidizes the hydrogen and carbon into water and carbon dioxide, respectively.

A typical biomass gasification process may include the following steps:

- Drying
- Thermal decomposition or pyrolysis
- Partial combustion of some gases, vapors, and char
- Gasification of decomposed products

Pyrolysis as explained in Chapter 5 is a thermal decomposition process that occurs in absence of any medium. Gasification, on the other hand, requires a gasifying medium like steam, air, or oxygen to rearrange the molecular structure of the feedstock in order to convert the solid feedstock

**Biomass Gasification, Pyrolysis and Torrefaction.**

into gases or liquids; it can also add hydrogen to the product. The use of a medium is essential for the gasification process, which is not the case for pyrolysis or torrefaction.

### 7.2.1 Gasifying Medium

Gasifying medium (also called "agent") reacts with solid carbon and heavier hydrocarbons to convert them into low-molecular-weight gases like CO and $H_2$. The main gasifying agents used for gasification are as follows:

- Oxygen
- Steam
- Air

Oxygen is a popular gasifying medium though it is primarily used for the combustion or the partial gasification in a gasifier. It may be supplied to a gasifier either in pure form or via air. The heating value and the composition of the gas produced in a gasifier are strong functions of the nature and amount of the gasifying agent used. A ternary diagram (Figure 3.12) of carbon, hydrogen, and oxygen demonstrates the conversion paths toward the formation of different products in a gasifier.

If oxygen is used as the gasifying agent, the conversion path moves toward the oxygen corner. Its products include CO for low amount of oxygen and $CO_2$ for high oxygen. When the amount of oxygen exceeds a certain (stoichiometric) amount, the process moves from gasification to combustion, and the product is "*flue gas*" instead of "*fuel gas*." The flue gas or the combustion product contains no residual heating value. A move toward the oxygen corner of the ternary diagram in a gasification process (Figure 3.12) results in low hydrogen content and an increase in carbon-based compounds such as CO and $CO_2$ in the product gas.

If steam is used as the gasification agent, the process moves upward toward the hydrogen corner in Figure 3.12. Then, the product gas contains more hydrogen per unit of carbon, resulting in a higher H/C ratio.

The choice of gasifying agent affects the heating value of the product gas as well. [For example, if air is used instead of oxygen, the nitrogen in it would dilute the product reducing the heating value of the product gas.] From Table 7.1, we can see that oxygen gasification has the highest heating value followed by steam and air gasification. Air, as the gasification medium, results in the lowest heating value in the product gas primarily due to the dilution effect of nitrogen.

## 7.3 THE GASIFICATION PROCESS

A typical gasification process generally follows the sequence of steps listed below (illustrated schematically in Figure 7.1).

**TABLE 7.1 Heating Values for Product Gas Based on Gasifying Medium**

| Medium | Heating Value ($MJ/Nm^3$) |
|---|---|
| Air | 4–7 |
| Steam | 10–18 |
| Oxygen | 12–28 |

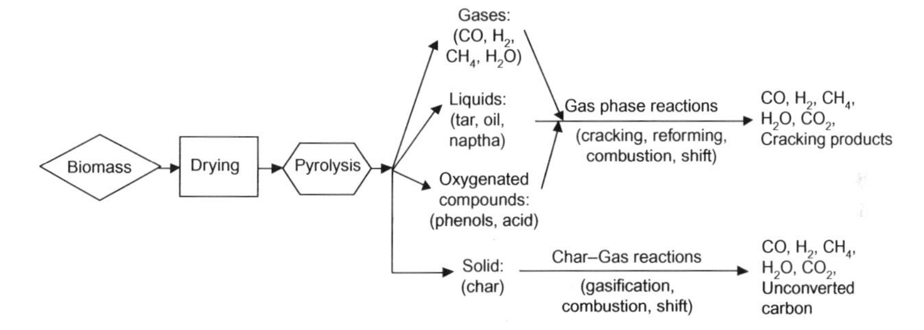


**FIGURE 7.1** Reaction sequence and potential paths for gasification.

- Preheating and drying
- Pyrolysis and or combustion
- Char gasification

Though these steps are frequently modeled in series, there is no sharp boundary between them, and they often overlap. The following paragraphs discuss the sequential phases of biomass gasification.

In a typical process, biomass is first heated (dried) and then it undergoes thermal degradation or pyrolysis. The products of pyrolysis (i.e., gas, solid, and liquid) react among themselves as well as with the gasifying medium to form the final gasification product. In most commercial gasifiers, the thermal energy necessary for drying, pyrolysis, and endothermic reactions comes from a certain amount of exothermic combustion reactions allowed in the gasifier. Table 7.2 lists some of the important chemical reactions taking place in a gasifier.

### 7.3.1 Drying

The typical moisture content of freshly cut woods ranges from 30% to 60%, and for some biomass, it can exceed 90% (see Table 3.11). Every kilogram of moisture in the biomass takes away a minimum of about 2242 kJ of extra energy from the gasifier to vaporize water, and that energy is not recoverable. For a high level of moisture, this loss is a concern, especially for

**TABLE 7.2 Typical Gasification Reactions at 25°C**

| Reaction Type | Reaction |
|---|---|
| **Carbon Reactions** | |
| R1 (Boudouard) | $C + CO_2 \leftrightarrow 2CO + 172$ kJ/mol[a] |
| R2 (water–gas or steam) | $C + H_2O \leftrightarrow CO + H_2 + 131$ kJ/mol[b] |
| R3 (hydrogasification) | $C + 2H_2 \leftrightarrow CH_4 - 74.8$ kJ/mol[b] |
| R4 | $C + 0.5\ O_2 \rightarrow CO - 111$ kJ/mol[a] |
| **Oxidation Reactions** | |
| R5 | $C + O_2 \rightarrow CO_2 - 394$ kJ/mol[b] |
| R6 | $CO + 0.5O_2 \rightarrow CO_2 - 284$ kJ/mol[c] |
| R7 | $CH_4 + 2O_2 \leftrightarrow CO_2 + 2H_2O - 803$ kJ/mol[d] |
| R8 | $H_2 + 0.5\ O_2 \rightarrow H_2O - 242$ kJ/mol[c] |
| **Shift Reaction** | |
| R9 | $CO + H_2O \leftrightarrow CO_2 + H_2 - 41.2$ kJ/mol[c] |
| **Methanation Reactions** | |
| R10 | $2CO + 2H_2 \rightarrow CH_4 + CO_2 - 247$ kJ/mol[c] |
| R11 | $CO + 3H_2 \leftrightarrow CH_4 + H_2O - 206$ kJ/mol[c] |
| R14 | $CO_2 + 4H_2 \rightarrow CH_4 + 2H_2O - 165$ kJ/mol[b] |
| **Steam-Reforming Reactions** | |
| R12 | $CH_4 + H_2O \leftrightarrow CO + 3H_2 + 206$ kJ/mol[d] |
| R13 | $CH_4 + 0.5O_2 \rightarrow CO + 2H_2 - 36$ kJ/mol[d] |

[a]*Higman and van der Burgt (2008), p. 12.*
[b]*Klass (1998), p. 276.*
[c]*Knoef (2005), p. 15.*
[d]*Higman and van der Burgt (2008), p. 3.*

energy applications. While we cannot do much about the inherent moisture residing within the cell structure, efforts may be made to drive away the external or surface moisture. A certain amount of predrying is thus necessary to remove as much moisture from the biomass as possible before it is fed into the gasifier. For the production of a fuel gas with a reasonably high heating value, most gasification systems use dry biomass with a moisture content of 10–20%.

The final drying takes place after the feed enters the gasifier, where it receives heat from the hot zone downstream. This heat preheats the feed and evaporates the moisture in it. Above 100°C, the loosely bound water that is in the biomass is irreversibly removed. As the temperature rises further, the low-molecular-weight extractives start volatilizing. This process continues until a temperature of approximately 200°C is reached.

### 7.3.2 Pyrolysis

In pyrolysis, no external agent is needed. As per the ternary diagram (Figure 3.12), a slow pyrolysis or torrefaction process moves the solid product toward the carbon corner, and thus more char is formed. The fast pyrolysis process, on the other hand, moves the product toward the C–H axis opposite to the oxygen corner (Figure 3.12). The oxygen is thereby largely diminished producing more liquid hydrocarbon.

As detailed in Chapter 5, pyrolysis involves the thermal breakdown of larger hydrocarbon molecules of biomass into smaller gas molecules (condensable and noncondensable) with no major chemical reaction with air, gas, or any other gasifying medium. This reaction generally precedes the gasification step.

One important product of pyrolysis is tar formed through condensation of the condensable vapor produced in the process. Being a sticky liquid, tar creates a great deal of difficulty in industrial use of the gasification product. A discussion of tar formation and ways of cracking or reforming it into useful noncondensable gases is presented in Chapter 6.

### 7.3.3 Char Gasification Reactions

The gasification step involves chemical reactions among the hydrocarbons in fuel, steam, carbon dioxide, oxygen, and hydrogen in the reactor, as well as chemical reactions among the evolved gases. Of these, char gasification is the most important. The biomass char produced through pyrolysis of biomass is not necessarily pure carbon. It contains a certain amount of hydrocarbon comprising hydrogen and oxygen.

Biomass char is generally more porous and reactive than coke produced through high temperature carbonization of coal. The porosity of biomass char is in the range of 40–50% while that of coal char is 2–18%. The pores of biomass char are much larger (20–30 μm) than those of coal char (~5 Å) (Encinar et al., 2001). Thus, its reaction behavior is different from that of chars derived from coal, lignite, or peat. For example, the reactivity of peat char decreases with conversion or time, while the reactivity of biomass char increases with conversion (Figure 7.2). This reverse trend can be attributed to the increasing catalytic activity of the biomass char's alkali metal constituents (Risnes et al., 2001).

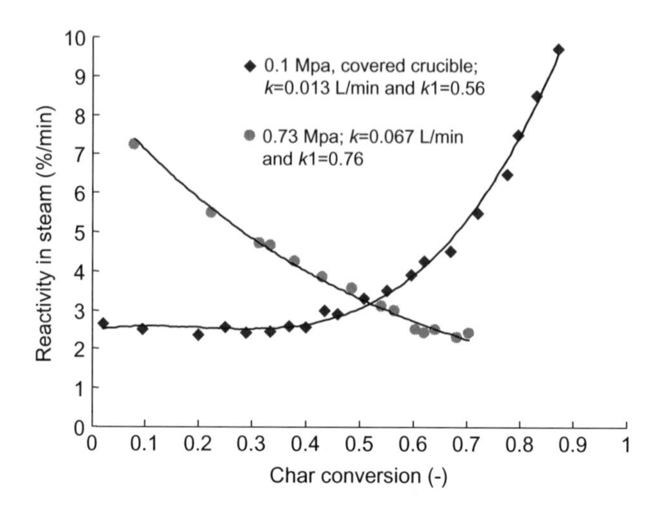


**FIGURE 7.2** Reactivities of peat char for gasification in steam decreases with conversion while that from hardwood increases with conversion. Source*: Data taken from Liliedahl and Sjostrom (1997).*

Gasification of biomass char involves several reactions between the char and the gasifying medium. The following is a description of some of those reactions with carbon, carbon dioxide, hydrogen, steam, and methane.

$$\text{Char} + O_2 \rightarrow CO_2 \text{ and } CO \tag{7.1}$$

$$\text{Char} + CO_2 \rightarrow CO \tag{7.2}$$

$$\text{Char} + H_2O \rightarrow CH_4 \text{ and } CO \tag{7.3}$$

$$\text{Char} + H_2 \rightarrow CH_4 \tag{7.4}$$

Equations (7.1)–(7.4) show how gasifying agents like oxygen, carbon dioxide, and steam react with solid carbon to convert it into low-molecular-weight gases like carbon monoxide and hydrogen. Some of the reactions are known by the popular names as listed in Table 7.2.

Gasification reactions are generally endothermic in nature, but some of them can be exothermic as well. For example, those of carbon with oxygen and hydrogen (R3, R4, and R5 in Table 7.2) are exothermic, whereas those with carbon dioxide and steam (reactions R1 and R2) are endothermic. The heat of reaction given in Table 7.2 for various reactions refers to a temperature of 25°C.

### *7.3.3.1 Speed of Char Reactions*

Chemical reactions take place at finite rates. The rate of gasification reactions of char (comprising of mainly carbon) depends primarily on its reactivity and the reaction potential of the gasifying medium. For example, amongst gasification medium, oxygen is most active, followed by steam and carbon dioxide. The rate of the char–oxygen reaction (R4: $C + 0.5O_2 \rightarrow CO$) is the fastest among the four reactions listed in Table 7.2 (R1, R2, R3, and R4). It is so fast that the reaction quickly consumes the entire oxygen, leaving hardly any for any other reactions.

The rate of the char–steam reaction (R2: $C + H_2O \rightarrow CO + H_2$) is three to five orders of magnitude slower than that of the char–oxygen reaction. The char–carbon dioxide reaction (R1: $C + CO_2 \rightarrow 2CO$), known as *Boudouard reaction*, is six to seven orders of magnitude slower (Smoot and Smith, 1985). The rate of water–steam gasification reaction (R2), known as *water–gas reaction*, is about two to five times faster than that of the Boudouard reaction (R1) (Blasi, 2009).

The char–hydrogen reaction (*hydrogasification reaction*) that forms methane ($C + 2H_2 \rightarrow CH_4$) is the slowest of all. Walker et al. (1959) estimated the relative rates of the above four reactions at a temperature of 800°C and one at a pressure of 0 kPa, as $10^5$ for oxygen, $10^3$ for steam, $10^1$ for carbon dioxide, and $3 \times 10^{-3}$ for hydrogen. The relative rates, $R$, may be shown as:

$$R_{C+O_2} \gg R_{C+H_2O} > R_{C+CO_2} \gg R_{C+H_2} \quad (7.5)$$

When steam reacts with carbon, it can produce CO and $H_2$. Under certain conditions, the steam and carbon reaction can also produce $CH_4$ and $CO_2$.

### 7.3.3.2 Boudouard Reaction

The gasification of char in carbon dioxide is popularly known as the *Boudouard reaction.*

$$C + CO_2 \leftrightarrow 2CO \text{ (reaction R1 in Table 7.2)} \quad (7.6)$$

Blasi (2009) described the Boudouard reaction through the following intermediate steps. In the first step, $CO_2$ dissociates at a carbon-free active site ($C_{fas}$), releasing carbon monoxide and forming a carbon–oxygen surface complex, C(O). This reaction being reversible can move in the opposite direction as well, forming a carbon active site and $CO_2$ in the second step. In the third step, the carbon–oxygen complex produces a molecule of CO.

$$\text{Step 1:} \quad C_{fas} + CO_2 \xrightarrow{k_{b1}} C(O) + CO \quad (7.7)$$

$$\text{Step 2:} \quad C(O) + CO \xrightarrow{k_{b2}} C_{fas} + CO_2 \quad (7.8)$$

$$\text{Step 3:} \quad C(O) \xrightarrow{k_{b3}} CO \quad (7.9)$$

where $k_i$ is the rate of the $i$th reaction.

The rate of the char gasification reaction in $CO_2$ is insignificant below 1000 K.

#### 7.3.3.3 Water–Gas Reaction

The gasification of char in steam, known as the *water–gas reaction*, is perhaps the most important gasification reaction.

$$C + H_2O \leftrightarrow CO + H_2 \text{ (R2 in Table 7.2)} \tag{7.10}$$

The first step involves the dissociation of $H_2O$ on a free active site of carbon ($C_{fas}$), releasing hydrogen and forming a surface oxide complex of carbon C(O). In the second and third steps, the surface oxide complex produces a new free active site and a molecule of CO.

$$\text{Step 1:} \quad C_{fas} + H_2O \xrightarrow{k_{w1}} C(O) + H_2 \tag{7.11}$$

$$\text{Step 2:} \quad C(O) + H_2 \xrightarrow{k_{w2}} C_{fas} + H_2O \tag{7.12}$$

$$\text{Step 3:} \quad C(O) \xrightarrow{k_{w3}} CO \tag{7.13}$$

Some models (Blasi, 2009) also include the possibility of hydrogen inhibition by C(H) or $C(H)_2$ complexes as below:

$$C_{fas} + H_2 \leftrightarrow C(H)_2 \tag{7.14}$$

$$C_{fas} + 0.5H_2 \leftrightarrow C(H) \tag{7.15}$$

The presence of hydrogen has a strong inhibiting effect on the char gasification rate in $H_2O$. For example, 30% hydrogen in the gasification atmosphere can reduce the gasification rate by a factor as high as 15 (Barrio et al., 2001). So an effective means of accelerating the water–gas reaction is continuous removal of hydrogen from the reaction site.

#### 7.3.3.4 Shift Reaction

Unlike the above reactions, shift reaction takes place between steam and an intermediate product of the gasification reaction. The other difference of this important reaction is that it is a gas-phase reaction. This reaction increases the hydrogen content of the gasification product at the expense of carbon monoxide. Some literature (Klass, 1998, p. 277) refers this reaction also as "water–gas shift reaction," though it is much different from the water–gas reaction (R2).

$$CO + H_2O \leftrightarrow CO_2 + H_2 \quad -41.2\ \text{kJ/mol (reaction R9 in Table 5.2)} \tag{7.16}$$

This is a prestep in syngas production in the downstream of a gasifier, where the ratio of hydrogen and carbon monoxide in the product gas is critical.

The shift reaction is slightly exothermic, and its equilibrium yield decreases slowly with temperature. Depending on temperature, it may be driven in either direction, that is, products or reactants. However, it is not sensitive to pressure (Petersen and Werther, 2005).

Above 1000°C, the shift reaction (R9) rapidly reaches equilibrium, but at a lower temperature, it needs heterogeneous catalysts (Figure 7.3). Probstein and Hicks (2006, p. 63) showed that this reaction has a higher equilibrium constant at a lower temperature, which implies a higher yield of $H_2$ at a lower temperature. With increasing temperature, the yield decreases, but the reaction rate increases. Optimum yield is obtained at about 225°C.

Because the reaction rate at such a low temperature is low, catalysts like chromium-promoted iron, copper–zinc, and cobalt–molybdenum are needed

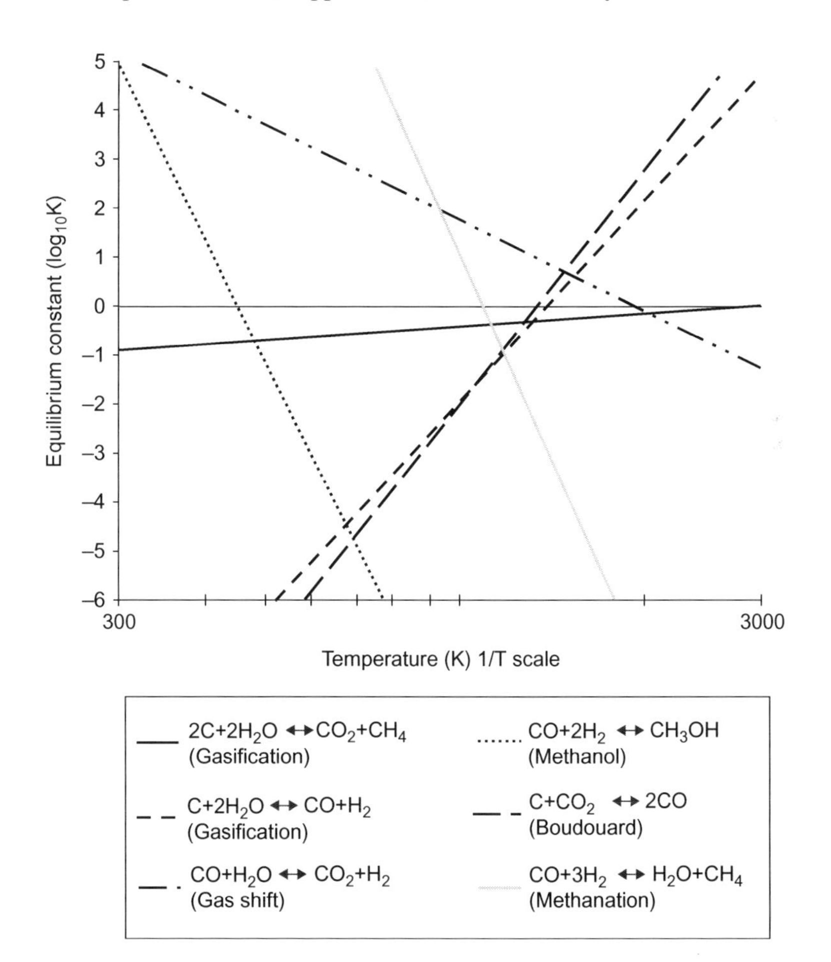


**FIGURE 7.3** Equilibrium constants for selected gasification reaction. Source: *From Probestein and Hicks (2006), p. 63.*

(Probstein and Hicks, 2006, p. 124). At higher temperatures (350–600°C) Fe-based catalysts may be employed. Pressure exerts no appreciable effect on the $H_2/CO$ ratio. Commercial shift conversions of CO uses the following catalysts (Boerrigter and Rauch, 2005):

- Copper-promoted catalyst, at about 300–510°C
- Copper–zinc–aluminum oxide catalyst, at about 180–270°C.

#### 7.3.3.5 Hydrogasification Reaction

This reaction involves the gasification of char in a hydrogen environment, which leads to the production of methane.

$$C + 2H_2 \rightarrow CH_4 \text{ (reaction R3 in Table 5.2)} \quad (7.17)$$

The rate of this reaction is much slower than that of the other reactions, and so it is not discussed here. It is of importance only when the production of synthetic natural gas is desired.

### 7.3.4 Char Combustion Reactions

Most gasification reactions are endothermic. To provide the required heat of reaction as well as that required for heating, drying, and pyrolysis, a certain amount of exothermic combustion reaction is allowed in a gasifier. Reaction R5 ($C + O_2 \rightarrow CO_2$) is the best reaction in this regard as it gives the highest amount of heat (394 kJ) per mole of carbon consumed. The next best is R4 ($C + \frac{1}{2}O_2 \rightarrow CO$), which also produces the fuel gas CO, but produces only 111 kJ/mol of heat. Additionally, the speed of R4 is also relatively slow.

When carbon comes in contact with oxygen, both R4 and R5 can take place, but their extent depends on temperature. A partition coefficient, $\beta$ may be defined to determine how oxygen will partition itself between the two. R4 and R5 may be combined and written as:

$$\beta C + O_2 \rightarrow 2(\beta - 1)CO + (2 - \beta)CO_2 \quad (7.18)$$

The value of the partition coefficient $\beta$ lies between 1 and 2 and depends on temperature. One of the commonly used expressions (Arthur, 1951) for $\beta$ is

$$\beta = \frac{[CO]}{[CO_2]} = 2400e^{-\left(\frac{6234}{T}\right)} \quad (7.19)$$

where $T$ is the surface temperature of the char.

Combustion reactions are generally faster than gasification reactions under similar conditions. Table 7.3 compares the rate of combustion and gasification for a biomass char at a typical gasifier temperature of 900°C. The combustion rates are at least one order of magnitude faster than the gasification reaction rate. Owing to pore diffusion resistance, finer char particles' combustion has a much higher reaction rate.

**TABLE 7.3 Comparison of the Effect of Pore Diffusion on Char Gasification and Combustion Rates**

| Particle Size (μm) | Combustion Rate ($min^{-1}$) | Gasification Rate ($min^{-1}$) | Combustion Rate/ Gasification Rate (−) |
|---|---|---|---|
| 6350 | 0.648 | 0.042 | 15.4 |
| 841 | 5.04 | 0.317 | 15.9 |
| 74 | 55.9 | 0.975 | 57.3 |

**Source:** Adapted from Reed (2002), pp. II−189.

Another important difference between char gasification and combustion reactions in a fluidized bed is that during gasification the temperature of the char particle is nearly the same as the bed temperature because of simultaneous exothermic and endothermic reactions on it (Gomez-Barea et al., 2008). In combustion, the char particle temperature can be much hotter than the bed temperature (Basu, 1977).

The availability of relative amounts of fuel, oxidant (air or oxygen), and steam (if used) govern the fraction of carbon or oxygen that enters R5 or R4 (Table 7.2). The presence of any more oxidant than that needed for the endothermic reaction will increase the gasifier temperature unnecessarily as well as reduce the quality of the product by diluting it with carbon dioxide. Example 7.1 illustrates how the heat balance works out in a gasifier.

**Example 7.1**

In an updraft gasifier, the water–gas gasification reaction ($C + H_2O \rightarrow CO + H_2 + 131$ kJ/mol) is to be carried out. Assume that drying and other losses in the system need 50% additional heat. Find a means to adjust the extent of the combustion reaction by controlling the supply of oxygen and carbon such that this need is met.

**Solution**

The reaction needs 131 kJ of heat for gasification of each mole of carbon. In oxygen-deficient or substoichiometric conditions like that present in a gasifier, the exothermic combustion reaction ($C + ½O_2 \rightarrow CO - 111$ kJ/mol) is more likely to take place than the more complete combustion reaction ($C + O_2 \rightarrow CO_2 - 394$ kJ/mol). If we adjust the feedstock such that for every mole of carbon gasified, only $p$ moles of carbon will be partially oxidized using $p/2$ mol of oxygen, the heat released by the combustion reaction will exactly balance the heat needed by the gasification reaction. In that case the reaction is

$$C + H_2O \rightarrow CO + H_2 + 131 \text{ kJ/mol} \quad \text{(i)}$$

- Heat required for endothermic reaction per mol of C = 131 kJ/mol
- Heat required for drying = 0.5 × 131 = 65.5 kJ
- Total heat required = 131 + 65.5 = 196.5 kJ.

If $p$ moles of carbon participate in the exothermic reaction, R4:

$$pC + 0.5pO_2 \rightarrow pCO - 111p \quad \text{(ii)}$$

Then, we have $111p = 196.5$ or $p = 1.77$.
Adding reactions (i) and (ii), we get the net reaction:

$$2.77C + H_2O + 0.88O_2 \rightarrow 1.77CO + H_2$$

Thus, for (2.77 × 12) kg of carbon, we need (2 + 16) kg of steam and (0.88 × 32) kg of oxygen. If we add more oxygen, the combustion reaction, R5, may take place and the temperature of the combustion zone may rise further.

### 7.3.5 Catalytic Gasification

Use of catalysts in the thermochemical conversion of biomass may not be essential, but it can help under certain circumstances. Two main motivations for catalyst's use are as follows:

1. Removal of tar from the product gas, especially if the downstream application or the installed equipment cannot tolerate it (see Chapter 6 for more details).
2. Reduction in methane content of the product gas, particularly when it is to be used as syngas (CO, $H_2$ mixture).

The development of catalytic gasification is driven by the need for tar reforming. When the product gas passes over the catalyst particles, the tar or condensable hydrocarbon can be reformed on the catalyst surface with either steam or carbon dioxide, thus producing additional hydrogen and carbon monoxide. The reactions may be written in simple form as:

Steam reforming reaction:

$$C_nH_m + nH_2O \xrightarrow{\text{catalyst}} (n + m/2)H_2 + nCO \quad (7.20)$$

Carbon dioxide (or dry) reforming reaction:

$$C_nH_m + nCO_2 \xrightarrow{\text{catalyst}} 2nCO + (m/2)H_2 \quad (7.21)$$

As we can see, instead of undesirable tar or soot, we get additional fuel gases through the catalytic tar-reforming reactions (Eq. (7.20)). Both gas yield and the heating value of the product gas improve.

The other option for tar removal is thermal cracking, but it requires high (>1100°C) temperature and produces soot; thus, it cannot harness the lost energy in tar hydrocarbon.

The second motivation for catalytic gasification is removal of methane from the product gas. For this, we can use either catalytic steam reforming or catalytic carbon dioxide reforming of methane. Reforming is very important for the production of syngas, which cannot tolerate methane and requires a precise ratio of CO and $H_2$ in the product gas. In steam reforming, methane reacts with steam at a temperature of 700–1100°C in the presence of a metal-based catalyst, and thus it is reformed into CO and $H_2$ (Li et al., 2007):

$$CH_4 + H_2O \xrightarrow{\text{catalyst}} CO + 3H_2 + 206\ \text{kJ/mol}\text{—steam reforming of methane} \tag{7.22}$$

This reaction is widely used in hydrogen production from methane, for which nickel-based catalysts are very effective.

The carbon dioxide reforming of methane is not as widely used commercially as steam reforming, but it has the special attraction of reducing two greenhouse gases ($CO_2$ and $CH_4$) in one reaction, and it can be a good option for removal of carbon dioxide from the product gas. The reaction is highly endothermic (Wang and Lu, 1996):

$$CH_4 + CO_2 \xrightarrow{\text{catalyst}} 2CO + 2H_2 + 247\ \text{kJ/mol}\text{—dry reforming of methane} \tag{7.23}$$

Nickel-based catalysts are also effective for the dry-reforming reaction (Liu et al., 2008).

#### *7.3.5.1 Catalyst Selection*

Catalysts for reforming reactions are to be chosen keeping in view their objective and practical use. Some important catalyst selection criteria for the removal of tar are as follows:

- Effectiveness
- Resistance to deactivation by carbon fouling and sintering
- Easily regenerated
- Strong and resistant to attrition
- Inexpensive

For methane removal, the following criteria are to be met in addition to those in the previous list:

- Capable of reforming methane
- Must provide the required $CO/H_2$ ratio for the syngas process

Catalysts can work in both in situ and postgasification reactions. The former may involve impregnating the catalyst in the biomass prior to gasification. It can be added directly in the reactor, as in a fluidized bed. Such application is effective in reducing the tar, but it is not effective in reducing methane (Sutton et al., 2001). In postgasification reactions, catalysts are placed in a secondary reactor downstream of the gasifier to convert the tar and methane formed. This has the additional advantage of being independent of the gasifier operating condition. The second reactor can be operated at temperatures optimum for the reforming reaction.

The catalysts in biomass gasification are divided into three groups:

1. *Earth metal catalysts*: Dolomite ($CaCO_3 \cdot MgCO_3$) is very effective for disposal of tar, and it is inexpensive and widely available, obviating the need for catalyst regeneration. It can be used as a primary catalyst by mixing it with the biomass or as a secondary catalyst in a reformer downstream, which is also called a *guard bed*. Calcined dolomite is significantly more effective than raw dolomite (Sutton et al., 2001). Neither, however, is very useful for methane conversion. The rate of the reforming reaction is higher with carbon dioxide than with steam.
1. *Alkali metal catalysts*: Potassium carbonate and sodium carbonate are important in biomass gasification as primary catalysts. $K_2CO_3$ is more effective than $Na_2CO_3$. Unlike dolomite, they can reduce methane in the product gas through a reforming reaction. Many biomass types have inherent potassium in their ash, so they can benefit from the catalytic action of the potassium with reduced tar production. However, potassium is notorious for agglomerating in fluidized beds, which offsets its catalytic benefit.
1. *Ni-based catalyst*: Nickel is highly effective as a reforming catalyst for reduction of tar as well as for adjustment of the $CO/H_2$ ratio through methane conversion. It performs the best when used downstream of the gasifier in a secondary bed, typically at 780°C (Sutton et al., 2001). Deactivation of the catalyst with carbon deposits is an issue. Nickel is relatively inexpensive and commercially available, though not as cheap as dolomite. Appropriate catalyst support is important for optimum performance.

### 7.3.6 Gasification Processes in Reactors

The sequence of gasification reactions depends to some extent on the type of gas–solid contacting reactors used. A brief description of this process as it occurs in some principal reactor types are discussed in the following sections.

#### 7.3.6.1 Moving-Bed Reactor

To explain the reaction process in moving-bed gasifiers, we take the example of a simple updraft gasifier reactor (Figure 7.4).

In a typical updraft gasifier, fuel is fed from the top; the product gas leaves from the top as well. The gasifying agent (air, oxygen, steam, or their mixture) is preheated and fed into the gasifier through a grid at the bottom.

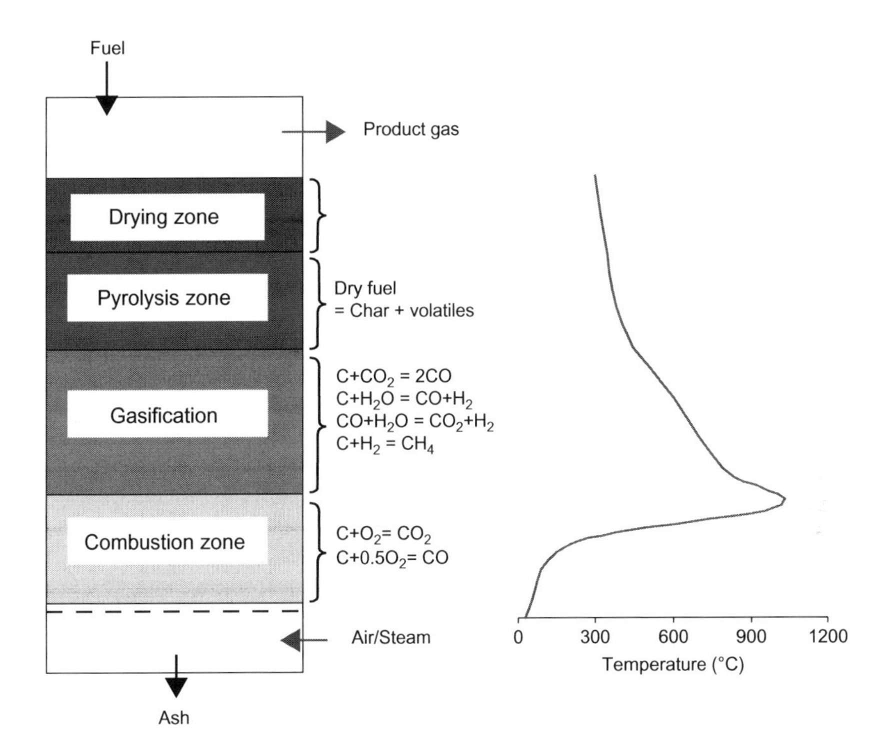


**FIGURE 7.4** Stages of gasification in an updraft gasifier.

The gas then rises through a bed of descending fuel or ash in the gasifier chamber.

The air (the gasifying medium), as it enters the bottom of the bed, meets hot ash and unconverted chars descending from the top (Figure 7.4). The temperature in the bottom layer well exceeds the ignition temperature of carbon, so the highly exothermic combustion reaction (Eq. (7.24)) takes place in the presence of excess oxygen. The released heat heats the upward-moving gas as well as the descending solids.

$$C + O_2 \rightarrow CO_2 - 394 \text{ kJ/mol} \tag{7.24}$$

The combustion reaction (Eq. (7.24)), being very fast, rapidly consumes most of the available oxygen. As the available oxygen is reduced further up, the combustion reaction changes into partial combustion, releasing CO and a moderate amount of heat.

$$C + 1/2O_2 \rightarrow CO - 111 \text{ kJ/mol} \tag{7.25}$$

The hot gas, a mixture of CO, $CO_2$, and steam (from the feed and the gasifying medium), from the combustion zone moves further up into the gasification zone, where char from the upper bed is gasified by Eq. (7.26).

The carbon dioxide concentration in the rising gas increases rapidly in the combustion zone, but once the oxygen is nearly depleted, the $CO_2$ enters the gasification reaction (Eq. (7.26)) with char, resulting in a decline in $CO_2$ concentration in the gasification zone.

$$\begin{aligned} C + CO_2 &\rightarrow 2CO + 172\ \text{kJ/mol} \\ C + H_2O &\rightarrow CO + H_2 + 131\ \text{kJ/mol} \end{aligned} \tag{7.26}$$

Sensible heat of the hot gas provides the heat for the two endothermic gasification reactions R1 and R2 (Table 7.2) in Eq. (7.26). These reactions are responsible for most of the gasification products like hydrogen and carbon monoxide. Because of their endothermic nature, the temperature of the gas reduces.

The zone above the gasification zone is for the pyrolysis of biomass. The residual heat of the rising hot gas heats up the dry biomass, descending from above. The biomass then decomposes (pyrolyzed) into noncondensable gases, condensable gases, and char. Both gases move up while the solid char descends with other solids.

The topmost zone dries the fresh biomass fed into it using the balance enthalpy of the hot product gas coming from the bottom. This gas is a mixture of gasification and pyrolysis products.

In a downdraft gasifier, biomass fed from the top descends, while air injected meets with the pyrolysis product, releasing heat (Figure 7.5). Thereafter, both product gas and solids (char and ash) move down in the downdraft gasifier. Here, a part of the pyrolysis gas may burn above the gasification zone. Thus, the thermal energy required for drying, pyrolysis, and gasification is supplied by the combustion of pyrolysis gas. This phenomenon is called *flaming pyrolysis*.

In downdraft gasifiers, the reaction regions are different from those for updraft gasifiers. Here, steam and oxygen or air is fed into a lower section of the gasifier (Figure 7.5) but biomass is fed at the top. The pyrolysis and combustion products flow downward. The hot gas then moves downward over the remaining hot char, where gasification takes place. Such an arrangement results in tar-free but low-energy-content gases.

#### 7.3.6.2 Fluidized-Bed Reactor

Unlike other types of reactors, a fluidized-bed gasifier contains nonfuel granular solids (bed solids) that act as a heat carrier and mixer. In a bubbling fluidized bed, the fuel fed from either the top or the sides mixes relatively fast over the whole body of the fluid bed (Figure 7.6). The gasifying medium (air, oxygen, steam, or their mixture) also serves as the fluidizing gas and so is sent through the bottom of the reactor.

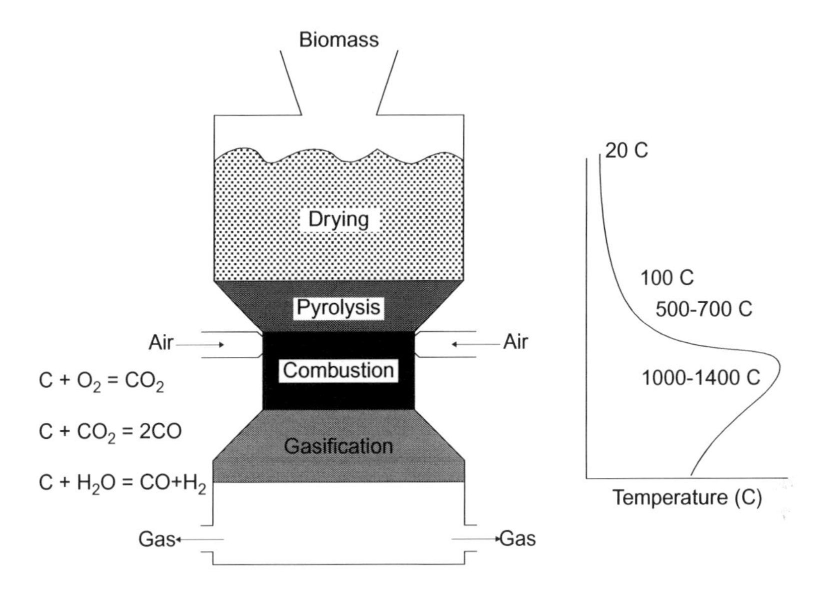


**FIGURE 7.5** Gasification reactions in a downdraft gasifier.

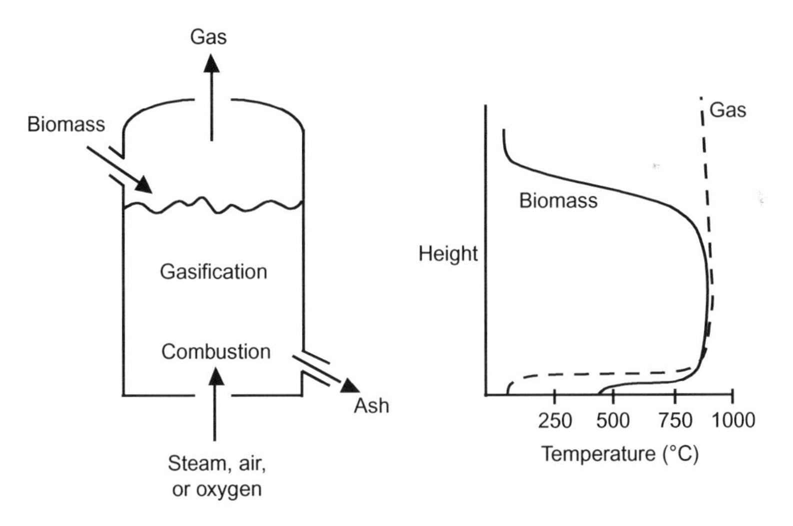


**FIGURE 7.6** Schematic of a bubbling fluidized-bed gasifier (Source: *Higman and Burgt, 2008, p. 106*).

In a typical fluidized-bed gasifier, fresh solid fuel particles are brought into contact with hot bed solids that quickly heat the particles to the bed temperature and make them undergo rapid drying and pyrolysis, producing char and gases.

Though the bed solids are well mixed, the fluidizing gas remains generally in plug-flow mode, entering from the bottom and leaving from the top. Upon entering the bottom of the bed, the oxygen goes into fast exothermic reactions (R4, R5, and R8 in Table 7.2) with char mixed with bed materials. The bed materials immediately disperse the heat released by these reactions to the entire fluidized bed. The amount of heat released near the bottom grid depends on the oxygen content of the fluidizing gas and the amount of char that comes in contact with it. The local temperature in this region depends on how vigorously the bed solids disperse heat from the combustion zone.

Subsequent gasification reactions take place further up as the gas rises. The bubbles of the fluidized bed can serve as the primary conduit to the top. They are relatively solids free. While they help in mixing, the bubbles can also allow gas to bypass the solids without participating in the gasification reactions. The pyrolysis products coming in contact with the hot solids break down into noncondensable gases. If they escape the bed and rise into the cooler freeboard, tar and char are formed.

A bubbling fluidized bed cannot achieve complete char conversion because of the back-mixing of solids. The high degree of solid mixing helps a bubbling fluidized-bed gasifier achieve temperature uniformity, but owing to the intimate mixing of fully gasified and partially gasified fuel particles, any solids leaving the bed contain some partially gasified char. Char particles entrained from a bubbling bed can also contribute to the loss in a gasifier. The other important problem with fluidized-bed gasifiers is the slow diffusion of oxygen from the bubbles to the emulsion phase. This encourages the combustion reaction in the bubble phase, which decreases gasification efficiency.

In a circulating fluidized bed (CFB), solids circulate around a loop that is characterized by intense mixing and longer solid residence time within its solid circulation loop. The absence of any bubbles avoids the gas-bypassing problem of bubbling fluidized beds.

Fluidized-bed gasifiers typically operate in the temperature range of 800–1000°C to avoid ash agglomeration. This is satisfactory for reactive fuels such as biomass, municipal solid waste (MSW), and lignite. Since fluidized-bed gasifiers operate at relatively low temperatures, most high-ash fuels, depending on ash chemistry, can be gasified without the problem of ash sintering and agglomeration.

Owing to the large thermal inertia and vigorous mixing in fluidized-bed gasifiers, a wider range of fuels or a mixture of them can be gasified. This feature is especially attractive for biomass fuels, such as agricultural residues and wood, that may be available for gasification at different times of the year. For these reasons, many developmental activities on large-scale biomass gasification are focused on fluidized-bed technologies.

### 7.3.6.3 Entrained-Flow Reactor

Entrained-flow gasifiers are preferred for the integrated gasification combined cycle plants. Reactors of this type typically operate at 1400°C and 20–70 bar pressure, where powdered fuel is entrained in the gasifying medium. Figure 7.7 shows two entrained-flow gasifier types. In the first one, oxygen, the most common gasifying medium, and the powdered fuel enter from the side; in the second one, they enter from the top.

In entrained-flow gasifiers, the combustion reaction, R5 (Eq. (7.24)), may take place right at the entry point of the oxygen, followed by reaction R4 (Eq. (7.25)) further downstream, where the excess oxygen is used up.

Powdered fuel (<75 μm) is injected into the reactor chamber along with oxygen and steam (air is rarely used). To facilitate feeding into the reactor, especially if it is pressurized, the fuel may be mixed with water to make a

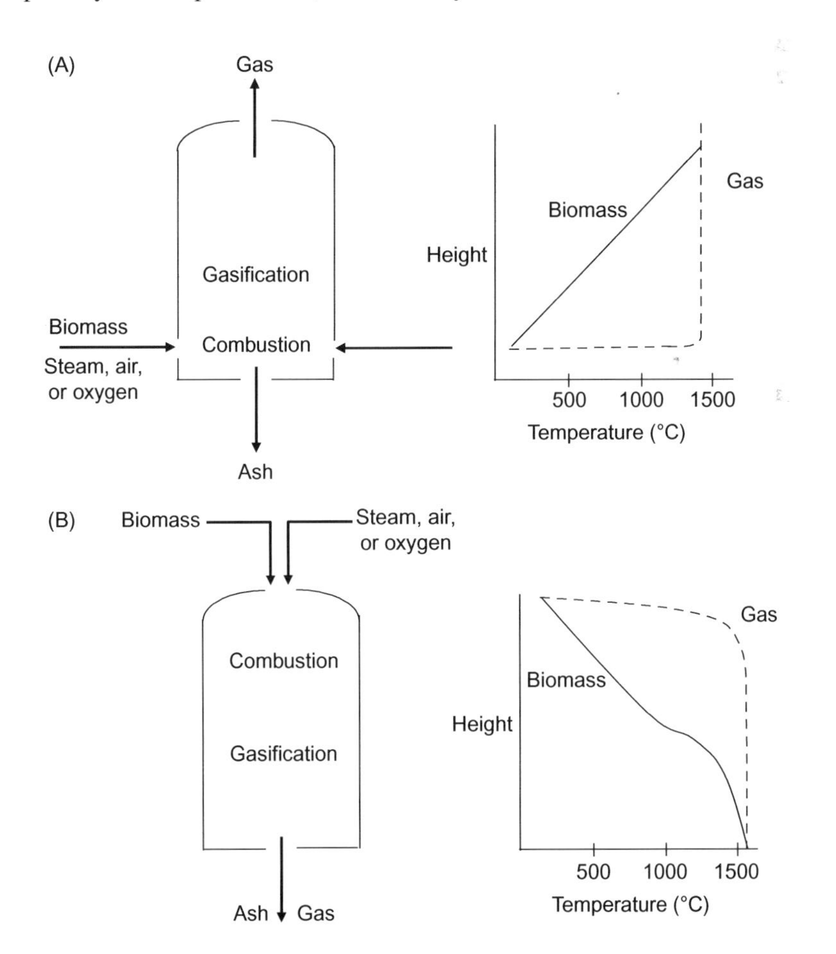


**FIGURE 7.7** Two main types of entrained-flow gasifiers.

slurry. The gas velocity in the reactor is sufficiently high to fully entrain the fuel particles. Slurry-fed gasifiers need additional reactor volume for evaporation of the large amount of water mixed with the fuel. Furthermore, their oxygen consumption is about 20% greater than that of a dry-feed system owing to higher blast requirements (Higman and van der Burgt, 2008).

Entrained-flow gasifiers are of two types depending on how and where the fuel is injected into the reactor. Chapter 8 discusses several types. In all of these designs, oxygen enters the reactor and reacts rapidly with the volatiles and char in exothermic reactions. These raise the reactor temperature well above the melting point of ash, resulting in complete destruction of tar. Such high temperatures should give a very high level of carbon conversion.

An entrained-flow gasifier may be viewed as a plug-flow reactor. Although the gas is heated to the reactor temperature rapidly upon entering, solids heat up less slowly along the reactor length because of the solid's larger thermal capacity and plug-flow nature, as shown in Figure 7.7. Some entrained-flow reactors are modeled as stirred tank reactors because of the rapid mixing of solids.

## 7.4 KINETICS OF GASIFICATION

Stoichiometric calculations (Section 3.6.6) can help determine the products of a completed reaction. Not all reactions in gasifier are instantaneous and completely convert reactants into products. Many of the chemical reactions discussed in the preceding sections proceed at a finite rate and to a finite extent.

To what extent a reaction progresses is determined by its equilibrium state. Its kinetic rates, on the other hand, determine how fast the reaction products are formed and whether the reaction completes within the gasifier chamber. A review of the basics of chemical equilibrium may be useful before discussing its results.

### 7.4.1 Chemical Equilibrium

Let us consider the reaction:

$$nA + mB \xrightarrow{k_{\text{for}}} pC + qD \tag{7.27}$$

where $n$, $m$, $p$, and $q$ are stoichiometric coefficients. The rate of this reaction, $r_1$, depends on $C_A$ and $C_B$, the concentration of the reactants $A$ and $B$, respectively as below:

$$r_1 = k_{\text{for}} C_A^n C_B^m \tag{7.28}$$

The reaction can also move in the opposite direction:

$$pC + qD \xrightarrow{k_{\text{back}}} nA + mB \tag{7.29}$$

The rate of the reverse reaction, $r_2$, is similarly written in terms of $C_C$ and $C_D$, the concentration of $C$ and $D$, respectively:

$$r_2 = k_{\text{back}}\, C_C^p C_D^q \tag{7.30}$$

When the reaction begins, the concentration of the reactants $A$ and $B$ is high and that of the product C and D is low. So the forward reaction rate $r_1$ is initially much higher than $r_2$, the reverse reaction rate, because the product concentrations are relatively low. The reaction in this state is not in equilibrium, as $r_1 > r_2$. As the reaction progresses, the forward reaction increases the buildup of products $C$ and $D$. This increases the reverse reaction rate. Finally, a stage comes when the two rates are equal to each other ($r_1 = r_2$). This is the equilibrium state. At equilibrium:

- There is no further change in the concentration of the reactants and the products.
- The forward reaction rate is equal to the reverse reaction rate.
- The Gibbs free energy of the system is at minimum.
- The entropy of the system is at maximum.

Under equilibrium state, we have

$$r_1 = r_2$$

$$k_{\text{for}} C_A^n C_B^m = k_{\text{back}} C_C^p C_D^q \tag{7.31}$$

### 7.4.1.1 Reaction Rate Constant

A rate constant, $k_i$, is independent of the concentration of reactants but is dependent on the reaction temperature, $T$. The temperature dependency of the reaction rate constant is expressed in Arrhenius form as:

$$k = A_0 \exp\left(-\frac{E}{RT}\right) \tag{7.32}$$

where $A_0$ is a preexponential constant, $R$ is the universal gas constant, and $E$ is the activation energy for the reaction.

The ratio of rate constants for the forward and reverse reactions is the equilibrium constant, $K_e$. From Eq. (7.31) we can write

$$K_e = \frac{k_{\text{for}}}{k_{\text{back}}} = \frac{C_C^p C_D^q}{C_A^n C_B^q} \tag{7.33}$$

The equilibrium constant, $K_e$, depends on temperature but not on pressure. Table 7.4 gives values of equilibrium constants and heat of formation of some gasification reactions (Probstein and Hicks, 2006, pp. 62–64).

#### 7.4.1.2 Gibbs Free Energy

Gibbs free energy, $G$, is an important thermodynamic function. Its change in terms of a change in entropy, $\Delta S$, and enthalpy, $\Delta H$, is written as:

$$\Delta G = \Delta H - T\Delta S \tag{7.34}$$

The change in enthalpy or entropy for a reaction system is computed by finding the enthalpy or entropy changes of individual gases in the system. It is explained in Example 7.2. An alternative approach uses the empirical equations given by Probstein and Hicks (2006). It expresses the Gibbs function (Eq. (7.35)) and the enthalpy of formation (Eq. (7.36)) in terms of temperature, $T$, the heat of formation at the reference state at 1 atm and 298 K, and a number of empirical coefficients, $a'$, $b'$, and so forth.

$$\Delta G^0_{f,T} = \Delta h^0_{298} - a'T\ln(T) - b'T^2 - \left(\frac{c'}{2}\right)T^3 - \left(\frac{d'}{3}\right)T^4 + \left(\frac{e'}{2T}\right) + f' + g'T \text{ kJ/mol} \tag{7.35}$$

$$\Delta H^0_{f,T} = \Delta h^0_{298} - a'T + b'T^2 + c'T^3 + d'T^4 + \left(\frac{e'}{T}\right) + f' \text{ kJ/mol} \tag{7.36}$$

The values of the empirical coefficients for some common gases are given in Table 7.5.

The equilibrium constant of a reaction occurring at a temperature $T$ may be known using the value of Gibbs free energy.

$$K_e = \exp\left(-\frac{\Delta G}{RT}\right) \tag{7.37}$$

**TABLE 7.4** Equilibrium Constants and Heats of Formation for Five Gasification Reactions

| | Equilibrium Constant ($\log_{10}$ K) | | | Heat of Formation (kJ/mol) | |
|---|---|---|---|---|---|
| **Reaction** | **298 K** | **1000 K** | **1500 K** | **1000 K** | **1500 K** |
| $C + ½O_2 \rightarrow CO$ | 24.065 | 10.483 | 8.507 | −111.9 | −116.1 |
| $C + O_2 \rightarrow CO_2$ | 69.134 | 20.677 | 13.801 | −394.5 | −395.0 |
| $C + 2H_2 \rightarrow CH_4$ | 8.906 | −0.999 | −2.590 | −89.5 | −94.0 |
| $2C + 2H_2 \rightarrow C_2H_4$ | −11.940 | −6.189 | −5.551 | 38.7 | 33.2 |
| $H_2 + ½O_2 \rightarrow H_2O$ | 40.073 | 10.070 | 5.733 | −247.8 | −250.5 |

**Source:** Data compiled from Probstein and Hicks (2006), p. 64.

**TABLE 7.5** Heat of Combustion, Gibbs Free Energy, and Heat of Formation at 298 K, 1 atm, and Empirical Coefficients from Eqs. (7.35) and (7.36)

| Product | HHV (kJ/mol) | $\Delta_f G_{298}$ (kJ/mol) | $\Delta_f H_{298}$ (kJ/mol) | Empirical Coefficients a′ | b′ | c′ | d′ | e′ | f′ | g′ |
|---|---|---|---|---|---|---|---|---|---|---|
| C | 393.5 | 0 | 0 | | | | | | | |
| CO | 283 | − 137.3 | − 110.5 | $5.619 \times 10^{-3}$ | $-1.19 \times 10^{-5}$ | $6.383 \times 10^{-9}$ | $-1.846 \times 10^{-12}$ | $-4.891 \times 10^{2}$ | 0.868 | $-6.131 \times 10^{-2}$ |
| $CO_2$ | 0 | − 394.4 | − 393.5 | $-1.949 \times 10^{-2}$ | $3.122 \times 10^{-5}$ | $-2.448 \times 10^{-8}$ | $6.946 \times 10^{-12}$ | $-4.891 \times 10^{2}$ | 5.27 | − 0.1207 |
| $CH_4$ | 890.3 | − 50.8 | − 74.8 | $-4.62 \times 10^{-2}$ | $1.13 \times 10^{-5}$ | $1.319 \times 10^{-8}$ | $-6.647 \times 10^{-12}$ | $-4.891 \times 10^{2}$ | 14.11 | 0.2234 |
| $C_2H_4$ | 1411 | 68.1 | 52.3 | $-7.281 \times 10^{-2}$ | $5.802 \times 10^{-5}$ | $-1.861 \times 10^{-8}$ | $5.648 \times 10^{-13}$ | $-9.782 \times 10^{2}$ | 20.32 | − 0.4076 |
| $CH_3OH$ | 763.9 | − 161.6 | − 201.2 | $-5.834 \times 10^{-2}$ | $2.07 \times 10^{-5}$ | $1.491 \times 10^{-8}$ | $-9.614 \times 10^{-12}$ | $-4.891 \times 10s^{2}$ | 16.88 | − 0.2467 |
| $H_2O$ (steam) | 0 | − 228.6 | − 241.8 | $-8.95 \times 10^{-3}$ | $-3.672 \times 10^{-6}$ | $5.209 \times 10^{-9}$ | $-1.478 \times 10^{-12}$ | 0 | 2.868 | − 0.0172 |
| $H_2O$ (water) | 0 | − 237.2 | − 285.8 | | | | | | | |
| $O_2$ | 0 | 0 | 0 | | | | | | | |
| $H_2$ | 285.8 | 0 | 0 | | | | | | | |

**Source**: Adapted from Probstein and Hicks (2006), pp. 55, 61.

Here, $\Delta G$ is the standard Gibbs function of reaction or free energy change for the reaction, $R$ is the universal gas constant, and $T$ is the gas temperature.

---

**Example 7.2**

Find the equilibrium constant at 2000 K for the reaction

$$CO_2 \rightarrow CO + 1/2O_2$$

**Solution**

Enthalpy change is written by taking the values for it from the NIST-JANAF thermochemical tables (Chase, 1998) for 2000 K:

$$\begin{aligned}\Delta H &= (h_f^o + \Delta h)_{CO} + (h_f^o + \Delta h)_{O_2} - (h_f^o + \Delta h)_{CO_2}\\ &= 1\ \text{mol}(-110,527 + 56,744)\ \text{J/mol} + 1/2\ \text{mol}\ (0 + 59,175)\ \text{J/mol}\\ &\quad - 1\ \text{mol}(-393,522 + 91,439)\ \text{J/mol} = 277,887\ \text{J}\end{aligned}$$

The change in entropy, $\Delta S$, is written in the same way as for taking the values of entropy change from the NIST-JANAF tables.

$$\begin{aligned}\Delta S &= 1 \times S_{CO} + 1/2 \times S_{O_2} - 1 \times S_{CO_2}\\ &= (1\ \text{mol} \times 258.71\ \text{J/mol}\ K) + (1/2\ \text{mol} \times 268.74\ \text{J/mol}\ K)\\ &\quad - (1\ \text{mol} \times 309.29\ \text{J/mol}\ K)\\ &= 83.79\ \text{J}/K\end{aligned}$$

From Eq. (7.34), the change in the Gibbs free energy can be written as:

$$\begin{aligned}\Delta G &= \Delta H - T\Delta S\\ &= 277.887\ \text{kJ} - (2000\ \text{K} \times 83.79\ \text{J}/K) = 110.307\ \text{kJ}\end{aligned}$$

The equilibrium constant is calculated using Eq. (7.37):

$$K_{2000\ K} = e^{-\frac{\Delta G}{RT}} = e^{-\left(\frac{110.307}{0.008314 \times 2000}\right)} = 0.001315 \tag{7.38}$$

---

### 7.4.1.3 *Kinetics of Gas–Solid Reactions*

The rate of gasification of char is much slower than the rate of pyrolysis of the biomass that produces the char. Thus, the volume of a gasifier is more dependent on the rate of char gasification than on the rate of pyrolysis. The char gasification reaction therefore plays a major role in the design and performance of a gasifier.

Typical temperatures of the gasification zone in downdraft and fluidized-bed reactors are in the range of 700–900°C. The three most common gas–solid reactions that occur in the char gasification zone are as follows:

$$\text{Boudouard reaction:}\quad (\text{R1:C} + CO_2 \rightarrow 2CO) \tag{7.39}$$

$$\text{Water} - \text{gas reaction:}\quad (\text{R2:C} + H_2O \leftrightarrow CO + H_2) \tag{7.40}$$

$$\text{Methanation reaction:}\quad (\text{R3:C} + 2H_2 \leftrightarrow CH_4) \tag{7.41}$$

The water–gas reaction, R2, is dominant in a steam gasifier. In the absence of steam, when air or oxygen is the gasifying medium, the Boudouard reaction, R1, is dominant. However, the steam gasification reaction rate is higher than the Boudouard reaction rate. Another important gasification reaction is the shift reaction, R9 ($CO + H_2O \leftrightarrow CO_2 + H_2$), which takes place in the gas phase. It is discussed in the next section.

A popular form of the gas–solid char reaction, $r$, is the $n$th-order expression:

$$r = \frac{1}{(1-X)^m}\frac{dX}{dt} = A_0 e^{-\frac{E}{RT}} P_i^n \ s^{-1} \tag{7.42}$$

where $X$ is the fractional carbon conversion, $A_0$ is the apparent preexponential constant ($s^{-1}$), $E$ is the activation energy (kJ/mol), $m$ is the reaction order with respect to the carbon conversion, $T$ is the temperature (K), and $n$ is the reaction order with respect to the gas partial pressure, $P_i$. The universal gas constant, $R$, is 0.008314 kJ/mol K.

### 7.4.1.4 Boudouard Reaction

Referring to the Boudouard reaction (R1) in Eq. (7.6), we can use the Langmuir–Hinshelwood rate, which takes into account CO inhibition (Cetin et al., 2005) to express the apparent gasification reaction rate, $r_b$:

$$r_b = \frac{k_{b_1} P_{CO_2}}{1 + (k_{b_2}/k_{b_3}) P_{CO} + (k_{b_1}/k_{b_3}) P_{CO_2}} \ s^{-1} \tag{7.43}$$

where $P_{CO}$ and $P_{CO_2}$ are the partial pressure of CO and $CO_2$, respectively, on the char surface (bar). The rate constants, $k_i$, are given in the form, $A$ exp ($-E/RT$), where $A$ is the preexponential factor ($bar^{-n}\ s^{-n}$). Barrio and Hustad (2001) gave some values of the preexponential factor and the activation energy for Birch wood (Table 7.6).

When the concentration of CO is relatively small, and when its inhibiting effect is not to be taken into account, the kinetic rate of gasification by the Boudouard reaction may be expressed by a simpler $n$th-order equation as:

$$r_b = A_b e^{-\frac{E}{RT}} P_{CO_2}^n \ s^{-1} \tag{7.44}$$

For the Boudouard reaction, the values of the activation energy, $E$, for biomass char are typically in the range of 200–250 kJ/mol, and those of the exponent, $n$, are in the range of 0.4–0.6 (Blasi, 2009). Typical values of $A$, $E$, and $n$ for char from birch, poplar, cotton, wheat straw, and spruce are given in Table 7.7.

The reverse of the Boudouard reaction has a major implication, especially in catalytic reactions, as it deposits carbon on its catalyst surfaces, thus deactivating the catalyst.

**TABLE 7.6** Activation Energy and Preexponential Factors for Birch Char Using the Langmuir–Hinshelwood Rate Constants for $CO_2$ Gasification

| Langmuir–Hinshelwood Rate Constants ($s^{-1}$ $bar^{-1}$) | Activation Energy $E$ (kJ/mol) | Preexponential Factor $A$ ($s^{-1}$ $bar^{-1}$) |
|---|---|---|
| $k_{b1}$ | 165 | $1.3 \times 10^5$ |
| $k_{b2}$ | 20.8 | 0.36 |
| $k_{b3}$ | 236 | $3.23 \times 10^7$ |

**Source**: Adapted from Barrio and Hustad (2001).

**TABLE 7.7** Typical Values for Activation Energy, Preexponential Factor, and Reaction Order for Char in the Boudouard Reaction

| Char Origin | Activation Energy $E$ (kJ/mol) | Preexponential Factor $A$ ($s^{-1}$ $bar^{-1}$) | Reaction Order, $n$ (−) | References |
|---|---|---|---|---|
| Birch | 215 | $3.1 \times 10^6\ s^{-1}\ bar^{-0.38}$ | 0.38 | Barrio and Hustad (2001) |
| Dry poplar | 109.5 | $153.5\ s^{-1}\ bar^{-1}$ | 1.2 | Barrio and Hustad (2001) |
| Cotton wood | 196 | $4.85 \times 10^8\ s^{-1}$ | 0.6 | DeGroot and Shafizadeh (1984) |
| Douglas fir | 221 | $19.67 \times 10^8\ s^{-1}$ | 0.6 | DeGroot and Shafizadeh (1984) |
| Wheat straw | 205.6 | $5.81 \times 10^6\ s^{-1}$ | 0.59 | Risnes et al. (2001) |
| Spruce | 220 | $21.16 \times 10^6\ s^{-1}$ | 0.36 | Risnes et al. (2001) |

$$2CO \rightarrow CO_2 + C - 172\ kJ/mol \tag{7.45}$$

The preceding reaction becomes thermodynamically feasible when $(P_{CO}^2/P_{CO_2})$ is much greater than that of the equilibrium constant of the Boudouard reaction (Littlewood, 1977).

### 7.4.1.5 Water–Gas Reaction

Referring to the water–gas reaction, the kinetic rate, $r_w$, may also be written in Langmuir–Hinshelwood form to consider the inhibiting effect of hydrogen and other complexes (Blasi, 2009).

$$r_w = \frac{k_{w_1} P_{H_2O}}{1 + (k_{w_1}/k_{w_3})P_{H_2O} + (k_{w_2}/k_{w_3})P_{H_2}} s^{-1} \tag{7.46}$$

where $P_i$ is the partial pressure of gas $i$ in bars.

Typical rate constants according to Barrio et al. (2001) for beech wood are

$$k_{w_1} = 2.0 \times 10^7 \exp(-199/RT) \text{ bar}^{-1} \text{ s}^{-1}$$

$$k_{w_2} = 1.8 \times 10^6 \exp(-146/RT) \text{ bar}^{-1} \text{ s}^{-1}$$

$$k_{w_3} = 8.4 \times 10^7 \exp(-225/RT) \text{ bar}^{-1} \text{ s}^{-1}$$

Most kinetic analysis, however, uses a simpler $n$th-order expression for the reaction rate:

$$r_w = A_w e^{-\frac{E}{RT}} P^n_{H_2O} s^{-1} \tag{7.47}$$

Typical values for the activation energy, $E$, for steam gasification of char for some biomass types are given in Table 7.8.

### 7.4.1.6 Hydrogasification Reaction

The hydrogasification reaction is as follows:

$$C + 2H_2 \Leftrightarrow CH_4 \tag{7.48}$$

With freshly devolatilized char, this reaction progresses rapidly, but graphitization of carbon soon causes the rate to drop to a low value. The reaction involves volume increase and so pressure has a positive influence on it. High pressure and rapid heating help this reaction. Wang and Kinoshita (1993) measured the rate of this reaction and obtained values of $A = 4.189 \times 10^{-3}$/s and $E = 19.21$ kJ/mol.

### 7.4.1.7 Steam Reforming of Hydrocarbon

For production of syngas ($CO$, $H_2$) direct reforming of hydrocarbon is an option. Here, a mixture of hydrocarbon and steam is passed over a nickel-based catalyst at 700–900°C. The final composition of the product gas depends on the following factors (Littlewood, 1977):

- H/C ratio of the feed
- Steam/carbon ratio
- Reaction temperature
- Operating pressure.

**TABLE 7.8 Activation Energy, Preexponential Factor, and Reaction Order for Char for the Water–Gas Reaction**

| Char Origin | Activation Energy $E$ (kJ/mol) | Preexponential Factor $A_w$ (s$^{-1}$ bar$^{-1}$) | Reaction Order $n$ (−) | References |
|---|---|---|---|---|
| Birch | 237 | $2.62\times10^8\ s^{-1}\ bar^{-n}$ | 0.57 | Barrio et al. (2001) |
| Beech | 211 | $0.171\times10^8\ s^{-1}\ bar^{-n}$ | 0.51 | Barrio et al. (2001) |
| Wood | 198 | $0.123\times10^8\ s^{-1}\ atm^{-n}$ | 0.75 | Hemati and Laguerie (1988) |
| Various biomass | 180–200 | | 0.04–1.0 | Blasi (2009) |

The mixture of CO and $H_2$ produced can be subsequently synthesized into required liquid fuels or chemical feedstock. The reactions may be described as:

$$C_mH_n + \frac{4m-n}{4}H_2O \Leftrightarrow \frac{4m+n}{8}CH_4 + \frac{4m-n}{8}CO_2 \tag{7.49}$$

$$CH_4 + H_2O \Leftrightarrow CO + 3H_2 \tag{7.50}$$

$$CO + H_2O \Leftrightarrow CO_2 + H_2 \tag{7.51}$$

The first reaction (Eq. (7.49)) is favorable at high pressure, as it involves an increase in volume in the forward direction. The equilibrium constant of the first reaction increases with temperature while that of the third reaction (Eq. (7.51)), which is also known as the shift reaction, decreases.

### *7.4.1.8 Kinetics of Gas-Phase Reactions*

Several gas-phase reactions play an important role in gasification. Among them, the shift reaction (R9), which converts carbon monoxide into hydrogen, is most important.

$$R9{:}CO + H_2O \xrightarrow{k_{for}} CO_2 + H_2 - 41.1\ kJ/mol \tag{7.52}$$

This reaction is mildly exothermic. Since there is no volume change, it is relatively insensitive to changes in pressure.

The equilibrium yield of the shift reaction decreases slowly with temperature. For a favorable yield, the reaction should be conducted at low temperature, but then the reaction rate will be slow. For an optimum rate, we need

**TABLE 7.9 Forward Reaction Rates, *r*, for Gas-Phase Homogeneous Reactions**

| Reaction | Reaction Rate (*r*) | Heat of Formation ($m^3$/mol/s) | References |
|---|---|---|---|
| $H_2 + ½O_2 \rightarrow H_2O$ | $K\, C_{H_2}^{1.5} C_{O_2}$ | $51.8\, T^{1.5} \exp(-3420/T)$ | Vilienskii and Hezmalian (1978) |
| $CO + ½O_2 \rightarrow CO_2$ | $K\, C_{CO} C_{O^2}^{0.5} C_{H_2O}^{0.5}$ | $2.238 \times 10^{12} \exp(-167.47/RT)$ | Westbrook and Dryer (1981) |
| $CO + H_2O \rightarrow CO_2 + H_2$ | $K\, C_{CO} C_{H_2O}$ | $0.2778 \exp(-12.56/RT)$ | Petersen and Werther (2005) |

*Note:* Here, the gas constant, *R*, is in kJ/mol K.

catalysts. Below 400°C, a chromium-promoted iron formulation catalyst ($Fe_2O_3-Cr_2O_3$) may be used (Littlewood, 1977).

Other gas-phase reactions include CO combustion, which provides heat to the endothermic gasification reactions:

$$R6{:}CO + \frac{1}{2}O_2 \xrightarrow{k_{for}} CO_2 - 284\ kJ/mol \tag{7.53}$$

These homogeneous reactions are reversible. The rate of forward reactions is given by the rate coefficients given in Table 7.9.

For the backward CO oxidation reaction ($CO + \frac{1}{2}O_2 \xleftarrow{k_{back}} CO_2$), the rate, $k_{back}$, is given by Westbrook and Dryer (1981) as:

$$k_{back} = 5.18 \times 10^8 \exp(-167.47/RT) C_{CO_2} \tag{7.54}$$

For the reverse of the shift reaction ($CO + H_2O \xleftarrow{k_{back}} CO_2 + H_2$), the rate is given as:

$$k_{back} = 126.2 \exp(-47.29/RT) C_{CO_2} C_{H_2}\ mol/m^3 \tag{7.55}$$

If the forward rate constant is known, then the backward reaction rate, $k_{back}$, can be determined using the equilibrium constant from the Gibbs free energy equation:

$$K_{equilibrium} = \frac{k_{for}}{k_{back}} = \exp\left(\frac{-\Delta G^0}{RT}\right) \text{ at 1 atm pressure} \tag{7.56}$$

$\Delta G^0$ for the shift reaction may be calculated (see Callaghan, 2006) from a simple correlation of:

$$\Delta G^0 = -32.197 + 0.031T - (1774.7/T),\ kJ/mol \tag{7.57}$$

where $T$ is in K.

### Example 7.3

For shift reaction $CO + H_2O \rightarrow CO_2 + H_2$, assume that the reaction begins with 1 mol of CO, 1 mol of $H_2O$, and 1 mol of nitrogen. Find:

- The equilibrium constant at 1100 K and 1 atm.
- The equilibrium mole fraction of carbon dioxide.
- Whether the reaction is endothermic or exothermic.
- If pressure is increased to 100 atm, the impact of the equilibrium constant at 1100 K.

**Solution**

Part (a): For the shift reaction, the Gibbs free energy at a certain temperature can be calculated from Eq. (7.57):

$$\Delta G^0 = -32.197 + 0.031T - (1774.7/T)$$

at 1100 K, $\Delta G^0 = 0.2896$ kJ/mol.

The equilibrium constant can be calculated from Eq. (7.56):

$$K_{equilibrium} = \frac{k_{for}}{k_{back}} = \exp\left(\frac{-\Delta G^0}{RT}\right)$$

$$K_{equilibrium} = \exp\left(\frac{-0.2896}{0.008314 \times 1100}\right)$$

$$K_{equilibrium} = 0.9688$$

Part (b): At equilibrium, the rate of the forward reaction will be equal to the rate of the backward reaction. So, using the definition of the equilibrium constant, we have

$$K_{equilibrium} = \frac{p_{CO_2} p_{H_2}}{p_{CO} p_{H_2O}} = 0.9688$$

where $p$ denotes the partial pressure of the various species. In this reaction, nitrogen stays inert and does not react. Thus, 1 mol of nitrogen comes out from it. If $x$ moles of CO and $H_2O$ react to form $x$ moles of $CO_2$ and $H_2$, then at equilibrium, $(1 - x)$ moles of CO and $H_2O$ remain unreacted. We can list the component mole fraction as:

| Species | Mole | Mole Fraction |
|---|---|---|
| CO | $(1-x)$ | $(1-x)/3$ |
| $H_2O$ | $(1-x)$ | $(1-x)/3$ |
| $CO_2$ | $x$ | $x/3$ |
| $H_2$ | $x$ | $x/3$ |
| $N_2$ | 1 | 1/3 |

The mole fraction $y$ is related to the partial pressure, $p$, by the relation $yP = p$, where $P$ stands for total pressure.

Substituting the values for the partial pressures of the various species, we get:

$$\frac{((x/3)P)((x/3)P)}{((1 - x/3)P)((1 - x/3)P)} = 0.9688$$

Solving for $x$, we get $x = 0.5$. Thus, the mole fraction of $CO_2$ at equilibrium $= (1 - x)/3 = 0.5/3 = 0.1667$.

Part (c): To determine if this reaction is exothermic or endothermic, the standard heats of formation of the individual components are taken from the NIST-JANAF thermochemical tables (Chase, 1998).

$$\Delta H = (h_f^0)_{CO_2} + (h_f^0)_{H_2} - [(h_f^0)_{CO} + (h_f^0)_{H_2O}]$$

$$\Delta H = -393.52 \text{ kJ/mol} - 0 \text{ kJ/mol} - [-110.53 \text{ kJ/mol} - 241.82 \text{ kJ/mol}]$$

$$\Delta H = -41.17 \text{ kJ/mol}$$

Since 41.17 kJ/mol of heat is given out, the reaction is exothermic.

Part (d): This reaction does not depend on pressure, as there is no volume change. The equilibrium constant changes only with temperature, so the equilibrium constant at 100 atm is the same as that at 1 atm for 1100 K. The equilibrium constant is 0.9688 at 100 atm for 1100 K.

### 7.4.2 Char Reactivity

Reactivity, generally a property of a solid fuel, is the value of the reaction rate under a well-defined condition of gasifying agent, temperature, and pressure. Proper values or expressions of char reactivity are necessary for all gasifier models. This topic has been studied extensively for more than 60 years, and a large body of information is available, especially for coal. These studies unearthed important effects of char size, surface area, pore size distribution, catalytic effect, and mineral content, pretreatment, and heating. The origin of the char and the extent of its conversion also exert some influence on reactivity.

Char can originate from any hydrocarbon—coal, peat, biomass, and so forth. An important difference between chars from biomass and those from fossil fuels like coal or peat is that the reactivity of biomass chars increases with conversion while that of coal or peat char decreases. Figure 7.3 plots the reactivity for hardwood and peat against their conversion (Liliedahl and Sjostrom, 1997). One can infer from here that while the conversion rate (at conversion 0.8) of hardwood char in steam is 9% per minute, that of peat char under similar conditions is only 1.5% per minute.

#### *7.4.2.1 Effect of Pyrolysis Conditions*

The pyrolysis condition under which the char is produced also affects the reactivity of the char. For example, vanHeek and Muhlen (1990) noted that the reactivity of char (in air) is much lower when produced above 1000°C compared to that when produced at 700°C. High temperatures reduce the number of active sites of reaction and the number of edge atoms. Longer residence times at peak temperature during pyrolysis also reduce reactivity.

### 7.4.2.2 Effect of Mineral Matter in Biomass

Inorganic materials in fuels can act as catalysts in the char–oxygen reaction (Zolin et al., 2001). In coal, inorganic materials reside as minerals, whereas in biomass they generally remain as salts or are organically bound. Alkali metals, potassium, and sodium are active catalysts in reactions with oxygen-containing species. Dispersed alkali metals in biomass contribute to the high catalytic activity of inorganic materials in biomass. In coal, CaO is also dispersed, but at high temperatures it sinters and vaporizes, blocking micropores.

Inorganic matter also affects pyrolysis, giving char of varying morphological characteristics. Potassium and sodium catalyze the polymerization of volatile matter, increasing the char yield; at the same time, they produce solid materials that deposit on the char pores, blocking them. During subsequent oxidation of the char, the alkali metal catalyzes this process. Polymerization of volatile matter dominates over the pore-blocking effect. A high pyrolysis temperature may result in thermal annealing or loss of active sites and thereby loss of char reactivity (Zolin et al., 2001).

### 7.4.2.3 Intrinsic Reaction Rate

Char gasification takes place on the surface of solid char particles, which is generally taken to be the outer surface area of the particle. However, char particles are highly porous, and the surface areas of the inner pore walls are several orders of magnitude higher than the external surface area. For example, the actual surface area (BET, named after Brunaeur, Emma, and Teller) of an internal pore of a 1 mm diameter beech wood char is 660 $cm^2$, while its outer surface is only 3.14 $cm^2$. Thus, if there is no physical restriction, the reacting gas can potentially enter the pores and react on their walls, resulting in a high overall char conversion rate. For this reason, two char particles with the same external surface area (size) may have widely different reaction rates because of their different internal structure.

From a scientific standpoint, it is wise to express the surface reaction rate on the basis of the actual surface on which the reaction takes place rather than the external surface area. The rate based on the actual pore wall surface area is the *intrinsic reaction rate*; the rate based on the external surface area of the char is the *apparent reaction rate*. The latter is difficult to measure, so sometimes it is taken as the reactive surface area determined indirectly from the reaction rate instead of the total pore surface area measured by the physical adsorption of nitrogen. This is known as the BET area (Klose and Wolki, 2005).

### 7.4.2.4 Mass-Transfer Control

For the gasification reaction to take place within the char's pores, the reacting gas must enter the pores. If the availability of the gas is so limited that it is entirely consumed by the reaction on the outer surface of the char, gasification is restricted to the external surface area. This can happen because of the limitation of the mass transfer of gas to the char surface. We can illustrate using the example of char gasification in $CO_2$:

$$C + CO_2 \rightarrow 2CO \tag{7.58}$$

Here, the $CO_2$ gas has to diffuse to the char surface to react with the active carbon sites. The diffusion, however, takes place at a finite rate. If the kinetic rate of this reaction is much faster than the diffusion rate of $CO_2$ to the char surface, all of the $CO_2$ gas molecules transported are consumed on the external surface of the char, leaving none to enter the pores and react on their surfaces. As the overall reaction is controlled by diffusion, it is called the *diffusion-* or *mass-transfer-controlled* regime of reaction.

On the other hand, if the kinetic rate of reaction is slow compared to the transport rate of $CO_2$ molecules, then the $CO_2$ will diffuse into the pores and react on their walls. The reaction in this situation is "kinetically controlled."

$$\begin{array}{ll} \text{Diffusion rate} \gg \text{kinetic rate} & \text{[kinetic control reaction]} \\ \text{Diffusion rate} \ll \text{kinetic rate} & \text{[diffusion control reaction]} \end{array} \tag{7.59}$$

Between the two extremes lie intermediate regimes. The relative rates of chemical reaction and diffusion determine the gas concentration profile in the vicinity of the char particle; how the reaction progresses; and how char size, pore distribution, reaction temperature, char gas relative velocity, and so forth influence overall char conversion. Figure 7.8 shows how the concentration profile of $CO_2$ around the particle changes with temperature. With a

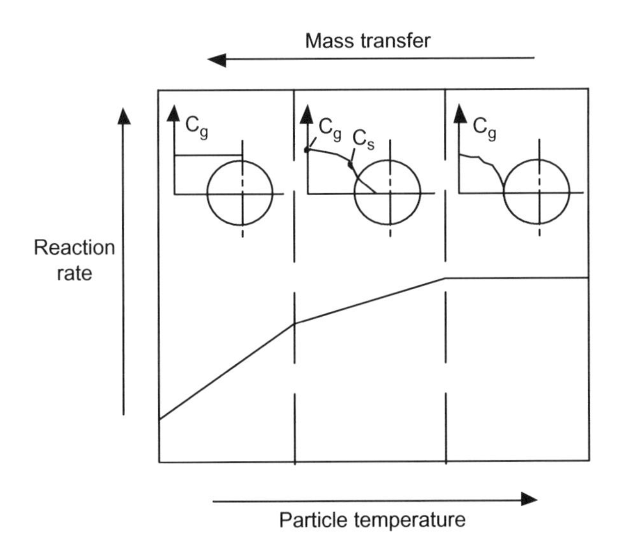


**FIGURE 7.8** Char gasification regimes in a porous biomass char particle.

rise in the surface temperature, the kinetic rate increases and therefore the overall reaction moves from the kinetic to the diffusion-controlled regime, resulting in less reaction within the pores.

The overall gasification rate of char particles, $Q$, when both mass-transfer and kinetic rates are important, may be written as:

$$Q = \frac{P_g}{(1/h_m) + (1/R_c)} \text{kg carbon/m}^2 \text{ s} \tag{7.60}$$

where $P_g$ is the concentration in partial pressure (bar) of the gasifying agent outside the char particle, $h_m$ is the mass-transfer rate (kg carbon (m$^2$ bar s)) to the surface, and $R_c$ is the kinetic rate of reaction: kg carbon (m$^2$ bar s).

## 7.5 GASIFICATION MODELS

Optimal conversion of chemical energy of the biomass or other solid fuel into the desired gas depends on proper configuration, sizing, and choice of gasifier operating conditions. In commercial plants, optimum operating conditions are often derived through trials on the unit or by experiments on pilot plants. Even though expensive, experiments can give more reliable design data than those can be obtained through modeling or simulation. There is, however, one major limitation with experimental data. If one of the variables of the original process changes, the optimum operating condition chosen from the specific experimental condition is no longer valid. Furthermore, an experimentally found optimum parameter can be size specific; that is, the optimum operating condition for one size of gasifier is not necessarily valid for any other size. The right choice between experiment and modeling, then, is necessary for a reliable design.

### 7.5.1 Simulation Versus Experiment

Simulation, or mathematical modeling, of a gasifier may not give a very accurate prediction of its performance, but it can at least provide qualitative guidance on the effect of design and operating or feedstock parameters. Simulation allows the designer or plant engineer to reasonably optimize the operation or the design of the plant using available experimental data for a pilot plant or the current plant.

Simulation can also identify operating limits and hazardous or undesirable operating zones, if they exist. Modern gasifiers, for example, often operate at a high temperature and pressure and are therefore exposed to extreme operating conditions. To push the operation to further extreme conditions to improve the gasifier performance may be hazardous, especially if it is done with no prior idea of how the gasifier might behave at those conditions. Modeling may provide a less expensive means of assessing the benefits and the associated risk.

Simulation can never be a substitute for good experimental data, especially in the case of gas–solid systems such as gasifiers. A mathematical model, however sophisticated, is useless unless it can reproduce real operation with an acceptable degree of deviation (Souza-Santos, 2004). Still, a good mathematical model can:

- Find optimum operating conditions or a design for the gasifier.
- Identify areas of concern or danger in operation.
- Provide information on extreme operating conditions (high temperature, high pressure) where experiments are difficult to perform.
- Provide information over a much wider range of conditions than one can obtain experimentally.
- Better interpret experimental results and analyze abnormal behavior of a gasifier, if that occurs.
- Assist scale-up of the gasifier from one successfully operating size to another, and from one feedstock to another.

### 7.5.2 Gasifier Simulation Models

Gasifier simulation models may be classified into the following groups:

- Thermodynamic equilibrium
- Kinetic
- Computational fluid dynamics (CFD)
- Artificial neural network (ANN)

The thermodynamic equilibrium model predicts the maximum achievable yield of a desired product from a reacting system (Li et al., 2001). In other words, if the reactants are left to react for an infinite time, they will reach equilibrium yield. The yield and composition of the product at this condition are given by the equilibrium model, which concerns the reaction alone without taking into account the geometry of the gasifier.

In practice, only a finite time is available for the reactant to react in the gasifier. So, the equilibrium model may give an ideal yield. For practical applications, we need to use the kinetic model to predict the product from a gasifier that provides a certain time for reaction. A kinetic model studies the progress of reactions in the reactor, giving the product compositions at different positions along the gasifier. It takes into account the reactor's geometry as well as its hydrodynamics.

The CFD models (Euler type) solve a set of simultaneous equations for conservation of mass, momentum, energy, and species over a discrete region of the gasifier. Thus, they give distribution of temperature, concentration, and other parameters within the reactor. If the reactor hydrodynamics is well known, a CFD model provides a very accurate prediction of temperature and gas yield around the reactor.

Neural network analysis is a relatively new simulation tool for modeling a gasifier. It works somewhat like an experienced operator, who uses his or her years of experience to predict how the gasifier will behave under a certain condition. This approach requires little prior knowledge about the process. Instead, the neural network *learns* by itself from sample experimental data (Guo et al., 1997).

#### 7.5.2.1 Thermodynamic Equilibrium Models

Thermodynamic equilibrium calculation is independent of gasifier design and so is convenient for studying the influence of fuel and process parameters. Though chemical or thermodynamic equilibrium may not be reached within the gasifier, this model provides the designer with a reasonable prediction of the maximum achievable yield of a desired product. However, it cannot predict the influence of hydrodynamic or geometric parameters, like fluidizing velocity, or design variables, like gasifier height.

Chemical equilibrium is determined by either of the following:

- Equilibrium constant (stoichiometric model)
- Minimization of the Gibbs free energy (non-stoichiometric model)

Prior to 1958, all equilibrium computations were carried out using the equilibrium constant formulation of the governing equations (Zeleznik and Gordon, 1968). Later, computation of equilibrium compositions by Gibbs free energy minimization became an accepted alternative.

This section presents a simplified approach to equilibrium modeling of a gasifier based on the following overall gasification reactions:

$$R1: CO_2 + C \rightarrow 2CO \tag{7.61}$$

$$R2: C + H_2O \rightarrow H_2 + CO \tag{7.62}$$

$$R3: C + 2H_2 \rightarrow CH_4 \tag{7.63}$$

$$R9: CO + H_2O \rightarrow CO_2 + H_2 \tag{7.64}$$

From a thermodynamic point of view, the equilibrium state gives the maximum conversion for a given reaction condition. The reaction is considered to be zero dimensional and there are no changes with time (Li et al., 2001). An equilibrium model is effective at higher temperatures (>1500 K), where it can show useful trends in operating parameter variations (Altafini et al., 2003). For equilibrium modeling, one may use stoichiometric or non-stoichiometric methods (Basu, 2006).

#### 7.5.2.2 Stoichiometric Equilibrium Models

In the stoichiometric method, the model incorporates the chemical reactions and species involved. It usually starts by selecting all species containing

C, H, and O, or any other dominant elements. If other elements form a minor part of the product gas, they are often neglected.

Let us take the example of 1 mol of biomass being gasified in $d$ moles of steam and $e$ moles of air. The reaction of the biomass with air (3.76 moles of nitrogen, 1 mol of oxygen) and steam may then be represented by:

$$\begin{aligned} CH_aO_bN_c + dH_2O + e(O_2 + 3.76N_2) \rightarrow n_1C + n_2H_2 + n_3CO \\ + n_4H_2O + n_5CO_2 + n_6CH_4 + n_7N_2 \end{aligned} \quad (7.65)$$

where $n_1, \ldots, n_7$ are stoichiometric coefficients. Here, $CH_aO_bN_c$ is the chemical representation of the biomass and $a$, $b$, and $c$ are the mole ratios (H/C, O/C, and N/C) determined from the ultimate analysis of the biomass. With $d$ and $e$ as input parameters, the total number of unknowns is seven.

An atomic balance of carbon, hydrogen, oxygen, and nitrogen gives:

$$\text{C:} \quad n_1 + n_3 + n_5 + n_6 = 1 \quad (7.66)$$

$$\text{H:} \quad 2n_2 + 2n_4 + 4n_6 + a + 2d \quad (7.67)$$

$$\text{O:} \quad n_3 + n_4 + 2n_5 = b + d + 2e \quad (7.68)$$

$$\text{N:} \quad n_7 = c + 7.52e \quad (7.69)$$

During the gasification process, reactions R1, R2, R3, and R9 (see Table 7.2) take place. The water–gas shift reaction, R9, can be considered a result of the subtraction of the steam gasification and Boudouard reactions, so we consider the equilibrium of reactions R1, R2, and R3 alone. For a gasifier pressure, $P$, the equilibrium constants for reactions $R_1$, $R_2$, and $R_3$ are given by:

$$K_{e_1} = \frac{y_{CO}^2 P}{y_{CO_2}} \quad \text{R1} \quad (7.70)$$

$$K_{e_2} = \frac{y_{CO} y_{H_2} P}{y_{H_2O}} \quad \text{R2} \quad (7.71)$$

$$K_{e_3} = \frac{y_{CH_4}}{y_{H_2}^2 P} \quad \text{R3} \quad (7.72)$$

where $y_i$ is the mole fraction for species $i$ of CO, $H_2$, $H_2O$, and $CO_2$.

The two sets of equations (stoichiometric and equilibrium) may be solved simultaneously to find the coefficients, ($n_1, \ldots, n_7$), and hence the product gas composition in an equilibrium state. Thus, by solving seven equations (Eqs. (7.66)–(7.72)) we can find seven unknowns ($n_1, \ldots, n_7$), which give both the yield and the product of the gasification for a given air/steam-to-biomass ratio. The approach is based on the simplified reaction path and the chemical formula of the biomass.

This is a greatly simplified example of the stoichiometric modeling of a gasification reaction. The complexity increases with the number of equations considered. For a known reaction mechanism, the stoichiometric equilibrium model predicts the maximum achievable yield of a desired product or the possible limiting behavior of a reacting system.

### *7.5.2.3 Nonstoichiometric Equilibrium Models*

In nonstoichiometric modeling, no knowledge of a particular reaction mechanism is required to solve the problem. In a reacting system, a stable equilibrium condition is reached when the Gibbs free energy of the system is at the minimum. So, this method is based on minimizing the total Gibbs free energy. The only input needed is the elemental composition of the feed, which is known from its ultimate analysis. This method is particularly suitable for fuels like biomass, the exact chemical formula of which is not clearly known.

The Gibbs free energy, $G_{\text{total}}$ for the gasification product comprising $N$ species ($i = 1, \ldots, N$) is given by:

$$G_{\text{total}} = \sum_{i=1}^{N} n_i \Delta G_{f,i}^{0} + \sum_{i=1}^{N} n_i RT \ln\left(\frac{n_i}{\sum n_i}\right) \tag{7.73}$$

where $\Delta G_{f,i}^{0}$ is the Gibbs free energy of formation of species $i$ at standard pressure of 1 bar.

Equation (7.73) is to be solved for unknown values of $n_i$ to minimize $G_{\text{total}}$, bearing in mind that it is subject to the overall mass balance of individual elements. For example, irrespective of the reaction path, type, or chemical formula of the fuel, the amount of carbon determined by ultimate analysis must be equal to the sum total of all carbon in the gas mixture produced. Thus, for each $j$th element we can write:

$$\sum_{i=1}^{N} a_{i,j} n_i = A_j \tag{7.74}$$

where $a_{i,j}$ is the number of atoms of the $j$th element in the $i$th species, and $A_j$ is the total number of atoms of element $j$ entering the reactor. The value of $n_i$ should be found such that $G_{\text{total}}$ will be minimum. We can use the Lagrange multiplier methods to solve these equations.

The Lagrange function (L) is defined as:

$$\mathrm{L} = G_{\text{total}} - \sum_{j=1}^{K} \lambda_j \left( \sum_{i=1}^{N} a_{ij} n_i - A_j \right) \text{ kJ/mol} \tag{7.75}$$

where $\lambda_\varphi$ is the Lagrangian multiplier for the $j$th element.

To find the extreme point, we divide Eq. (7.75) by $RT$ and take the derivative:

$$\left(\frac{\partial L}{\partial n_i}\right) = 0 \tag{7.76}$$

Substituting the value of $G_{\text{total}}$ from Eq. (7.73) in Eq. (7.75), and then taking its partial derivative, the final equation is of the form given by:

$$\left(\frac{\partial L}{\partial n_i}\right) = \frac{\Delta G^0_{f,i}}{RT} + \sum_{i=1}^{N} \ln\left(\frac{n_i}{n_{\text{total}}}\right) + \frac{1}{RT}\sum_{j=1}^{K} \lambda_j \left(\sum_{i=1}^{N} a_{ij} n_i\right) = 0 \tag{7.77}$$

### 7.5.2.4 Kinetic Models

Gas composition measurements for gasifiers often vary significantly from those predicted by equilibrium models (Kersten, 2002; Li et al., 2001; Peterson and Werther, 2005). This shows the inadequacy of equilibrium models and underscores the need of kinetic models to simulate gasifier behavior.

A kinetic model gives the gas yield and product composition a gasifier achieves after a finite time (or in a finite volume in a flowing medium). Thus, it involves parameters such as reaction rate, residence time of particles, and reactor hydrodynamics. For a given operating condition and gasifier configuration, the kinetic model can predict the profiles of gas composition and temperature inside the gasifier and overall gasifier performance.

The model couples the hydrodynamics of the gasifier reactor with the kinetics of gasification reactions inside the gasifier. At low reaction temperatures, the reaction rate is very slow, so the residence time required for complete conversion is long. Therefore, kinetic modeling is more suitable and accurate at relatively low operating temperatures ($<800°C$) (Altafini et al., 2003). For higher temperatures, where the reaction rate is faster, the equilibrium model may be of greater use.

Kinetic modeling has two components: reaction kinetics and reactor hydrodynamics.

### 7.5.2.5 Reaction Kinetics

Reaction kinetics must be solved simultaneously with bed hydrodynamics and mass and energy balances to obtain the yields of gas, tar, and char at a given operating condition.

As the gasification of a biomass particle proceeds, the resulting mass loss is manifested either through reduction in size with unchanged density or reduction in density with unchanged size. In both cases the rate is expressed in terms of the external surface area of the biomass char. Some models, where the reaction is made up of char alone, can define a reaction rate based on reactor volume. There are thus three ways of defining the char gasification reaction for biomass: (i) shrinking core model, (ii) shrinking particle model, and (iii) volumetric reaction rate model.

### 7.5.2.6 Reactor Hydrodynamics

The kinetic model considers the physical mixing process and therefore requires knowledge of reactor hydrodynamics. The hydrodynamics may be defined in terms of the following types with increasing sophistication and accuracy:

- Zero dimensional (stirred tank reactor)
- One dimensional (plug flow)
- Two dimensional
- Three dimensional

Unlike other models, the kinetic model is sensitive to the gas–solid contacting process involved in the gasifier. Based on this process, the model may be divided into three groups: (i) moving or fixed bed, (ii) fluidized bed, and (iii) entrained flow. Short descriptions of these are given in Section 7.6.

### 7.5.2.7 Neural Network Models

An alternative to the sophisticated modeling of a complex process, especially for one not well understood, is an ANN. An ANN model mimics the working of the human brain and provides some human characteristics in solving models (Abdulsalam, 2005). It cannot produce an analytical solution, but it can give numerical results. This technique has been used with reasonable success to predict gas yield and composition from gasification of bagasse, cotton stem, pine sawdust, and poplar in fluidized beds (Guo et al., 1997); in MSW; and also in a fluidized bed (Xiao et al., 2009).

The ANN model can deal with complex gasification problems. It uses a high-speed architecture of three hidden layers of neurons (Kalogirou, 2001): one to receive the input(s), one to process them, and one to deliver output(s). Figure 7.9 shows the arrangement of neuron layers and the connection patterns between them. Kalogirou (2001) suggested the following empirical formula to estimate the number of hidden neurons:

$$\text{Number of hidden neurons} = \tfrac{1}{2}(\text{inputs} + \text{outputs}) + \sqrt{\text{number of training patterns}} \tag{7.78}$$

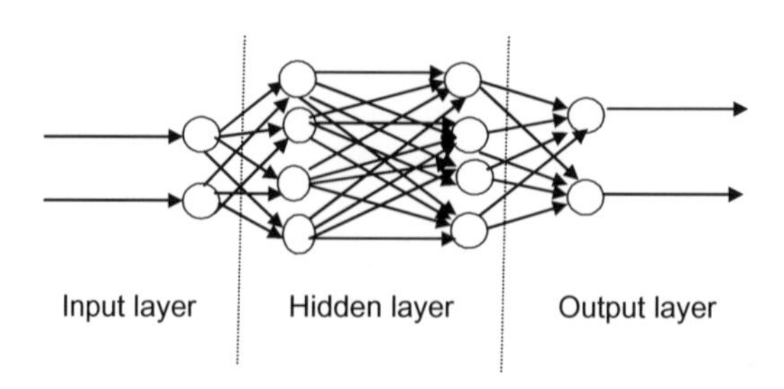


**FIGURE 7.9** Schematic diagram of a multilayer feed-forward neural network (*Source: Kalogirou, 2001*).

The input layer has two values associated with it: inputs and weights. Weights are used to transfer data from layer to layer. In the first step, the information is processed at the nodes and then added up (summation); the result is passed through an activation function. The outcome is the node's "activation value," which is multiplied by the specific weight and transferred to the next node.

#### *7.5.2.8 Network Training*

Training modifies the connection weights in some orderly fashion using learning methods (Guo et al., 2001). It begins with a set of data (with inputs and outputs targeted); the weights are adjusted until the difference between the neural network output and the corresponding target is minimum (Kalogirou et al., 1999). When the training process satisfies the required tolerance, the network holds the weights constant and uses the network to make output predictions. After training, the weights contain meaningful information. A back-propagation algorithm is used to train the network. Multilayer feed-forward neural networks are used to approximate the function.

A neural network may return poor results for data that differ from the original data it was trained with. This happens sometimes when limited data are available to calibrate and evaluate the constants of the model (Hajek and Judd, 1995). After structuring the neural network, information starts to flow from the input layer to the output layer according to the concepts described here.

#### *7.5.2.9 CFD Models*

CFD can have an important role in the modeling of a fluidized-bed gasifier. A CFD-based code involves a solution of conservation of mass, momentum, species, and energy over a defined domain or region. The equations can be written for an element, where the flux of the just-mentioned quantities moving in and out of the element is considered with suitable boundary conditions.

A CFD code for gasification typically includes a set of submodels for the sequence of operations such as the vaporization of a biomass particle, its pyrolysis (devolatilization), the secondary reaction in pyrolysis, and char oxidation (Babu and Chaurasia, 2004a,b; Di Blasi, 2008). Further sophistications such as a subroutine for fragmentation of fuels during gasification and combustion are also developed (Syred et al., 2007). These subroutines can be coupled with the transport phenomenon, especially in the case of a fluidized-bed gasifier.

The hydrodynamic or transport phenomenon for a laminar flow situation is completely defined by the Navier–Stokes equation, but in the case of turbulent flow, a solution becomes difficult. A complete time-dependent solution of the instantaneous Navier–Stokes equation is beyond today's computation capabilities (Wang and Yan, 2008), so it is necessary to assume some models for the turbulence. The Reynolds-averaged Navier–Stokes

($k$-$\varepsilon$) model or large eddy simulation filters are two means of accounting for turbulence in the flow.

For a fluidized bed, the flow is often modeled using the Eulerian–Lagrange concept. The discrete phase is applied to the particle flow; and the continuous phase to the gas. Overmann et al. (2008) used the Euler–Euler and Euler–Lagrange approaches to model wood gasification in a bubbling fluidized bed. Their preliminary results found both to have comparable agreement with experiments. If the flow is sufficiently dilute, the particle–particle interaction and the particle volume in the gas are neglected.

A two-fluid model is another CFD approach. Finite difference, finite element, and finite volume are three methods used for discretization. Commercial software such as ANSYS, ASPEN, Fluent, Phoenics, and CFD2000 are available for solution (Miao et al., 2008). A review and comparison of these codes is given in Xia and Sun (2002) and Norton et al. (2007).

Recent progress in numerical solution and modeling of complex gas–solid interactions has brought CFD much closer to real-life simulation. If successful, it will be a powerful tool for optimization and even design of thermochemical reactors like gasifiers (Wang and Yan, 2008). CFD models are most effective in modeling entrained-flow gasifiers, where the gas–solid flows are less complex than those in fluidized beds and the solid concentration is low.

Models developed by several investigators employ sophisticated reaction kinetics and complex particle–particle interaction. Most of them, however, must use some submodels, fitting parameters or major assumptions into areas where precise information is not available. Such weak links in the long array make the final result susceptible to the accuracy of those "weak links." If the final results are known, we can use them to back-calculate the values of the unknown parameters or to refine the assumptions used.

The CFD model can thus predict the behavior of a given gasifier over a wider range of parameters using data for one situation, but this prediction might not be accurate if the code is used for a different gasifier with input parameters that are substantially different from the one for which experimental data are available.

## 7.6 KINETIC MODEL APPLICATIONS

This section briefly discusses how kinetic models can be applied to the three major gasifier types.

### 7.6.1 Moving-Bed Gasifiers

A basic moving-bed or fixed-bed gasifier can use the following assumptions:

- The reactor is uniform radially (i.e., no temperature or concentration gradient exists in the radial direction).
- The solids flow downward (in an updraft gasifier) as a plug flow.

- The gas flows upward as a plug flow.
- The interchange between two phases takes place by diffusion.

The mass balance of a gas species, $j$, can be written (Souza-Santos, 2004, p. 134) as:

$$u_g \frac{d\rho_{g,j}}{dz} = D_{g,j} \frac{d^2\rho_{g,j}}{dz^2} + R_{m,j} \tag{7.79}$$

where $u_g$ is the superficial gas velocity, $z$ is the distance, $\rho_{g,j}$ is the density of the $j$th gas, and $D_{g,j}$ is the diffusivity of the $j$th gas. $R_{m,j}$, the production or consumption of the $j$th gas element, is related to $Q_{gasification}$ heat generation or absorption.

Similarly, an energy balance equation can be written for a d$z$ element as:

$$\rho_g C_{pg} u_g \frac{dT}{dz} = \lambda_g \frac{d^2T}{dz^2} + Q_{gasification} + Q_{conv} + Q_{rad} + Q_{mass} \tag{7.80}$$

where, $Q_{gasification}$, $Q_{conv}$, $Q_{rad}$, and $Q_{mass}$ are the net heat flow into the element due to gasification, convection, radiation, and mass transfer, respectively. These terms can be positive or negative. $\rho_g$, $C_{pg}$, and $\lambda_g$ are the density, specific heat, and thermal conductivity of the bulk gas, respectively.

Equations (7.79) and (7.80) can be solved simultaneously with appropriate expression for the reaction rate $R_{m,j}$.

### 7.6.2 Fluidized-Bed Gasifiers

The kinetic modeling of fluidized-bed gasifiers requires several assumptions or submodels. It takes into account how the fluidized-bed hydrodynamics is viewed in terms of heat and mass transfer, and gas flow through the fluidized bed. The bed hydrodynamics defines the transport of the gasification medium through the system, which in turn influences the chemical reaction on the biomass surface. Each of these is subject to some assumptions or involves submodels.

One can use several versions of the fluidization model:

- Two-phase model of bubbling fluidized bed: bubbling and emulsion phases.
- Three-phase model of bubbling fluidized bed: bubbling, cloud, and emulsion phases.
- Fluidized bed divided into horizontal sections or slices.
- Core-annulus structure.

Gas flow through the bed can be modeled as:

- Plug flow in the bubbling phase; ideally mixed gas in the emulsion phase.
- Ideally mixed gases in both phases.
- Plug flow in both phases (there is exchange between phases).

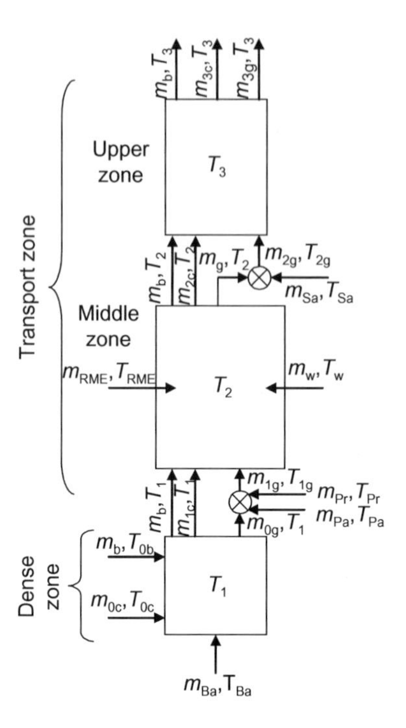


**FIGURE 7.10** Model of a CFB gasifier.

- Plug flow through the bubble and emulsion phases without mass transfer between phases.
- Plug flow of gas upward in the core and solid backflow in the annulus.

The following sections present the essentials of a model for a CFB combustor and one for a bubbling fluidized-bed gasifier (Kaushal et al., 2008). A typical one-dimensional steady-state model of a CFB combustor, as shown in Figure 7.10, assumes gases as ideal and in the plug-flow regime. The riser is divided into three hydrodynamic zones: lower dense bed zone, intermediate middle zone, and top dilute zone. The solids are assumed uniform in size with no attrition. Char is a homogeneous matrix of carbon, hydrogen, and oxygen.

A bubbling fluidized-bed gasifier is divided into several zones with different hydrodynamic characteristics: dense zone and freeboard zone for bubbling beds and core-annulus for circulating beds. The dense zone additionally deals with the drying and devolatilization of the introduced feed. Superheated steam is introduced at the lower boundary of the dense zone. Each zone is further divided into cells, which individually calculate their local hydrodynamic and thermodynamic state using chosen equations or correlations. The cells are solved sequentially from bottom to top, with the output of each considered the input for the next. The conservation equations for carbon, bed material, and energy are evaluated not in each cell but across the entire zone. Therefore, each zone shows a homogeneous char concentration in the bed material and a uniform temperature. Additional input parameters to the model are geometric data, particle properties, and flow-rates.

#### *7.6.2.1 Hydrodynamic Submodel (Bubbling Bed)*

The dense zone (assumed to be the bubbling bed) is modeled according to the modified two-phase theory. Bubble size is calculated as a function of bed height (Darton and LaNauze, 1977), and it is assumed that all bubbles at any cross-section are of uniform size:

$$d_b = 0.54\frac{(U-U_{mf})^{0.4}}{g^{0.2}}\left(z+4\sqrt{\frac{A}{N_{or}}}\right)^{0.8} \tag{7.81}$$

where $(N_{or}/A)$ is the number of orifices per unit of cross-section area of the bed.

The interphase mass transfer between bubbles and emulsion, essential for the gas–solid reactions, is modeled semiempirically using the specific bubble surface as the exchange area, the concentration gradient, and the mass-transfer coefficient. The mass-transfer coefficient, $K_{BE}$, based on the bubble–emulsion surface area (Sit and Grace, 1978), is:

$$K_{BE} = \frac{U_{mf}}{4} + \sqrt{\frac{4\varepsilon_{mf}D_r U_B}{\pi d_B}} \tag{7.82}$$

where $U_{mf}$ and $\varepsilon_{mf}$ are, respectively, minimum fluidization velocity and voidage at a minimum fluidizing condition, $D_r$ is the bed diameter, and $U_B$ is the rise velocity of a bubble of size $d_B$.

The axial mean voidage in the freeboard is calculated using an exponential decay function.

#### *7.6.2.2 Reaction Submodel*

Gasification reactions proceed at a finite speed; this process is divided into three steps: drying, devolatilization, and gasification. The time taken for drying and devolatilization of the fuel is much shorter than the time taken for gasification of the remaining char. Some models assume instantaneous drying and devolatilization because the rate of reaction of the char, which is the slowest, largely governs the overall process.

The products of devolatilization are $CO_2$, CO, $H_2O$, $H_2$, and $CH_4$. The gases released during drying and devolatilization are not added instantaneously to the upflowing gas stream, but are added along the height of the gasifier in a predefined pattern. The total mass devolatilized, $m_{volatile}$, is therefore the sum of the carbon, hydrogen, and oxygen volatilized from the solid biomass.

$$m_{volatile} = m_{char} + m_{hydrogen} + m_{oxygen} \tag{7.83}$$

Char gasification, the next critical step, may be assumed to move simultaneously through reactions R1, R2, and R3 (Table 7.2). As these three reactions occur simultaneously on the char particle, reducing its mass, the overall rate is given as:

$$m_{\text{char}} = m_{\text{Boudouard}} + m_{\text{steam}} + m_{\text{methanation}} \tag{7.84}$$

The conversion of the porous char particle may be modeled assuming that the process follows shrinking particle (diminishing size), shrinking core (diminishing size of the unreacted core), or progressive conversion (diminishing density). The shift reaction is the most important homogenous reaction followed by steam reforming. The bed materials may catalyze the homogeneous reactions, but only in the emulsion phase, because the bubble phase is assumed to be free of solids.

### 7.6.3 Entrained-Flow Gasifiers

Extensive work on the modeling of entrained-flow gasifiers is available in the literature. CFD has been successfully applied to this gasifier type. This section presents a simplified approach to entrained-flow gasification following the work of Vamvuka et al. (1995).

The reactor is considered to be a steady-state, one-dimensional plug-flow reactor in the axial direction and well mixed radially—similar to that shown in Figure 7.11. Fuel particles shrink as they are gasified. Five gas–solid reactions (R1–R5 in Table 7.2) can potentially take place on the char particle surface. The reduction in the mass of char particles is the sum of these individual reactions, so if there are $N_c$ char particles in the unit gas volume, the total reduction, $W_c$, in the plug flow is as shown in the equation that follows the figure.

$$dW_c = -(N_c A\, dz)\sum_{k=1}^{5} r_k(T_s, L_T) \tag{7.85}$$

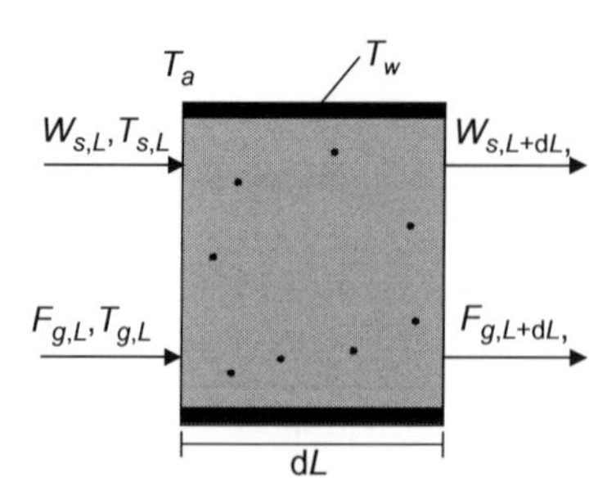


**FIGURE 7.11** One-dimensional entrained-flow model.

where $r_k(T_s, L_r)$ is the surface reaction rate of the $k$th reaction (one of R1–R5) at the reactor's surface temperature, $T_s$, and length, $L_r$. $A$ is its cross-section area.

Gaseous reactants diffuse to the char surface to participate in $k$ reactions. Thus, if $a_{jk}$ is the mass of the $j$th gas, required for the $k$th reaction, the overall diffusion rate of this gas from free stream concentration, $y_j$, to the char surface, $y_{js}$, may be related to the total of all reactions consuming the $j$th gas as follows:

$$\sum_{k=1}^{5} a_{jk} r_k(T_s, L_r) = 4\pi r_c^2 \left[ \frac{D_{gj} P}{R T_g r_c} (y_j - y_{js}) \right] \tag{7.86}$$

where $y_{js}$ and $y_j$ are mole fractions of gas on the char surface and in the bulk gas, respectively; $P$ is the reactor pressure; and $D_{gj}$ is the diffusion coefficient of the $j$th gas in the mixture of gases.

The surface reaction rate, $r_k(T_s, L_r)$, may be written in $n$th-order form as:

$$r_k(T_s, L_r) = 4\pi r_c^2 K_{sk}(T_s)(P y_{js})^n \ \text{mol/s} \tag{7.87}$$

where $n$ is the order of reaction, and $K_{sk}(T_s)$ is the surface reaction rate constant at temperature $T_s$. For conversion of gaseous species, we can write:

$$\frac{dF_{gj}}{dZ} = \pm N_c A \sum_{k=1}^{5} a_{j,k} r_k(T_s, L_r) \tag{7.88}$$

where $a_{j,k}$ is the stoichiometric coefficient for the $j$th gas in the $k$th reaction.

The total molar flow-rate of the $j$th gas is found by adding the contribution of each of nine gas–solid and gas–gas reactions:

$$F_{gj} = F_{gj0} + \sum a_{jk} \xi_k \tag{7.89}$$

where $F_{gj0}$ is the initial flow-rate of the gas.

#### 7.6.3.1 Energy Balance

Some of the five equations (reactions R1–R5) are endothermic while some are exothermic. The overall heat balance of reacting char particles is known from a balance of a particle's heat generation and heat loss to the gas by conduction and radiation.

$$\begin{aligned} \frac{d(W_c C_{pc} T_s)}{dz} &= -N_c A \left[ \sum_{k=1}^{5} r_k(T_s, L_r) \right] \Delta H_k(T_s) \\ &\quad + 4\pi r_c^2 \left[ \frac{\lambda_g}{r_c} (T_s - T_g) + e_p \sigma (T_g^4 - T_c^4) \right] \end{aligned} \tag{7.90}$$

where $C_{pc}$ is the specific heat of the char, $\Delta H_k$ is the heat of reaction of the $k$th reaction at the char surface at temperature $T_s$, $e_p$ is the emissivity of the char particle, $\lambda_g$ is the thermal conductivity of the gas, and $\sigma$ is the Stefan–Boltzmann constant.

A similar heat balance for the gas in an element d$z$ in length can be carried out as:

$$d\left(\frac{\sum_j F_{gj} C_{pg} T_g}{dZ}\right) = -A\left[\sum_{k=6}^{9} \xi_k \Delta H_k(T_g)\right] - 4\pi r_c^2 N_c A\left[\frac{\lambda_g}{r_c}(T_g - T_c) + e_p\sigma(T_g^4 - T_c^4)\right] - [h_{conv}(T_g - T_w) + e_w\sigma(T_g^4 - T_w^4)]\pi D_r \quad (7.91)$$

where $\xi_k$ is the extent of the gas-phase $k$th reaction with the heat of reaction, $\Delta H_k$ $(T_g)$; $h_{conv}$ is the gas-wall convective heat transfer coefficient; and $D_r$ is the reactor's internal diameter.

The first term on the right of Eq. (7.91) is the net heat absorption by the gas-phase reaction, the second is the heat transfer from the gas to the char particles, and the third is the heat loss by the gas at temperature $T_g$ to the wall at temperature $T_w$.

The equations are solved for an elemental volume, $A_r dL_r$, with boundary conditions from the previous upstream cell. The results are then used to solve the next downstream cell.

## SYMBOLS AND NOMENCLATURE

| | |
|---|---|
| $A$ | cross-sectional area of bed or reactor ($m^2$) |
| $A_0$ | preexponential coefficient in Eq. (7.42) ($s^{-1}$) |
| $A_b$, $A_w$ | preexponential coefficients in Eqs. (7.44) and (7.47), respectively ($bar^{-n}$ $s^{-1}$) |
| $A_j$ | total number of atoms of element $j$ entering the reactor (−) |
| $a_{i,j}$ | number of atoms of $j$th element in $i$th species (−) |
| $a_{jk}$ | mass of $j$th gas, required for the $k$th reaction (kg) |
| $C_i$ | molar concentration of $i$th gas ($mol/m^3$) |
| $C_{pc}$ | specific heat of char (kJ/kg K) |
| $C_{pg}$ | specific heat of the bulk gas |
| $D_r$ | internal diameter of the reactor(m) |
| $D_{g,j}$ | diffusion coefficient of the $j$th gas in the mixture of gases ($m^2/s$) |
| $d_b$ | diameter of the bubble (m) |
| $E$ | activation energy (kJ/mol) |
| $e_p$ | emissivity of char particle (−) |
| $F_{gl0}$ | initial flow-rate of the gas (mol/s) |
| $F_{gl}$ | molar flow-rate of the $l$th gas (mol/s) |

| | |
|---|---|
| $G_{total}$ | total Gibbs free energy (kJ) |
| $g$ | acceleration due to gravity, 9.81 (m/s$^2$) |
| $\Delta H_k$ | heat of reaction of $k$th reaction at char surface (kJ/mol) |
| $\Delta H$ | enthalpy change (kJ) |
| $h_i^0$, $h_f^0$ | heat of formation at reference state (kJ/mol) |
| $h_{conv}$ | gas-wall convective heat transfer coefficient (kW/m$^2$ K) |
| $h_m$ | mass-transfer coefficient (kg carbon/m$^2$ bar$^2$ s) |
| $k$ | first-order reaction rate constant (s$^{-1}$) |
| $k_0$ | preexponential factor (s$^{-1}$) |
| $k_{liq}$ | rate constant for the liquid yield of pyrolysis (s$^{-1}$) |
| $k_{BE}$ | bubble–emulsion mass exchange coefficient (m/s) |
| $k_c$ | rate constant for the char yield of pyrolysis (s$^{-1}$) |
| $k_g$ | rate constant for the gas yield of pyrolysis (s$^{-1}$) |
| $k_1$ | rate constant of three primary pyrolysis reactions taken together (s$^{-1}$) |
| $K$ | number of element in Eq. (7.77) |
| $K_{e1}$, $k_{e2}$, $k_{e3}$ | rate constants in Eq. (7.70)–(7.72) (bar$^{-1}$ s$^{-1}$) |
| $K_{sk}$ | surface reaction rate constant for $k$th reaction, mol/m$^2$ bar$^n$ |
| $K_e$, $K_{equilibrium}$ | equilibrium constant (–) |
| $l$ | number of gaseous reactants (–) |
| $L_r$ | length of the reactor (m) |
| L | Lagrangian function (–) |
| $m_b$ | mass of the biomass in the primary pyrolysis process (kg) |
| $m_0$ | initial mass of the biomass (kg) |
| $m_c$ | mass of the biomass remaining after complete conversion (kg) |
| $m$ | reaction order with respect to carbon conversion in Eq. (7.42) (–) |
| $m$, $n$, $p$, $q$ | stoichiometric coefficients in Eqs. (7.27)–(7.29) |
| $n$ | reaction order with respect to the gas partial pressure, Eq. (7.44) (–) |
| $N$ | number of species present (–) |
| $N_c$ | number of char particles in unit gas volume (–) |
| $N_{or}$ | number of orifices in a bed of area ($A_r$) |
| $P_g$ | partial pressure of gasifying agent outside the char particle (bar) |
| $P_i$ | partial pressure of the species $i$ (bar) |
| $P$ | total pressure of the species (bar) |
| $Q$ | char gasification rate (kg Carbon/m$^2$ s) |
| $Q_{gasification}$, $Q_{conv}$, $Q_{rad}$, and $Q_{mass}$ | energy transfer due to gasification, convection, radiation, and mass transfer, respectively (kW/m$^3$ of bed) |
| $R$ | gas constant (8.314 J/mol K or 8.314 × 10$^{-5}$ m$^3$ bar/mol K) |
| $R_c$ | chemical kinetic reaction rate (kg carbon/m$^2$ bar$^2$ s) |
| $R_{m,j}$ | rate of production or consumption of gas species $j$ (kg/m$^3$ s) |
| $r_i$ | reaction rate of the $i$th reaction (s$^{-1}$) |
| $r_c$ | char particle radius (m) |
| $T$ | temperature (K) |
| $T_s$ | surface temperature of char particles (K) |
| $T_g$ | gas temperature (K) |
| $T_w$ | wall temperature (K) |

| | |
|---|---|
| $t$ | time (s) |
| $u_g$ | superficial gas velocity in Eq. (7.79) (m/s) |
| $U$ | fluidization velocity (m/s) |
| $U_B$ | bubble rise velocity (m/s) |
| $U_{mf}$ | minimum fluidization velocity (m/s) |
| $X$ | fractional change in the carbon mass of the biomass (−) |
| $y$ | mole fraction of a species (−) |
| $y_l$ | mole fraction of gas in the bulk (−) |
| $y_{ls}$ | mole fraction of gas on the char surface (−) |
| $z$ | height above grid or distance along a reactor from fuel entry (m) |
| $\alpha_{l,k}$ | stoichiometric coefficient for *l*th gas in *k*th reaction (−) |
| $\beta$ | partition coefficient (−) |
| $\lambda$ | Lagrangian multiplier (−) |
| $\lambda_g$ | thermal conductivity of gas (kJ/m K) |
| $\sigma$ | Stefan–Boltzmann constant ($5.67 \times 10^{-8}$ W/m$^2$/K$^4$) |
| $\Delta G, \Delta G^0$ | change in Gibbs free energy (kJ) |
| $\Delta G^0_{fi}$ | change in Gibbs free energy of formation of species $i$ (kJ) |
| $\Delta\xi_k$ | extent of gas-phase *k*th reaction (−) |
| $\rho_j$ | density of *j*th gas (kg/m$^3$) |
| $\varrho_{mf}$ | voidage at minimum fluidization condition |
| $\rho_g$ | density of the bulk gas |
| $\Delta S$ | entropy change (kJ/K) |

Chapter 8

# Design of Biomass Gasifiers

## 8.1 INTRODUCTION

A gasification plant includes the gasifier reactor as well as its auxiliary or support equipment. So, the design of a gasification plant would involve design of individual units like:

- Gasifier reactor
- Biomass-handling system
- Biomass-feeding system
- Gas-cleanup system
- Ash or solid residue-removal system

This chapter deals with the design of the gasifier reactor alone. Chapter 12 discusses the design of the handling and feeding systems. Gas-cleaning systems are briefly discussed in Chapter 6.

As with most process plant equipment, the design of a gasifier is generally done in the following three major phases:

*Phase 1.* Process design and preliminary sizing.
*Phase 2.* Optimization of design.
*Phase 3.* Detailed mechanical design.

For cost estimation and/or for submission of initial bids, most manufacturers use the first step of sizing the gasifier. The second step is considered only for a confirmed project—that is, when an order is placed and the manufacturer is ready for the final stage of detailed mechanical or manufacturing design. The detailed mechanical design begins after the design is optimized and actual manufacturing is to begin.

This chapter mainly concerns the first phase and, briefly, the second phase (design optimization). To set the ground for design methodologies, a short description of different gasifier types is presented, followed by a discussion of design considerations and design methodologies.

### 8.1.1 Gasifier Types

Gasifiers are classified mainly on the basis of their gas–solid contacting mode and gasifying medium. Based on the gas–solid contacting mode,

**Biomass Gasification, Pyrolysis and Torrefaction.**

**TABLE 8.1 Comparison of Characteristics of Some Commercial Gasifiers**

| Parameters | Fixed/ Moving Bed | Fluidized Bed | Entrained Bed |
|---|---|---|---|
| Feed size | <51 mm | <6 mm | <0.15 mm |
| Tolerance for fines | Limited | Good | Excellent |
| Tolerance for coarse | Very good | Good | Poor |
| Gas exit temperature | 450–650°C | 800–1000°C | >1260°C |
| Feedstock tolerance | Low-rank coal | Low-rank coal and excellent for biomass | Any coal including caking but unsuitable for biomass |
| Oxidant requirements | Low | Moderate | High |
| Reaction zone temperature | 1090°C | 800–1000°C | 1990°C |
| Steam requirement | High | Moderate | Low |
| Nature of ash produced | Dry | Dry | Slagging |
| Cold-gas efficiency | 80% | 89% | 80% |
| Application | Small capacities | Medium-size units | Large capacities |
| Problem areas | Tar production and utilization of fines | Carbon conversion | Raw-gas cooling |

**Source:** Data compiled from Basu (2006).

gasifiers are broadly divided into three principal types (Table 8.1): (i) fixed or moving bed, (ii) fluidized bed, and (iii) entrained-flow bed. Each is further subdivided into specific commercial types as shown in Figure 8.1.

One particular gasifier type is not necessarily suitable for the full range of gasifier capacities. There is an appropriate range of application for each. For example, the moving-bed (updraft and downdraft) type is used for smaller units ($<10\ MW_{th}$); the fluidized-bed type is more appropriate for intermediate units ($5-100\ MW_{th}$); entrained-flow reactors are used for large-capacity units ($>50\ MW_{th}$). Figure 8.2 developed with data from Maniatis (2001) and Knoef (2005) shows the overlapped range of application for different types of gasifiers. Downdraft gasifiers are for the smallest size while entrained-flow gasifiers are the largest size.

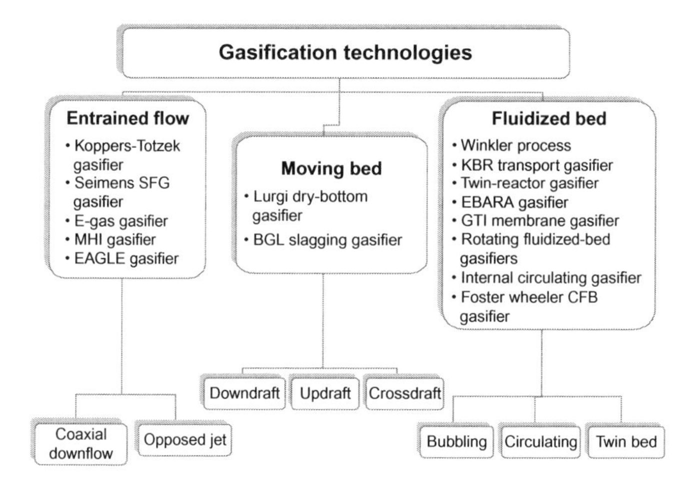


**FIGURE 8.1** Gasification technologies and their commercial suppliers.

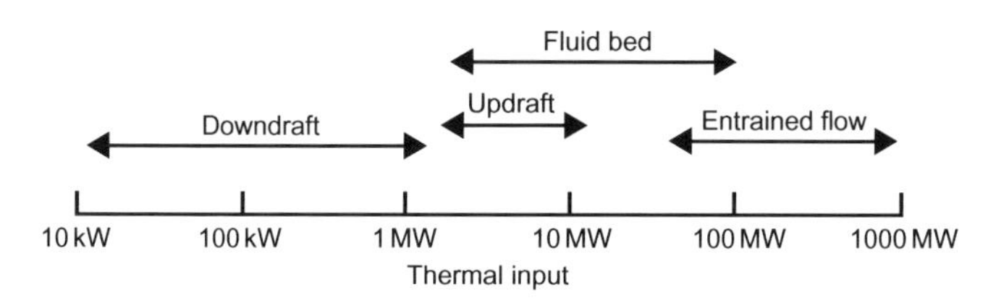


**FIGURE 8.2** Range of applicability for biomass gasifier types.

## 8.2 FIXED-BED/MOVING-BED GASIFIERS

In entrained-flow and fluidized-bed gasifiers, the gasifying medium conveys the fuel particles through the reactor, but in a *fixed-bed* (also known as *moving-bed*) gasifier, the fuel is supported on a grate (hence its name). This type is also called *moving bed* because the fuel moves down in the gasifier as a plug. Fixed-bed gasifiers can be built inexpensively in small sizes, which is one of their major attractions. For this reason, large numbers of small-scale moving-bed biomass gasifiers are in use around the world.

Both mixing and heat transfer within the moving (fixed) bed are rather poor, which makes it difficult to achieve uniform distribution of fuel, temperature, and gas composition across the cross-section of the gasifier. Thus, fuels that are prone to agglomeration can potentially form agglomerates during gasification. This is why fixed-bed gasifiers are not very effective for biomass fuels or coal with a high caking index in large-capacity units.

There are three main types of fixed- or moving-bed gasifier: (i) updraft, (ii) downdraft, and (iii) crossdraft. Table 8.2 compares their characteristics.

**TABLE 8.2 Characteristics of Fixed-Bed Gasifiers**

| Fuel (wood) | Updraft | Downdraft | Crossdraft |
|---|---|---|---|
| Moisture wet basis (%) | 60 max | 25 max | 10–20 |
| Dry-ash basis (%) | 25 max | 6 max | 0.5–1.0 |
| Ash-melting temperature (°C) | >1000 | >1250 | |
| Size (mm) | 5–100 | 20–100 | 5–20 |
| Application range (MW) | 2–30 | 1–2 | |
| Gas exit temperature (°C) | 200–400 | 700 | 1250 |
| Tar (g/N $m^3$) | 30–150 | 0.015–3.0 | 0.01–0.1 |
| Gas LHV (MJ/N $m^3$) | 5–6 | 4.5–5.0 | 4.0–4.5 |
| Hot-gas efficiency (%) | 90–95 | 85–90 | 75–90 |
| Turn-down ratio (–) | 5–10 | 3–4 | 2–3 |
| Hearth load (MW/$m^2$) | <2.8 | | |

**Source:** Adapted from Knoef (2005), p. 26.

### 8.2.1 Updraft Gasifiers

An updraft gasifier is one of the oldest and simplest of all designs. Here, the gasification medium (air, oxygen, or steam) travels upward while the bed of fuel moves downward, and thus the gas and solids are in countercurrent mode. The product gas leaves from near the top of the gasifier as shown in Figure 8.3. The gasifying medium enters the bed through a grate or a distributor, where it meets with the hot bed of ash. The ash drops through the grate, which is often made moving (rotating or reciprocating), especially in large units to facilitate ash discharge. Chapter 7 describes this process in more detail.

Updraft gasifiers are suitable for high-ash (up to 25%), high-moisture (up to 60%) biomass. They are also suitable for low-volatile fuels such as charcoal. Tar production is very high (30–150 g/$nm^3$) in an updraft gasifier, which makes it unsuitable for high-volatility fuels. On the other hand, as a countercurrent unit, an updraft gasifier utilizes combustion heat very effectively and achieves high cold-gas efficiency (Section 8.11.1). Updraft is more suitable for direct firing, where the gas produced is burnt in a furnace or boiler with no cleaning or cooling required. Here, the tar produced does not have to be cleaned.

Updraft gasifiers find commercial use in small units like improvised cooking stoves in villages and in large units like South African Synthetic

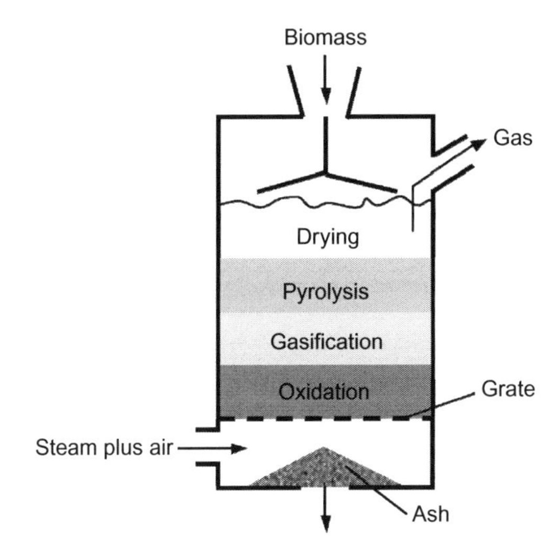


**FIGURE 8.3** Schematic of an updraft gasifier.

Oils (SASOL) for production of gasoline from coal. The following is a brief description of two important large-scale commercial updraft gasifier technologies.

### *8.2.1.1 Dry-Ash Gasifier*

Lurgi, a process development company, developed a pressurized dry-ash updraft gasifier. It is called *dry ash* because the ash produced is not molten. One that produces molten ash is called a *slagging* gasifier.

Though the peak temperature (in the combustion zone) is 1200°C, the maximum gasification temperature is 700–900°C. The reactor pressure is in the neighborhood of 3 MPa, and the residence time of coal in the gasifier is between 30 and 60 min (Ebasco Services Inc., 1981). The gasification medium is a mixture of steam and oxygen, steam and air, or steam and oxygen-enriched air. It uses a relatively high steam/fuel carbon ratio (~1.5).

The coal is first screened to between 3 and 40 mm (Probstein and Hicks, 2006, p. 162) and then fed into a lock hopper. The gasifying agent moves upward in the gasifier while the solids descend. The reactor is a double-walled pressure vessel. Between the two walls lies water that quickly boils into steam under pressure, utilizing the heat loss from the reactor. As the coal travels down the reactor, it undergoes drying, devolatilization, gasification, and combustion. Typical residence time in the gasifier is about an hour (Probstein and Hicks, 2006, p. 162). In a dry-ash gasifier, the temperature is lower than the melting point of ash, so the solid residue dries and is removed from the reactor by a rotating grate.

The dry-ash technology has been used at SASOL in South Africa, the world's biggest gasification complex. SASOL produces 55 million $nm^3$/day of syngas, which is used to produce 170,000 bbl/day of Fischer–Tropsch liquid fuel (Figure 11.3).

### *8.2.1.2 Slagging Gasifier*

The British Gas/Lurgi consortium developed a moving-bed gasifier that works on the same principle as the dry-ash gasifier, except a much higher temperature (1500–1800°C) is used in the combustion zone to melt the ash (hence its name *slagging gasifier*). Such a high temperature requires a lower steam-to-fuel ratio (~0.58) than that used in dry-ash units (Probstein and Hicks, 2006, p. 169).

Coal crushed to 5–80 mm is fed into the gasifier through a lock hopper system (Minchener, 2005). The gasifier's tolerance for coal fines is limited, so briquetting is used in places where the coal carries too many of them. Gasification agents, oxygen and steam, are introduced into the pressurized (~3 MPa) gasifier vessel through sidewall-mounted tuyers (lances) at the elevation where combustion and slag formation occur.

The coal introduced at the top gradually descends through several process zones. The feed is first dried in the top zone and then devolatilized as it descends. The descending coal is transformed into char and then passes into the gasification (reaction) zone. Below this zone, any remaining carbon is oxidized, and the ash content melts, forming slag. Slag is withdrawn from the slag pool through an opening in the hearth plate at the bottom of the gasifier vessel. The product gas leaves from the top, typically at 400–500°C (Minchener, 2005).

## 8.2.2 Downdraft Gasifiers

A downdraft gasifier is a cocurrent reactor where air enters the gasifier at a certain height below the top. The product gas flows downward (giving the name *downdraft*) and leaves from lower section of the gasifier through a bed of hot ash (Figures 8.4 and 8.5). Since it passes through the high-temperature zone of hot ash, the tar in the product gas finds favorable conditions for cracking (see Chapter 6). For this reason, a downdraft gasifier, of all types, has the lowest tar production rate.

Air from a set of nozzles, set around the gasifier's periphery, flows downward and meets with pyrolyzed char particles, developing a combustion zone (zone III shown schematically in Figure 8.5 and described in the discussion of throatless downdraft gasifiers that follows) of about 1200–1400°C. Then the gas descends further through the bed of hot char particles (zone IV), gasifying them. The ash produced leaves with the gas, dropping off at the bottom of the reactor.

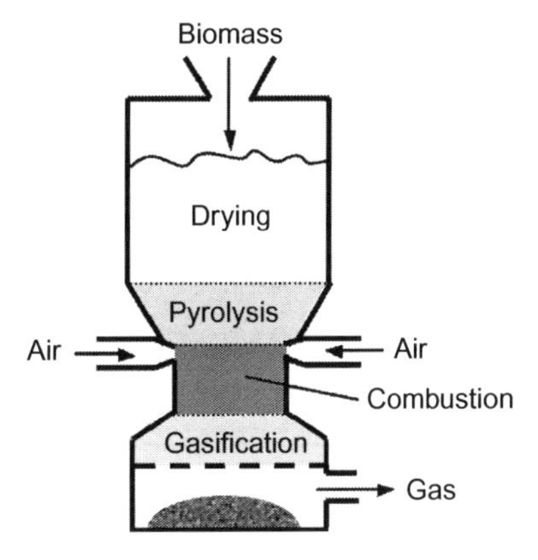


**FIGURE 8.4** Schematic of a throated-type downdraft gasifier.

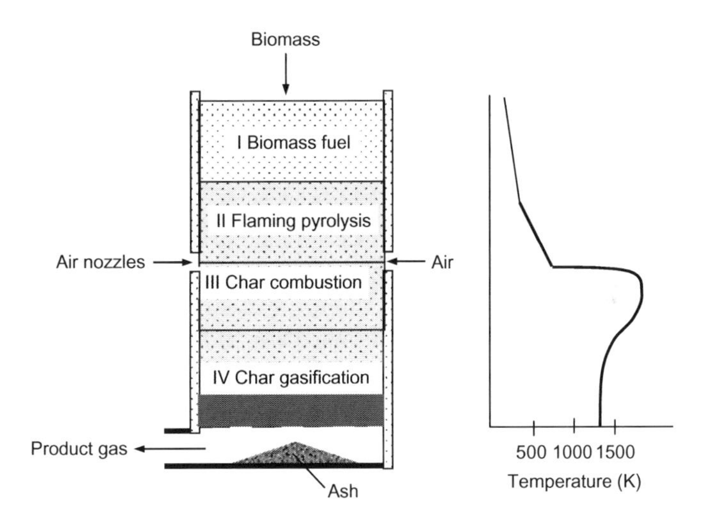


**FIGURE 8.5** Schematic of the operation of a throatless downdraft gasifier. Temperature gradient along the height shown at the right.

Downdraft gasifiers work well with internal-combustion engines that need cleaner gas. The engine suction draws air through the bed of fuel, and gas is produced at the end. Low tar content (0.015–3 g/nm$^3$) in the product gas is another motivation for their use with internal-combustion engines. A downdraft gasifier requires a shorter time (20–30 min) to ignite and bring the plant up to working temperature compared to the time required by an updraft gasifier.

There are two principal types of downdraft gasifier: throatless and throated. The throatless (or open-core) type is illustrated in Figure 8.5.

Reactions in different zones and at different temperatures are plotted on the right. The throated (or constricted) type is shown in Figure 8.4.

#### *8.2.2.1 Throatless Gasifier*

This gasifier type is also called *open top* or *stratified throatless*. Here, the top is exposed to the atmosphere, and there is no constriction in the gasifier vessel because the walls are vertical. Figure 8.5 shows that a throatless design allows unrestricted movement of the biomass down the gasifier, which is not possible in the throated type shown in Figure 8.4. The absence of a throat avoids bridging or channeling. Open core is another throatless design, but here air is not added from the middle as in other types of downdraft gasifiers. Air is drawn into the gasifier from the top by the suction created downstream of the gasifier. Such gasifiers are suitable for finer or lighter fuels. Rice husk is an example of such biomass.

The followings are some of the shortcomings of a downdraft gasifier:

1. It operates best on pelletized fuel instead of fine light biomass.
2. The moisture in the fuel must not exceed 25%.
3. A large amount of ash and dust remain in the product gas.
4. As a result of its high exit temperature, it has a lower gasification efficiency.

#### Operating Principle

Because an open top, or a throatless, gasifier is simple in construction, it is used to describe the gasification process in the downdraft gasifier (Figure 8.5). The throatless process can be divided into four zones (Reed and Das, 1988, p. 39). The first, or uppermost, zone receives raw fuel from the top that is dried in the air drawn through the first zone. The second zone receives heat from the third zone principally by thermal conduction.

During its journey through the first zone, the biomass heats up (zone I in Figure 8.5). Above 350°C, it undergoes pyrolysis, breaking down into charcoal, noncondensable gases ($CO$, $H_2$, $CH_4$, $CO_2$, and $H_2O$), and tar vapors (condensable gases). The pyrolysis product in zone II receives only a limited supply of air from below and burns in a fuel-rich flame. This is called *flaming pyrolysis*. Most of the tar and char produced burn in zone III, where they generate heat for pyrolysis and subsequent endothermic gasification reactions (Reed and Das, 1988, p. 28).

Zone III contains ash and pyrolyzed char produced in zone II. While passing over the char, hot gases containing $CO_2$ and $H_2O$ undergo steam gasification and Boudouard reactions, producing $CO$ and $H_2$. The temperature of the downflowing gas reduces modestly, owing to the endothermic gasification reactions, but it is still above 700°C.

The bottommost layer (zone IV) consists of hot ash and/or unreacted charcoal, which crack any unconverted tar in this layer. Figure 8.5 shows the reactions and temperature distribution along the gasifier height. In one version of the throatless downdraft gasifier, the *open-core* type, the air enters from the top along with the feed. This type is free from some of the problems of other downdraft gasifiers.

### 8.2.2.2 Throated Gasifier

The cross-sectional area of a throated (also called *constricted*) gasifier is reduced at the throat and then expanded as shown in Figure 8.4. The purpose of the constriction is for the oxidation (combustion) zone to be at the narrowest part of the throat and to force all of the pyrolysis gas to pass through this narrow passage. Air is injected through nozzles just above the constriction. The height of the injection is about one-third of the way up from the bottom (Reed and Das, 1988, p. 33).

The movement of the entire mass of pyrolysis products through this hot and narrow zone results in a uniform temperature distribution over the cross-section and allows most of the tar to crack there. In the 1920s, a French inventor, Jacques Imbert, developed the original design, which is popularly known as an *Imbert gasifier* (Figure 8.6).

The fuel, fed at the top, descends along a cylindrical section that serves as storage. After the biomass is pyrolyzed, the air burns the pyrolysis product and/or some charcoal produced from pyrolysis. The hot char and the pyrolysis product pass through the throat, where most of the tar is cracked and the char is gasified. Figure 8.6 shows a flat-type throat construction, but it can be a V-type like in Figure 8.4.

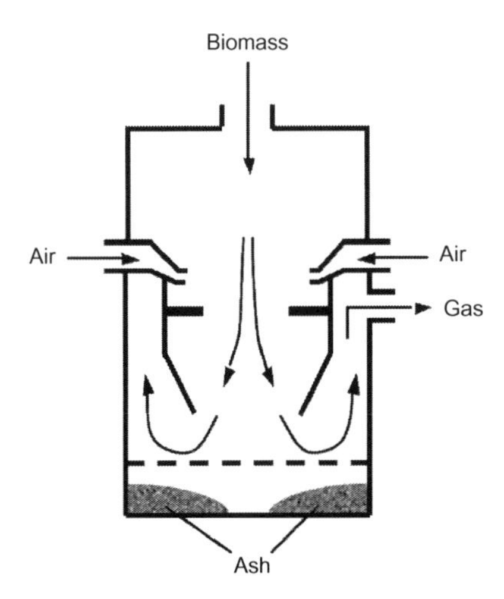


**FIGURE 8.6** Constricted downdraft gasifier (Imbert type). Air/oxygen is added through nozzles around the vessel just above the constriction. Source: *Adapted from Reed and Das (1988), p. 39.*

Throated downdraft gasifiers are not suitable for scale-up to larger sizes because they do not allow for uniform distribution of flow and temperature in the constricted area. To above 1 $MW_{th}$ capacity, an annular constriction can be employed, but this has not been the practice to date.

### 8.2.3 Crossdraft Gasifiers

A crossdraft gasifier is a cocurrent moving-bed reactor, in which the fuel is fed from the top and air is injected through a nozzle from the side (Figure 8.7) of the gasifier. It is primarily used for gasification of charcoal with very low ash content. Unlike the downdraft and updraft types, it releases the product from its sidewall opposite to the entry point of the air for gasification. Because of this configuration, the design is also referred to as *sidedraft*. High-velocity air enters the gasifier through a nozzle set at a certain height above the grate. Excess oxygen in front of the nozzles facilitates combustion (oxidation) of part of the char, creating a very-high-temperature (>1500°C) zone. The remaining char is then gasified to CO downstream in the next zone (Figure 8.7). The product gas exits from the opposite side of the gasifier. Heat from the combustion zone is conducted around the pyrolysis zone, so the fresh biomass is pyrolyzed while passing through it.

This type of gasifier is generally used in small-scale biomass units. One of its important features is a relatively small reaction zone with low thermal capacity, which gives a faster response time than that of any other moving-bed type. Moreover, start-up time (5–10 min) is much shorter than in downdraft and updraft units. These features allow a sidedraft gasifier to respond well to load changes when used directly to run an engine. Because its tar production is low (0.01–0.1 $g/nm^3$), a crossdraft gasifier requires a relatively simple gas-cleaning system.

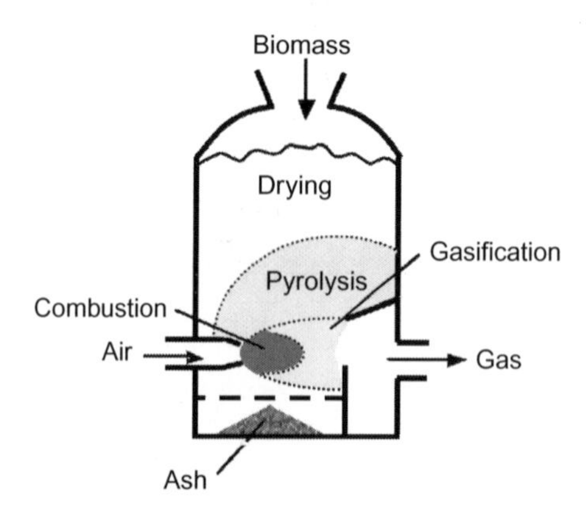


**FIGURE 8.7** Schematic of a crossdraft gasifier.

Crossdraft gasifiers can be very light and small (<10 $kW_e$). Since layers of fuel and ash insulate the walls from the high-temperature zone, the gasifier vessel can be constructed of ordinary steel with refractory linings on the nozzle and gas exit zone.

The crossdraft design is less suitable for high-ash or high-tar fuels, but it can handle high-moisture fuels if the top is open so that the moisture can escape. Particle size should be controlled, as unscreened fuel runs the risk of bridging and channeling. Crossdraft gasifiers work better with charcoal or pyrolyzed fuels. For unpyrolyzed fuels, the height of the air nozzle above the grate becomes critical (Reed and Das, 1988, p. 32).

## 8.3 FLUIDIZED-BED GASIFIERS

Fluidized-bed gasifiers are noted for their excellent mixing and temperature uniformity. A fluidized bed is made of granular solids called *bed materials*, which are kept in a semi-suspended condition (*fluidized state*) by the passage of the gasifying medium through them at the appropriate velocities. The excellent gas–solid mixing and the large thermal inertia of the bed make this type of gasifier relatively insensitive to the fuel's quality (Basu, 2006). Along with this, the temperature uniformity greatly reduces the risk of fuel agglomeration.

The fluidized-bed design has proved to be particularly advantageous for gasification of biomass. Its tar production lies between that for updraft (~50 $g/nm^3$) and downdraft gasifiers (~1 $g/nm^3$), with an average value of around 10 $g/nm^3$ (Milne et al., 1998, p. 14). There are two principal fluidized-bed types: bubbling and circulating.

### 8.3.1 Bubbling Fluidized-Bed Gasifier

The bubbling fluidized-bed gasifier, developed by Fritz Winkler in 1921, is perhaps the oldest commercial application of fluidized beds; it has been in commercial use for many years for the gasification of coal (Figure 8.8). For biomass gasification, it is one of the most popular options. A fairly large number of bubbling fluidized-bed gasifiers of varying designs have been developed and are in operation (Lim and Alimuddin, 2008; Narváez et al., 1996).

Because they are particularly suitable for medium-size units (<25 $MW_{th}$), many biomass gasifiers operate on the bubbling fluidized-bed regime. Depending on operating conditions, bubbling-bed gasifiers can be grouped as low-temperature and high-temperature types. They can also operate at atmospheric or elevated pressures.

In the most common type of fluidized bed, biomass crushed to less than 10 mm is fed into a bed of hot materials. These bed materials are fluidized with steam, air, or oxygen, or their combination, depending on the choice of gasification medium. The ash generated from either the fuel or the inorganic

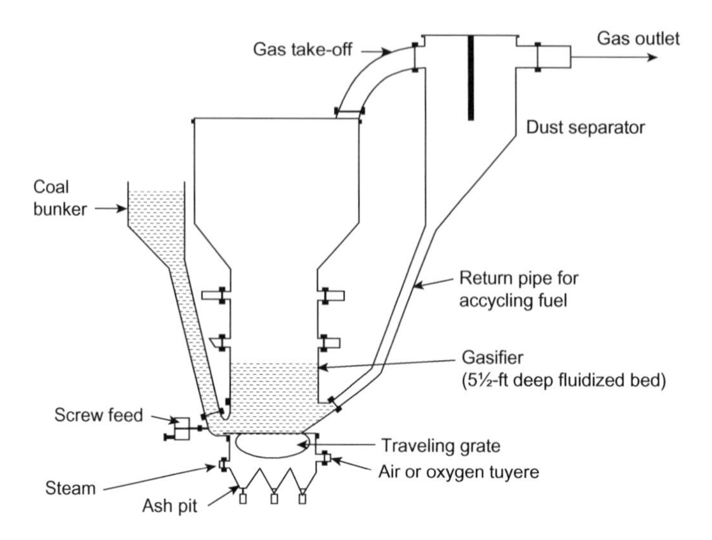


**FIGURE 8.8** A sketch of the original Winkler bubbling fluidized-bed gasifier.

materials associated with it are drained easily from the bottom of the bed. The bed temperature is normally kept below 980°C for coal and below 900°C for biomass to avoid ash fusion and consequent agglomeration.

The gasifying medium may be supplied in two stages. The first-stage supply is adequate to maintain the fluidized bed at the desired temperature; the second-stage supply, added above the bed, converts entrained unreacted char particles and hydrocarbons into useful gas.

High-temperature Winkler (HTW) gasification is an example of high-temperature, high-pressure bubbling fluidized-bed gasification for coal and lignite. Developed by Rheinbraun AG of Germany, the process employs a pressurized fluidized bed operating below the ash-melting point. To improve carbon conversion efficiency, small char particles in the raw gas are separated by a cyclone and returned to the bottom of the main reactor (Figure 8.9).

The gasifying medium (steam and oxygen) is introduced into the fluidized bed at different levels as well as above it. The bed is maintained at a pressure of 10 bar while its temperature is maintained at about 800°C to avoid ash fusion. The overbed supply of the gasifying medium raises the local temperature to about 1000°C to minimize production of methane and other hydrocarbons.

The HTW process produces a better-quality gas compared with the gas that is produced by traditional low-temperature fluidized beds. Though originally developed for coal, it is suitable for lignite and other reactive fuels like biomass and treated municipal solid waste (MSW).

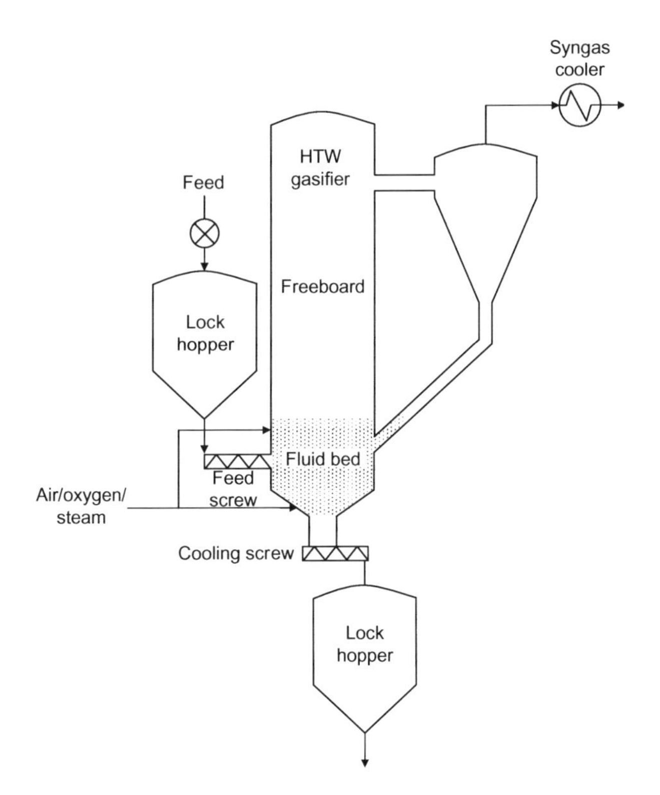


**FIGURE 8.9** HTW bubbling fluidized-bed gasifier.

## 8.3.2 Circulating Fluidized-Bed Gasifier

A circulating fluidized-bed (CFB) gasifier has a special appeal for biomass gasification because of the long gas residence time it provides. It is especially suitable for fuels with high volatiles. A CFB reactor typically comprises a riser, a cyclone, and a solid recycle device (Figure 8.10). The riser serves as the gasifier reactor.

Although the HTW process (Figure 8.9) appears similar to a CFB, it is only a bubbling bed with limited solid recycle. The circulating and bubbling fluidized beds are significantly different in their hydrodynamic behavior. In a CFB, the solids are dispersed all over the tall riser, allowing a long residence time for the gas as well as for the fine particles. The fluidization velocity in a CFB is much higher (3.5–5.5 m/s) than that in a bubbling bed (0.5–1.0 m/s). Also, there is large-scale migration of solids out of the CFB riser. These are captured and continuously returned to the riser's base. The recycle rate of the solids and the fluidization velocity in the riser are sufficiently high to maintain the riser in a special hydrodynamic condition known

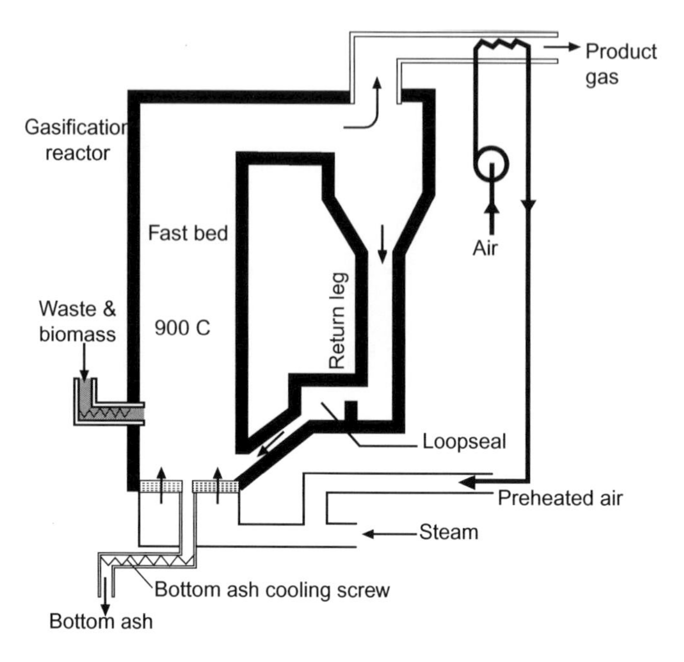


**FIGURE 8.10** Circulating fluidized bed gasifier.

as *fast-fluidized bed*. Depending on the fuel and the application, the riser operates at a temperature of 800–1000°C.

The hot gas from the gasifier passes through a cyclone, which separates most of the solid particles associated with it, and the loop seal returns the particles to the bottom of the gasifier. Foster Wheeler developed a CFB gasifier where an air preheater is located in the standpipe below the cyclone to raise the temperature of the gasification air and indirectly raise the gasifier temperature (Figure 8.10).

Many commercial gasifiers of this type have been installed in different countries. One of the biggest is a 140 MW CFB gasifier attached to a 560 $MW_e$ pulverized coal (PC) fired unit at Vaasa, Finland, for biomass cofiring. It provides a cheap supplementary fuel by gasifying wood, peat and straw replacing upto 40% coal. Several manufacturers around the world have developed versions of the CFB gasifier that work on the same principle and vary only in engineering details.

### *8.3.2.1 Transport Gasifier*

This type of gasifier has the characteristics of both entrained-flow and fluidized-bed reactors. The hydrodynamics of a transport gasifier is similar to that of a fluid catalytic cracking reactor. A transport gasifier operates at

**TABLE 8.3 Comparison of Hydrodynamic Operating Conditions of a Commercial Transport Gasifier and CFB of Fluid Catalyst Cracking Units**

| | Reactor type | | |
|---|---|---|---|
| Parameter | Transport (Smith et al., 2002) | CFB (Petersen and Werther, 2005) | Fluid catalytic cracker (Zhu and Venderbosch, 2005) |
| Particle size (μm) | 200–350 | 180–230 | 20–150 |
| Riser velocity (m/s) | 12–18 | 3.5–5.0 | 6–28 |
| Circulation rate (kg/$m^2$ s) | 730–3400 | 2.5–9.2[a] | 400–1200 |
| Riser temperature (°C) | 910–1010 | 800–900 | 500–550 |
| Riser pressure (bar) | 140–270 psig | 1 bar | 150–300 kPa |
| Operation as | KBR gasifier | CFB gasifier | FCC cracker |

[a]*Computed from comparable units. KBR - Kellon, Brown and Root*

circulation rates, velocities, and riser densities considerably higher than those of a conventional CFB. This results in higher throughput, better mixing, and higher mass and heat-transfer rates. The fuel particles are also very fine (Basu, 2006) and as such it requires a pulverizer or a hammer mill. A comparison of typical hydrodynamic operating conditions in a transport gasifier, CFB, and fluid catalytic cracking unit is given in Table 8.3.

A transport gasifier consists of a mixing zone, a riser, a gas–solid separator, a standpipe, and a J-leg. Coal, sorbent (for sulfur capture), and air are injected into the reactor's mixing zone. The gas–solid disengager removes the larger carried-over particles, and the separated solids return to the mixing section through the J-valve located at the base of the standpipe (Figure 8.11). Most of the remaining finer particles are removed by a cyclone located downstream of the disengager from which the gas exits the reactor. The reactor can use either air or oxygen as the gasification medium.

Use of oxygen as the gasifying medium avoids nitrogen, the diluting agent in the product gas. For this purpose, air is more suitable for power generation, while oxygen is more suitable for chemicals production. The transport gasifier has proved to be effective for gasification of coal, but it is yet to be proven for biomass.

### *8.3.2.2 Twin Reactor System*

One of the major problems in air gasification of coal or biomass is the dilution of its product gas by the nitrogen in the air. This air is essential for the

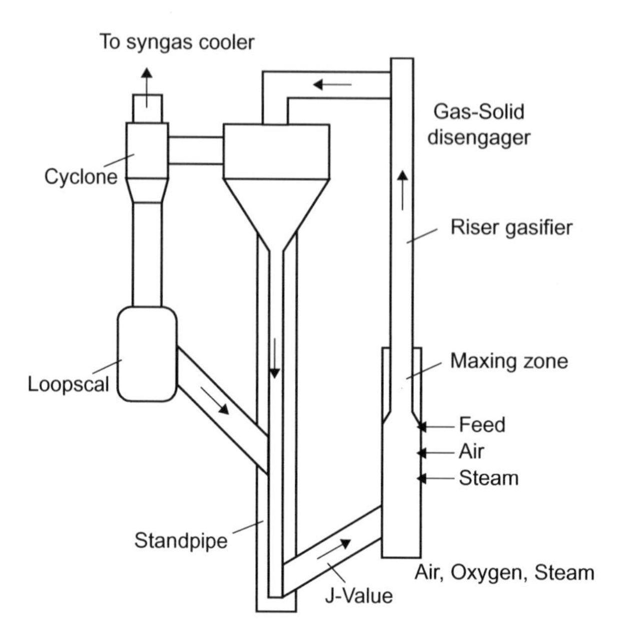


**FIGURE 8.11** A sketch of a typical transport fluidized-bed gasifier.

exothermic combustion reaction necessary for a self-sustained gasifier. To avoid the dilution, oxygen could be used instead, but oxygen gasification is expensive and highly energy intensive (see Example 8.5). A twin reactor (e.g., a dual fluidized bed) overcomes this problem by separating the combustion reactor from the gasification reactor such that the nitrogen released in the air combustion does not dilute the product gas. Twin reactor systems are used for coal and biomass. They are either externally or internally circulating.

This type of system has some limitations; for example, Corella et al. (2007) identified two major design issues with the twin or dual fluidized-bed system:

1. Biomass contains less char than coal contains; however, if this char is used for gasification, the amount of char available may not be sufficient to provide the required endothermic heat to the gasifier reactor to maintain a temperature above 900°C. This external heating may be necessary.
2. Though the gasifier runs on steam, only a small fraction (<10%) of the steam participates in the gasification reaction; the rest of it simply leaves the gasifier, consuming a large amount of heat and diluting the product gas.

The Technical University of Vienna used an externally circulating system to gasify various types of biomass in an industrial plant in Gussing, Austria. The system is comprised of a bubbling fluidized-bed gasifier and a CFB combustor (Figure 8.12). The riser in a CFB that is fluidized by air operates

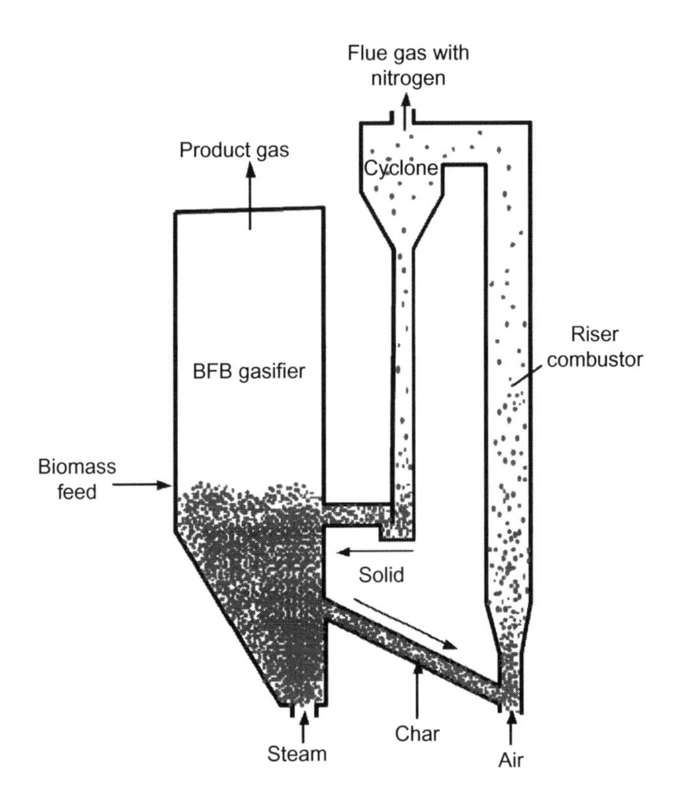


**FIGURE 8.12** Twin reactor (dual fluidized-bed) gasifier.

as a combustor; the bubbling fluidized bed in the return leg operates as a gasifier. Pyrolysis and gasification take place in the bubbling fluidized bed, which is fluidized by superheated steam. Unconverted char and tar move to the riser through a nonmechanical valve.

Tar and gas produced during pyrolysis are combusted in the riser's combustion zone. Heat generated by combustion raises the temperature of the inert bed material to around 900°C. This material leaves the riser and is captured by the cyclone at the riser exit. The collected solids drop into a standpipe and are then circulated into the bubbling fluidized-bed reactor to supply heat for its endothermic reactions. The char is gasified in the bubbling bed in the presence of steam, producing the product gas. This system overcomes the problem of tar by burning it in the combustor. In this way, a product gas relatively free of tar can be obtained.

Rentech-Silvagas process is another technology based on the externally circulating principle. Here, both the combustor and the gasifier work on CFB principles.

Besides these, there is an internally circulating system. In the internally circulating design, the fluidized-bed reactor is divided into two chambers

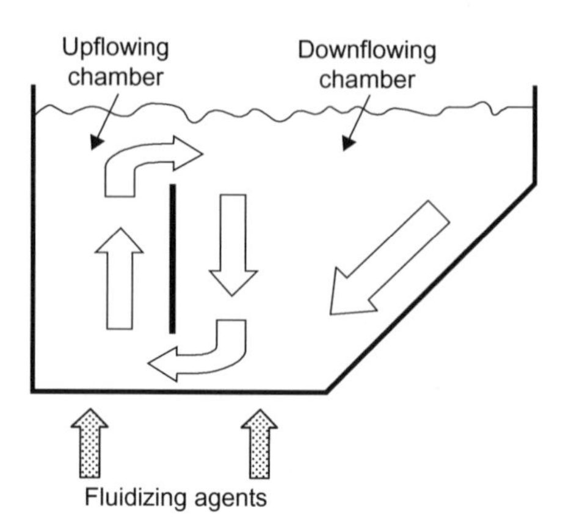


**FIGURE 8.13** Internally circulating dual fluidized-bed gasifier.

and connected by a window at the bottom of the division wall separating them. The chambers are fluidized at different velocities (Figure 8.13), which result in their having varying bed densities. As the bed height is the same in both, the hydrostatic pressure at the bottom of the two chambers is different. The biomass and sand, thus, flow from the higher-density chamber to the lower-density chamber, creating a continuous circulation of bed materials similar to the natural circulation in a boiler. This helps increase the residence time of solids in the fluidized bed.

Such an arrangement can provide a more uniform distribution of biomass particles in the reactor, with increased gasification yield and decreased tar and fine solids (char) in the syngas (Freda et al., 2008). A special feature of the twin reactor is that more air or oxygen can be added in one part of the bed to encourage combustion, and more steam can be added in another part to encourage gasification.

### *8.3.2.3 Chemical Looping Gasifier*

Primary motivation of chemical looping gasification is production of two separate streams of gases—a product gas rich in hydrogen and a gas stream rich in carbon dioxide ($CO_2$) such that the latter can be sequestrated while the hydrogen can be used for applications that require hydrogen-rich gas. The system uses calcium oxide as a carrier of carbon dioxide between two reactors: a gasifier (bubbling fluidized bed) and a regenerator (CFB). The $CO_2$ produced during gasification is captured by the CaO and released in a second reactor during sorbent regeneration.

Figure 8.14 is a schematic of the chemical looping process. Biomass is fed into the gasifier that receives calcium oxide from the regenerator and superheated steam from an external source. During gasification, the $CO_2$

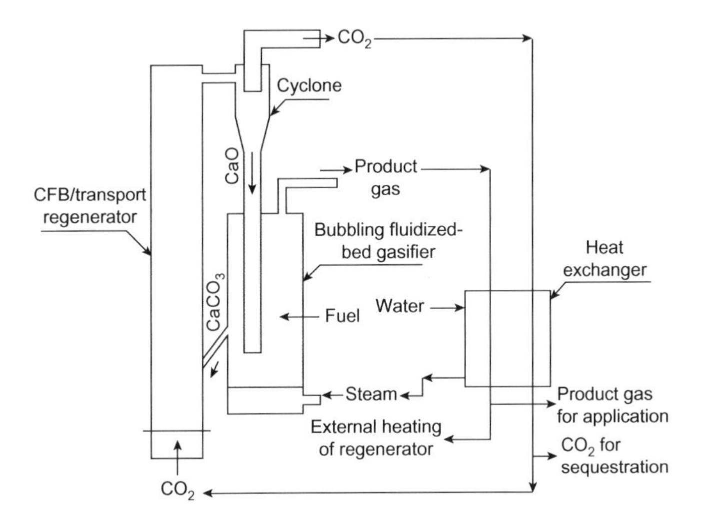


**FIGURE 8.14** Chemical looping gasification with CaO as the carrier of $CO_2$ between the gasifier and the regenerator.

produced is captured by the calcium oxide that makes up the bubbling fluidized bed (Acharya et al., 2009) as follows:

$$\text{Gasification reaction: } C_nH_hO_o + (2n-p)H_2O + n\text{CaO} \leftrightarrow n\text{CaCO}_3 + \left(\frac{h}{2} + 2n - o\right)H_2 \quad (8.1)$$

$$CO + H_2O \leftrightarrow CO_2 + H_2 \quad (8.2)$$

$$CO_2 \text{ removal reaction: CaO} + CO_2 \rightarrow \text{CaCO}_3 \quad (8.3)$$

The removal of the reaction product, $CO_2$, from the system as it is produced increases the rate of forward reaction (Eq. (8.2)), enhancing the water–gas shift reaction, therefore yielding more hydrogen in the product gas. The calcium carbonate formed in the gasifier (Eq. (8.3)) is transferred to a circulating/transport regenerator, where it is calcined into calcium oxide and carbon dioxide.

$$\text{Regeneration: CaCO}_3 \rightarrow \text{CaO} + CO_2 + 178.3\ \text{kJ/mol} \quad (8.4)$$

The carbon dioxide and the product gas leave the regenerator and gasifier, respectively, at a high temperature. The hot product can be used for generation of steam needed for gasification. The extent of calcination of calcium carbonate depends on several factors including the fluidizing medium in the regenerator section, temperature, and residence time. If the medium is carbon dioxide as shown in Figure 8.14, the conversion is relatively low

**TABLE 8.4** Effect of Medium, Temperature, and Residence Time on the Calcination of $CaCO_3$ in a Lime-Based Chemical Looping Gasifier

| Medium | $CO_2$ | | $H_2O$ | |
|---|---|---|---|---|
| Temperature (°C) | Conversion (%) | Time (min) | Conversion (%) | Time (min) |
| 600 | | | 8.78 | 30 |
| 700 | | | 73.22 | 30 |
| 800 | 7.58 | 30 | 96.94 | 30 |
| 900 | 20 | 30 | 100 | 25 |
| 950 | 72.89 | 30 | 100 | 19.16 |
| 1000 | 92.95 | 30 | 100 | 10 |

**Source:** Data taken from Acharya, et al. (2012).

(Table 8.4). If it is, on the other hand steam, a very high conversion is achieved even at lower temperature.

## 8.4 ENTRAINED-FLOW GASIFIERS

It is the most successful and widely used type of gasifier for large-scale gasification of coal, petroleum coke, and refinery residues. Entrained-flow gasifier is ideally suited to most types of coal except low-rank coal, which, like lignite and biomass, is not attractive because of its large moisture content. High-ash coal is also less suitable because cold-gas efficiency decreases with increasing ash content. For slurry-fed coal, the economic limit is 20% ash; for dry feed it is 40% (Higman and Burgt, 2008, p. 122).

The suitability of entrained-flow gasification for biomass is questionable for a number of reasons. Owing to a short residence time (a few seconds) in entrained-flow reactors, the fuel needs to be very fine, but grinding fibrous biomass into such fine particles is difficult. Entrained-flow gasifiers need ash to be molten. For biomass with high CaO and low alkali metals (Na, K), the ash-melting point is high (Mettanant et al., 2009), and therefore to provide for such high combustion temperature, a higher amount of oxygen is required. On the other hand, for biomass with high alkali content, the ash-melting point is much lower. This reduces the oxygen required to raise the temperature of the ash above its melting points. However, molten biomass ash is highly aggressive, which greatly shortens the life of the gasifier's refractory lining.

For these reasons, entrained-flow reactors are not preferred for biomass gasification. Still, they have the advantage of easily destroying tar, which is very high in biomass and is a major problem in biomass gasification.

Entrained-flow gasifiers are essentially cocurrent plug-flow reactors, where gas and fuel travel. The hydrodynamics is similar to that of the well-known PC boiler, where the coal is ground in a pulverizing mill to sizes below 75 μm and then conveyed by part of the combustion air to a set of burners suitably located around the furnace. The reactor geometry of the entrained-flow gasifier is much different from the furnace geometry of a PC boiler. Additionally, an entrained-flow gasifier works in a sub-stoichiometric supply of oxygen, whereas a PC boiler requires excess oxygen.

The gasification temperature of an entrained-flow gasifier generally well exceeds 1000°C. This allows production of a gas that is nearly tar-free and has very low methane content. A properly designed and operated entrained-flow gasifier can have a carbon conversion rate close to 100%. The product gas, being very hot, must be cooled in downstream heat exchangers that produce the superheated steam required for gasification.

Figure 8.15 describes the working principle of an entrained-flow gasifier by means of a simplified sketch. The high-velocity jet forms a recirculation zone near the entry point. Fine fuel particles are rapidly heated by radiative heat from the hot walls of the reactor chamber and from the hot gases downstream and start burning in excess oxygen. The bulk of the fuel is consumed near the entrance zone through devolatilization; here, the temperature may rise to as high as 2500°C.

The combustion reaction consumes nearly all of the oxygen feed, so the residual char undergoes gasification reactions in $CO_2$ and $H_2O$ environments downstream of this zone. These reactions are relatively slow compared to the devolatilization reaction, so the char takes much longer to complete its conversion to gases. For this reason, a large reactor length is required.

Entrained-flow gasifier design may be classified into two broad groups: (i) the top-fed downflow (used by GE Energy and Siemens SFG), shown in

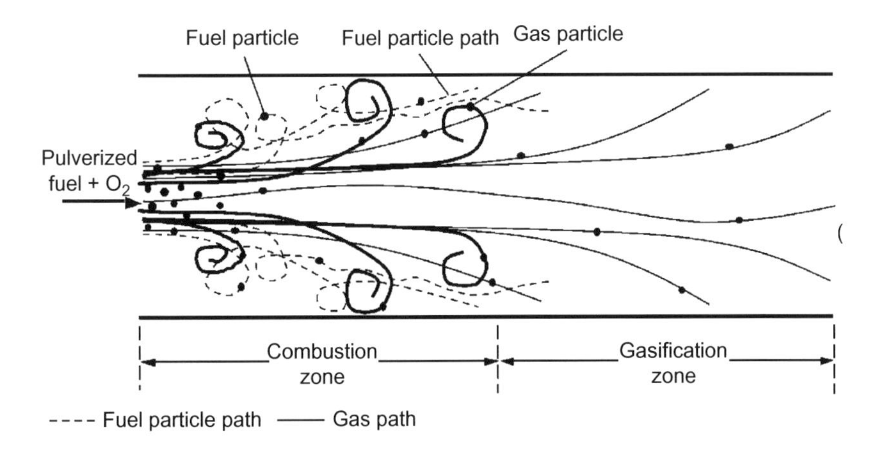


**FIGURE 8.15** Simplified sketch of gas–solid flow in an entrained-flow gasifier.

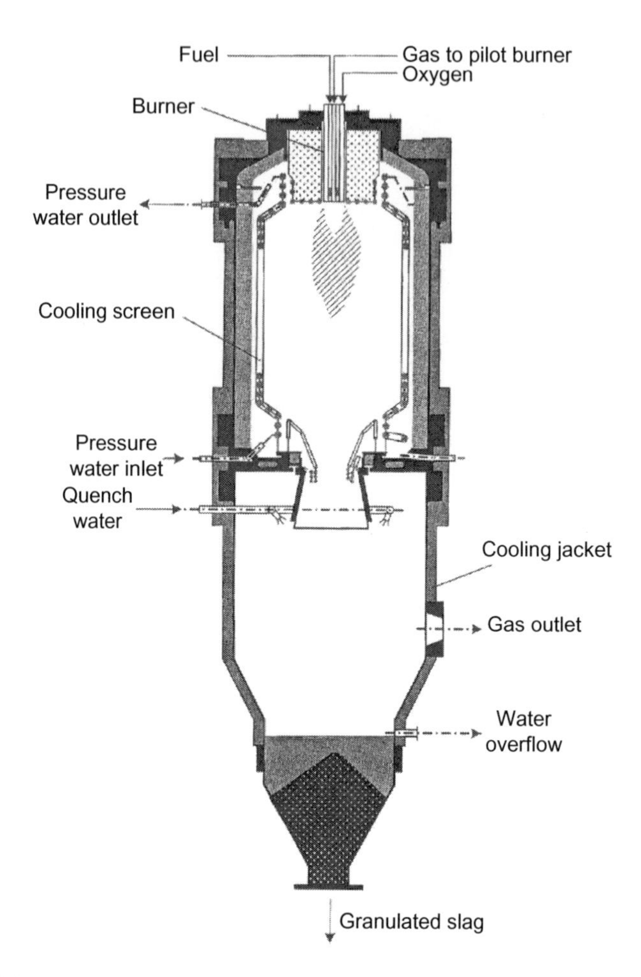


**FIGURE 8.16** A schematic of a top-fed downflow entrained-flow gasifier.

Figure 8.16, and (ii) the side-fed upflow (used by Koppers-Totzek, the Shell gasification process, Prenflo, and the Lurgi multipurpose), shown in Figure 8.17.

## 8.4.1 Top-Fed Gasifier

Top-fed gasifiers use a vertically cylindrical reactor vessel into which pulverized fuel (biomass or coal) and gasifying agent(s) are conveyed by oxygen and injected from the top. This vessel resembles a vertical furnace with a downward burner (Figure 8.16). The fuel and the gasifying agent(s) are injected into the reactor through a jet that generally sits at the reactor's middle section.

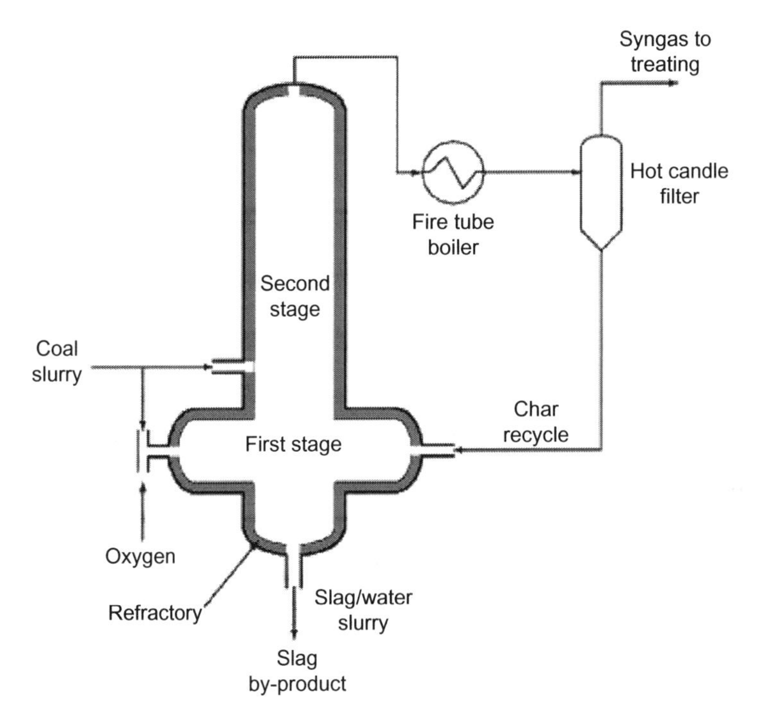


**FIGURE 8.17** A schematic of a side-fed entrained-flow gasifier.

The fuel gasifier process of Siemens uses a top-fired reactor design, in which the reactants are introduced through the single centrally mounted burner. This has several advantages. First, it is of an axisymmetric construction, reducing equipment costs; second, the flow of reactant occurs from a single burner, reducing the number of burners to be controlled; finally, the product gas and the slag flow in the same direction, which reduces any potential blockage in a slag trap (Higman and van der Burgt, 2008, p. 132).

## 8.4.2 Side-Fed Gasifier

In side-fed gasifiers, powdered fuel is injected through horizontal nozzles set opposite each other in the reactor's lower section (Figure 8.17). Jets of fuel and gasifying agents form a stirred-tank reactor characterized by a high degree of mixing. The product gas moves upward and exits through the top. Because of the high oxygen availability in this mixing zone, rapid exothermic reactions take place, raising the gas temperature to well above the ash-melting point (>1400°C). Thus, the ash, instead of traveling up, is separated in this zone as slag from the fuel and drained. Some gasifier designs (e.g., E-gas, MHI, and Eagle) inject additional fuel further downstream from the main reaction zone.

The Koppers-Totzek atmospheric pressure gasifier also uses side feeding. It consists of two side-mounted burners where a mixture of coal and oxygen is injected. The gas leaves from the top of the gasifier at temperatures around 1500°C and is quenched with water downstream. The reactor has a steam jacket to protect its shell from high temperatures (Higman and van der Burgt, 2008, p. 129).

The E-gas gasifier is a side-fed two-stage entrained-flow slagging gasifier with a coal–water slurry feed. It is designed to use sub-bituminous coal (Figure 8.17). The coal slurry is fed at the nonslagging stage, where the upflowing gas heats it. Thus, the gas exits at a lower temperature and then passes through a fire-tube boiler and is filtered in a hot candle filter. The char, separated out by the filter, is taken back to the slagging zone. The slag is quenched in a water bath at the bottom of the slagging reactor.

### 8.4.3 Advantages of Entrained-Flow Gasifiers

Entrained-flow gasifiers have several advantages over other types:

- Low tar production
- A range of acceptable feed
- Ash produced as slag
- High-pressure, high-temperature operation
- Very high conversion of carbon
- Low methane content well suited for synthetic gas production.

### 8.4.4 Entrained-Flow Gasification of Biomass

For thermal gasification of the refractory components of biomass (those difficult to gasify) such as lignin, the minimum temperature requirement is similar to that for coal (~900°C) (Higman and van de Burgt, 2008, p. 147). Entrained-flow gasification of biomass is therefore rather limited and has not been seen on a commercial scale for the following reasons:

- The residence time in the reactor is very short. For the reactions to complete, the biomass particles must be finely ground. Being fibrous, biomass cannot be pulverized easily.
- Molten ash from biomass is highly aggressive because of its alkali compounds and can corrode the gasifier's refractory or metal lining.

Given these shortcomings, entrained-flow gasifiers are not popular for biomass. However, there is at least one successful entrained-flow biomass gasifier known as the Choren process.

#### *8.4.4.1 Choren Process*

The Choren entrained-flow biomass gasifier is comprised of three stages (Figure 8.18). The first stage receives biomass in a horizontal stirred-type

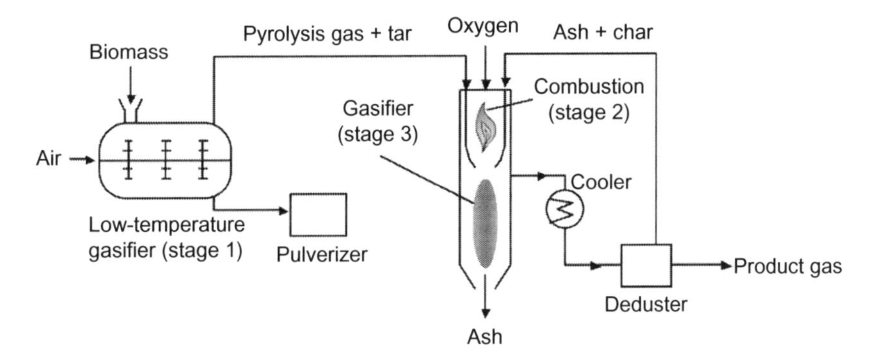


**FIGURE 8.18** Choren process. The biomass is gasified in an entrained-flow gasifier, facilitated by a rotary-type partial gasifier (stage 1).

low-temperature reactor for pregasification at 400–500°C in a limited supply of air. This produces solid char and a tar-rich volatile product. The latter flows into the second chamber (stage 2), an entrained-flow combustor, where oxygen and the product gas from the first stage are injected downward into the reactor. Combustion raises the temperature to 1300–1500°C and completely cracks the tar. The hot combustion product flows into the third chamber (stage 3), where the char is gasified.

The solid char received from the first stage is pulverized and fed into the third stage of the Choren process. It is gasified in the hot gasification medium produced in the second stage. Endothermic gasification reactions reduce the temperature to about 800°C. Char and ash from the product gas are separated and recycled into the second-stage combustor. The ash melts at the high temperature in the combustor and is drained from the bottom. Now the molten ash solidifies, forming a layer on the membrane wall that protects the wall against the corrosive action of fresh molten biomass ash. The product gas is processed downstream for Fischer–Tropsch synthesis or other applications.

## 8.5 PLASMA GASIFICATION

In plasma gasification, high-temperature plasma helps gasify biomass hydrocarbons. It is especially suitable for MSW and other waste products. This process may also be called "plasma pyrolysis" because it essentially involves thermal disintegration of carbonaceous material into fragments of compounds in an oxygen-starved environment. The heart of the process is a plasma gun, where an intense electric arc is created between two electrodes spaced apart in a closed vessel through which an inert gas is passed (Figure 8.19).

Though the temperature of the arc is extremely high (~13,000°C), the temperature downstream, where waste products are brought in contact with

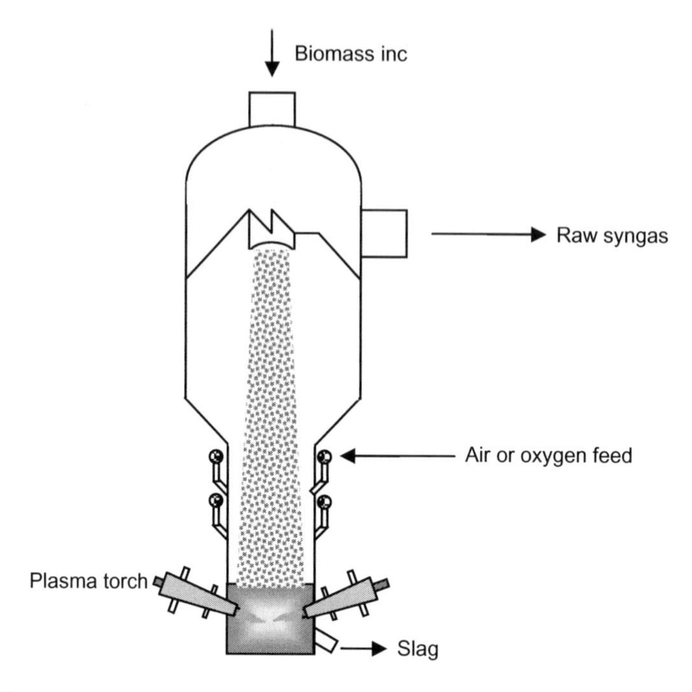


**FIGURE 8.19** Plasma gasification of solid waste.

it, is much lower (2700–4500°C). The downstream temperature is still sufficiently high, however, to pyrolyze complex hydrocarbons into simple gases such as CO and $H_2$. Simultaneously, all inorganic components (e.g., glass, metals, silicates, and heavy metals) are fused into a volcanic-type lava, which after cooling forms an inert basaltic slag. The product gas leaves the gasifier at very high temperatures (1000–1200°C).

A typical plasma reactor provides exceptionally high temperature that cause the tar products to be cracked and harmful products like dioxin and furan to be destroyed.

Owing to the high reactor temperature and the presence of chlorine in wastes, the life of the reactor liner is an issue. However, an attractive feature is that plasma gasification is relatively insensitive to the quality of the feedstock. This is the result of an independent energy source run by electricity instead of partial combustion of the gasification product.

## 8.6 PROCESS DESIGN

The design of a gasifier involves both process and hardware. The process design gives the type and yield of the product, operating conditions, and the basic size of the reactor. The hardware design involves structural and mechanical components, such as grate, main reactor body, insulation, cyclone, and others, that are specific to the reactor type. This section focuses on gasifier process design.

### 8.6.1 Design Specification

For any design, specification of the plant is very important. The input includes the specification of the fuel, gasification medium, and product gas. A typical fuel specification will include proximate and ultimate analysis, operating temperatures, and ash properties. The specification of the gasifying medium is based on the selection of steam, oxygen, and/or air and their proportions.

These parameters could influence the design of the gasifier as follows:

1. The desired heating value of the product gas dictates the choice of gasification medium. Table 7.1 gives typical ranges of heating value of product gases for different mediums. If air is the gasification medium, the lower heating value (LHV) of gas is in the range of 4–7 MJ/$m^3$, while in cases of oxygen- and steam-based gasifiers, it is in the range of 10–20 MJ/$m^3$ (Ciferno and Marano, 2002, p. 4). It may be noted that when the feedstock is biomass, the heating value is lower due to its high oxygen and moisture content.
2. Hydrogen can be maximized with steam, but if it is not a priority, oxygen or air is a better option, as it reduces the energy used in generating steam and the energy lost through unutilized steam.
3. If nitrogen in the product gas is not acceptable, air cannot be chosen.
4. Capital cost is lowest for air, followed by steam. A much larger investment is needed for an oxygen plant, which also consumes a large amount of auxiliary power.
5. Equivalence ratio (ER) has a major influence on carbon conversion efficiency.

For the product gas, the specification includes:

a. Desired gas composition
b. Desired heating value
c. Desired production rate (N $m^3$/s or $MW_{th}$ produced)
d. Yield of the product gas per unit fuel consumed
e. Required power output of the gasifier, $Q$.

The outputs of process design include geometric and operating and performance parameters. The geometric or basic size includes reactor configuration, cross-sectional area, and height (hardware design). Important operating parameters are (i) reactor temperature, (ii) preheat temperature of steam, air, or oxygen, and (iii) amount (i.e., steam/biomass ratio) and relative proportion of the gasifying medium (i.e., steam/oxygen ratio). Performance parameters of a gasifier include carbon conversion and cold-gas efficiency.

A typical process design starts with a mass balance followed by an energy balance. The following subsections describe the calculation procedures for these.

## 8.6.2 Mass Balance

Basic mass and energy balance is common to all types of gasifiers. It involves calculations for product gas flow and fuel feed rate.

### 8.6.2.1 Product Gas Flow-Rate

The gasifier's required power output, $Q$ ($MW_{th}$), is an important input parameter specified by the client. Based on this, the designer makes a preliminary estimation of the amount of fuel to be fed into the gasifier and the amount of gasifying medium. The volume flow-rate of the product gas, $V_g$ (N $m^3$/s). For a desired $LHV_g$ (MJ/N $m^3$) is found by:

$$V_g = \frac{Q}{LHV_g} \text{ N m}^3/\text{s} \tag{8.5}$$

The net heating value or LHV of producer gas ($LHV_g$) can be calculated from its composition. The composition may be predicted by the equilibrium calculations, described later, or by more sophisticated kinetic modeling of the gasifier, as discussed in Chapter 7. In the absence of these, a reasonable guess can be made either from published data on similar fuels in similar gasification conditions or from the designer's experience.

For example, for air-blown fluidized-bed biomass gasifiers, the $LHV_g$ is in the range 3.5–6 MJ/N $m^3$ (Enden and Lora, 2004). For oxygen gasification, it is in the range 10–15 MJ/N $m^3$ (Ciferno and Marano, 2002). So, for an air-blown gasifier, we start with a value of 5 MJ/N $m^3$ as a reasonable guess (Quaak et al., 1999).

### 8.6.2.2 Fuel Feed Rate

To find the biomass feed rate, $M_f$, the required power output is divided by the LHV of the biomass ($LHV_{bm}$) and by the gasifier efficiency, $\eta_{gef}$.

$$M_f = \frac{Q}{LHV_{bm}\eta_{gef}} \tag{8.6}$$

The LHV may be related to the higher heating value (HHV) and its hydrogen and moisture contents (Quaak et al., 1999) as:

$$LHV_{bm} = HHV_{daf} - 20,300 \times H_{daf} - 2260 \times M_{daf} \tag{8.7}$$

Here, $H_{daf}$ is the hydrogen mass fraction in the fuel, $M_{daf}$ is the moisture mass fraction, and $HHV_{daf}$ is the HHV in kJ/kg on a dry on moisture-ash-free basis. By using the definition of these, one can relate the HHV on moisture-ash-free basis to that on only dry-basis value as:

$$\mathrm{HHV_{daf}} = \mathrm{HHV_d}\left(\frac{1-M}{1-\mathrm{ASH}-M}\right) \tag{8.8}$$

where the subscripts d and daf refer to dry and moisture-ash-free basis, respectively; $M$ is the moisture fraction; and ASH is the ash fraction in fuel on a raw-fuel basis.

On a dry basis, $\mathrm{HHV_d}$ is typically in the range 18–21 MJ/kg (Van Loo and Koppejan, 2003, p. 48). It may be calculated from the ultimate analysis for the biomass using the following equation (Van Loo and Koppejan, 2003, p. 29):

$$\begin{aligned}\mathrm{HHV_d} = {} & 0.3491\mathrm{C} + 1.1783\mathrm{H} + 0.1005\mathrm{S} - 0.0151\mathrm{N} \\ & - 0.1034\mathrm{O} - 0.0211\mathrm{ASH}\end{aligned} \tag{8.9}$$

where C, H, S, N, O, and ASH are the mass fraction of carbon, hydrogen, sulfur, nitrogen, oxygen, and ash in the fuel on a dry basis.

#### *8.6.2.3 Flow-Rate of Gasifying Medium*

The amount of gasification medium has a major influence on yield and composition of the product gas. This section discusses methods for choosing that amount.

##### Air

The theoretical air requirement for complete combustion of a unit mass of a fuel, $m_{th}$, is an important parameter. It is known as the *stoichiometric air* requirement. Its calculation is shown in Eq. (3.32). For an air-blown gasifier operating, the amount of air required, $M_a$, for gasification of unit mass of biomass is found by multiplying it by another parameter equivalence ratio (ER):

$$M_a = m_{th}\mathrm{ER} \tag{8.10}$$

For a fuel feed rate of $M_f$, the air requirement of the gasifier, $M_{fa}$, is:

$$M_{fa} = m_{th}\mathrm{ER} \times M_f \tag{8.11}$$

For a biomass gasifier, 0.25 may be taken as a first-guess value for ER. A more detailed discussion of this is presented next.

#### *8.6.2.4 Equivalence Ratio*

Equivalence ratio (ER) is an important design parameter for a gasifier. It is the ratio of the actual air–fuel ratio to the stoichiometric air–fuel ratio. This definition is the same as that of excess air (EA) used for a combustion system, except that it is used only for air-deficient situations, such as those found in a gasifier.

$$\mathrm{ER}(<1.0)_{\mathrm{gasification}} = \frac{\text{actual air}}{\text{stoichiometric air}} = \mathrm{EA}(>1.0)_{\mathrm{combustion}} \quad (8.12)$$

In a combustor, the amount of air supplied is determined by the stoichiometric (or theoretical) amount of air and its *excess air coefficient.* In a gasifier, the air supply is only a fraction of the stoichiometric amount. The stoichiometric amount of air may be calculated based on the ultimate analysis of the fuel.

ER dictates the performance of the gasifier. For example, pyrolysis takes place in the absence of air and hence the ER is zero; for gasification of biomass, it lies between 0.2 and 0.3.

Downdraft gasifiers give the best yield for ER—0.25 (Reed and Das, 1988, p. 25). With a lower ER value, the char is not fully converted into gases. Some units deliberately operate with a low ER to maximize their charcoal production. A lower ER gives rise to higher tar production. So, updraft gasifiers, which typically operate with an ER of less than 0.25, have higher tar content. With an ER above 0.25, some product gases are also burnt, increasing the temperature.

The quality of gas obtained from a gasifier strongly depends on the value of ER, which must be significantly below 1.0 to ensure that the fuel is gasified rather than combusted. However, an excessively low ER value (<0.2) results in several problems, including incomplete gasification, excessive char formation, and a low heating value of the product gas. On the other hand, too high an ER (>0.4) results in excessive formation of products of complete combustion, such as $CO_2$ and $H_2O$, at the expense of desirable products, such as CO and $H_2$. This causes a decrease in the heating value of the gas. In practical gasification systems, the ER's value is normally maintained within the range of 0.20–0.30. Figure 8.20 shows the variation in carbon

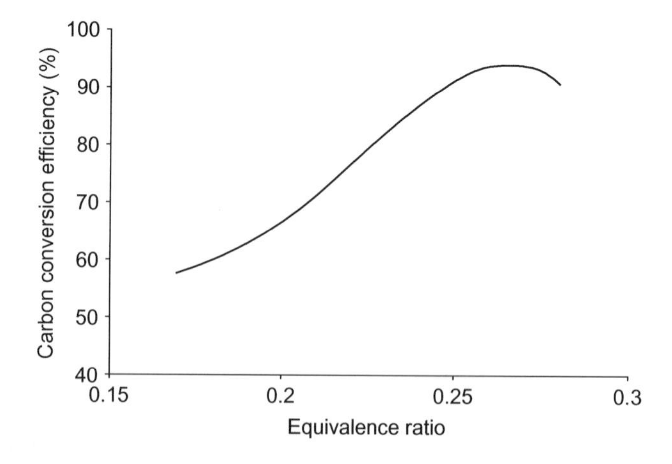


**FIGURE 8.20** Effect of ER on carbon conversion in a fluidized-bed gasifier.

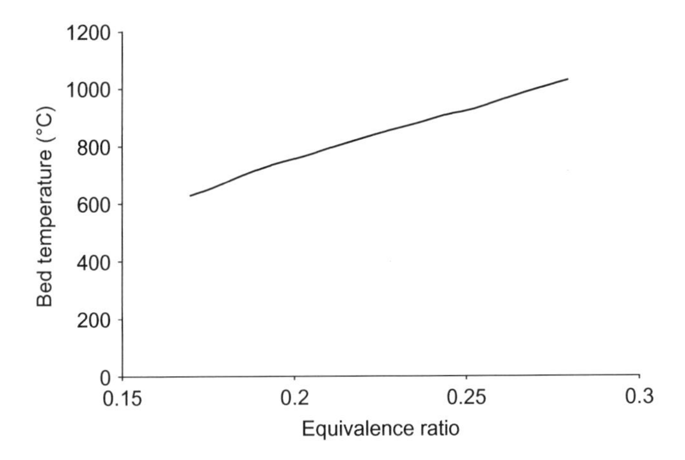


**FIGURE 8.21** Gasifier temperature in a CFB riser increases with ER.

conversion efficiency of a CFB gasifier for wood dust against the ER. The efficiency increases with equivalent ratio and then it starts declining. The optimum value here is 0.26, but it may change depending on many factors.

The bed temperature of a fluidized-bed gasifier increases with the ER because the higher the amount of air, the greater the extent of the combustion reaction and the higher the amount of heat released (Figure 8.21). Example 8.1 illustrates the calculation procedure for ER.

## Oxygen

Oxygen is used primarily to provide the thermal energy needed for the endothermic gasification reactions. The bulk of this heat is generated through the following partial and/or complete oxidation reactions of carbon:

$$C + 0.5O_2 \rightarrow CO - 111\ \mathrm{kJ/mol} \tag{8.13}$$

$$C + O_2 \rightarrow CO_2 - 394\ \mathrm{kJ/mol} \tag{8.14}$$

It can be seen that for the oxidation of 1 mol of carbon to $CO_2$, the oxygen requirement is $(2 \times 16)/12 = 2.66$ mol, while that for carbon to CO is $(16/12) = 1.33$ mol. Thus, the reaction in Eq. (8.13) is more likely to take place in oxygen-deficient regions.

Besides supplying the energy for the endothermic gasification reactions, the gasifier must provide energy to raise the feed and gasification medium to the reaction temperature, as well as to compensate for the heat lost to the reactor walls. For a self-sustained gasifier, part of the chemical energy in the biomass provides the heat required. The total heat necessary comes from the oxidation reactions. The energy balance of the gasifier is thus the main consideration in determining the oxygen-to-carbon (O/C) ratio.

Equilibrium calculations can show that as the O/C ratio in the feed increases, $CH_4$, CO, and hydrogen in the product decreases but $CO_2$ and $H_2O$ in the product increases. Beyond a O/C ratio of 1.0, hardly any $CH_4$ is produced.

When air is the gasification medium, as is the case for 70% of all gasifiers (Ciferno and Marano, 2002), the nitrogen in it dilutes the product gas. The heating value of the gas is therefore relatively low (4–6 MJ/m$^3$). When pure oxygen from an air-separation unit is used, the heating value is higher, in the range 12–28 MJ/m$^3$, but a large amount of energy (~2.18 MJ/kg $O_2$) is spent in separating the oxygen from the air (Grezin and Zakharov, 1988).

Either atmospheric air or oxygen from an air-separation unit can meet the oxygen requirement of a gasifier.

### Steam

Superheated steam is used as a gasification medium either alone, with air, or with oxygen. It contributes to the generation of hydrogen.

$$C + H_2O \rightarrow CO + H_2 \tag{8.15}$$

The quantity of steam, $M_{fh}$, is known from the steam-to-carbon (S/C) molar ratio.

$$\text{Steam flow-rate, } M_{fh} = 18\frac{M_f C}{12}(S/C) \text{ kg steam/kg fuel} \tag{8.16}$$

where $M_f$ is the fuel feed rate and $C$ is the carbon mass fraction in the fuel.

The S/C mole ratio has an important influence on product composition as the ER has. Both hydrogen and CO increase with an increasing S/C ratio for a given temperature and O/C molar ratio. The production of these two gases increases with decreasing pressure, decreasing oxygen, and decreasing S/C ratio. However, there is only a marginal gain in increasing the S/C molar ratio above 2–3, as the excess steam simply leaves the gasifier unreacted (Probstein and Hicks, 2006, p. 119). So a value in the range of 2.0–2.5 can give a reasonable starting value.

---

**Example 8.1**

A moving-bed gasifier 4 m in diameter operates at 25 bar of pressure and consumes 750 kg/min (dry-ash-free basis) of bituminous coal, 1930 kg/min of steam, and 280 N m$^3$/min of oxygen to produce a product gas that contains 1000 N m$^3$ of syngas (a mixture of $H_2$ and CO). The mean gasifier temperature is 1000°C. The volumetric composition of the product gas is:

$CO_2$: 32%
$H_2S$: 0.4%
CO: 15.2%
$H_2$: 42.3%
$CH_4$: 8.6%
$C_2H_4$: 0.8%
$N_2$: 0.7%

The ultimate analysis of the coal on a moisture-ash-free basis is:

C: 77.3%

H: 5.9%

S: 4.3%

N: 1.4%

O: 11.1%

Find

1. The S/C molar ratio
2. The O/C molar ratio
3. The ER
4. The hearth load in energy produced per unit of grate area and space velocity

The heating values of the product gas constituents may be taken from Table C.2 in Appendix C.

**Solution**

From the feed rate of coal, steam, and oxygen, we can find the molar feed rate by dividing the mass rate by the molecular weight as here:

Carbon moles: $750 \times 0.773/12 = 48.31$ kmol/min

Steam moles: $1930/18 = 107.22$ kmol/min

Oxygen moles: $280/22.4 = 12.5$ kmol/min

From these, we can calculate the following:

1. S/C molar ratio $= 107.22/48.31 =$ **2.22**
2. O/C molar ratio $= 12.5/48.31 =$ **0.26**
3. To find the stoichiometric oxygen requirement, the oxygen required to oxidize carbon to $CO_2$, hydrogen to $H_2O$, and sulfur to $SO_2$ has to be calculated.

To produce 1 mol of $CO_2$, 12 kg of carbon (1 mol) reacts with 32 kg of oxygen (1 mol):

$$C + O_2 = CO_2$$

Therefore, the oxygen required for 1 kg of carbon is 32/12.

To produce 1 mol of $SO_2$, 32 kg of sulfur (1 mol) reacts with 32 kg of oxygen (1 mol):

$$S + O_2 = SO_2$$

Therefore, the oxygen required for 1 kg of sulfur is $32/32 = 1$.

Similarly, 4 kg of hydrogen reacts with 32 kg of oxygen to produce $H_2O$:

$$2H_2 + O_2 = 2H_2O$$

Therefore, the oxygen required for 1 kg of hydrogen is $32/4 = 8$.

$$\text{Stoichiometric oxygen requirement} = \frac{32C}{12} + 8H + S - O$$

$$= \frac{32 \times 0.773}{12} + 8 \times 0.059 + 0.043 - 0.111$$

$$= 2.465 \text{ kg of } O_2/\text{kg of fuel}$$

The total $O_2$ required is:

$$750 \times 2.465 = 1848.75 \text{ kg of } O_2/\text{min}$$

The $O_2$ supplied is:

$$\text{Moles of } O_2 \times 32 = 12.5 \times 32 = 400 \text{ kg of } O_2/\text{min}$$

From this, we can calculate:

$$ER = 400/1848.75 = 0.22$$

$$ER = 280/1848.75 = 0.15$$

The syngas constituents in the total product gas are CO (15.2%) and $H_2$ (42.3%). So, to produce 1000 N m$^3$/min of syngas, the amount of product gas, $Q_{pr}$, is:

$$Q_{pr} = 1000/(0.152 + 0.423) = 1739 \text{ N m}^3/\text{min}$$

The cross-sectional area of the gasifier reactor, $A$, is:

$$A = \pi\, 4^2/4 = 12.56 \text{ m}^2$$

Assuming the operating temperature to be 1000°C and the pressure to be 25 bar, the volumetric flow-rate of product gas is:

$$Q'_{pt} = Q_{pt}\left(\frac{1}{25}\right)\left(\frac{1273}{273}\right) = 324 \text{ m}^3/\text{min}$$

The space velocity of the gas flow $V_g$ is $Q'_{pr}/A = 324/(12.56 \times 60) = 0.43$ m/s.

The energy produced per N m$^3$ of product gas is found by multiplying the volume fraction by the heating value of each constituent, which is taken from Table C.2 in Appendix C. Adding together the contribution of all product gas constituents, gives the total heating value, HHV, as:

$$\begin{aligned} \text{HHV} = {} & 0.004 \times 25.1 + 0.152 + 0.152 \times (282.99/22.4) + 0.423 \\ & \times (285.84/22.4) + 0.086 \times (890.36/22.4) + 0.008 \times 63.4 = 11.33 \text{ MJ/N m}^3 \end{aligned}$$

Thus, the total energy produced, $E_{total}$, is $Q_{pr} \times \text{HHV} = 1739 \times 11.33/60 = 328.3$ MW$_{th}$.

The hearth load is:

$$E_{total}/A = 328.3/12.56 = 26.14 \text{ MW/m}^2$$

### 8.6.3 Energy Balance

Unlike combustion reactions, most gasification reactions are endothermic. Thus, heat must be supplied to the gasifier for these reactions to take place at the designed temperature. In laboratory units, this is not an issue because the heat is generally supplied externally. In commercial units, it is a major issue, and it must be calculated and provided for. The amount of external heat supplied to the gasifier depends on the heat requirement of the endothermic reactions as well as on the gasification temperature. The latter is a design choice, and it is discussed next.

#### 8.6.3.1 Gasification Temperature

The choice of gasification temperature is an important process choice. Because lignin, a refractory component of biomass, does not gasify well at lower temperatures, thermal gasification of lignocellulosic biomass prefers a minimum gasification temperature in the range 800–900°C. For biomass, an entrained-flow gasifier typically maintains a peak temperature well exceeding 900°C. For coal, the minimum is 900°C for most gasifier types (Higman and van der Burgt, 2008, p. 163).

A higher peak gasification temperature is chosen for an entrained-flow gasifier. A higher ash-melting temperature requires a higher choice of the gasifier temperature. This temperature is raised through the gasifier's exothermic oxidation reactions, so a high reaction temperature also means a high oxygen demand.

In entrained-flow gasifiers, the peak gasification temperature is typically in the range of 1400–1700°C, as it is necessary to melt the ash; however, the gas exit temperature is much lower. The peak temperature of a fluidized-bed gasifier is in the range of 700–900°C to avoid softening of bed materials. It is about the same as the gas exit temperature in a fluidized-bed gasifier. In a crossdraft gasifier, the mean gasification temperature is about 1250°C, whereas the peak temperature is about 1500°C. The gas exit temperature of a downdraft gasifier is about 700°C, but its peak gasifier temperature at the throat is 1000°C. The updraft gasifier has the lowest gas exit temperature (200–400°C), while its gasification temperature may be up to 900°C (Knoef, 2005). Once the gasification temperature is known, the designer can turn to the heat balance on this basis.

#### 8.6.3.2 Heat of Reaction

Heat of reaction is the heat gained or lost in a chemical reaction. To calculate it for gasification, we consider an overall gasification reaction where 1 mol of biomass ($C_aH_bO_c$) is gasified in $\alpha$ moles of steam and $\beta$ moles of oxygen. The overall equation is:

$$\begin{aligned} C_aH_bO_c + \alpha H_2O + \beta O_2 = {} & A' \cdot C + B' \cdot CO_2 + C' \cdot CO + D' \cdot CH_4 \\ & + E' \cdot H_2O + F' \cdot H_2 + Q \end{aligned} \tag{8.17}$$

The equilibrium analysis of Section 7.5.2 gives the mole fraction $A'$, $B'$, $C'$, $D'$, $E'$, and $F'$ in the flue gas for given values of $\alpha$ and $\beta$. The chosen S/B ratio defines $\alpha$ while the ER defines $\beta$. The heat of reaction, $Q$, for the overall gasification reaction (Eq. (8.17)) may be found from the heat of formation of the products and reactants:

$$\begin{aligned}\text{Heat of reaction} &= \text{heat of formation of product} - \text{heat of formation of reactant}\\ &= \text{heat of formation of } [A'\cdot \mathrm{C} + B'\mathrm{CO_2} + C'\cdot \mathrm{CO}\\ &\quad + D'\cdot \mathrm{CH_4} + E'\cdot \mathrm{H_2O} + F'\cdot \mathrm{H_2}] - \text{heat of formation of}\\ &\quad [\alpha\cdot \mathrm{H_2O} + \beta\cdot \mathrm{O_2} + \text{biomass}]\end{aligned} \tag{8.18}$$

The heat of formation at 25°C, or 298K, is available in Table C.6 (Appendix C). The heat of formation at any other temperature, $T$, in Kelvin, can be found from the relation:

$$\Delta \mathrm{H}_T^0 = \Delta \mathrm{H}_{298}^0 + \sum\left(\int_{298}^{T} A' C_{\mathrm{p},j}\,\mathrm{d}T\right)_{\text{product}} - \sum\left(\int_{298}^{T} \alpha C_{\mathrm{p},j}\,\mathrm{d}T\right)_{\text{reactants}} \tag{8.19}$$

where $C_{p,i}$ is the specific heat of a substance $i$ at temperature $T$ Kelvin, and $A'$,$B'$,$C'$, $D'$, $E'$, $D'$, $\alpha$, and $\beta$ are the stoichiometric coefficients of the products and reactants, respectively. The specific heat of gases as a function of temperature is given in Table C.4 (Appendix C).

The net heat, $Q_{\text{gasification}}$, to be supplied to the reactor is the algebraic sum of heat of reactions.

$$Q_{\text{gasification}} = \Delta \mathrm{H}_T \ \mathrm{kJ/mol} \tag{8.20}$$

This expression takes into account both exothermic combustion and endothermic gasification reactions. If the value of $Q_{\text{gasification}}$ works out to be negative, the overall process is exothermic, and so no net heat for the reactions is required.

---

**Example 8.2**

Find the heat of reaction for the following reaction at 1100K:

$$\mathrm{C} + \mathrm{H_2O(gas)} = \frac{1}{2}\mathrm{CH_4} + \frac{1}{2}\mathrm{CO_2}$$

**Solution**

Taking values at the reference temperature, 298K from Table C.6, we have

Heat of formation at 298 K for C for $CH_4 = 0$; $H_2O\ (g) = -241.8$ kJ/mol;
$= -74.8$ kJ/mol; $CO_2 = -393.5$ kJ/mol

$$\begin{aligned}\text{Total } \Delta \mathrm{H}_{298}^0 &= \text{product} - \text{reactant}\\ &= \left[\frac{1}{2}(-74.8 - 393.5) - (-241.8)\right] = 7.65\ \mathrm{kJ/mol}\end{aligned}$$

Now, to find the value at 1100K, we use Eq. (8.19):

$$\Delta H^0_{1100} = \Delta H^0_{298} + \sum\left(\int_{298}^{1100} (C_{p,CH_4} + C_{p,CO_2})dT\right)_{product} - \sum\left(\int_{298}^{1100} C_{p,H_2O}dT\right)_{reactants} \quad (8.21)$$

The specific heats of gases are taken from Table C.4 (Appendix C) as:

$$C_{p,CH_4} = 22.35 + 0.0481T \text{ kJ/kmol}$$

$$C_{p,CO_2} = 43.28 + 0.0114T - 818363/T^2 \text{ kJ/mol}$$

$$C_{p,H_2O} = 34.4 + 0.00062T + 0.0000056\ T^2 \text{ kJ/mol}$$

The integrations of respective gas components are:

$$C_{p,CH_4} = \left[22.35T + \frac{0.0481T^2}{2}\right]_{298}^{1100} = 22.35 \times (1100 - 298) + \frac{0.0481}{2} \times (1100^2 - 298^2) = 44.8895 \text{ kJ/mol}$$

$$C_{p,CO_2} = \left[43.28T + \frac{0.0114T^2}{2} + \frac{818,363}{T}\right]_{298}^{1100} = 43.28 \times (1100 - 298) + \frac{0.0114}{2} \times (1100^2 - 298^2) + \frac{818,363}{1100 - 298} = 42.1218 \text{ kJ/mol}$$

$$C_{p,H_2O} = \left[34.4T + \frac{0.000628T^2}{2} + \frac{0.0000056T^3}{3}\right]_{298}^{1100} = 34.4 \times (1100 - 298) + \frac{0.000628}{2} \times (1100^2 - 298^2) + \frac{0.0000056}{3}(1100^3 - 298^3) = 30.376 \text{ kJ/mol}$$

Substituting these values and integrating the above expression, we get:

$$\Delta H^0_{1100} = 7.65 + 104.58 - 33.578 = 78.65 \text{ kJ/mol}$$

Thus, this reaction is endothermic and it is written as:

$$C + H_2O(gas) \rightarrow \frac{1}{2}CH_4 + \frac{1}{2}CO_2 + 78.65 \text{ kJ/mol}$$

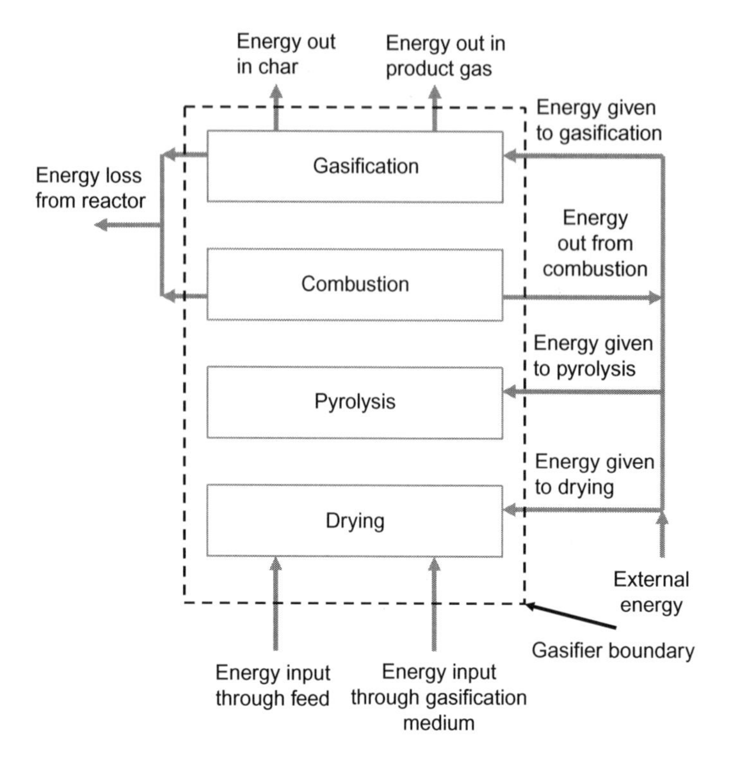


**FIGURE 8.22** Energy flow in and out of a gasifier.

Figure 8.22 shows the energy flow in and out of a gasifier. Biomass enters with its chemical energy and sensible heat. The gasifying agents enter with sensible heat at the reference temperature. External heat is added for heating the feeds to the gasification temperature for meeting any shortfall in the reaction heat requirement and for wall losses from the reactor. The product gas, with its chemical energy, leaves at the gasifier temperature. Unburnt char leaves with a potential energy in it. The unutilized steam and other gases also leave at the gasification temperature.

The overall energy balance may be written as:

*Energy input:* Enthalpy of (biomass + steam + oxygen) at reference temperature + heating value of biomass + external heat

*Energy output:* Enthalpy of product gas at gasifier temperature + heating value of product gas + heat in unconverted char + heat loss from the reactor.

If $A$ is the amount of air needed and $W$ is the total steam (from moisture or otherwise) needed to gasify $F$ kg of fuel to produce 1 N m$^3$ of product gas, we can write the energy balance of the gasifier taking 0°C as the reference:

$$\begin{aligned} ACp_aT_0 + FCp_fT_0 + WH_0 + F \times \text{HHV} + Q_{\text{ext}} &= (C_{\text{CO}}V_{\text{CO}} + C_{\text{CO}_2}V_{\text{CO}_2} \\ &+ C_{\text{CH}_4}V_{\text{CH}_4} + C_{\text{H}_2}V_{\text{H}_2} + C_{\text{O}_2}V_{\text{O}_2} + C_{\text{N}_2}V_{\text{N}_2})T_{\text{g}} + (1 - X_{\text{g}})WH_{\text{g}} \\ &+ P_{\text{c}}q_{\text{c}} + Q_{\text{gasification}} + Q_{\text{loss}} + Q_{\text{product}} \end{aligned} \quad (8.22)$$

where $H_0$ and $H_g$ are the enthalpies of steam at the reference temperature and the gasifier exit temperature; $C_i$ and $V_i$ are the volumetric specific heat and the volume of the gas species, $i$, at temperature $T_g$ leaving the gasifier; $(1 - X_g)W$ is the net amount of steam remaining in the product gas of the gasification reaction; $P_c$ is the amount of char produced; and $q_c$ is the heating value of the char. Here, $Q_{loss}$ is the total heat loss through the wall, radiation from the bed surface, ash drain, and entrained solids, corresponding to 1 N m$^3$ of gas generation. This allows computation of external heat addition, $Q_{ext}$ kJ/N m$^3$ of product gas to the system. $Q_{product}$ is the amount of energy in the product gas and $Q_{gasification}$ is the net heat of reaction.

## 8.7 PRODUCT GAS PREDICTION

A typical gasifier design starts with a desired composition of the product gas. Equilibrium and other calculations are carried out to check how closely that targeted composition can be achieved through a choice of design parameters.

The product of combustion reactions is predominantly made up of carbon dioxide and steam, the percentages of which can be estimated with a fair degree of accuracy from simple stoichiometric calculations. For gasification reactions, this calculation is not straightforward; as the fraction of the fuel gasified and the compositions of the product gas need to be estimated carefully. Unlike combustion reactions, gasification reactions do not always reach equilibrium, so only a rough estimate is possible through an equilibrium calculation. Still, this can be a reasonable start for the design until detailed kinetic modeling is carried out in the design optimization stage.

### 8.7.1 Equilibrium Approach

An equilibrium calculation ideally predicts the product of gasification if the reactants are allowed to react in a fully mixed condition for an infinite period of time. There are two types of equilibrium model. The first one is based on equilibrium constants (stoichiometric model). The specific chemical reactions used for the calculations have to be defined, so this model is not suitable for complex reactions where the chemical formulae of the compounds, the reaction path, or the reaction equations are unknown. This requires the second model type (nonstoichiometric model), which involves minimization of the Gibbs free energy. This process is more complex but it is advantageous because the chemical reactions are not needed.

#### 8.7.1.1 Stoichiometric Model

The stoichiometric model requires a selection of appropriate chemical reactions and information concerning the values of the equilibrium constants. Section 7.5.2.1 in Chapter 7 explains the calculation procedure, so it is not repeated here.

#### 8.7.1.2 Nonstoichiometric Model

The nonstoichiometric model is based on the premise that at an equilibrium stage the total Gibbs free energy has to be minimized. It is described briefly in Chapter 7. Using Eq. (7.77), we can write the Gibbs free minimization equation for five gas species as follows:

$$CH_4: \frac{(\Delta \overline{G^o} CH_4)}{RT} + \ln\left(\frac{n_{CH_4}}{n_{total}}\right) + \frac{1}{RT}\lambda_C + \frac{4}{RT}\lambda_H = 0 \tag{8.23}$$

$$CO_2: \frac{(\Delta \overline{G^o}_{CO_2})}{RT} \ln\left(\frac{n_{CO_2}}{n_{total}}\right) + \frac{1}{RT}\lambda_C + \frac{2}{RT}\lambda_O = 0 \tag{8.24}$$

$$CO: \frac{(\Delta \overline{G^o}_{CO})}{RT} + \ln\left(\frac{n_{CO}}{n_{total}}\right) + \frac{1}{RT}\lambda_C + \frac{1}{RT}\lambda_O = 0 \tag{8.25}$$

$$H_2: \frac{(\Delta \overline{G^o}_{H_2})}{RT} + \log\left(\frac{n_{H_2}}{n_{total}}\right) + \frac{2}{RT}\lambda_H = 0 \tag{8.26}$$

$$H_2O: \frac{(\Delta \overline{G^o}_{H_2O})}{RT} + \ln\left(\frac{n_{H_2O}}{n_{total}}\right) + \frac{2}{RT}\lambda_H + \frac{1}{RT}\lambda_O = 0 \tag{8.27}$$

The five molar fractions of gases, such as ($nCH_4/n_{total}$), and the three Lagrangian constants, $\lambda_H$, $\lambda_O$, and $\lambda_C$, can be solved from the five equations and the three mass balance equations for C, H, and O derived from Eq. (7.74). Thus, for given feed and gasification medium and temperature, we can obtain the composition of the product gas.

Equilibrium models have some limitations. The effect of tar is not considered here, even though tar can be a major problem in the gasification process and can affect plant operation. An equilibrium model may, for example, result in overestimation of the hydrogen production. Kinetics, heat, and mass transfer determine the actual extent of chemicals participation in the chemical equilibrium in a given time or space domain (Florin and Harris, 2008). Furthermore, the equilibrium model assumes infinite speed of reaction and that all reactions will complete; these assumptions are not valid for most practical gasifiers. Nevertheless, equilibrium calculations give a good starting point, providing basic process parameters.

## 8.8 GASIFIER SIZING

The process design described in the previous section determines such operating parameters as gasification temperature, feed rates of fuel, and gasification medium. Now we can move to the next step, which involves the choice of gasifier configuration and type. Section 8.1.1 discusses the choice of gasifier. Table 8.5 compares the choices by their strength and weaknesses. Table 8.6 gives a range of product gas composition and heating values obtained in several commercial or large demonstration plants of generic types. The data of a fixed bed may not be typical as it is the value from one manufacturer gasifying MSW. Fixed-bed gasifiers for biomass are known for high tars unless it uses a downdraft system. Another table of typical composition is given in Table 2.6. By carefully examining these along with the type of plant to be designed, we can make a rational choice of gasifier type.

Once the gasifier type has been chosen, the designer can then proceed with the geometric design, where the basic sizes (the geometric dimensions of critical components) of the reactor are determined. At this stage, the designer decides on the geometric configuration of the reactor and its preliminary size. Both configuration and size depend on the reactor technology used.

### 8.8.1 Moving-Bed Gasifiers

A moving-bed gasifier may be designed on the basis of characteristic design parameters such as specific grate gasification rate, hearth load, and space velocity.

Specific grate gasification rate is the mass of fuel gasified per unit of cross-sectional area in unit time. The hearth load of a gasifier may be

**TABLE 8.5** Comparison of Strength and Weaknesses of Different Types of Gasifiers

| Class | Types | Strength/Weakness | Power Production |
|---|---|---|---|
| Fixed bed | Downdraft | Low heating value, moderate dust, low tar | Small to medium scale |
| | Updraft | Higher heating value, moderate dust, high tar | |
| | Crossdraft | Low heating value, moderate dust | |
| Fluidized bed | Bubbling | Higher than fixed-bed throughput, improved mass and heat transfer from fuel, higher heating value, higher efficiency | Medium scale |

**TABLE 8.6** Sample Performance of Some Commercial Gasifiers

| Reactor Type | BFB | CFB | Fixed Bed (updraft) | Entrained Flow |
|---|---|---|---|---|
| Feedstock | Biomass | Biomass | MSW | Coal |
| $H_2$ | 5–26 | 7–20 | 23 | 24 |
| CO | 13–27 | 9–22 | 39 | 67 |
| $CO_2$ | 12–40 | 11–16 | 24 | 4 |
| $H_2O$ | <18 | 10–14 | Dry | 3 |
| $CH_4$ | 3–11 | <9 | 5 | 0.02 |
| Higher hydrocarbon | <3 | <4 | 5 | 0 |
| $H_2S$ | ~0 | ~0 | 0.05 | 1 |
| $O_2$ | <0.2 | 0 | – | 0 |
| $N_2$ | 13–56 | 46–52 | – | 1 |
| $NH_3$ | 0 | 0 | – | 0.04 |
| Tars | <0.11 | <1 | Included in higher hydrocarbon | 0 |
| $H_2$/CO ratio | 0.2–1.6 | 0.6–1.0 | 0.6 | 0.36 |
| Heating value (MJ/m$^3$) | 4–13 | 4–7.5 | – | 9.5 |
| Throughput (ton/day) | 4.5–181 | 9–108 | 181 | 2155 |
| Pressure (bar) | 1–35 | 1–19 | 1 | 30 |
| Temperature (°C) | 650–950 | 800–1000 | – | 1400 |
| Gasification medium | $O_2$/air/ steam | Air | $O_2$ | $O_2$/steam |

Biomass includes corn stover, wood, and pulp sludge; –, unknown.
**Source:** Data compiled from Ciferno and Marano (2002).

expressed in terms of the fuel gasified, the volume of gas that is produced, or the energy throughput.

$$\text{Hearth load } (\text{kg}/s\ \text{m}^2) = \frac{\text{mass of fuel gasified}}{\text{hearth cross-sectional area}}$$

$$\text{Hearth load } (N\,\text{m}^3/s\,\text{m}^2) = \frac{\text{volumetric gas production rate}}{\text{hearth cross-sectional area}}$$

or

$$\text{Hearth load } (\text{MW}/m^2) = \frac{\text{energy throughput in product gas}}{\text{hearth cross-sectional area}} \tag{8.28}$$

The hearth load in volume flow-rate of gas per unit of cross-sectional area is also known as *superficial gas velocity* or *space velocity*, as it has the unit of velocity (at reference temperature and pressure).

The following section discusses type-specific design considerations.

#### 8.8.1.1 Updraft Gasifier

Updraft gasifiers are one of the simplest and most common types of gasifier for biomass. The maximum temperature increases when the feed of air or oxygen increases. Thus, the amount of oxygen feed for the combustion reaction is carefully controlled such that the temperature of the combustion zone does not reach the slagging temperature of the ash, causing operational problems. The gasification temperature may be controlled by mixing steam and/ or flue gas with the gasification medium.

The hearth load of an updraft gasifier is generally limited to 2.8 $\text{MW/m}^2$ or 150 $\text{kg/m}^2\text{/h}$ for biomass (Overend, 2004). For coal, it might be higher. In an oxygen-based coal gasifier, for example, the hearth load of a moving bed can be greater than 10 $\text{MW/m}^2$. A higher hearth load increases the space velocity of gas through the hearth, fluidizing finer particles in the bed. Probstein and Hicks (2006, p. 148) quoted space velocities for coal on the order of 0.5 m/h for steam–air gasification and 5.0 m/h for steam–oxygen gasification. Excessive heat generation in such a tightly designed gasifier may cause slagging. Based on the characteristics of some commercial updraft coal gasifiers, Rao et al. (2004) suggested a specific grate gasification rate as 100–200 kg $\text{fuel/m}^2\text{h}$ for RDF pellets, with the gas-to-fuel ratio in the range 2.5–3.0. Carlos (2005) obtained a rate of 745–916 $\text{kg/m}^2$ h with air–steam and air preheat at temperatures of 350°C and 830°C, respectively.

For an updraft gasifier, the height of the moving bed is generally greater than its diameter. Usually, the height-to-diameter ratio is more than 3:1 (Chakraverty et al., 2003). If the diameter of a moving bed is too large, there may be a material flow problem, so it should be limited to 3–4 m in diameter (Overend, 2004).

#### 8.8.1.2 Downdraft Gasifier

As we saw in Figures 8.4 and 8.6, the cross-sectional area of a downdraft gasifier may be nonuniform; it is narrowest at the throat. The hearth load is, therefore, based on the cross-sectional area of the throat for a throated gasifier, and for a throatless or stratified downdraft gasifier, it is based on the gasifier cross-sectional area. The actual velocity of gas is, however, significantly higher than the designed space velocity because much of the flow passage is occupied by fuel particles. The velocity is higher in the throat also because of the higher temperature there. Table 8.7 gives some characteristic values of these parameters.

In a downdraft gasifier, the gasification air is injected by a number of nozzles from the periphery (refer to Figure 8.6). The total nozzle area is typically 7–4% of the throat area. The number of nozzles should be an odd number so that the jet from one nozzle does not hit a jet from the opposite side, leaving a dead space in between. To ensure adequate penetration of nozzle air into the hearth, the diameter of a downdraft gasifier is generally limited to 1.5 m. This naturally restricts the size and capacity of a downdraft gasifier.

Table 8.8 lists typical sizes for the Imbert-type downdraft gasifier and shows the relation between throat size and air nozzle diameter and the unit output.

### 8.8.2 Fluidized-Bed Gasifiers

No established design method for sizing a fluidized-bed gasifier is available in the literature because, though nearly a century old, this type is still

**TABLE 8.7** Maximum Values of Hearth Load Based on Throat Area for Downdraft Gasifiers

| Plant | Gasifier Type | Medium | $D_{throat}$ (m) | $D_{air\ entry}$ (m) | Superficial Velocity at Throat (m/s) | Hearth Load[a] ($MW/m^2$) |
|---|---|---|---|---|---|---|
| Gengas | Imbert | Air | 0.15 | 0.3 | 2.5 | 15 |
| Biomass Corp. | Imbert | Air | 0.3 | 0.61 | 0.95 | 5.7 |
| SERI | Throatless | Air | 0.15 | | 0.28 | 1.67 |
| Buck Rogers | Throatless | Air | 0.61 | | 0.23 | 1.35 |
| Buck Rogers | Throatless | Air | 0.61 | | 0.13 | 0.788 |
| Syngas | Throatless | Air | 0.76 | | 1.71 | 10.28 |
| Syngas | Throatless | Oxygen | 0.76 | | 1.07 | 12.84 |
| SERI | Throatless | Oxygen | 0.15 | | 0.24 | 1.42 |

[a]*Based on throat area.*
**Source:** Data compiled from Reed and Das (1988), p. 36.

**TABLE 8.8** Sizes of Imbert-Type Gasifiers of Different Capacities

| $d_r/d_h$(−) | $d_h$ (mm) | $d_r$ (mm) | $d_{r'}$(mm) | $h$ (mm) | $H$ (mm) | $R$ (mm) | $A$ (no.) | $d_m$ (mm) | Range of Gas Output (N $m^3$/h) | Maximum Wood Consumption (kg/h) | Air Blast Velocity (m/s) |
|---|---|---|---|---|---|---|---|---|---|---|---|
| 268/60 | 60 | 268 | 150 | 80 | 256 | 100 | 5 | 7.5 | 4–30 | 14 | 22.4 |
| 300/100 | 100 | 300 | 208 | 100 | 275 | 115 | 5 | 10.5 | 10–77 | 36 | 29.4 |
| 400/130 | 130 | 400 | 258 | 110 | 370 | 155 | 7 | 10.5 | 17–120 | 57 | 32.6 |
| 400/150 | 135 | 400 | 258 | 120 | 370 | 155 | 7 | 12 | 21–150 | 71 | 32.6 |
| 400/175 | 175 | 400 | 308 | 130 | 370 | 155 | 7 | 13.5 | 26–190 | 90 | 31.4 |
| 400/200 | 200 | 400 | 318 | 145 | 370 | 153 | 7 | 16 | 33–230 | 110 | 31.2 |

Variables not defined in the figure are defined as follows: $d_m$ = inner diameter of the tuyere and $A$ = number of tuyeres.
**Source:** Data compiled from Reed and Das (1988).

evolving. This section presents a tentative method for determining size based on available information.

#### 8.8.2.1 Cross-Sectional Area

The inside cross-sectional area of the fluidized-bed gasifier, $A_b$, is found by dividing the volumetric flow-rate of the product gas flow, $V_g$, by the chosen superficial gas or fluidization velocity through it, $U_g$, at the operating temperature and pressure.

$$A_b = \frac{V_g}{U_g} \tag{8.29}$$

The volume of gas $V_g$ at the operating temperature and pressure is estimated from the mass of air (or other medium), $M_{fa}$, required for gasification. This gas flow-rate should also be appropriate for fluidization. Thus, $V_g$ is necessarily the gas passing through the grate and the bed.

In some designs, part of the gasifying medium is injected above the distributor grid. In that case, $V_g$ is only the amount that passes through the grid. We can use the mass of gasification medium, $M_{fa}$, required for gasification for the computation of $V_g$:

$$V_g = \frac{M_{fa}}{\rho_g} \tag{8.30}$$

where $\rho_g$ is the density of the medium at the gasifier's operating temperature and pressure.

Equation (8.29) requires choosing an appropriate value for the superficial gas (fluidizing) velocity, $U_g$, through the gasification zone. This is critical as it must be within acceptable limits for the selected particle size to ensure satisfactory fluidization and to avoid excessive entrainment.

#### 8.8.2.2 Fluidization Velocity

The range of fluidizing velocity, $U_g$, in a bubbling bed depends on the mean particle size of the bed materials. The choice is made in the same way as for a fluidized-bed combustor. The range should be within the minimum fluidization and terminal velocities of the mean bed particles. The particle size may be within group B or group D of Geladart's powder classification (see Basu, 2006, Appendix I). The typical fluidization velocity for silica sand (~1 mm mean diameter) may, for example, vary between 1.0 and 2.0 m/s.

If the gasifier reactor is a CFB type, the fluidization velocity in its riser (Figure 8.12) must be within the limits of fast fluidization, which favors groups A or group B particles. Typical fluidization velocity for particle size in the range 150–350 μm is 3.5–5.0 m/s in a CFB. This type of bed has another important operating condition to be satisfied for operation in the

CFB regime. Solids, captured in the gas–solid separator at the gasifier exit, must be recycled back to the gasifier at a rate sufficiently high to create a "fast-fluidized" bed condition in the riser. Additional details about this are available in Basu (2006) or Kunii and Levenspiel (1991).

### 8.8.2.3 Gasifier Height

Since gasification involves only partial oxidation of the fuel, the heat released inside a gasifier is only a fraction of the fuel's heating value, and part of it is absorbed by the gasifier's endothermic reactions. Thus, it is undesirable to extract any further heat from the main gasifier column. For this reason, the height of a fluidized-bed gasifier is not determined by heat-transfer considerations as for fluidized-bed boilers. Instead, solid residence times are major considerations. For coal it could be about an hour (Probstein and Hicks, 2006, p. 154).

The total height of the gasifier is made up of the height of the fluidized bed and that of the freeboard above it:

$$\text{Total gasifier height} = \text{bubbling bed height (depth)} + \text{freeboard height} \tag{8.31}$$

### 8.8.2.4 Fluidized-Bed Height

The bed height (or depth) of a bubbling fluidized-bed gasifier is an important design parameter. Gas–solid gasification reactions are slower than combustion reactions, so a bubbling-bed gasifier is necessarily deeper than a bubbling-bed combustor, which is typically 1.0–1.5 m deep for units larger than 1 m in diameter. Besides pilot plant data or design experience, there is presently no simple means of deciding the bed depth. A deeper bed allows longer gas residence time, but the depth should not be so great compared to its diameter as to cause slugging. The selection of bed height depends on economics. A higher bed height means a higher pressure drop and also a taller reactor. It also should provide a longer residence time for better carbon conversion.

The gasification agent, $CO_2$ or $H_2O$, entering the grid takes a finite time to react with char particles to produce the gas. The bulk of the gasifying agent travels up through the bubbles but very little reaction takes place in the bubble phase. Rather, the reaction takes place mostly in the emulsion phase. The extent to which oxygen or steam is converted into fuel gases thus depends on the gas exchange rate between the bubble and emulsion phases as well as on the char–gas reaction rate in the emulsion phase. This is best computed through a kinetic model of the gasifier as described in Section 7.6.2. An alternative is to use an approach based on residence time, as described next.

### Residence-Time-Based Design Approach

A bubbling fluidized bed must be sufficiently deep to provide reactants the time to complete the gasification reactions. This is why residence time is an important consideration for determination of bed height. An approach based on residence time, developed primarily for coal gasification, can be used for biomass char gasification, which gives at least a first estimate of the bed height for a biomass-fueled bubbling fluidized-bed gasifier.

The residence time approach is based on the assumption that the conversion of char into gases is the slowest of all gasifier processes, so the reactor should provide adequate residence time for the char to complete its conversion to the desired level. Here is a simplified method.

Given the following assumption:

1. The reactivity factor $f_o = 1$ (which lies between $0 < f_o \leq 1$).
2. The solid is in a perfectly mixed condition (i.e., continuous stirred-tank reactor, CSTR).

Then, the volume of the fluidized bed, $V$, is calculated using the equation:

$$V = \frac{W_{out}\theta}{\rho_b} \tag{8.32}$$

where $W_{out}$ is the char moving out; kg/s = $(1 - X)\ W_{in}$; $X$ is the fraction of the char in the converted feed; $\rho_b$ is the bed density, which can be estimated theoretically from fluidization hydrodynamics and regime (kg/m$^3$); and $\theta$ is the residence time of the char in the bed or reaction time (s).

The residence time approach assumes that the water–gas reaction, ($C + H_2O \rightarrow CO + H_2$), as written in Eq. (8.33) is the main gasification reaction, where the char is consumed primarily by the steam gasification reaction for $n$th-order kinetics:

$$\frac{1}{m}\frac{dC}{dt} = k[H_2O]^n \tag{8.33}$$

where $m$ is the initial mass of the biomass and C is the total amount of carbon gasified in time, $t$. Taking a logarithm of this:

$$\ln\left(\frac{1}{m}\frac{dC}{dt}\right) = \ln(k) + n\ \ln[H_2O] \tag{8.34}$$

Experiments can be carried out taking a known weight of the biomass and measuring the change in carbon conversion at different time intervals for a given temperature, steam flow, and pressure. Using these data, graphs are plotted between $\ln((1/m)(dC/dt))$ and $\ln[H_2O]$. The $y$ intercept in this graph will give the value of $k$, and the slope will give the value of $n$. An example of such a plot is shown in Figure 8.23. The experiment is carried out for

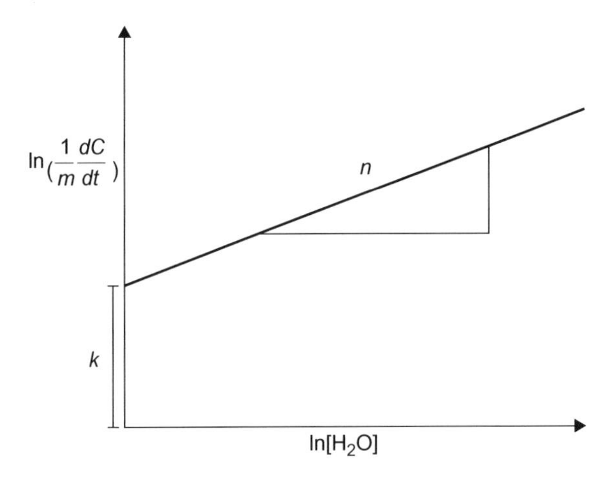


**FIGURE 8.23** Plot of Eq. (8.34) for determination of residence time.

different operating temperatures such as 700°C, 800°C, and 900°C; so, for each temperature, one $k$ value is obtained.

Now $k$ can be expressed as:

$$k = k_0 \exp\left(-\frac{E_a}{RT}\right)$$
$$\ln k = \ln k_0 \frac{E_a}{RT} \quad (8.35)$$

This shows that if we plot a graph between ln $k$ and $1/T$, the $y$ intercept will give the value of $k_0$ and the slope will give the value of $(-E_a/R)$.

The reaction rate for the steam gasification of biomass is given by:

$$\frac{dC}{dt} = k_0 m \exp\left(-\frac{E_a}{RT}\right)[H_2O]^n \quad (8.36)$$

This gives the generalized reaction rate that shows the dependence of the gasification rate on temperature, mass of carbon or char, and concentration of steam/air/oxygen.

From a knowledge of the reaction rate, the residence time, $\theta$, can be calculated as:

$$\theta = C_0 \frac{X}{r} \quad (8.37)$$

where $C_0$ is the initial carbon in the biomass particle (kg), $X$ is the required carbon conversion (−), and $r$ is the steam gasification reaction rate (kg/s). We can avoid such experiments if there is a suitable expression for the rate of steam gasification of the designed biomass char (Sun et al., 2007).

From knowledge of the required solid residence time, $\theta$, then, the bed volume, $V_{bed}$, is:

$$V_{bed} = \frac{F[C]\theta}{(1-\varepsilon)\rho_s x_{char}} \tag{8.38}$$

where $F[C]$ is the char feed rate into the gasifier and $\rho_s$ is the density of the bed solids. In a typical bubbling bed, the bed voidage is ~0.7. The bed generally contains 5–8% (by weight) of reacting char ($x_{char}$); the remaining solids are inert bed materials.

The bed height, $H_{bed}$, is known by dividing bed volume by the bed area, $A_b$, which is known from chosen superficial velocity.

$$H_{bed} = \frac{V_{bed}}{A_b} \tag{8.39}$$

Design charts for residence time, $\theta$, of test coals for different feed conversions and S/C or O/C ratios are given in the *Coal Conversion Systems Technical Data Book* (U.S. DOE, 1978). The residence time may be adjusted for the reactivity of the char in question and for the reactivity of its partial gasification before it enters the gasifier.

## Other Considerations

Although virgin biomass contains little or no sulfur, some waste biomass fuels do. For these, limestone is fed into the fluidized-bed gasifier for in-bed sulfur removal. The height of the gasifier (freeboard and bed) should be adequate to allow the residence time needed for the desired sulfur capture.

The tar produced should be thermally cracked inside the gasifier as far as possible. Therefore, the depth of the gasifier should be such that the gas residence time is adequate for the desired tar conversion/cracking.

The deeper the bed, the higher the pressure drop across it and the higher the pumping cost of air. Because bubble size increases with bed height, a deeper bed gives larger bubbles with reduced gas–solid mixing. Furthermore, if the bubble size becomes comparable to the smallest dimension of the bed cross-section, a highly undesirable slugging condition is reached. This imposes another limit on how deep the dense section of a fluidized bed can be.

Some biomass char, like that from wood, is fine and easily undergoes attrition in a fluidized bed. In such cases, a deeper bed may not guarantee a longer residence time (Barea, 2009). Here, special attention must be paid to capturing the char and either combusting it in a separate chamber to provide heat required by the gasifier or re-injecting it at an appropriate point in the bed where solids are descending.

A kinetic model (*n*th-order, shrinking particle, and shrinking core) may also be used to determine the residence time, the net solid holdup, and therefore the height of the dense bed.

#### *8.8.2.5 Freeboard Height*

Entrainment of unconverted fine char particles from the bubbling bed is a major source of carbon loss. The empty space above the bed, the *freeboard*, allows entrained particles to drop back into it. A bubbling, turbulent, or spouted fluidized bed must have such a freeboard section to help avoid excessive loss of bed materials through entrainment and to provide room for conversion of finer entrained char particles. The freeboard height must be sufficient to provide the required residence time for char conversion. It can be determined from experience or through kinetic modeling.

A larger cross-sectional area and a taller freeboard increase the residence time of gas/char and reduce entrainment. From an entrainment standpoint, the freeboard height need not exceed the transport disengaging height of a bed because no further reduction in entrainment is achieved beyond this.

## 8.9 ENTRAINED-FLOW GASIFIER DESIGN

Because the gas residence time in an entrained-flow reactor is very short—on the order of a few seconds—to complete the reactions, the biomass particles must be ground to extremely fine sizes (less than 1 mm). The residence time requirement for the char is thus on the order of seconds. Section 8.9.1 describes some important considerations for entrained-flow gasifier design.

Although an entrained-flow gasifier is ideally a plug-flow reactor, in practice this is not necessarily so. The side-fed entrained-flow gasifier, for example, behaves more like a CSTR. As we saw in Figure 8.15, at a certain distance from the entry point, fuel particles may have different residence times depending on the path they took to arrive at that section. Some may have traveled a longer path and so have a longer residence time. For this reason, a plug-flow assumption may not give a good estimate of the residence time of char.

### 8.9.1 Gasifier Chamber

Most commercial entrained-flow gasifiers operate under pressure and therefore are compact in size. Table 8.9 gives data on some of these operating in the United States and China.

A typical downflow entrained-flow gasifier is a cylindrical pressure vessel with an opening at the top for feed and another at the bottom for discharge of ash and product gas. The walls are generally lined with refractory and insulating materials, which serve three purposes: (i) they reduce heat

**TABLE 8.9 Characteristic Sizes of Some Entrained-Flow Gasifiers**

| Gasifier | Volume ($m^3$) | Reactor External Diameter (m) | Reactor Internal Diameter (m) | Reactor Height (m) |
|---|---|---|---|---|
| Tennessee Eastman | 12.7 | 2.79 | 1.67 | 4.87 |
| Cool water | 17 | 3.17 | 2.13 | 3.73 |
| Cool water | 25.5 | 3.17 | 2.13 | 6 |
| Cool water | 12.7 | 2.79 | 1.67 | 4.62 |
| Shandong fertilizer | 12.7 | 2.79 | 1.67 | 4.87 |
| Shanghai Chemical | 12.7 | 2.79 | 1.67 | 4.87 |
| Harbei fertilizer | 12.7 | 2.79 | 1.67 | 4.87 |

**Source:** Data compiled from Zen (2006).

loss through the wall, (ii) they act as thermal storage to help ignition of fresh feed, and (iii) they prevent the metal enclosure from corrosion.

The thickness of the refractory and insulation used is to be chosen with care. For example, biomass ash melts at a lower temperature and is more corrosive than most coal ash, so special care needs to be taken in designing the gasifier vessel for biomass feedstock.

The construction of a side-fed gasifier is more complex than that of a top-fed gasifier, as the reactor vessel is not entirely cylindrical and requires numerous openings. The bottom opening is for the ash drain, the top opening is for the product gas, and the side ports are for the feed. Additional openings may also be required depending on the design. Because of the complexity in the design of a pressure vessel operating at 30–70 bar and temperatures exceeding 1000°C, any additional openings or added complexity in the reactor configuration must be weighed carefully against perceived benefits and manufacturing difficulties.

## 8.10 AUXILIARY ITEMS

The following subsections discuss the design of auxiliary systems in fluidized-bed gasifiers.

### 8.10.1 Position of Biomass-Feeding Position

The feed points for the biomass should be such that entrainment of any particles in the product gas is avoided. This can happen when the feed points are located too close to the expanded bed surface of a bubbling fluidized bed. If they are in close proximity to the distributor plate, excessive combustion of the volatiles in the fluidizing air produced can occur. To avoid this, they should be some distance further above the grate.

Nascent tar is released close to the feed point, so tar cracking can be important for some designs. If tar is a major concern, the feed port should be close to the bottom of the gasifier so that the tar has adequate residence time to crack (Barea, 2009).

### 8.10.2 Distributor Plate

The distributor plate of a fluidized bed supports the bed materials. It is no different from that used for a fluidized combustor or boiler. The ratio of pressure drop across the bed and that across the distributor plate must be chosen appropriately to arrive at the plate design. More details are available in books on distributor plate design, including Basu (2006, chapter 11). The typical open area in the air distributor grate is only a few percentage points.

### 8.10.3 Bed Materials

For the process design of a fluidized-bed gasifier, the choice of bed materials is crucial. These comprise mostly granular inorganic solids and some (<10%) fuel particles. For biomass, sand or other materials are used (as explained next); coal gasification requires granular ash produced from the gasification process. Sometimes limestone is added with coal particles to remove sulfur. At different stages of calcination and sulfurization, the limestone can also form a part of the bed material.

Biomass has very little ash (less than 1% for wood), so silica sand is normally used as the inert bed material. This is a natural choice because silica is inexpensive and the most readily available granular solid. One major problem with silica sand is that it can react with the potassium and sodium components of the biomass to form eutectic mixtures having low melting points, thereby causing severe agglomeration. To avoid this, the following alternative materials can be used:

- Alumina ($Al_2O_3$)
- Magnesite ($MgCO_3$)
- Feldspar (a major component of Earth's crust)
- Dolomite ($CaCO_3 \cdot MgCO_3$)
- Ferric oxide ($Fe_2O_3$)
- Limestone ($CaCO_3$)

Magnesite (MgO) was successfully used in the first biomass-based integrated gasification combined cycle plant in Värnamo, Sweden (Ståhl et al., 2001).

Tar is a mixture of higher-molecular-weight (higher than benzene) chemical compounds that condense on downstream metal surfaces at lower temperatures. It can plug the passage and/or make the gas unsuitable for use. The bed materials, besides serving as a heat carrier, can catalyze the gasification reaction by increasing the gas yield and reducing the tar. Bed materials that act as a catalyst for tar reduction are an attractive option. Some are listed here (Pfeifer et al., 2005; Ross et al., 2005):

- Olivine
- Activated clay (commercial)
- Acidified bentonite
- Raw bentonite
- House brick clay

Common house brick clay can be effectively used in a CFB gasifier to reduce tar emission and enhance hydrogen production. The alkalis deposited on the bed materials from biomass may potentially behave as catalysts if their agglomerating effect can be managed (Ross et al., 2005).

Tar production can be reduced using olivine. The Fe content of olivine is catalytically active, and that helps with tar reforming (Hofbauer, 2002). Nickel-impregnated olivine gives even better tar reduction because nickel is active for the steam tar reforming reaction (Pfeifer et al., 2005).

Bingyan et al. (1994) reported using ash from the fuel itself (sawmill dust) as the bed material in a CFB gasifier. This riser is reportedly operated at a very low velocity of 1.4 m/s, which is 3.5 times the terminal velocity of the biomass particles. Chen et al. (2005) tried to operate a 1-MWe CFB gasifier with rice husk alone, but the system had difficulty with fluidization in the loop seal because of the low sphericity of the husk ash; however, the main riser reportedly operated in the fast bed regime without major difficulty.

## 8.11 DESIGN OPTIMIZATION

Design optimization generally starts after the preliminary design is complete and actual project execution is set to begin. It has two aspects: (i) process and (ii) engineering.

Process optimization tells the designer if the preliminary design will give the best performance in terms of efficiency and gas yield, and how this is related to the operation and design parameters. Commercial simulation programs (mathematical models) or computational fluid dynamics codes are the most effective tools for this purpose. Engineering optimization involves optimizing the reactor configuration to enhance its operability, maintainability, and cost reduction.

### 8.11.1 Process Optimization

Process optimization enhances gasifier performance in terms of the following indicators:

- Cold- and hot-gas efficiency
- Unconverted carbon and tar concentration in the product gas
- Composition and heating value of the product gas

One can approach optimization either through experiments or through kinetic modeling.

Experiments are the best and most reliable means of optimizing process parameters, as they are based on the actual or prototype gasifier. However, they have several limitations and are expensive. Furthermore, practical difficulties may not allow all operational parameters to be explored. An alternative is to conduct tests in a controlled laboratory-scale unit and to calibrate the resulting data to the full-scale unit. This allows the scale-up of data from the laboratory to the full-scale unit with a reasonable degree of confidence.

#### *8.11.1.1 Optimization Through Kinetic Modeling*

With a kinetic model, we can predict the performance of a gasifier already designed because it utilizes both configuration and dimensions of the reactor. Kinetic modeling can help optimize or fine-tune the operating parameters for best performance in a given situation. Section 7.6 describes a kinetic model for gasifiers.

## 8.12 PERFORMANCE AND OPERATING ISSUES

Gasifier performance is measured in terms of both quality and quantity of gas produced. The amount of biomass converted into gas is expressed by gasification efficiency. The product quality is measured in terms of heating value as well as amount of desired product gas.

### 8.12.1 Gasification Efficiency

The efficiency of gasification is expressed as cold-gas efficiency, hot-gas efficiency, or net gasification efficiency. These are described in the following subsections:

#### 8.12.1.1 Cold-Gas Efficiency

Cold-gas efficiency is the potential energy output over the energy input. If $M_f$ kg of solid fuel is gasified to produce $M_g$ kg of product gas with an LHV of $Q_g$, the efficiency is expressed as:

$$\eta_{cg} = \frac{Q_g M_g}{\mathrm{LHV}_f M_f} \tag{8.40}$$

where $\mathrm{LHV}_f$ is the LHV of the solid fuel.

---

**Example 8.3**

Air – steam gasifier data includes the mass composition of the feedstock:

C: 66.5%
O: 7%
H: 5.5%
N: 1%
Moisture: 7.3%
Ash: 12.7%
LHV: 28.4 MJ/kg

and the volume composition of the product gas:

CO: 27.5%
$CO_2$: 3.5%
$CH_4$: 2.5%,
$H_2$: 15%
$N_2$: 51.5%

The dry air supply rate is 2.76 kg/kg of feed, the steam supply rate is 0.117 kg/kg of feed, the moisture content is 0.01 kg of $H_2O$ per kg of dry air, and the ambient temperature is 20°C.

Find:

- The amount of gas produced per kg of feed
- The amount of moisture in the product gas
- The carbon conversion efficiency
- The cold-gas efficiency

**Solution**

Table C.3 (Appendix C) shows the mass fraction of $N_2$ and $O_2$ in air as 0.755 and 0.232, respectively. The nitrogen supply from air is:

$$0.755 \times 2.76 = 2.08 \text{ kg } N_2/\text{kg feed}$$

The total nitrogen supplied by the feed air and the fuel feed, which carry 1% nitrogen, is:

$$2.08 + 0.01 = 2.09 \text{ kg } N_2/\text{kg feed} = (2.09/28) = 0.0747 \text{ kmol } N_2/\text{kg feed}$$

noting that volume percent equals molar percent in a gas mixture.

Since the product gas contains 51.5% by volume of nitrogen, the amount of the product gas per kilogram of feed is:

$$0.0747/0.515 = 0.145 \text{ kmol gas/kg feed}$$

Similarly, the oxygen from the air flow to the gasifier is:

$$0.232 \times 2.76 = 0.640 \text{ kg/kg feed}$$

The steam supplied per kilogram of fuel is 0.117 kg, so the oxygen associated with the steam supply is:

$$0.117 \times (8/9) = 0.104 \text{ kg/kg feed}$$

Oxygen also enters through the 7.3% moisture in the fuel and the 1% moisture in the air feed. The total oxygen from moisture is:

$$\begin{aligned} 0.073 \times (8/9) + 0.01 \times 2.76 \times (8/9) &= 0.065 + 0.0245 \\ &= 0.0895 \text{ kg/kg feed} \end{aligned}$$

The total oxygen flow to the gasifier, including the 7% that comes with the fuel, is:

$$0.640 + 0.104 + 0.0895 + 0.07 = 0.9035 \text{ kg } O_2\text{/kg feed}$$

*Hydrogen Balance*

The total hydrogen inflow to the gasifier with fuel, steam, and moisture in the fuel and moisture in the air is:

$$0.055 + 0.117/9 + 0.073/9 + 0.0276/9 = 0.0792 \text{ kg/kg feed}$$

The hydrogen leaving with $H_2$ and $CH_4$ in dry gas, noting that 1 mol of $CH_4$ contributes 2 mol of $H_2$, is:

$$\begin{aligned} (0.15 + 2 \times 0.025) \times 0.145 &= 0.029 \text{ kmol/kg feed} = 0.029 \times 2 \\ &= 0.058 \text{ kg hydrogen/kg feed} \end{aligned}$$

To find the moisture in the product gas, we deduct the hydrogen in the dry gas from the total hydrogen inflow obtained earlier, using the hydrogen balance:

$$\begin{aligned} \text{Hydrogen inflow} - \text{hydrogen out through dry product gas} &= 0.0792 - 0.058 \\ &= 0.0212 \text{ kg/kg feed} \end{aligned}$$

The steam or moisture associated with this hydrogen in the gas is:

$$0.0212 \times (18/2) = 0.1908 \text{ kg/kg feed}$$

*Carbon Balance*

The carbon-bearing gases—CO, $CO_2$, and $CH_4$ in the dry gas each contains 1 mol of carbon. So the total carbon in 0.145 kmol/kg of fuel product gas is:

$$\begin{aligned} (0.275 + 0.035 + 0.025) \times 0.145 &= 0.0485 \text{ kg mol/kg feed} = 0.0485 \times 12 \\ &= 0.583 \text{ kg/kg feed} \end{aligned}$$

The carbon input, as found from the composition of the feed, is 0.665 kg/kg feed. The carbon conversion efficiency is found by dividing the carbon in the product gas by that in the fuel is equal to:

$$(0.583/0.665) \times 100 = 87.6\%$$

***Energy Balance***

The heats of combustion for different gas constituents are taken from Table C.2 (Appendix C). They are:

CO: 12.63 $MJ/nm^3$

Hydrogen: 12.74 $MJ/nm^3$

Methane: 39.82 $MJ/nm^3$

We note that 1 kg of feed produces 0.145 kmol of gas, the volumetric composition of which is:

CO: 27.5%

$CO_2$: 3.5%

$CH_4$: 2.5%

$H_2$: 15%

$N_2$: 51.5%

By multiplying the heating value of the appropriate constituents of the product gas, we can find the total heating value of the product gas (the volume of 1 kmol of any gas is 22.4 $nm^3$):

$$(12.63 \times 0.275 + 12.74 \times 0.15 + 39.82 \times 0.025)\text{MJ/nm}^3$$
$$\times 0.145\ \text{kmol/kg feed} \times 22.4\ \text{nm}^3/\text{kmol} = \mathbf{20.6\ MJ/kg\ feed}$$

The total energy input is equal to the heating value of the feed, which is 28.4 MJ/kg.

From Eq. (8.40), the cold-gas efficiency is:

$$(20.6/28.4) \times 100 = 72.5\%$$

---

### 8.12.1.2 Hot-Gas Efficiency

Sometimes gas is burned in a furnace or boiler without being cooled, creating a greater utilization of the energy. Therefore, by taking the sensible heat of the hot gas into account, the hot-gas efficiency, $\eta_{hg}$, can be defined as:

$$\eta_{hg} = \frac{Q_g M_g + M_g C_p (T_f - T_0)}{\text{LHV}_f M_f} \tag{8.41}$$

where $T_f$ is the gas temperature at the gasifier exit or at the burner's entrance and $T_0$ is the temperature of the fuel entering the gasifier. The hot-gas efficiency assumes the heating of the unconverted char to be a loss.

---

**Example 8.4**

The gas produced by the gasifier in Example 8.3 is supplied directly to a burner at the gasifier exit temperature, 900°C, to be burnt for cofiring in a boiler. Find the hot-gas efficiency of the gasifier.

**Solution**

The product gas enters the burner at 900°C (1173K). To find the enthalpy of the product gas, we add the enthalpies of its different components. Specific heats of individual components are calculated using the relations from Table C.4 (Appendix C). For example, the specific heat of CO at 1173K is:

$$27.62 + 0.005 \times 1173 = 33.48\ \text{kJ/kmol K}$$

From Example 8.3, the amount of product gas is 0.145 kmol/kg fuel. The enthalpy of CO in the product gas that contains 27.5% CO above the ambient temperature, 25°C or 298K, is:

$$(0.145 \times 0.275)\ \text{kmol/kg feed} \times 33.48\ \text{kJ/kmol K} \times (1173 - 298)\text{K} \times 10^{-3}\ \text{MJ/kJ}$$
$$= 1.168\ \text{MJ/kg feed}$$

Similarly enthalpy of other products,

$CO_2$: $(0.145 \times 0.035) \times 56.06 \times (1173 - 298) \times 10^{-3} = 0.249$ MJ/kg feed
H2: $(0.145 \times 0.15) \times 31.69 \times (1173 - 298) \times 10^{-3} = 0.603$ MJ/kg feed
$N_2$: $(0.145 \times 0.515) \times 32.13 \times (1173 - 298) \times 10^{-3} = 2.099$ MJ/kg feed
$CH_4$: $(0.145 \times 0.025) \times 78.65 \times (1173 - 298) \times 10^{-3} = 0.249$ MJ/kg feed

The amount of steam in the flue gas was calculated as 0.1908 kg/kg of feed. To find the enthalpy of this steam above 298 K, we take values of the steam enthalpy at 1 bar of pressure at 1173 and 298 K. The values are 4398.05 and 104.93 kJ/kg, respectively, so the enthalpy in water is:

$$H_2O{:}\,0.1908 \times (4398.05 - 104.93) \times 10^{-3} = 0.819\ \text{MJ/kg feed}$$

Adding these, we get the total enthalpy of the product gas at 900°C.

$$1.168 + 0.249 + 0.603 + 2.099 + 0.249 + 0.819 = 4.368\ \text{MJ/kg feed}$$

The total thermal energy is:

$$\text{Heating value} + \text{enthalpy} = 20.6 + 4.368 = 24.968\ \text{MJ/kg coal}$$

The total gasifier efficiency is:

$$\text{Total thermal energy/heat in feedstock} = \frac{24.968}{28.4} \times 100\%$$
$$= 87.92\%$$

---

#### *8.12.1.3 Net Gasification Efficiency*

The enthalpy or energy content of the gasification medium can be substantial, and so, for a rigorous analysis, these inputs should be taken into consideration. At the same time, part of the input energy is returned (energy credit) by the tar or oil produced as well as by any recovery of the heat of vaporization in the product gas. A more rigorous energy balance may thus be written as:

- Total gross energy input = fuel energy content + heat in gasifying mediums
- Net energy input = total energy input − energy recovered through burning tar, oil, and condensation of steam in the gas

The net gasification efficiency can be written as:

$$\eta_{net} = \frac{\text{net energy in the product gas}}{(\text{total energy input to the gasifer} - \text{credits})} \tag{8.42}$$

---

**Example 8.5**

In most steam-fed gasifiers, a large amount of steam remains unutilized. For the given problem, find the amount of unutilized steam. Also find the cold-gas and net gasification efficiency of a fixed-bed gasifier that uses steam and oxygen to gasify grape wastes (HHV = 21,800 kJ/kg). The product gas composition (mass basis) is:

CO: 31.8%
$H_2$: 3.1%
$CO_2$: 38.2%
$CH_4$: 1.2%
$C_3H_8$: 0.9%
$N_2$: 1%
$H_2O$: 44.8%

The HHV of the product gas is 8.78 MJ/kg.

The ultimate and proximate analyses of the biomass are as given in Table 8.10. The total fuel feed rate is 25 kg/s; the oxygen feed rate is 5.3 kg/s. The steam is fed into the gasifier at a rate of 27 kg/s at 180°C and 5 bar of pressure. The product contains dry gas, condensable moisture, and tar. The tar production rate is 1.3 kg/s and is analyzed to contain 85% carbon and 15% hydrogen by weight. The heating value of the tar is 42,000 kJ/kg. The oxygen is produced from air using an oxygen-separation unit that consumes 4000 kJ of energy/kg of the oxygen produced (assume full conversion of char).

Find the amount of product gas produced and the fraction of steam that remains unutilized.

**Solution**

Hydrocarbon hydrogen from the ultimate analysis is $5.83 \times (1 - 0.04) = 5.6\%$. Additional hydrogen also in the moisture is $0.04 \times (2/18) = 0.44\%$. Thus, the total hydrogen, on an as-received basis, is $5.6 + 0.44 = 6.04\%$. The feed rate of the total hydrogen through the fuel is $25 \times 6.04/100 = 1.51$ kg/s.

A mass balance between input and output helps determine the production rate of the gas. Output equals input, so:

Product + ash + tar and oil = fuel + oxygen + steam
Product + $(25 \times 0.042) + 1.3 = 25 + 5.3 + 27$
Product = 54.95 kg/s

The product contains gas, the composition of which was given previously ($M_{gas}$), as well as the condensate, $M_{cond}$. To find the gas, we carry out a carbon balance from its measured composition.

Part (a): Carbon Balance

The total carbon in the gas (%) is:

$$\text{Molecular weight of carbon} \times \sum \left( \frac{\text{percent of species with carbon in product gas}}{\text{molecular weight of the species}} \right)$$

$$= 12 \times \left( \frac{31.8}{28} + \frac{38.2}{44} + \frac{1.2}{16} + \frac{0.9}{44} \right) = 25.19\%$$

The carbon balance gives:

Carbon in gas + carbon in tar and oil = carbon in fuel

$$\begin{aligned} M_{gas} \times 25.19\% + 1.3 \times 85\% &= 25 \times 55.59\% \\ M_{gas} &= 50.78\ \text{kg/s} \\ \text{Total product} &= 54.95 = M_{gas} + M_{cond} \\ M_{cond} &= 4.17\ \text{kg/s} \end{aligned}$$

Part (b): Water Balance

Water enters the gasifier through the steam as well as through the moisture in the fuel, so:

Water in steam + water in fuel = water used in gasification + water leaving as waste steam/water

The water used in gasification is:

$$27 + 25 \times 0.04 - 50.78 \times 0.448 = 5.25\ \text{kg/s}$$

Therefore, the percent of steam not utilized is:

$$1 - 5.25/27 = 19.44\%$$

**TABLE 8.10 Analyses of Ultimate and Proximate of the Biomass feedstock for Example 8.5**

| | Proximate Analysis in Mass (%) | | Ultimate Analysis (dry basis) in Mass (%) |
|---|---|---|---|
| Ash | 4.2 | Carbon | 55.59 |
| Volatile matter | 70.4 | Hydrogen | 5.83 |
| Fixed carbon | 21.4 | Nitrogen | 2.09 |
| Moisture | 4.0 | Sulfur | 0.21 |
| | | Oxygen | 32.08 |
| | | Ash | 4.2 |

### 8.12.2 Operational Considerations

A large number of operational issues confront a biomass gasifier. Universal to all gasifier types are problems related to biomass handling and feeding. Bridging of biomass over the exit of a hopper is common for plants that use low-shape-factor (flaky) biomass such as leaves and rice husk. This problem is discussed in more detail in Chapter 12.

#### *8.12.2.1 Fixed-Bed Gasifier*

Charcoal particles become porous and finer during their residence in the gasification zone. Thus, in a downdraft gasifier, when fine charcoal drops into the ash pit, the product gas can easily carry the particles as dust. Escaping particles can be a source of carbon loss, and they often plug downstream equipment.

Under steady state, the rate of drying, pyrolysis, or gasification at any layer must be equal to the feed into the section. Otherwise, the conversion zone will move either up or down in a moving-bed gasifier. Thus, dry fuel is fed into an updraft gasifier designed for wet fuel, the pyrolysis zone may travel upward faster, thus consuming the layer of fresh fuel above and leading to premature pyrolysis. The gas lost in this way may result in lower gasification efficiency.

On the other hand, if the fuel is more wet than designed, its pyrolysis may be delayed. This may move the pyrolysis zone downward. In the extreme case, the cooler pyrolysis zone may sink sufficiently to extinguish the gasification and combustion reaction. Clearly, a proper balance of rates of fuel flow and air flow is required for stabilization of each of these zones in respective places.

#### *8.12.2.2 Fluidized-Bed Gasifier*

The start-up of a fluidized-bed gasifier is similar to the start-up of a fluidized bed combustor. The inert bed materials are preheated either by an overbed burner or by burning gas in the bed. Once the bed reaches the ignition temperature of the fuel, the feed is started. Combustion in bed is allowed to raise the temperature. Once it reaches the required temperature, the air/oxidizer-to-fuel ratio is slowly adjusted to switch from combustion to gasification mode.

One major problem with fluidized-bed gasifiers is the entrainment (escape) of fine char with the product gas. The superficial velocity in a fluidized bed is often sufficiently high to transport small and light char particles, contributing to major carbon loss. A tall freeboard can reduce the problem, but that has a cost penalty. Instead, most fluidized-bed gasifiers use a cyclone and a recycle system to return the entrained char particles back to the gasifier.

#### 8.12.2.3 Entrained-Flow Gasifier

The start-up procedure for an entrained-flow gasifier takes a long time because a start-up burner must heat up the reactor vessel wall that is lined with heavy refractory. During this time, the reactor vessel is not pressurized. Once oil or gas flame heats up the thick refractory wall to $\sim 1100°C$, the start-up burner is withdrawn and the fuel is injected along with the oxidizer (Weigner et al., 2002). The hot reactor wall serves as a thermal storage and igniter for the fuel, which once ignited the fuel continues to burn in the combustion zone, consuming the oxygen. For this reason, the fuel injector in an entrained-flow reactor is also called the "burner." The reactor is pressurized slowly once the main fuel is ignited.

The gasifying medium is rarely premixed with the fuel. The fuel and the medium are often injected coaxially, as in a pulverized coal (PC) burner in a boiler or furnace. They immediately mix on entering the reactor. The operation of a gasifier's "burner" is similar to that of conventional burners, so design methods for PC or oil burners can be used for a rough and an initial sizing. The use of a separate start-up burner involves replacing it with a fuel injector. This is especially difficult for water-cooled walls because their lower thermal inertia cannot hold the wall temperature long enough. Integration of the start-up burner in the existing fuel injector is the best option.

#### 8.12.2.4 Tar Cracking

Several options for tar control and destruction are available. In fixed-bed gasifiers, thermal cracking or burning has been used with success. In one such design, as shown in Figure 8.24, the air entering the gasifier passes through an aspirator that entrains the tar vapor. The mixture is then burnt in the combustion zone. The aspirator can be outside or inside the gasifier. Fluidized-bed gasifiers can use appropriate bed materials to crack or reduce tars. More details are discussed in Section 8.10.3.

## SYMBOLS AND NOMENCLATURE

| | |
|---|---|
| $A_b$ | cross-sectional area of the fluidized bed ($m^2$) |
| **ASH** | fractional of ash in the fuel in dry basis (–) |
| **C** | fractional of carbon in the fuel in dry basis (–) |
| $C_i$ | volumetric specific heat of gas $i$ (kJ/N $m^3$ K) |
| $C_0$ | initial carbon in the biomass (kg) |
| $C_p$ | specific heat of the gas (kJ/kg C) |
| $E_a$ | activation energy (kJ/mol) |
| **EA** | excess air coefficient (–) |
| **ER** | equivalence ratio (–) |
| $F$ | amount of dry fuel required to obtain 1 N $m^3$ of product gas (kg/N $m^3$) |
| $F[C]$ | char feed rate into the gasifier (kg/s) |
| **H** | fractional of hydrogen in the fuel in dry basis (–) |

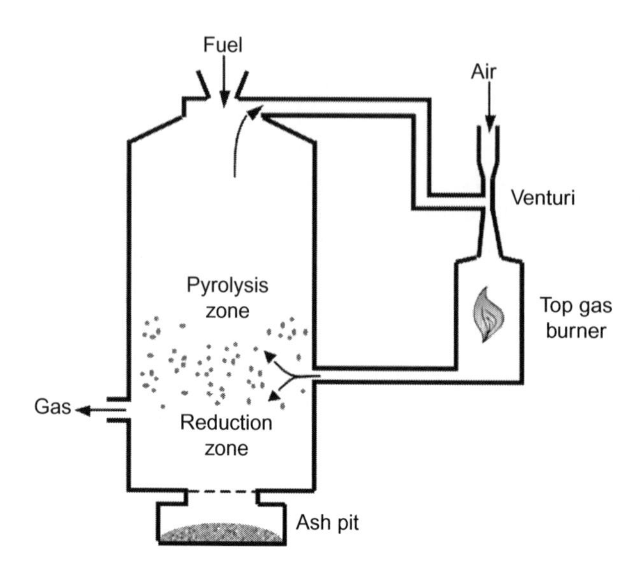


**FIGURE 8.24** Gasifier with an aspirator for cracking tar. Fresh air picks up the tar from the gasifier and injects it into the high-temperature combustion zone.

| | |
|---|---|
| **HHV** | higher heating value (kJ/kg) |
| $\mathbf{HHV_d}$ | higher heating value of biomass on dry basis (MJ/kg) |
| $\mathbf{HHV_{daf}}$ | higher heating value of biomass on dry-ash-free basis (MJ/kg) |
| $H_{bed}$ | height of the bed (m) |
| $H_g$ | enthalpy of steam at gasification temperature (kJ/kg) |
| $H_{in}$ | heat of the input gas (kJ) |
| $[H_2O]$ | concentration of steam (−) |
| $k$ | rate constant ($s^{-1}$) |
| $k_0$ | pre-exponential constant in the Arrhenius equation ($s^-1$) |
| $\mathbf{LHV_{bm}}$ | lower heating value of the biomass (MJ/kg) |
| $\mathbf{LHV_{daf}}$ | lower heating value of biomass on dry-ash-free basis (MJ/kg) |
| $\mathbf{LHV_f}$ | lower heating value of the solid fuel (MJ/N $m^3$) |
| $\mathbf{LHV_g}$ | lower heating value of the produced gas (MJ/N $m^3$) |
| $m$ | mass-flow rate of carbon or char (kg/s) |
| $m_{th}$ | theoretical air requirement for complete combustion of a unit of biomass (kg/kg) |
| $M_a$ | amount of air required for gasification of unit mass of biomass (kg/kg) |
| $M$ | fractional of moisture in the fuel (−) |
| $M_{daf}$ | moisture based on dry-ash-free basis |
| $M_f$ | fuel flow-rate (kg/s) |
| $M_{fh}$ | quantity of steam (kg/s) |
| $M_g$ | gas produced (kg/s) |
| $n$ | order of reaction (−) |
| $n_i$ | number of moles of species $i$ (−) |
| **N** | fractional of nitrogen in the fuel in dry basis (−) |
| $n_{total}$ | total number of moles |

| | |
|---|---|
| **O** | fractional of oxygen in the fuel in dry basis (−) |
| $P_c$ | amount of char produced per N $m^3$ of product gas (kg/N $m^3$) |
| $q_c$ | heating value of char (kJ/kg) |
| $Q$ | power output of the gasifier ($MW_{th}$) |
| $Q_{ext}$ | external heat addition to the system (kJ/N $m^3$) |
| $Q_g$ | lower heating value of the product gas from gasification (MJ/N $m^3$) |
| $Q_{gasification}$ | heat supplied to gasify 1 mol of biomass (kJ/mol) |
| $Q_{loss}$ | heat loss from the gasifier (kJ/N $m^3$) |
| $r$ | steam gasification reaction rate (kg/s) |
| $R$ | universal gas constant (0.008314 kJ/mol K) |
| **S** | fractional of sulfur in the fuel in dry basis (−) |
| **SC** | steam-to-carbon molar ratio (−) |
| $t$ | time (s) |
| $T$ | temperature (K) |
| $T_f$ | gas temperature at the exit (°C) |
| $T_g$ | gas temperature (°C) |
| $T_0$ | gas temperature at the entrance (°C) |
| $U_g$ | fluidizing velocity (m/s) |
| $V$ | volume of the fluidized bed ($m^3$) |
| $V_{bed}$ | volume of the bed ($m^3$) |
| $V_{daf}$ | volatile based on dry-mass-free basis |
| $V_g$ | gas generation rate ($m^3$/s) |
| $V_g$ | volumetric flow rate of product gas (N $m^3$/s) |
| $V_i$ | volumetric fraction of gas species $i$ (−) |
| $W$ | total steam needed in Eq. (8.22) (kg/s) |
| $W_{in}$ | rate of the char moving in (kg/s) |
| $W_{out}$ | rate of the char moving out (kg/s) |
| $x_{char}$ | weight of the reacting char (kg) |
| $X$ | fraction of char in the feed converted (−) |
| $X_c$ | fixed carbon fraction in the fuel (kg carbon/kg dry fuel) |
| $X_{char}$ | char fraction in bed (−) |
| $X_g$ | fraction of steam used up in gasification |
| $\varepsilon$ | voidage of the bed (−) |
| $\lambda_I$ | Lagrangian multiplier for species $i$ (−) |
| $\rho_g$ | density of air at the opening temperature and pressure of the gasifier (kg/$m^3$) |
| $\theta$ | residence time of char in bed or reactor (s) |
| $\rho_b$ | bed density (kg/$m^3$) |
| $\rho_s$ | density of bed solids (kg/$m^3$) |
| $\eta_{gef}$ | gasifier efficiency (−) |
| $\eta_{ceff}$ | cold-gas efficiency (−) |
| $\eta_{cg}$ | cold-gas efficiency of the gasifier (−) |
| $\eta_{hg}$ | hot-gas efficiency of the gasifier (−) |
| $\eta_{net}$ | net gasification efficiency of the gasifier (−) |
| $\Delta H_T$ | heat of formation at temperature $T$ (kJ/mol) |

Chapter 9

# Hydrothermal Gasification of Biomass

## 9.1 INTRODUCTION

This chapter deals with conversion of biomass into liquid or gaseous products in hydrothermal medium. Here the hydrothermal medium is a water-rich phase above about 200°C at sufficiently high pressures to keep the water in either a liquid or supercritical state (Peterson et al., 2008).

In the mid-1970s, Sanjay Amin, a graduate student working at the Massachusetts Institute of Technology (MIT), was studying the decomposition of organic compounds in hot water (steam reforming):

$$C_6H_{10}O_5 + 7H_2O \rightarrow 6CO_2 + 12H_2 \tag{9.1}$$

While conducting an experiment in subcritical water, he observed that in addition to producing hydrogen and carbon dioxide, the reaction was producing much char and tars. Herguido et al. (1992) also made similar observations in the steam gasification of biomass at atmospheric pressure.

Sanjay interestingly noted that when he raised the water above its "critical state," the tar that formed in the subcritical state disappeared entirely (Amin et al., 1975). This important finding kick-started research and development on supercritical water oxidation (SCWO), for disposal of organic waste materials (Tester et al., 1993), which has now become a commercial option for disposal of highly contaminated organic wastes (Shaw and Dahmen, 2000).

Biomass in general contains substantially more moisture than do fossil fuels like coal. Some aquatic species, such as water hyacinth, or waste products, such as raw sewage, can have water contents exceeding 90%. Thermal gasification, where air, oxygen, or subcritical steam is the gasification medium, is very effective for dry biomass, but it becomes very inefficient for a high-moisture biomass because the moisture must be substantially driven away before thermal gasification can begin; in addition, a large amount of the extra energy ($\sim$2242 kJ/kg moisture) is consumed in its evaporation. For example, Yoshida et al. (2003) saw the efficiency of their thermal gasification system reduce from 61% to 27%, while the water content of

**Biomass Gasification, Pyrolysis and Torrefaction.**

the feed increased from 5% to 75%. So, for gasification of very wet biomass, some other means such as anaerobic digestion (see Section 3.2.2) and hydrothermal gasification in high-pressure hot water are preferable because the water in these processes is not a liability as it is in thermal gasification. Instead, it serves as a reaction medium and a reactant.

The efficiencies of hydrothermal or anaerobic processes do not decrease with moisture content of the biomass. For anaerobic digestion and supercritical gasification, Yoshida et al. (2003) found the gasification efficiency to remain nearly unchanged, at 31% and 51%, respectively, even when the moisture in the biomass increased from 5% to 75%. A major limitation of anaerobic digestion is, however, that it is very slow and most importantly, it produces methane only, no hydrogen. If hydrogen is the desired product, as is often the case, an additional step of steam reforming the methane ($CH_4 + H_2O = CO + 3H_2$) must be added to the anaerobic digestion process.

Hydrothermal gasification involves gasification in an aqueous medium at a very high temperature and pressure exceeding or close to its critical value. While subcritical water (pressure, $P$ and temperature, $T$ are below their critical values) has been used effectively for hydrothermal reaction, supercritical water (SCW) has attracted more attention owing to its unique features. SCW ($P > P_c$; $T > T_c$) offers rapid hydrolysis of biomass, high solubility of intermediate reaction products, including gases, and a high ion product near (but below) the critical point that helps ionic reaction. These features make SCW an excellent reaction medium for gasification, oxidation, and synthesis reactions.

This chapter deals primarily with hydrothermal gasification of biomass in SCW, while explaining the properties of SCW and the biomass conversion process in it. The effects of different parameters on SCW gasification and design considerations for the SCW gasification plants are also presented.

## 9.2 SUPERCRITICAL WATER

Water above its critical temperature (374.29°C) and pressure (22.089 MPa) is said to be in *supercritical* (Figure 9.1) state or simply as SCW. Water or steam below this pressure and temperature is called *subcritical*. The term *water* in a conventional sense may not be applicable to SCW except for its chemical formula, $H_2O$, because above the critical temperature SCW is neither water nor steam. It has a water-like density but a steam-like diffusivity. Table 9.1 compares the properties of subcritical water and steam with those of SCW, indicating that SCW's properties are intermediate between the liquid and gaseous states of water in subcritical pressure.

Figure 9.1 shows that the higher the temperature, the higher the pressure required for water to be in its liquid phase. Above a critical point, the line separating the two phases disappears, and thereby the division between

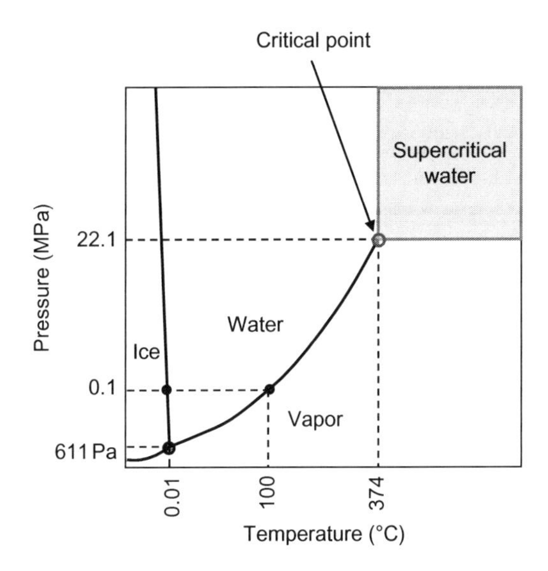


**FIGURE 9.1** Phase diagram of water showing the supercritical region.

**TABLE 9.1 Properties of Supercritical and Subcritical Water**

| Properties | Subcritical Water | Supercritical Water | Supercritical $CO_2$ | Subcritical Steam |
|---|---|---|---|---|
| Temperature (°C) | 25 | 400 | 55 | 150 |
| Pressure (MPa) | 0.1 | 30 | 28 | 0.1 |
| Density (kg/m$^3$) | 997 + | 358 + | 835 | 0.52 + |
| Dynamic viscosity (μ) (kg/m s) | $890.8 \times 10^{-6}$ | $43.83 \times 10^{-6+}$ | $0.702 \times 10^{-6}$ | $14.19 \times 10^{-6+}$ |
| Diffusivity of small particles (m$^2$/s) | $\sim 1.0 \times 10^{-9\#}$ | $\sim 1.0 \times 10^{-8\#}$ | | $\sim 1.0 \times 10^{-5\#}$ |
| Dielectric constant$^a$ | 78.46 | 5.91 | | 1.0 |
| Thermal conductivity (λ) (W/m k) | $607 \times 10^{-3+}$ | $330 \times 10^{-3+}$ | | $28.8 \times 10^{-3+}$ |
| Prandtl number ($C_p\mu/\lambda$) | 6.13 | 3.33 | | 0.97 |

*$^a$Uematsu and Franck (1980), # Serani et al. (2008), + Haar et al. (1984).*

the liquid and vapor phases disappears. Temperature and pressure at this point are known as *critical temperature* and *critical pressure*, respectively. Above these values, water attains supercritical state and hence is called SCW.

*Subcritical water* ($T < T_{sat}$; $P < P_c$): When the pressure is below its critical value, $P_c$, and the temperature is below its critical value, $T_c$, the fluid is called *subcritical*. If the temperature is below its saturation value, the fluid is known as *subcritical water*, as shown in the lower left block of Figure 9.1.

*Subcritical steam* ($T > T_{sat}$; $P < P_c$. Note: $T$ may be above $T_c$): When water (below critical pressure) is heated, it experiences a drop in density and an increase in enthalpy; this change is very sharp when the temperature of the water just exceeds it saturation value, $T_{sat}$. Above the saturation temperature, but below the critical value, the fluid ($H_2O$) is called *subcritical steam*. This regime is shown below the saturation line in Figure 9.1.

*Supercritical water* ($T > T_c$; $P > P_c$): When heated above its critical pressure, $P_c$, water experiences a continuous transition from a liquid-like state to a vapor-like state. The vapor-like, supercritical, state is shown in the upper right block in Figure 9.1. Unlike in the subcritical stage, no heat of vaporization is needed for the transition from liquid-like to vapor-like. Above the critical pressure, there is no saturation temperature separating the liquid and vapor states. However, there is a temperature, called *pseudo-critical temperature*, that corresponds to each pressure ($>P_c$) above which the transition from liquid-like to vapor-like takes place. The pseudo-critical temperature is characterized by a sharp rise in the specific heat of the fluid.

The pseudo-critical temperature depends on the pressure of the water. It can be estimated within 1% accuracy by the following empirical equation (Malhotra, 2006):

$$\begin{aligned} T^* &= (P*)^F \\ F &= 0.1248 + 0.01424P* - 0.0026(P*)^2 \\ T^* &= \frac{T_{sc}}{T_c};\ P* = \frac{P}{P_c} \end{aligned} \tag{9.2}$$

where $T_{sat}$ is the saturation temperature at pressure $P$; $P_{sat}$ is the saturation pressure at temperature $T$; $P_c$ is the critical pressure of water, 22.089 MPa; $T_c$ is the critical temperature of water, 374.29°C; and $T_{sc}$ is the pseudo-critical temperature at pressure $P$ ($P > P_c$).

## 9.2.1 Properties of SCW

The critical point marks a significant change in the thermophysical properties of water (Figure 9.2). There is a sharp rise in the specific heat (less than 5 to higher than 90 kJ/kg K) near the critical temperature followed by a similar drop (Figure 9.2). The peak value of specific heat decreases with system pressure. The thermal conductivity of water also drops from 0.330 at 400°C to 0.176 W/m K at 425°C. The drop in molecular viscosity is also significant,

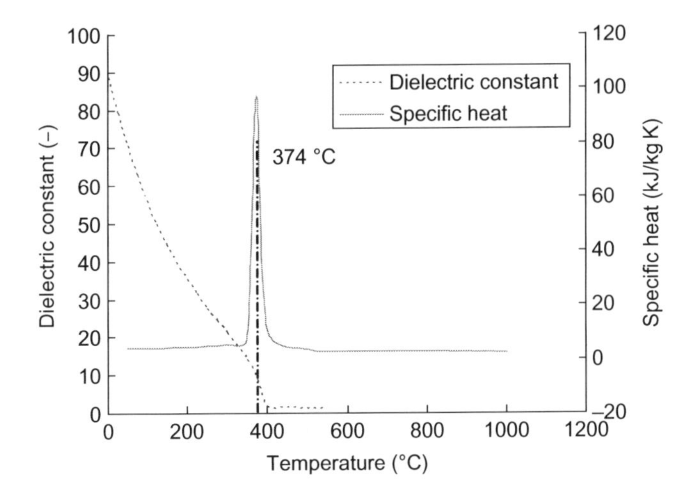


**FIGURE 9.2** Specific heat of water above its critical pressure shows a peak at it "pseudo-critical" temperature. Dielectric constant at 22.1 MPa, also plotted on this graph, shows rapid decline closer to the critical temperature.

although the viscosity starts rising with temperature above the critical value. Above this critical point, water experiences a dramatic change in its solvent nature primarily because of its loss of hydrogen bonding. The dielectric constant of the water drops from a value of about 80 in the ambient condition to about 10 at the critical point (Figure 9.2). This changes the water from a highly polar solvent at an ambient condition to a nonpolar solvent, like benzene, in a supercritical condition.

The change in density in SCW across its pseudo-critical temperature is much more modest. For example, at 25 MPa it can drop from about 1000 to 200 kg/m$^3$ while the water moves from a liquid-like to a vapor-like state (Figure 9.3). At subcritical pressure, however, there is an order of magnitude drop in density when the water goes past its saturation temperature. For example, at 0.1 MPa or 1 atm of pressure, the density reduces from 1000 to 0.52 kg/m$^3$ as the temperature increases from 25°C to 150°C (refer to Table 9.1).

The most important feature of SCW is that we can "manipulate" and control to a certain degree its properties around its critical point simply by adjusting the temperature and pressure. SCW possesses a number of special properties that distinguish it from ordinary or subcritical water. Some of those properties relevant to gasification are as follows:

- The solvent property of water can be changed much near or above its critical point as a function of temperature and pressure.
- Subcritical water is polar, but SCW is nonpolar because of its low dielectric constant. This makes it a good solvent for nonpolar organic

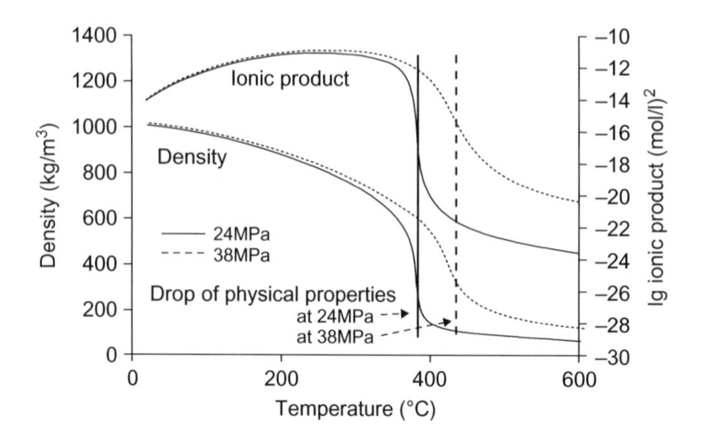


**FIGURE 9.3** At a given temperature, both density and ion product of water increase with pressure. Data plotted for 24 MPa show that these values reduce fast at the critical temperature, but that at 38 MPa it is more gentle. (Note: ion product is plotted as negative log kw, from Kritz (2004).)

compounds but a poor one for strongly polar inorganic salts. SCW can be a solvent for gases, lignin, and carbohydrates, which show low solubility in ordinary (subcritical) water. Good miscibility of intermediate solid organic compounds as well as gaseous products in liquid SCW allows single-phase chemical reactions during gasification, removing the interphase barrier of mass transfer.

- SCW has a high density compared to subcritical steam at the same temperature. This feature favors the forward reaction between cellulose and water to produce hydrogen.
- Near its critical point, water has higher ion products ($[H^+][OH^-] \sim 10^{-11}$ $(mol/l)^2$) than it has in its subcritical state at ambient conditions ($\sim 10^{-14}$ $(mol/l)^2$) (Figure 9.3). Owing to this high $[H^+]$ and $[OH^-]$ ion, the water can be an effective medium for acid- or base-catalyzed organic reactions (Serani et al., 2008). Above the critical point, however, the ion product drops rapidly ($\sim 10^{-24}$ $(mol/l)^2$ at 24 MPa), and the water becomes a poor medium for ionic reactions.
- Most ionic substances, such as inorganic salts, are soluble in subcritical water but nearly insoluble under typical conditions in SCW. As the temperature rises past the critical point, the density as well as the ionic product decreases (Figure 9.3). Thus, highly soluble common salt (NaCl) becomes insoluble at higher temperatures above the critical point. This tunable solubility property of SCW makes it relatively easy to separate the salts as well as the gases from the product mixture in an SCW gasifier.
- Gases, such as oxygen and carbon dioxide, are highly miscible in SCW, allowing homogeneous reactions with organic molecules either for

oxidation or for gasification. This feature makes SCW an ideal medium for destruction of hazardous chemical waste through SCWO.

- SCW possesses excellent transport properties. Its density is lower than that of subcritical water but much higher than that of subcritical steam. This, along with other properties like low viscosity, low surface tension (surface tension of water reduces from $7.2 \times 10^{-2}$ at 25°C to 0.07 at 373°C), and high diffusivity greatly contribute to the SCW's good transport property, which allows it to easily enter the pores of biomass for effective and fast reactions.
- Reduced hydrogen bonding is another important feature of SCW. The high temperature and pressure break the hydrogen-bonded network of water molecules.

Table 9.1 compares some of these properties of water under subcritical and supercritical conditions.

### 9.2.2 Application of SCW in Chemical Reactions

Chemical reactions involve the mixing of reactants. If the mixing is incomplete, the reaction will be incomplete, even if the right amounts of reactant and the right temperature are available. The mixing is better when all reactants are either in the gas phase or in the liquid phase compared to that when one reactant is in the solid phase and the other is in the gas or liquid phase. The absence of interphase resistance in a monophase reaction medium greatly improves the mixing. The conventional thermal gasification of solid biomass in air or steam involves heterogeneous mixing, and therefore the gas–solid interphase resistance limits the conversion reactions.

SCW allows reactions to take place in a single phase, as most organic compounds and gases are completely miscible in it. It is thus a superior reaction medium. Because the absence of interphase mass transfer resistance facilitates better mixing and therefore higher conversion, SCW can be an excellent medium for the following three types of reactions:

1. *Hydrothermal gasification of biomass*: SCW is an ideal medium for gasification of very wet biomass, such as aquatic species and raw sewage, which ordinarily have to be dried before they can be gasified economically. SCW gasification produces gas at high pressure and thus obviates the need for an expensive product gas compression step for transport or use in combustion.
2. *Synthesis reactions*: A variety of organic reactions like hydrolysis and molecular rearrangement can be effectively carried out in SCW, which serves as a solvent, a reactant, and sometimes a catalyst. There is no need for acid or base solvents, the disposal of which is often a problem.
3. *Supercritical water oxidation*: Complete miscibility of oxygen in SCW helps harmful organic compounds to be easily oxidized and degraded. Thus, SCW is an attractive means of turning pollutants into harmless oxides.

### 9.2.3 Advantages of SCW Gasification over Conventional Thermal Gasification

The following are two broad routes for the production of energy or chemical feedstock from biomass:

1. *Biological:* Direct biophotolysis, indirect biophotolysis, biological reactions, photofermentation, and dark fermentation are the five major biological processes.
2. *Thermochemical:* Combustion, pyrolysis, liquefaction, and gasification are the four main thermochemical processes.

Thermal conversion processes are relatively fast, taking minutes or seconds to complete, while biological processes, which rely on enzymatic reactions, take much longer, on the order of hours or even days. Thus, for commercial use, thermochemical conversion is preferred.

Gasification may be carried out in air, oxygen, subcritical steam, or water near or above its critical point. This chapter concerns hydrothermal gasification of biomass above or very close to the water's critical point to produce energy and/or chemicals.

Conventional thermal gasification faces major problems from the formation of undesired tar and char. The tar can condense on downstream equipment, causing serious operational problems, or it may polymerize to a more complex structure, which is undesirable for hydrogen production. Char residues contribute to energy loss and operational difficulties. Furthermore, very wet biomass can be a major challenge to conventional thermal gasification because it is difficult to economically convert if it contains more than 70% moisture. The energy used in evaporating fuel moisture (2257 kJ/kg), which effectively remains unrecovered, consumes a large part of the energy in the product gas.

Supercritical water gasification (SCWG) can largely overcome these shortcomings, especially for very wet biomass or organic waste. For example, the efficiency of thermal gasification of a biomass containing 80% water in conventional steam reforming is only 10%, while that of hydrothermal gasification in SCW can be as high as 70% (Dinjus and Kruse, 2004). Gasification in near or SCW therefore offers the following benefits:

- Tar production is low. The tar precursors, such as phenol molecules, are completely soluble in SCW and so can be efficiently reformed in SCW gasification.
- SCWG achieves higher thermal efficiency for very wet biomass.
- SCWG can produce in one step a hydrogen-rich gas with low CO, obviating the need for an additional shift reactor downstream.
- Hydrogen is produced at high pressure, making it ready for downstream commercial use.

- Carbon dioxide can be easily separated because of its much higher solubility in high-pressure water.
- Char formation is low in SCWG.
- Heteroatoms like S, N, and halogens leave the process with aqueous effluent, avoiding expensive gas cleaning. Inorganic impurities, being insoluble in SCW, are also removed easily.
- The product gas of SCWG automatically separates from the liquid containing tarry materials and char if any.

## 9.3 BIOMASS CONVERSION IN SCW

There are three major routes for SCW-based conversion of biomass into energy which are as follows:

1. *Liquefaction:* Formation of liquid fuels above critical pressure (22.1 MPa) but near critical temperature (300–400°C).
2. *Gasification to* $CH_4$*:* Conversion in SCW in a low-temperature range (350–500°C) in the presence of a catalyst.
3. *Gasification to* $H_2$*:* Conversion in SCW with or without catalysts at higher ($>600°C$) temperatures.

Here we discuss only the last two gasification options.

### 9.3.1 Gasification

Supercritical biomass gasification takes place typically at around 500–750°C in the absence of catalysts, and at an even lower (350–500°C) temperature with catalysts. The biomass decomposes into char, tar, gas, or other intermediate compounds, which are reformed into gases like CO, $CO_2$, $CH_4$, and $H_2$. The process is schematically shown in Figure 9.4. If the biomass is represented by the general formula $C_6H_{12}O_6$, the gasification process may be described by the following overall reaction:

$$mC_6H_{12}O_6 + nH_2O \rightarrow wH_2 + xCH_4 + yCO + zCO_2 \tag{9.3}$$

Gasification in SCW involves, among other reactions, hydrolysis and oxidation reactions. A brief description of these reactions follows.

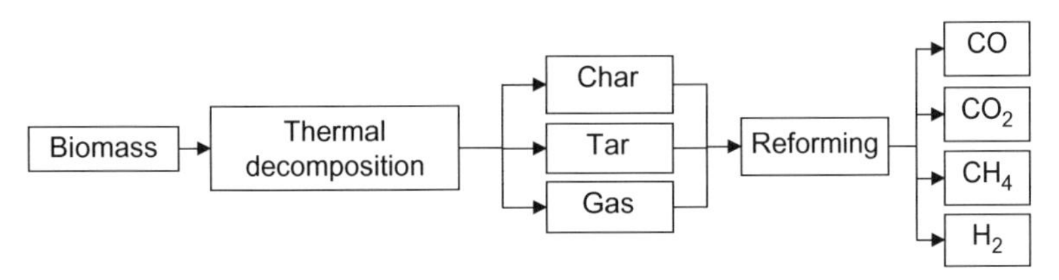


**FIGURE 9.4** Biomass conversion process.

### 9.3.2 Hydrolysis

Hydrolysis (meaning "splitting with water") is the reaction of an organic compound with water. Here, a bond of an organic molecule is broken, and the water molecule is also broken into [$H^+$] and [$OH^-$]. The organic molecules are cleaved into two parts by the water molecule: One part gains the [$H^+$] ion; the other part, the [$OH^-$] ion. Acid or base catalysts generally catalyze hydrolysis reactions. Water near its critical point (at high temperature and pressure) has a high ion product, so the water itself catalyzes the hydrolysis reaction.

A simplified representation of the reaction scheme is shown in Figure 9.5B with polyethylene terephthalate (PET) as an example. The hydrolysis of PET into terephthalic acid and ethylene glycol in SCW is a better option than other reactions (e.g., methanolysis or glycololysis) because it does not require solvents and catalysts like others. Here, water near its critical point is used to accomplish this reaction in a shorter time. Additionally, SCW avoids the need to recover and dispose of external solvents or catalysts. Figure 9.5A shows the

(A)

(B) Pet(polyethylene terephthalate)



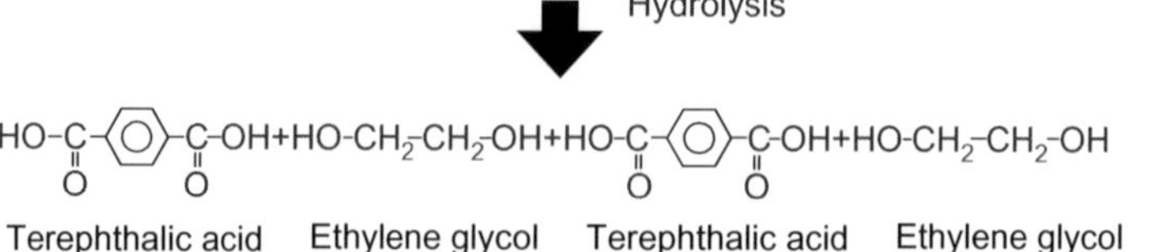



**FIGURE 9.5** (A) The photograph shows PET in SCW and SCW after hydrolysis. (B) Hydrolysis of PET in SCW. Source: *From Kobe Steel (2010).*

photograph of PET in ordinary water before and after hydrolysis in SCW into fine particles of teraphthalic acid in ethylene glycol solution.

### 9.3.3 SCW Oxidation

SCW that exhibits complete miscibility with oxygen is a homogeneous reaction medium for the oxidation of organic molecules. This feature of SCW allows oxidation of harmful or toxic substances at low temperature in a process known as SCWO or cold combustion. In a typical SCWO unit, the entire mixture (water, oxygen, and waste) remains as a single fluid phase with no interphase transport limitations. This allows very rapid and complete (>99.9%) oxidation of the organic wastes to harmless lower-molecular-weight compounds like $H_2O$, $N_2$, and $CO_2$. Unlike thermal incineration, SCWO does not produce toxic by-products such as dioxin. This method of waste treatment is especially attractive for highly dilute toxic wastes in water.

One important shortcoming of this process is the production of highly corrosive liquid effluents because chlorine, sulfur, and phosphorous, if present in the waste, are converted into their corresponding acids (Serani et al., 2008). The destruction of polychlorinated biphenyls (PCBs) in SCW, producing carbon dioxide and hydrochloric acid, may be represented by the following simple reaction:

$$C_{12}H_{10-m}Cl_m(\mathrm{PCB}) + (19+m)/2\ O_2 + (5-m)\ H_2O = 12CO_2 + m\mathrm{HCl} \tag{9.4}$$

Conventional thermal incineration uses very high temperature to destroy by-products like dioxin, but results in the production of another pollutant, NOx. This is not the case with SCWO owing to its low-temperature operation (450–600°C).

### 9.3.4 Scheme of an SCWG Plant

A typical SCWG plant includes the following key components:

- Feedstock pumping system
- Feed preheater
- Gasifier/reactor
- Heat-recovery (product-cooling) exchanger
- Gas–liquid separator
- Optional product-upgrading equipment

The feed-preheating system is very elaborate and accounts for the majority (~60%) of the capital investment in an SCW gasification plant.

Figure 9.6 explains the SCWG process using the example of an SCWG plant for gasifying sewage sludge. Biomass is made into a slurry for feeding.

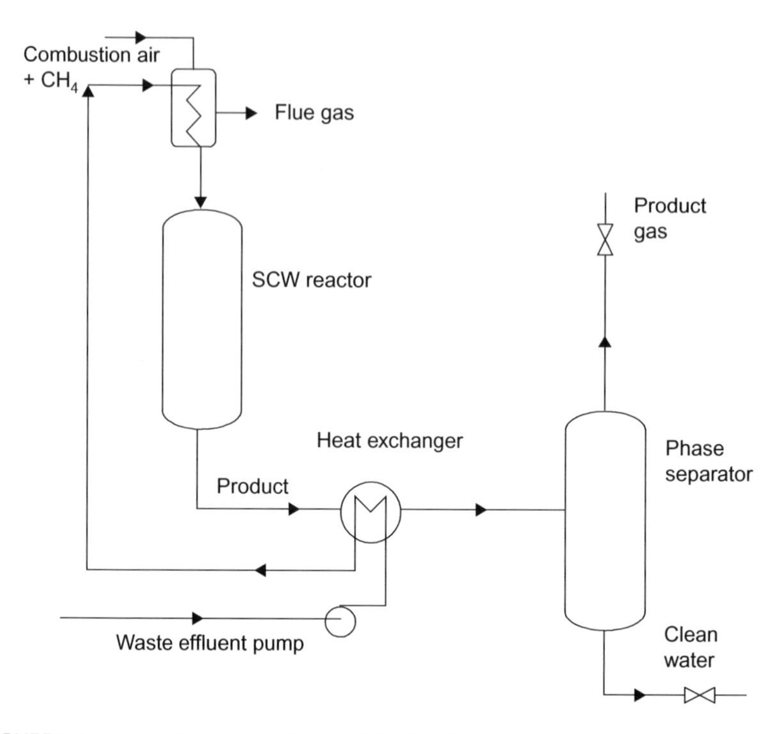


**FIGURE 9.6** Schematic of a pilot plant for SCWG of biomass.

It is then pumped to the required supercritical pressure. Alternatively, water may be pressurized separately and the biomass fed into it. In any case, the feedstock needs to be heated to the designed inlet temperature for the gasifier, which must be above the critical temperature and well above the designed gasification temperature because the enthalpy of the water provides the energy required for the endothermic gasification reactions. This temperature is a critical design parameter.

The sensible heat of the product of gasification may be partially recovered in a waste heat-recovery exchanger and used for partial preheating of the feed (Figure 9.6). For complete preheating, additional heat may be obtained from one of the following:

- Externally fired heater (Figure 9.6)
- Burning of a part of the fuel gas produced to supplement the external fuel
- Controlled burning of unconverted char in the reactor system (refer to Figure 9.12 later in this chapter)

After gasification, the product is first cooled in the waste heat-recovery unit. Thereafter, it cools to room temperature in a separate heat exchanger by giving off heat to an external coolant.

The next step involves separation of the reaction products. The solubility of hydrogen and methane in water at low temperature but high pressure is considerably low, so they are separated from the water after cooling while the carbon dioxide, because of its high solubility in water, remains in the liquid phase. For complete separation of $CO_2$, the gas may be scrubbed with additional water (refer to Figure 9.15 later in this chapter). The gaseous hydrogen is separated from the methane in a pressure swing adsorber. The $CO_2$-rich liquid is depressurized to the atmospheric pressure, separating the carbon dioxide from the water and unconverted salts.

## 9.4 EFFECT OF OPERATING PARAMETERS ON SCW GASIFICATION

The product of gasification is defined by its yield and composition which are influenced by a number of gasifier design and operating parameters. For proper design and operation of an SCW gasifier, a good understanding of the influence of the following parameters is important:

- Reactor temperature
- Catalyst use
- Residence time in the reactor
- Solid concentration in the feed
- Heating rate
- Feed particle size
- Reactor pressure
- Reactor type

### 9.4.1 Reactor Temperature

Temperature has an important effect on the conversion, the product distribution, and the energy efficiency of an SCW gasifier, which typically operates at a maximum temperature of nearly 600°C. The overall carbon conversion increases with temperature; at higher temperatures hydrogen yield is higher while methane yield is lower. Figure 9.7 shows the temperature dependence of gasification efficiency and product distribution in a reactor operated at 28 MPa (30-s residence, 0.6-M[1] glucose) (Lee et al., 2002). We see that the hydrogen yield increases exponentially above 600°C, while the CO yield, which rises gently with temperature, begins to drop above 600°C owing to the start of the shift reaction Eq. (7.52).

Gasification efficiency is measured in terms of hydrogen or carbon in the gaseous phase as a fraction of that in the original biomass. Carbon conversion efficiency increases continually with temperature, reaching close to

[1]M - mol/liter

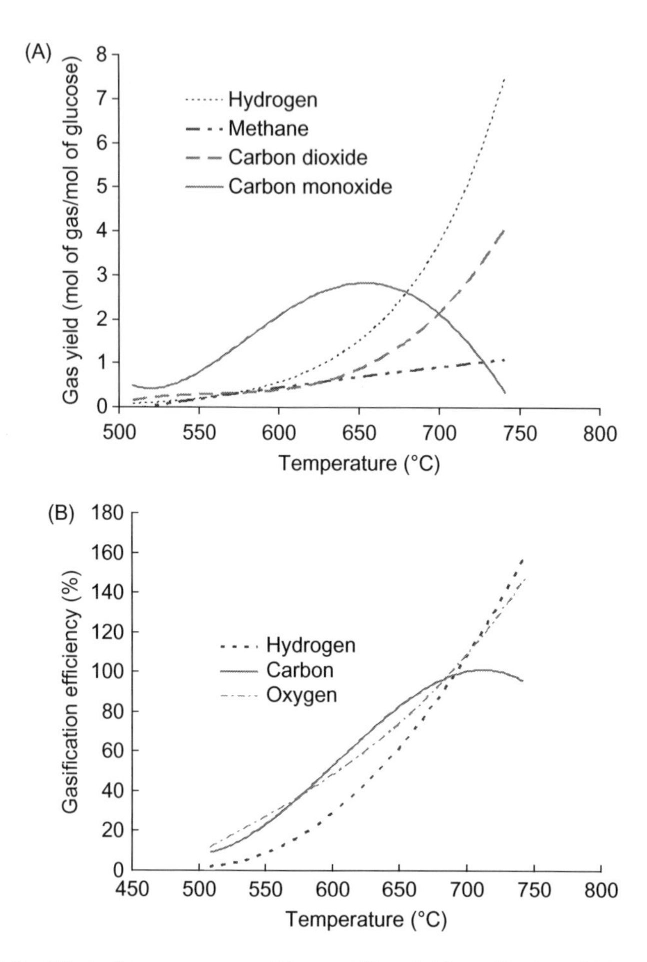


**FIGURE 9.7** Effect of temperature on (A) gas yield and (B) gasification efficiency. Source: *From Lee et al. (2002).*

100% above 700°C. Hydrogen conversion efficiency (the fraction of hydrogen in glucose converted into gas) also increases with temperature. It appears strange that at 740°C, the hydrogen conversion efficiency exceeds 100%, reaching 158%. This clearly demonstrates that the extra hydrogen comes from the water, confirming that water is indeed a reactant in the SCWG process as well as a reaction medium.

Hydrothermal gasification of biomass has been divided into three broad temperature categories: high, medium, and low with their desired products (Peterson et al., 2008). Table 9.2 shows that the first group targets production of hydrogen at a relatively high temperature (>500°C); the second targets production of methane at just above the critical temperature (~374.29°C) but below 500°C; and the third gasifies at subcritical temperature, using only

**TABLE 9.2 Three Categories of Hydrothermal Gasification of Biomass Based on the Target Product ($T_c$ ~374.29°C)**

| Temperature Range | Catalyst Use | Desired Product |
|---|---|---|
| High temperature (>500°C) | Not needed | Hydrogen-rich gas |
| Medium temperature ($T_c$ to 500°C) | Needed | Methane-rich gas |
| Low temperature (<$T_c$) | Essential | Other gases from smaller organic molecules |

simple organic compounds as its feedstock. The last two groups, because of their low-temperature operation, need catalysts for reactions.

## 9.4.2 Catalysts

An effective degradation of biomass and the gasification of intermediate products of thermal degradation into lower-molecular-weight gases like hydrogen require the SCW reactor to operate in the high-temperature range (>600°C). The higher the temperature, the better the conversion, especially for production of hydrogen, but lower the SCW's energy efficiency. A lower gasification temperature is therefore desirable for higher thermodynamic efficiency of the process.

Catalysts help gasify the biomass at lower temperatures, thereby retaining, at the same time, high conversion and high thermal efficiency. Additionally, some catalysts also help gasification of difficult items like the lignin in biomass. Watanabe et al. (2003) noted that the hydrogen yield from lignin at 400°C and 30 MPa is doubled when a metal oxide ($ZrO_2$) catalyst is used in the SCW. The yield increases four times with a base catalyst (NaOH) compared to gasification without a catalyst. The three principal types of catalyst used so far for SCW gasification are: (1) alkali, (2) metal, and (3) carbon-based.

An important positive effect of catalysts in SCWG is the reduction in required gasification temperature for a given yield. Minowa et al. (1998) noted a significant reduction in unconverted char while gasifying cellulose with a $Na_2CO_3$ catalyst at 380°C. Base catalysts (e.g., NaOH and KOH) offer better performance, but they are difficult to recover from the effluent. Some alkalis (e.g., NaOH, KOH, $Na_2CO_3$, $K_2CO_3$, and $Ca(OH)_2$) are also used. They, too, are difficult to recover.

The special advantage of metal oxide catalysts is that they can be recovered, regenerated, and reused. Commercially available nickel-based catalysts are effective in SCW biomass gasification. Among them, Ni/MgO (nickel supported on an MgO catalyst) shows high-catalytic activity, especially for biomass (Minowa et al., 1998).

Metal catalysts have a severe corrosion effect at the temperatures needed to secure high yields of hydrogen. To overcome this problem, Antal et al. (2000) used carbon (e.g., coal-activated and coconut shell-activated carbon and macadamia shell and spruce wood charcoal). The carbon catalysts resulted in high yields of gas without tar formation.

### 9.4.3 Residence Time

A longer residence of the reactants in the reactor gives a better yield. Lu et al. (2006) experimented with 2% (by weight) sawdust and 2% carboxymethyl cellulose (CMC) in a flow reactor at 650°C and 25 MPa. Mettanant et al. (2009b) experimented with 2% rice husk in a batch reactor under the same conditions. Both found a steady increase in hydrogen and a moderate increase in methane (Figure 9.8) when the residence time was increased by three times and six times, respectively. Total organic carbon in the liquid product decreases with residence time, whereas carbon and hydrocarbon gasification efficiencies increase. This implies that a longer residence time is favorable for SCW biomass gasification. The optimum residence time, beyond which no further improvement in conversion efficiency is possible, depends on several factors. At a higher temperature, the residence time required for a given conversion is shorter.

### 9.4.4 Solid Concentration in Feedstock

Unlike in other gasification methods, solids fraction in the feed have an important effect on the gasification in SCW. Thermodynamic calculations

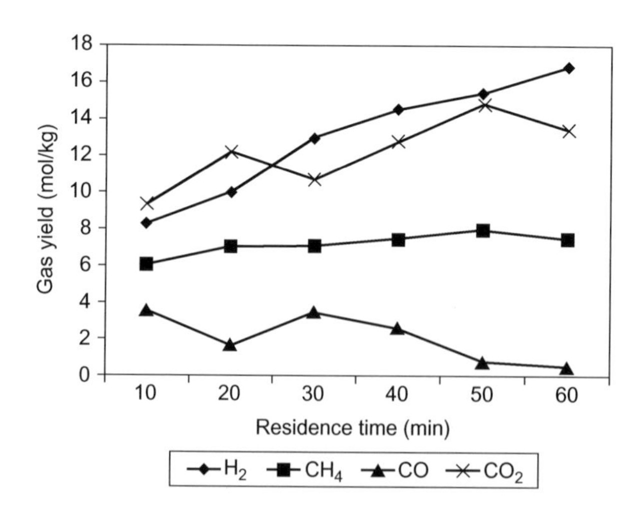


**FIGURE 9.8** Effect of residence time on the gasification of 2% rice husk in SCW at 650°C, 30 MPa in a batch reactor.

suggest that the conversion of carbon to gases in SCW declines rapidly when the solid content in a liquid feed exceeds 50% (Prins et al., 2005), but experimental results show this to occur for a much lower concentration. Experimental data (Mettanant et al., 2009b; Schmieder et al., 2000) show that gasification efficiency starts to decline when the solid concentration exceeds a value as low as 2%.

Table 9.3 presents data (Mozaffarian et al., 2004) that show the effect of solid content in feed. Although experimental conditions and feedstock vary, we can broadly classify these results into groups of low, medium, and high solid feedstock. For a lower feed concentration (<2%), carbon conversion efficiency is in the range 100% to 92% and reduces to 60–90% for an intermediate concentration (2–10%) and to 68–80% for a >10% concentration. An SCW gasifier, thus, needs a very low solid concentration in the feed for high carbon conversion efficiency. This requires higher pumping costs and liquid effluent disposal which may be a major impediment in commercialization of SCWG.

The reactor type also influences how solid concentration affects gasification efficiency. For example, Kruse et al. (2003) noted that a stirred reactor shows opposite results—that is, higher gasification efficiency at higher solid content (1.8–5.4%) in feed. This contrasts with data from Schmieder et al. (2000) from tumbling and tubular reactors that indicate a decrease in gasification efficiency with solid content (0.2–0.6 M). In stirred reactors, reactants are very well mixed, resulting in a heating rate that is faster than that achieved in other reactor types. This may be the explanation for the higher gasification efficiency where there is a higher solid content. The exact reason for this decrease is not clear and is a major issue in the development of commercial SCW gasifiers. Catalysts, high gasification temperatures, and high heating rates can avoid the drop in conversion of a high-solid-content feedstock (Lu et al., 2006).

### 9.4.5 Heating Rate

Limited data obtained by Sinag et al. (2004) suggest that at a higher heating rate the yield of hydrogen, methane, and carbon dioxide increases while that of carbon monoxide decreases. Matsumura et al. (2006) noted some improvement on g carbon gasification efficiency.

### 9.4.6 Feed Particle Size

The effect of biomass particle size is not well researched. With limited data, Lu et al. (2006) showed that smaller particles result in a slightly improved hydrogen yield and higher gasification efficiency. However, Mettanant et al. (2009b) did not observe any effect when they varied the size of rice husk particles in the range of 1.25 to 0.5 mm. Even if the size effect is

**TABLE 9.3** Effect of Different Operating Parameters Including Solid Concentration in Feed on Gasification

| | | $C<2$ wt% | | $2<C<10$ wt% | | | $C>10$ wt% | |
|---|---|---|---|---|---|---|---|---|
| **Reference** | **Holgate (1995)** | **Yu (1993)** | **Kruse (1999)** | **Hao (2003)** | **Xu (1996)** | **Kruse (2003)** | **Yu (1993)** | **Xu (1996)** |
| Feedstock | Glucose | Glucose | Wood | Glucose | Formic acid | Baby food | Glucose | Glucose |
| Conc. (wt%) | 0.01 | 1.8 | 1 | 7.2 | 2.8 | 5.4 | 14.4 | 22 |
| $P$ (bar) | 246 | 345 | 350 | 250 | 345 | 300 | 345 | 345 |
| $T$ (°C) | 600 | 600 | 450 | 650 | 600 | 500 | 600 | 600 |
| Reactor type | Flow reactor | Tubular flow reactor | Autoclave | Tubular flow reactor −9 mm | Tubular flow reactor | SCTR | Tubular flow reactor | Tubular flow reactor 1 |
| Residence time (s) | 6 | 34 | 7200 | 210 | 34 | 300 | 34 | 34 |
| Carbon conversion efficiency (%) | 100 | 90 | 91.8 | 89.6 | 93 | 60 | 68 | 80 |
| **Gas composition** | | | | | | | | |
| $H_2$ | 61.3 | 61.6 | 28.9 | 21.5 | 49.2 | 44 | 25 | 11 |
| $CO_2$ | 36.8 | 29 | 48.4 | 35.5 | 48.1 | 41 | 16.6 | 5.7 |
| CO | – | 2 | 3.3 | 18.3 | 1.7 | 0.4 | 41.6 | 62.3 |
| $CH_4$ | 1.8 | 7.2 | 19 | 15.8 | 1 | 14.6 | 16.7 | 16.5 |
| $C_{2,3}$ | – | | – | 5.3 | – | – | – | 4.5 |

$X_c$—carbon conversion.
$C$—initial feed concentration.
$C$—concentration of solid in feed.
**Source:** Compiled from Mozaffarian et al. (2004).

confirmed with further data, it remains to be seen if the extra energy required for grinding is worth the improvement.

### 9.4.7 Pressure

Experiments by Van Swaaij et al. (2003) in their microreactor over the range of 28–34.5 MPa 710°C, those by Kruse et al. (2003) in a stirred tank (30–50 MPa, 500°C), and those by Lu et al. (2006) in a plug-flow reactor (18–30 MPa, 625°C) showed no major effect of pressure on carbon conversion or product distribution. Nor did Mettanant et al. (2009b) see much effect in their temperature and pressure range, although they noted a clear positive effect of pressure at 700°C.

### 9.4.8 Reactor Type

The reactors used so far for SCWG research have been either batch or continuous (flow). Depending on their type of mixing, they can be further divided as follows:

- Autoclave
- Tubular steel
- Stirred tank
- Quartz capillary tube
- Fluidized bed

A batch reactor is simple, does not require a high-pressure pump and can be used for almost all biomass feedstock. However, its reaction processes are not isothermal and it needs time to heat up and cool down. During heat up many reactions occur that cause transformation of the feedstock; this does not happen in a continuous-flow reactor.

Reactor type has an important effect on the influence of feed concentration. The drop in gasification efficiency with feed concentration, noted in tubular reactors, was not found in the stirred-tank reactor studied by Matsumura et al. (2005). However, the reactor used was exceptionally small (1.0 mm in diameter), so validation of this finding in a reasonably large reactor (Matsumura and Minowa, 2004) is necessary. The process development of SCW gasifiers is lagging laboratory research because of engineering difficulties and the high cost of pilot plant construction.

## 9.5 APPLICATION OF BIOMASS CONVERSION IN SCWG

Three major areas of application for biomass SCWG are: (1) energy conversion, (2) waste remediation, and (3) chemical production.

### 9.5.1 Energy Conversion

All three of the following important feedstock for the energy industry can be produced by biomass conversion in SCW:

- *Bio-oil:* Potential use in the transport sector
- *Methanol:* Though a chemical feedstock, may be used for combustion
- *Hydrogen:* Potential use in fuel cells

The overall efficiency of an energy conversion system depends on the technology route, on the wetness of the biomass, and on many other factors. Yoshida et al. (2003) compared the effect of moisture content on the net efficiency of seven options for electricity generation, including an SCWG combined cycle. Interestingly, the SCWG-based system shows a total efficiency independent of moisture content, while for all other systems, total efficiency decreases with increasing moisture. Total electricity generation efficiency is even higher than that for conventional combustion-based systems. Integrated gasification combined cycle (IGCC) efficiency is higher than that of SCWG for biomass containing less than 40% moisture. Above 40%, its efficiency drops below that of SCWG (Figure 9.9).

Yoshida et al. (2003) also compared the total heat utilization efficiency of seven energy conversion processes:

1. Direct combustion of biomass
2. Combustion of biomass-oil produced by liquefaction or pyrolysis
3. Combustion of methanol produced by thermal gasification
4. Combustion of methanol produced by SCWG
5. Combustion of biogas produced by thermal gasification

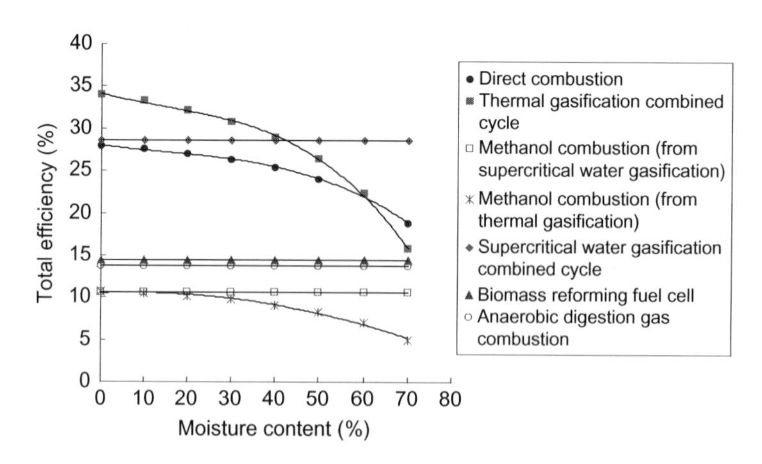


**FIGURE 9.9** Dependence of the net electricity generation efficiency of different biomass-based processes on the biomass moisture content (Yoshida et al., 2003).

6. Combustion of methanol produced by SCWG
7. Anaerobic digestion

SCWG has the distinction of easily separating $CO_2$ from the product gas. This makes it an optimal technology for generation of electricity and heat from biomass when $CO_2$ emission limits become binding.

Fuel cells have the highest energy conversion efficiency for electricity generation, but they need hydrogen as their fuel. For hydrogen production, from very wet biomass, SCW gasification could be an attractive route. However, the capital costs of a fuel cell and that of a gasification plant have an important bearing on the economic viability of this generation option.

### 9.5.2 Waste Remediation

Waste treatment is another SCWG application. As explained in Section 9.3.3, in SCW even highly toxic wastes can be oxidized to harmless disposable residues. The agricultural industry produces large volumes of nontoxic but unhealthy products such as animal extracts and farm wastes that need to be disposed of productively. Many of these contain so much moisture that economical combustion or thermal gasification is not possible. Anaerobic digestion is a widely used alternative, especially in developing countries for production of useful gas (mostly methane) from animal extracts. Along with methane, anaerobic digestion produces fermentation sludge, which can be used as fertilizer.

Nevertheless, anaerobic gasification is orders of magnitude slower than thermal and other gasification processes, even with the use of catalysts. As a result, this makes large-scale commercial operation of anaerobic digesters difficult. Furthermore, the attractiveness of this method depends on the price of fertilizer, which can vary as a result of over- or undersupply in the market (Matsumura, 2002).

SCWG or SCWO is an alternative suitable for waste treatment because it does not depend on the production of sludge and is much faster than anaerobic digestion. Matsumura (2002) noted that SCWG has better energy efficiency, cheaper gas production, and faster $CO_2$ payback time (64.8%, 3.05 yen/MJ, and 4.19 years, respectively) in comparison with biomethanation (49.3%, 3.74 yen/MJ, and 5.05 years, respectively).

### 9.5.3 Chemical Production

Solvents are an important component of many chemical reactions. SCW acts as a solvent, but can also be a reactant and/or a catalyst. Ordinary subcritical water is popular as a solvent for reactions, especially because it is inexpensive and easily disposed of. Many organics, however, do not react efficiently in it. For these reactions, acid or base solvents are needed, which are good for synthesis reactions but, unless they can be efficiently recovered, are expensive and hazardous to dispose of.

Owing to its unique properties, SCW can act as a solvent for some reactions. Based on their studies of the following reactions Krammer et al. (1999) noted that many hydration, dehydration, as well as hydrolysis reactions can take place in SCW with good selectivity and high space/time yield, with no acids or bases as support materials.

- Dehydration of 1,4-butandiol and glycerine
- Hydrolysis of ether acetate, acetonitrile, and acetamide
- Reaction of acetone cayanohydrine

Production of useful chemicals from biomass is another use for SCW gasification. During its degradation in SCW, biomass produces phenols. Phenol production increases with feed concentration (Kruse et al., 2003). Because phenol is an important feedstock for the green resin, wood composite, and laminate industries, SCW provides an effective medium for green chemistry.

## 9.6 REACTION KINETICS

Limited information is available on the global kinetics of SCW gasification. Lee et al. (2002) studied the kinetics of glucose (used as the model biomass) in SCWG with a plug-flow reactor.

$$C_6H_{12}O_6 + 6H_2O = 6CO_2 + 12H_2 \tag{9.5}$$

We define the reaction rate, $r$, as the depletion of the biomass carbon fraction, $C$, with time. Assuming pseudo-first-order kinetics, we can write:

$$r = -\frac{dC}{d\tau} = k_g C \tag{9.6}$$

where $k_g$ is the reaction rate constant.

The fraction of carbon converted into gas, $X_c$, may be related to the current carbon fraction, $C$, and the initial carbon fraction, $C_0$, in the fuel:

$$X_c = 1 - \frac{C}{C_0} \tag{9.7}$$

Now replacing the carbon fraction in Eq. (9.6) and integrating, we get:

$$k_g = -\frac{\ln(1 - X_c)}{\tau} \tag{9.8}$$

Table 9.4 presents some data on the global kinetics for SCWG of model compounds. The rates measured by Mettanant et al. (2009b), Lee et al. (2002), and Kabyemela et al. (1997) show how the reaction rate decreases with increasing solid carbon in the feed.

**TABLE 9.4 Global Kinetic Rate of Gasification of Model Compound in SCW**

| | Blasi et al. (2007) | Mettanant et al. (2009b) | Lee et al. (2002) | Kabyemela et al. (1997) |
|---|---|---|---|---|
| Feed | Waste water from wood gasifier/TOC | Rice husk | Glucose/ COD | Glucose |
| Reactor | Plug | Batch | Plug | Plug |
| Temperature (K) | 723–821 | 673–873 | 740–1023 | 573–673 |
| Residence time (s) | 60–120 | 3600 | 16–50 | 0.02–2 |
| Solid content | 7–15 g(Toc)/l | 9.4 mol/l | 0.6 mol/l | 0.007 mol/l |
| Preexponential, $A$ ($s^{-1}$) | 897 | 184 | 897 ± 29 | |
| Activation energy, $E$ (kJ/mol) | 76 | 77.4 | 71 ± 3.9 | 96 |
| $k_g$ ($s^{-1}$) | 0.002–0.003 | 0.0002–0.006 | 0.01–0.55 | 0.15–9.9 |

## 9.7 REACTOR DESIGN

Because SCWG is likely to enter the market once its major development barriers are removed this section discusses important considerations for design of an SCWG reactor. The discussion is based on limited information available in laboratory units and on the design of thermal gasifiers (see Chapter 8).

The major design parameters for an SCWG reactor are temperature, residence time, pressure, catalysts, and feed concentration. Important design considerations for auxiliary or support equipment are (1) waste heat-recovery exchanger and feed-preheating system, (2) the biomass feed system, and (3) product separation. The following subsections present a brief discussion of some of the design parameters.

### 9.7.1 Reactor Temperature

The temperature and pressure of an SCWG must be above the critical value of 374.21°C and 22.089 MPa, respectively. As explained in Section 9.4.7, pressure has a minor effect on biomass conversion, but the effect of gasification temperature is a major one (see also Section 9.4.1).

Because feedstock (biomass and water) must be heated to the reaction temperature using energy from an external source, the lower the designed reactor temperature, the lower the energy required for feed preheat and the more efficient the process. The gasification temperature should be above 600°C for a reasonable hydrogen yield, but it can be lower if catalysts are used.

For synthetic natural gas (SNG) production, high methane, and low hydrogen are required; therefore, we can choose a reaction temperature of 350–500°C, but catalysts are necessary for a reasonable yield. With catalysts, methane-rich gas may be produced even just below the critical temperature (~350°C) (Mozaffarian et al., 2004).

### 9.7.2 Catalyst Selection

The choice of catalyst influences reactor temperature, product distribution, and plugging potential. Section 9.4.2 discussed the catalysts used in SCW gasification. They are selected on the basis of the desired product. Catalyst deactivation is an issue assigned to most catalyzed reactions because the deactivated catalysts must be regenerated. If they are deactivated because of carbon deposits, as happens in a fluid catalytic cracker (FCC), they can be combusted by adding oxygen, preferably in a separate chamber. The combustion reaction reactivates the catalysts and can additionally provide enough heat for preheating the feed.

### 9.7.3 Reactor Size

Consider a simple reactor receiving $W_f$ of feed while producing $W_p$ of product per unit of time. The product comprises a number of hydrocarbon components represented by species $i$. The total carbon in the product gas is sum of carbon in the individual gaseous hydrocarbons:

$$\text{Total carbon production in the product gas} = \sum W_p C_i \alpha_i \text{ kmol/s} \quad (9.9)$$

where $\alpha_i$ is the number of carbon atoms in component $i$ in the gas product; $C_i$ is the mole fraction of $i$ in the gas product; and $W_p$ is the product gas flow-rate (kmol/s). The amount of carbon in the feed is known from the feed rate, $W_f$ (kg/s), and its carbon fraction, $F_c$. The carbon gasification yield, $Y$, is defined as the ratio of gasified carbon to the carbon in the feed:

$$Y = \frac{\sum_i 12 W_p \alpha_i C_i}{W_f F_c} \quad (9.10)$$

where 12 is the carbon's molecular weight (kg/kmol).

From Eq. (9.8) the reaction rate is given in terms of conversion as:

$$k_g = \frac{\ln(1 - X_c)}{\tau} \quad (9.11)$$

where $\tau$ is the residence time in a reactor of volume $V$.

For a continuous stirred-tank reactor,

$$\tau = \frac{V}{\text{Volume flow-rate of feed at reactor condition}} \text{ s} \quad (9.12)$$

Thus, for a known reaction rate, $k_g$, and a desired conversion, $X_c$, we can estimate the reactor volume required for gasification.

### 9.7.4 Heat-Recovery Heat-Exchanger Design

A feedstock preheater is the second most important part of an SCW gasifier system. The heat required to preheat the feedstock (water and biomass) is a significant fraction of the potential heating value of the product gas. Without efficient recovery of heat from the product gas, the external energy needed for gasification may exceed the energy produced, making the gasifier a net energy consumer. The feedstock should therefore utilize as much enthalpy as possible from the sensible heat of the product. This is one of the most important aspects of SCW plant design.

Figure 9.10 compares the capital costs of different components of an SCWG plant. We can see that the heat-recovery exchanger represents 50–60% of the total capital cost of the plant which makes it a critical component.

Efficient heat exchange between the feed and the product is the primary goal of an SCWG heat-recovery system. However, for SCW system intended for hazardous waste reduction (SCWO) or synthesis reaction (SCWS), heat exchanger may not be all that important since the primary purpose of these systems is the production of chemicals, not energy as in a supercritical gasifier.

The heat-exchange efficiency, $\eta$, defines how much of the available heat in the product stream can be picked up by the feed stream.

$$\eta = \frac{H_{\text{product-out}} - H_{\text{product-in}}}{H_{\text{feed-in}} - H_{\text{product-in}}} \tag{9.13}$$

where $H$ is the enthalpy, and the subscripts define the fluid/solid it refers to.

Theoretically, the heat-exchange efficiency can be 100% if no heat of vaporization is required to heat the feed and an infinite heat-exchange

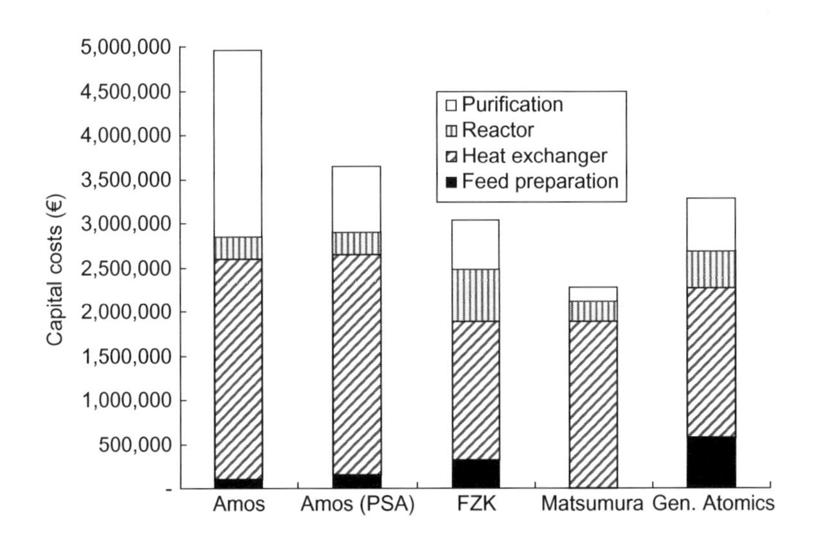


**FIGURE 9.10** Investment cost of different SCW plant design based on a throughput of 5000 kg/h of sewage sludge. Source: *From Gasafi et al. (2008).*

surface area is available. Of course, these conditions are not possible. Figure 9.11 shows variations in heat-exchange efficiency with changes in tube surface area and water pressure.

The specific heat of water rises sharply close to its critical point and then drops equally as the temperature increases (Figure 9.2). Thus, around the critical point we may expect a modest temperature rise along the heat-exchanger length.

Thermal conductivity in SCW is lower than that in subcritical water because SCW's intermolecular space is greater than that in liquid. A slight increase in conductivity is noticed as the fluid approaches the critical point. This increase is due to an increase in the agitation of molecules when the change from a liquid-like to a gas-like state SCW takes place. Above the critical point, thermal conductivity decreases rapidly with temperature.

The heat-transfer coefficient varies with temperature near its pseudo-critical value (see Section 9.2) because of variations in the thermophysical properties of water. As the temperature approaches the pseudo-critical value, conductivity and viscosity decrease but specific heat increases. The drop in viscosity and the peak of specific heat at the pseudo-critical temperature overcome the effect of decreased thermal conductivity so as to increase the overall heat-transfer rate.

As the temperature further increases, beyond the pseudo-critical point, the specific heat decreases sharply; the drop in thermal conductivity continues as well, and therefore the heat-transfer coefficient reduces. For a given heat flux, the wall temperature rises for the drop in heat-transfer coefficient. Generally, for high heat flux and low mass flux, the heat transfer deteriorates, leading to

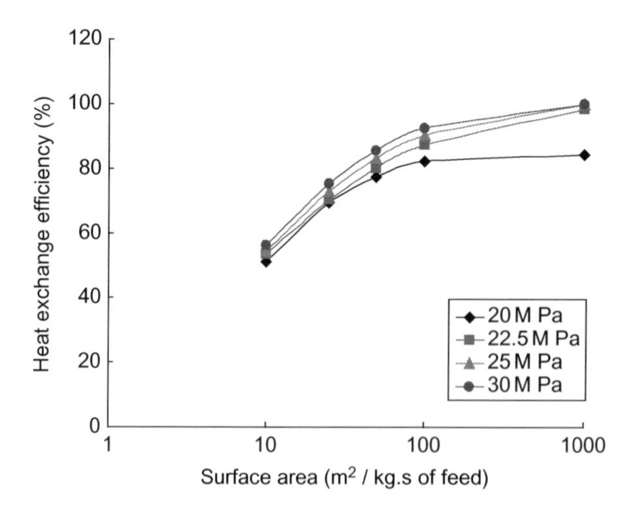


**FIGURE 9.11** Calculated heat exchange efficiency for water–water counter current heat exchanger at different pressures when feed and product entered at 24°C and 600°C, respectively. Source: *Data taken from Knoef (2005).*

hot spots in the tube. In horizontal tubes while bottom part experience the sharp rise in heat transfer coefficient was the pseudo critical temperature the upper part does not experience the change to that extent due to buoyancy effect (Shang et al. 2008).

#### *9.7.4.1 Heat Transfer in SCW*

Table 9.5 illustrates the designed performance of a typical heat-recovery exchanger for SCWG. The data are taken from a large operating near-SCWG plant. The fluid-to-wall heat-transfer coefficient in clean SCW in the vertical tube may be calculated by the correlation of Mokry et al. (2011):

$$N_u = 0.061\ \mathrm{Re}_b^{0.904}\mathrm{Pr}_b^{-0.684}\left(\frac{p_w}{p_b}\right)^{0.564} \tag{9.14}$$

Based on experiments and reviews of 15 correlations Jager et al. (2011) recommended the following correlations for horizontal tubes in SCW.

$$N_u = 0.0069\ \mathrm{Re}_b^{0.9}\mathrm{Pr}_b^{-0.66}\left(\frac{p_w}{p_b}\right)^{0.43}\left(1+\frac{2.4}{x/d}\right) \tag{9.15}$$

where $x$ is length and $d$ is diameter of tube.

Heat transfer in SCWG may vary because of solids in the fluid. Thus, applicability of these equations to SCWG is uncertain. Information on this aspect of heat transfer is presently unavailable.

### 9.7.5 Carbon Combustion System

Because gasification and pyrolysis reactions are endothermic, heat from some external source is required for operation of the reactor. In thermal gasification systems, the reaction temperature is very high (800–1000°C), so a large amount of energy is required for production of fuel gases from biomass or other feedstock. This heat is generally provided by allowing a part of the

**TABLE 9.5** Sample Data from Product to Feed Heat Exchanger Pilot Plant (VERENA)

| Flow-Rate kg/h (Methanol%) | Product In (°C) | Product Out (°C) | Feed In (°C) | Feed Out (°C) | Reactor Temperature (°C) |
|---|---|---|---|---|---|
| 100 (10%) | 561 | 168 | 26 | 405 | 582 |
| 90 (20%)[a] | 524 | 155 | 22 | 388 | 537 |

[a] *Heat exchanger surface area—1.1 m$^2$, heat transfer coefficient—920 W/m$^2$C.*
**Source:** Boukis et al. (2005).

hydrocarbon or carbon in the feed to combust in the gasifier, but then a part of the energy in the feedstock is lost.

An SCW gasifier operates at a much lower (450–650°C) temperature and thus requires a much lower but finite amount of heat. Thermodynamically, the heat recovered from the gasification product is inadequate to raise the feed to the gasification temperature (450–600°C) and provide the required reaction heat. This shortfall is made up either by an external source or by combustion of part of the product gas in a heater.

Both options are expensive. For example, a study of an SCWG design for gasification of 120 t/day (5000 kg/h) of sewage sludge with 80% water showed that 122 kg/h of natural gas is required to provide the gasification heat. This, along with an electricity consumption of 541 kW, constitutes 23% of the total revenue requirement for the plant (Gasafi et al., 2008). A better alternative would be controlled combustion of the unconverted char upstream of the gasifier, which would make SCWG energy self-sufficient.

Although SCWG is known for its low char and tar production, in practice we expect some char formation. A low gasification temperature is thermodynamically more efficient, but raises the char yield as (Figure 9.7), gasification efficiency is low at lower temperatures. If this char can be combusted in SCW, it can provide the extra heat needed for preheating the feed, thereby improving the efficiency of the overall system.

Combustion of char offers an additional benefit for an SCWG that sometimes uses solid catalysts which are deactivated after being coated with unconverted char in the gasifier. A combustor can burn the deposited carbon and regenerate the catalyst. The generated heat is carried to the gasifier by both solid catalysts and the gasifying medium (SCW and $CO_2$).

Recycling of solid catalysts is an issue for plug-flow reactors. Special devices such as fluidized beds may be used for these, as shown in Figure 9.12. Here, the catalysts or their supports are granular solids, which are separated from the product fluid leaving the reactor in a hydrocyclone operating in an SCW state. The separated solids drop into a bubbling fluidized bed combustor, where oxygen or air is injected to facilitate burning of the deposited carbon. The bed is fluidized by pressurized water already heated above its critical temperature in a heat-recovery heat exchanger.

Under supercritical conditions, oxidation or combustion reactions occur in a homogeneous phase where carbon is converted to carbon dioxide.

$$C + O_2 = CO_2 - 393.8\ \text{kJ/mol} \tag{9.16}$$

Because these reactions are exothermic, the process can become thermally self-sustaining with the appropriate concentration of oxygen. Heated water from the combustor carries the regenerated catalysts to the gasification reactor, into which the biomass is fed directly.

Under supercritical conditions, water acts as a nonpolar solvent. As a result, the SCW fully dissolves oxygen gas. The mass transfer barrier between

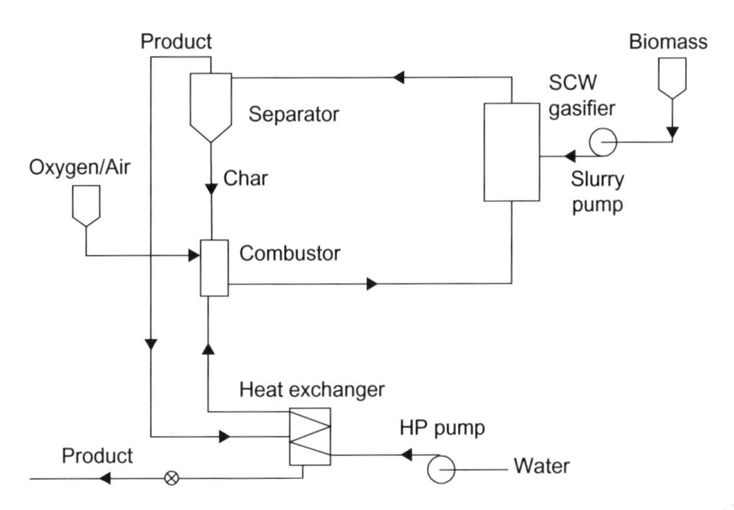


**FIGURE 9.12** A conceptual system for combustion of residual carbon deposited on solid catalysts to provide heat for SCWG of biomass.

dissolved oxygen and solid char may be lower than that between gas and char. This, along with its high-density feature, may allow the SCW to conduct the combustion reaction quickly and efficiently. Another advantage of low-temperature combustion is that it avoids formation of toxic by-products.

## 9.7.6 Design of Gas–Liquid Separator System

In an SCWG system, the product gas mixture is separated from water in two stages. In the first stage, initial separation takes place in a high-pressure but low-temperature separator. In the second stage, final separation occurs under low pressure and low temperature.

At low temperatures (25–100°C), hydrogen or methane has very low solubility (0.001–0.006) in water, even at high pressure (Figure 9.13). So the bulk of the hydrogen is separated from the water when cooled. From Figure 9.14 one notes that the solubility of carbon dioxide is an order of magnitude higher (0.01–0.03) than that of hydrogen (Figure 9.13) at this low temperature and high pressure. Figure 9.15 shows one scheme where hydrogen separated using its low solubility features. Other gases like $CO_2$ are also separated from the water but to a limited extent.

This feature can be exploited to separate the hydrogen from the carbon dioxide, but the $CO_2$'s equilibrium concentration may not be sufficient to dissolve it entirely in the high-pressure water. Additional water may be necessary to dissolve all of these gases except hydrogen so that the hydrogen alone remains in the gas phase (S1, Figure 9.14). The equilibrium concentration of these gases in water can be calculated from the equation of state, such as Peng–Robinson or SAFT.

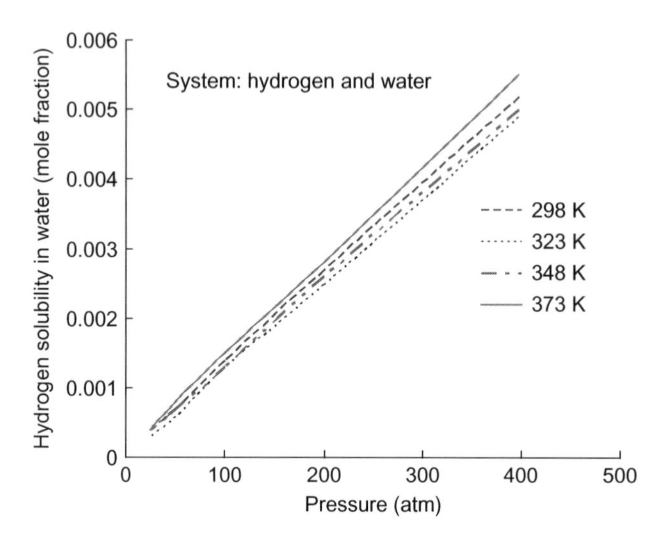


**FIGURE 9.13** Hydrogen solubility in water. Source: *From Ji et al. (2006).*

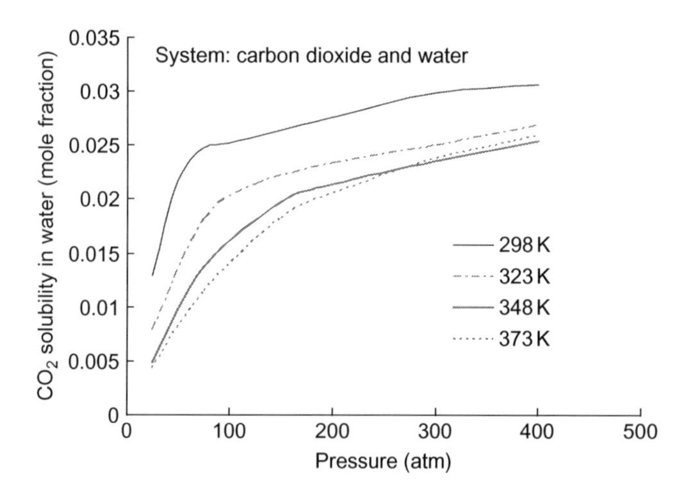


**FIGURE 9.14** Solubility of carbon dioxide in water. Source: *From Ji et al. (2006).*

The liquid mixture is next depressurized through a pressure regulator before it enters the second separator (S2, Figure 9.15). The solubility of most gases reduces with a decrease in pressure, so the second unit separates the rest of the $CO_2$ from the gas.

Feng et al. (2004a,b) calculated the phase equilibrium of different gases in water for a plant using different relations. Values calculated using SAFT equilibrium showed the best agreement with experimental results. These

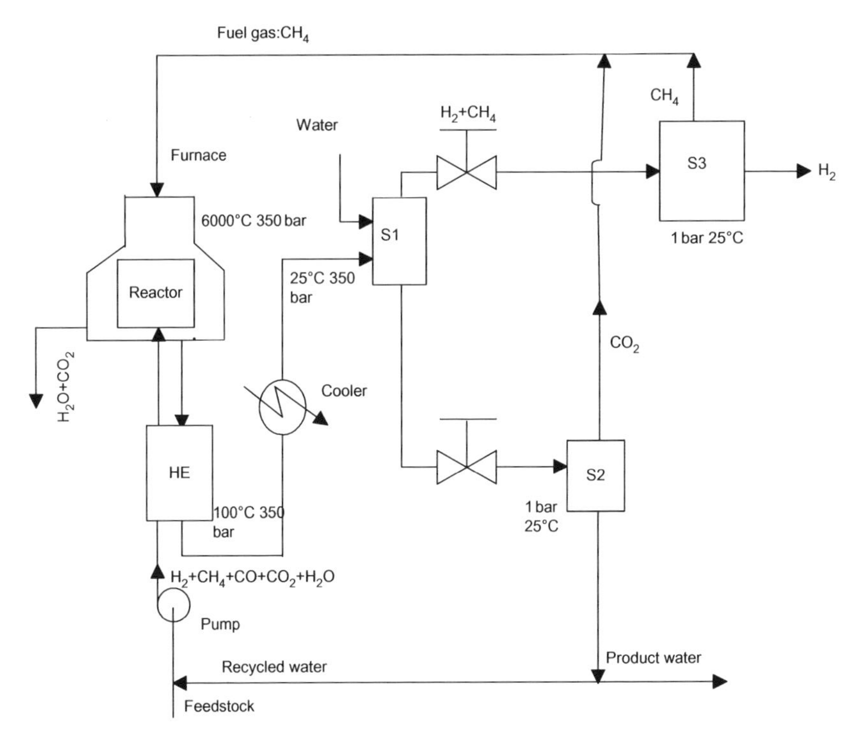


**FIGURE 9.15** A gas–liquid separation scheme for an SCW gasifier plant. HE, waste heat recovery heat exchanger; S1, hydrogen separator; S2, carbon dioxide separator; S3, pressure swing adsorber for separation of methane from hydrogen.

**TABLE 9.6 Solubility of Some Gases in Water at 25°C and Different Temperature**

| | Pressure (Bar) | | | | | | | | |
|---|---|---|---|---|---|---|---|---|---|
| | 60 | 90 | 120 | 140 | 200 | 300 | 400 | 600 | 1000 |
| $CH_4$ ($cm^3/g\ H_2O$) | 1.8 | 2.34 | 2.9 | 3.3 | | | | | |
| $H_2$ ($cm^3/g\ H_2O$) | 1.0 | 1.5 | 2.0 | 2.1 | 3.0 | 4.5 | 7.9 | 9.0 | 15 |
| $CO_2$ ($cm^3/g\ H_2O$) | 27 | | 32 | | 33 | | 39 | | |

**Source:** Collected from experimental and calculated values of Feng et al. (2004b).

results are shown in Table 9.6 to illustrate the process. It is apparent that at 25°C the solubility of $CO_2$ is orders of magnitude higher than that of methane and hydrogen. The solubility of methane and hydrogen is similar at nearly all pressures. For their separation, then, it is necessary to use a system such as a pressure swing adsorber (S3), as shown in Figure 9.15.

An important consideration is the additional water required to keep the carbon dioxide dissolved while the hydrogen is being separated. The amount, which may be considerable, can be expressed as the ratio of water to gaseous product ($R$) on a weight basis. When pressure and $R$ increase, the purification of hydrogen increases but the amount of hydrogen in the gas phase decreases. Therefore, we can recover more hydrogen with less purity or less hydrogen with more purity. This depends on an adjustment of the pressure and $R$. Example 9.1 illustrates the computation.

---

**Example 9.1**

Design a separator to produce 79% pure hydrogen from an SCWG operating at 250 bars of pressure. Assume the following overall gasification equation, which produces hydrogen, methane, carbon dioxide, and carbon monoxide.

$$C_6H_{10}O_5 + 4.5H_2O = 4.5CO_2 + 7.5H_2 + CH_4 + 0.5CO$$

**Solution**

We use the carbon dioxide solubility curve in Figure 9.15 to design the separator. Here, at 250 bars of pressure and 25°C, we find the solubility of $CO_2$ to be 0.028 mole fraction. This implies that 1 mol of water is needed to dissolve 0.028 mol of carbon dioxide.

To separate gaseous hydrogen from liquid water, we reduce the ambient temperature to 25°C. From Figure 9.13 we find that the hydrogen solubility is only 0.0031 at 250 bars and 25°C, so (1−0.0031) or 0.9969 fraction of hydrogen produced will be in the gas phase here. The gas may, however, contain other gases, so to ensure that the hydrogen is 79% pure, we need to add water to the separator. If we know the operating temperature, pressure, and weight ratio of the water to the gas mixture, the amount of product in the liquid and vapor phases can be calculated according to an equation of state. Here, we use Figure 9.16 computed by the Peng–Robinson equation. For 250 bars of pressure and a mole fraction of 79% in the gas phase, we get $R = 80$. Thus, the amount of water required is 80 × (the mass of product gas).

From the overall gas equation, the mass of product gas is $4.5 \times 44 + 7.5 \times 2 + 1 \times 16 + 0.5 \times 28 = 243$ g/mol of biomass. The mass of water is $80 \times 243 = 19{,}440$ g = 19.4 kg.

From the property table of water, we get the density of water at 25°C and 250 bars, which is 1008.5 kg/m³. The volume of water is 19.4 kg/1008.5 kg/m³ = 0.0192 m³.

$$\begin{aligned}\text{Volume of product gas} &= \sum(nRT/P) \\ &= (4.5 + 7.5 + 1 + 0.5) \times 10^{-3}\text{kmol} \times (8.314\ \text{kPa m}^3 \\ &\quad \text{kmol}^{-1}\ \text{K}^{-1} \times 298\ \text{K})/(250 \times 10^2\ \text{kPa}) \\ &= 0.00134\ \text{m}^3\end{aligned}$$

Therefore, the total volume of biomass that is gasified is $0.0192 + 0.00134 = 0.0205$ m³/mol.

---

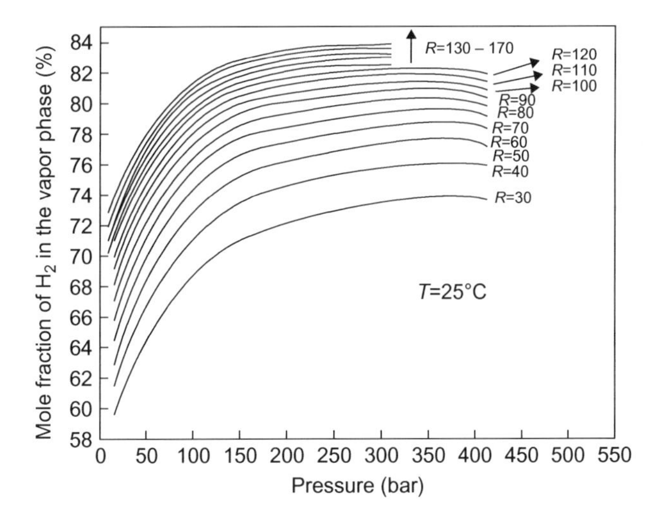


**FIGURE 9.16** Equilibrium mole fraction of hydrogen in gas phase for different water to product gas ratios, *R*. Source: *From Ji et al. (2006).*

## 9.7.7 Biomass Feed System

The feeding of biomass into a high-pressure ($>22$ MPa) reactor is a formidable challenge for an SCW gasifier. If the feed is a dilute stream of organics, the problem is not so severe, as pumps can handle light slurries. However, if it is fibrous solid granular biomass that needs to be pumped against high pressure, the problem is especially difficult for the reasons that follow:

- The irregular size and the low shape factor of biomass makes it difficult to flow.
- Pulverization is necessary for pumping the biomass, but it is very difficult to pulverize. Pretreatment of the feedstock is necessary.
- Fibrous by nature, biomass does not flow well through an augur or gear pump, and it is difficult to make a uniform slurry for pumping through impellers.

Most of the research work on SCWG generally used model water-soluble biomass such as glucose, digested sewage sludge, and wastewater (Blasi, 2007), which are easy to pump. For other types of biomass, Antal et al. (2000) used additives or emulsifiers such as corn starch gel, sodium CMC, and xanthan to make pumpable slurries. In an industrial application, large-scale use of emulsifiers is impractical.

A sludge pump was successfully used in a 100 kg/h pilot plant; however, the solids had to be ground to less than 1 mm particles and pretreated before pumping. Even then grass and fibrous materials clogged the membrane pump's vents (Boukis et al., 2005). Cement pumps have been suggested but, to date, have not been tried for pumping biomass in an SCW gasifier (Knoef, 2005).

Another important problem, is plugging of the feed line during the preheating stage, in which the feed being heated can start breaking down. Char and other intermediate products can deposit on the tube walls, blocking the passage, and thereby creating a dangerous situation.

Carbon buildup on the reactor wall has an adverse effect. It reduces the gas yield when the reactor is made of metals that have catalytic effects, although it is not associated with the feed system. Lu et al. (2006) showed that gas yields, gasification efficiency, and carbon efficiency are reduced by 3.25 mol/kg, 20.35%, and 17.39%, respectively, when carbon builds up on the reactor wall compared to when the reactor is clean. Similar results were found by Antal et al. (2000).

## 9.8 CORROSION

In an SCWG or SCWO, where the temperature can go as high as 600°C and the pressure can be in excess of 22.089 MPa, water becomes highly corrosive. SCWG and SCWO plants, work with organic compounds, which react with oxygen in SCWO to produce mostly $CO_2$ and $H_2O$, or hydrolyze in SCWG. Halogen, sulfur, and phosphorous in the feed are converted into mineral acids such as HCl, $H_2SO_4$, or $H_3PO_4$. High-temperature water containing these acids along with oxygen is extremely corrosive to stainless steels and nickel–chromium alloys (Friedrich et al., 1999).

After oxidation of neutral or acidic feeds, the pH of SCWO solutions is low, making it as corrosive as hydrochloric acid (Boukis et al., 2001). Chlorine is especially corrosive in SCW. Interestingly, a supercritical steam boiler, which is one of the most common uses of SCW, is relatively free from corrosion because the water used in the boiler is well treated and contains no corrosive species such as salts and oxygen or has only very low concentrations.

The following sections briefly describe the mechanism and the prevention of corrosion in biomass SCWG plants. More details are available in reviews presented by Kritz (2004) and Marrone and Hong (2008).

### 9.8.1 Mechanism of Corrosion

An oxide layer that forms on metal surfaces generally protects them and guards against further attack from corrosive elements. This protective layer can be destroyed through chemical or electrochemical dissolution, where the protective layer is removed by a chemical process using either an acidic or an alkaline solution depending on the pH value in the local region. In electrochemical dissolution, depending on the electrochemical potential, the metal can undergo either transpassive or active dissolution. All forms of electrochemical corrosion, require the presence of aggressive ionic species (as reactants, products, or both), which in turn requires the existence of an aqueous environment capable of stabilizing them.

Stainless and nickel–chromium alloys experience high corrosion rates at supercritical pressure but not at subcritical temperatures because of transpassive dissolution (Friedrich et al., 1999), where the nickel or iron cannot form a stable insoluble oxide that protects the alloy. Under supercritical conditions, the acids are not dissociated and ionic corrosion products cannot be dissolved by the solution because of the solvent's low polarity. Consequently, corrosion drops down to low values.

Electrochemical corrosion requires the presence of ionic species like halides, nickel-based alloys, and compounds. These show high corrosion rates which decrease at higher temperatures. High-pressure water in an SCW reactor provides favorable conditions for this, but once the water enters the supercritical domain the solubility and concentration of ionic species in it decreases, although the reaction rate continues to be higher because of higher temperatures. The total corrosion reduces because of decreased concentration of the reacting species. Thus, corrosion in a plant increases with temperature, reaching a peak just below the critical temperature, and then reduces when the temperature is supercritical. The corrosion rate increases downstream, where the temperature drops into the subcritical region.

At a relatively low supercritical pressure (e.g., 25 MPa), the salt NaCl is not soluble. Thus, in an SCW a reaction that produces NaCl, the salt can precipitate on the reactor wall. Sometimes water and brine trapped between the salt deposit and the metal can create a local condition substantially different from conditions in the rest of the reactor in terms of corrosion. This is known as *under deposit corrosion.*

In general, a reaction environment that is characterized by high density, high temperature, and high ion concentration (e.g., acidic) is most conducive to corrosion in an SCW reactor. Rather than the severity of corrosion in terms of whether the flow is supercritical or subcritical, the density of the water should be the major concern.

### 9.8.2 Prevention of Corrosion

According to Marrone and Hong (2008), corrosion prevention in an SCW unit is broadly classified in four ways: (1) contact avoidance, (2) corrosion-resistant barriers, (3) process adjustments, and (4) corrosion-resistant materials.

#### 9.8.2.1 Contact Avoidance

The following are some innovative options that may be used to reduce contact between corroding species and the reactor wall:

- A transpiring wall on which water constantly washes down, preventing any corroding material's contact with the wall surface.
- A centrifugal motion created in the reactor to keep lighter reacting fluids away from the wall.
- In a fluidized bed, neutralizing or retaining of the corrosive species by the fluidized particles.

#### 9.8.2.2 Corrosion-Resistant Barriers

Corrosion-resistant liners are used inside the reactor to protect the vessel wall. These are required to withstand the reactor's high temperature but not its high pressure. Titanium is corrosion resistant, but in large quantities, such as required for the reactor shell, it is not recommended because of the risk of fire if it comes in contact with high concentrations of oxidant, particularly when pure oxygen is used in an SCWO. In much smaller quantities, titanium can be as a liner; alternatively, some type of sacrificial liner can be used.

#### 9.8.2.3 Process Adjustments

Changes in process conditions may reduce or even avoid corrosion in some cases, but they may not be practical in many situations. For example, if the corrosion is as a result of acidic reaction, the addition of a base to the feed may preneutralize the reactant. Since most of the corrosion occurs just below critical temperature, the water without the feed may be preheated to a sufficiently high temperature such that on mixing with the cold feed the reaction zone quickly reaches the design reactor's temperature; then the biomass may be fed directly into the reactor to reduce the corrosion in the feed preheat section.

#### 9.8.2.4 Corrosion-Resistant Materials

If corrosion cannot be avoided altogether, it can be reduced by the use of highly corrosion-resistant materials. Choosing one of these as the primary construction material in an SCWO system is the simplest and most basic means of corrosion control. The following materials have been tried in supercritical environments. Of course, no single material can meet all design requirements, so some optimization is required. The materials listed are arranged in the order of least-to-most corrosion resistant.

- Stainless steel
- Nickel-based alloys
- Titanium
- Tantalum
- Niobium
- Ceramics

## 9.9 ENERGY CONVERSION EFFICIENCY

Matsumura (2002) estimated the energy required for SCWG of water hyacinth. His analysis came up with a high overall efficiency. Gasafi et al. (2008) carried out a similar analysis for sewage sludge that came up with a much lower efficiency. The energy consumption of these two biomass types is compared in Table 9.7. We note that the energy required to pump and preheat the feed is a substantial fraction of the energy produced in an SCW plant.

Overall efficiency may depend on the type of feedstock used. Yoshida et al. (2003) studied options for electricity generation from biomass, including SCWG combined cycle, thermal gasification, and direct combustion. They concluded that the SCWG combined cycle offers the highest efficiency for high-moisture biomass, but it does not for low-moisture fuels.

## 9.10 MAJOR CHALLENGES

Commercialization of SCW biomass gasification must overcome the following major challenges:

- SCWG requires a large heat input for its endothermic reactions and for maintenance of its moderately high reaction temperature. This heat requirement greatly reduces energy conversion efficiency unless most of the heat is recovered from the sensible heat of the reaction product. For this reason, the efficiency of the heat exchanger and its capital cost greatly affect the viability of SCWG.

**TABLE 9.7** Energy Consumption for Gasification of Biomass

| Investigators | Matsumura et al. (2002) | Gasafi et al. (2008) |
|---|---|---|
| Feedstock | Watery hyacinth | Sewage sludge |
| Potential energy in feed (MW) | 4.44 | 1.44 |
| Energy in product gas (MW) | 3.32 | 1.38 |
| Electricity consumption in pumping and others (MW) | 0.54 | 0.05 |
| External energy used for feed preheating (MW) | 1.69 | 0.33 |
| Net energy production (MW) | 1.09 | 0.99 |
| Overall efficiency (%) | 24.5 | 68.6 |

- The feeding of wet solid biomass, which is fibrous and widely varying in composition, is another major challenge. A slurry pump has been used to feed solid slurry into high-pressure reactors, but it has not been tested for feeding biomass slurry into a supercritical reactor with ultrahigh pressure.
- The drop in gasification efficiency and gas yield with an increase in dry solids in the feed may be a major obstacle to commercial SCWG. Efforts are being made to improve this ratio using different catalysts, but a cost-effective method has yet to be discovered.
- Separation of carbon dioxide from other gases may require the addition of large amounts of water at high pressure (see Section 9.7.6). This can greatly increase the system's cost and reduce its overall energy efficiency.
- The heating of biomass slurry in the heat exchanger and reactor is likely to cause fouling or plugging because of the tar and char produced during the preheating stage. Further research is required to address this important challenge. A final problem that might inhibit commercialization of SCWG is the corrosion of the reactor wall.

## SYMBOLS AND NOMENCLATURE

$A$ cross-sectional area ($m^2$)
$C_i$ mole fraction of the component $i$ in the gas product
$F_c$ carbon fraction in feed
$k_g$ reaction rate ($s^{-1}$)
$H$ enthalpy of products for product-out, product-in, and feed-in (kJ)
$L$ length of gasifier reactor (m)
$Q'$ volume flow-rate through reactor ($m^3/s$)
$V$ volume of reactor ($m^3$)
$W_p$ product gas flow-rate (kmol/s)
$W_f$ feed rate (kg/s)
$X_c$ carbon conversion fraction
$Y$ gasification yield
$\propto$ viscosity (N s/$m^2$)
$\propto_i$ number of carbon atoms of component $i$ in the gas product
$\eta$ heat-exchange efficiency
$\tau$ residence time (s)
$\rho_b$ density in bulk, kg/$m^3$

Chapter 10

# Biomass Cofiring and Torrefaction

## 10.1 INTRODUCTION

Direct combustion is an important option for biomass energy conversion, and it has been used since the dawn of human civilization when man discovered fire. Direct combustion of biomass is in use in many parts of the world including Nepal, where it is still the primary source of energy in rural areas. Greatest use of biomass is in small-scale applications like a domestic stove, where biomass is used as firewood. Large-scale commercial use, though growing especially for heating and for electricity production, is still not the dominant application of biomass. Of late, the motivation for use of biomass to replace fossil fuels in steam power plants, cement industries, and iron making is growing because it could reduce carbon footprint of those industries. Owing to the large difference in combustion properties of biomass and fossil fuels, it is difficult to replace a fossil fuel entirely in a fossil fuel fired system with biomass, without any major performance penalty. An acceptable practical option is partial replacement by cofiring biomass in an existing fossil fuel fired combustion plant because it reduces the extent of incompatibility. This chapter examines cofiring of biomass for partial replacement of fossil fuels in the above industries.

The interest in biomass cofiring is rising because of the growing need for immediate reduction in greenhouse gas (GHG) emission from large power plants. Although extensive research on carbon capture and storage for sequestration (CCS) is being conducted, and many countries have committed major funds for demonstration plants, wide-scale commercial use of CCS for reduction in carbon dioxide is not likely to happen in the short term because of the large number of technical and scale-up issues this technology faces. Even if CCS overcomes all these issues, the electricity generation cost would still be high.

Utility industries and regulators around the world have recognized that for immediate reduction in GHG emission from fossil fuel fired power plants, one could take an incremental step by cofiring $CO_2$ neutral biomass in an existing fossil fuel fired power plant. This being proven and less expensive could be the best near-term solution for GHG reduction. As the amount

Biomass Gasification, Pyrolysis and Torrefaction.

of biomass cofiring is relatively small (<10% by mass), much of the existing infrastructure of an existing power plant could be utilized for cofiring. For the same reason unlike CCS, cofiring would have a minimum performance penalty on the existing power plant.

Several excellent references (Tumuluru et al., 2011a,b; Van Loo and Koppejan, 2008) are available for direct combustion of biomass for cofiring. So, this chapter would not discuss much on direct combustion of biomass in boilers. It will instead discuss cofiring in general and an improved option of biomass cofiring, where biomass is torrefied before it is fired in the boiler.

## 10.2 BENEFITS AND SHORTCOMINGS OF BIOMASS COFIRING

The option of cofiring a small amount of biomass in an existing fossil fuel fired power plant has several advantages over complete switch over to biomass:

1. It is one of the most cost-effective practical means for GHG reduction.
2. Modern fossil fuel fired steam plants, because of its high-temperature and pressure, are much more efficient than smaller conventional biomass energy conversion systems. Cofiring that rides on such plants naturally offers much higher energy conversion efficiency for biomass.
3. Cofiring biomass in existing coal-fired boilers is among the lowest generation unit ($/kWh) cost among biomass-based power production options.
4. Combustion technology being conventional, cofiring has the lowest technical risk and is ready for immediate implementation in large scale.
5. The carbon dioxide abatement cost ($/ton $CO_2$) of cofiring is much lower than that in CCS (Figure 10.1).
6. Biomass cofiring may have some synergistic effect on corrosion. For example, sulfur and aluminum silicate in coal/peat could combine with the alkali in biomass forming alkali silicate/sulfate preventing the formation of corrosive alkali chloride compound in biomass-fired plant (Kasman and Berg, 2006).

Since biomass cofiring rides on a highly efficient steam power plant equipped with advanced pollution control systems, the biomass energy conversion efficiency is very high and cleanest. Figure 10.1 shows that cofiring could reduce the $CO_2$ abatement cost considerably. One should, however, be careful about this comparison as it depends much on the price of biomass. For example, the cost of $CO_2$ abatement by cofiring may even exceed that by CCS if the biomass price exceeds $120/dry ton (Ortiz et al., 2011, p. xvii).

One shortcoming of biomass cofiring is a modest reduction in the thermal efficiency of the plant. Using data from three commercial plants, Tillman (2000) correlated empirically the loss in plant efficiency with the amount of biomass fired on percentage mass basis ($Z$). This equation may not be

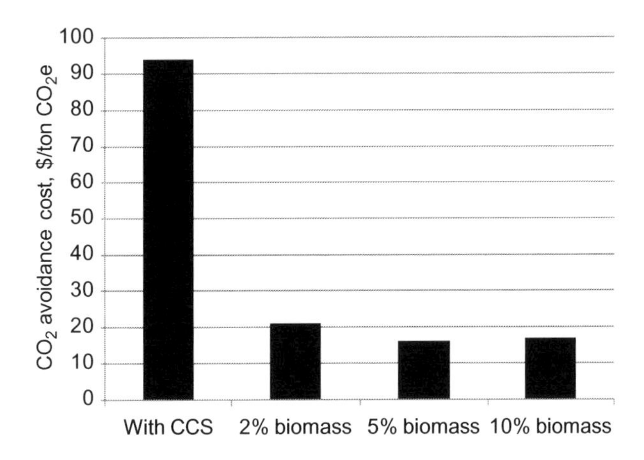


**FIGURE 10.1** Comparison of $CO_2$ capture cost by cofiring different percentage of biomass with that of carbon capture and storage from a PC-fired boiler for biomass price $40 per dry ton. Source*: Plotted with data from Ortiz (2011), p. 61 and NETL (2010).*

applicable to any plant as it was developed for one particular commercial plant for one set of parameters, but it shows the relationship in a qualitative term:

$$\text{Efficiency loss} = 0.0044Z^2 + 0.0055Z\% \tag{10.1}$$

## 10.3 EMISSION REDUCTION THROUGH BIOMASS COFIRING

Emissions of $NO_x$, $SO_2$, fly ash, and mercury from a coal-fired power plant are of particular concern as they contribute to near-term local air pollution, and as such their reduction is desirable. Biomass because of its inherent property emits much less or none of these pollutants. Thus, cofiring could bring a positive emission reduction from an existing power plant.

The ash content of biomass is generally much lower than that in coal (Table 10.1). As a result, there is an overall reduction in fly ash production with associated particulate reduction when some biomass replaces coal in a power plant.

Biomass contains lesser amounts of nitrogen and sulfur than those in coal. So, through cofiring, one could bring about a modest reduction in $NO_x$ and $SO_2$ emission. Many biomasses contain calcium, which is very effective in absorbing the sulfur released from the coal during cofiring.

The higher volatile and higher hydrogen content of biomass could be exploited in $NO_x$ reduction procedures such as air staging and reburning in a cofired boiler. The improvement in $NO_x$ reduction increases with the

**TABLE 10.1** Comparison of Properties of Coal with Some Biomass

| | Upper Freeport | Spring Creek | Bryan | | Biomass | | | |
|---|---|---|---|---|---|---|---|---|
| | Bituminous | Subbituminous | Lignite | Peat | Pine bark[a] | Forest Residue[b] | Bagasse | Rapeseed Expeller[c] |
| Heating value (db), MJ/kg | 31 | 28 | 14 | – | 21 | – | ~20 | |
| Ash in fuel (db), % | 13.4 | 5.7 | 50.4 | NS | 2.9 | NS | NS | NS |
| $SiO_2$ | 59.6 | 32.6 | 62.4 | 32.1 | 39.0 | 11.6 | 73 | 0 |
| $Al_2O_3$ | 27.4 | 13.4 | 21.5 | 17.3 | 14.0 | 2.0 | 5.0 | 0 |
| $Fe_2O_3$ | 4.7 | 7.5 | 3.0 | 18.8 | 3.0 | 1.8 | 2.5 | 0.3 |
| CaO | 0.62 | 15.1 | 3 | 15.1 | 25.5 | 40 | 6.2 | 15.0 |
| $Na_2O$ | 0.42 | 7.41 | 0.59 | 0.5 | 1.3 | 0.6 | 0.3 | 0 |
| $K_2O$ | 2.47 | 0.87 | 0.92 | 1.4 | 6 | 9.2 | 3.9 | 22.8 |
| $P_2O_5$ | 0.42 | 0.44 | – | 3.7 | | 4.4 | 1.0 | 41.1 |
| Others | Rest | Rest | Rest | Rest | Rest | Rest | Rest | Rest |
| ID (oxidizing) °C | 1404 | 1510 | 1354 | | 1210 | | | |

[a] *High Ca, K, and low silica.*
[b] *High silica and low Ca and K.*
[c] *High Ca, K, and potassium and low silica.*
ID, initial deformation temperature of ash. NS, not specified
**Source:** Kitto and Stultz (2005) and Hiltunen et al. (2008).

percentage amount of biomass fired (*Z*). For the set of plant for which Eq. (10.1) was developed, it was correlated as (Tillman 2000):

$$\text{Reduction in NO}_x\text{ emission} = 0.75Z\% \tag{10.2}$$

Reduction is measured in kg/GJ input and *Z* is in mass percentage of biomass in feed.

Adding biomass to the coal could also decrease the nitrous oxide ($N_2O$) emission, a strong GHG (about 300 times more potent than $CO_2$), from the boiler due to the higher oxygen–carbon (O/C) ratio of biomass and probably due to catalytic effects of relatively high amount of calcium, potassium, and sodium in biomass (EUBIONET, 2003, p. 21).

### 10.3.1 $CO_2$ Reduction

Biomass being a carbon neutral fuel, the $CO_2$ released from the combustion of biomass in a cofired boiler does not make any net contribution to the global $CO_2$ inventory (see Section 1.3.2). Additionally, owing to its high hydrogen–carbon (H/C) molar ratio, biomass combustion releases lesser amount of $CO_2$ than that by lower H/C ratio fuels like coal or oil.

## 10.4 CARBON CAPTURE AND STORAGE (CCS) VERSUS BIOMASS FIRING

Generation of power without much or no addition to the atmospheric inventory of $CO_2$ can be achieved through one of the following two options:

1. Carbon dioxide capture and its sequestration from existing or new coal-fired power plants (CCS).
2. Conversion of existing coal-based power plant into 100% biomass firing.

The first option (CCS) is most talked about and researched as it provides a lasting and complete solution to the emission of GHG from coal-fired plants. In this option, efforts are being made to retrofit existing coal-fired units with a CCS plant as a $CO_2$ scrubbing system. The CCS option allows generation of electric power from coal in conventional means, while largely avoiding the release of the $CO_2$ to the atmosphere. This is done by removing carbon dioxide from the flue gas by amine scrubbing or other techniques and sequestering it appropriately. One option for $CO_2$ scrubbing involves separation of carbon dioxide by using oxygen instead of air for combustion. This option, known as "oxy-firing," requires complete change in the entire firing system of the boiler such that the flue gas is free from nitrogen and it contains mostly $CO_2$. The flue gas being primarily made of $CO_2$ it permits easy compression and its eventual sequestration in appropriate storage.

The CCS technology when fully developed would require additional energy for separating $CO_2$ (or for generating $O_2$ for oxy-firing) from flue gas and for pumping and transporting $CO_2$ to storage locations. Such additional power requirement will naturally account for additional $CO_2$ release.

The cost of CCS is also a major factor. The additional carrying charge of a relatively large capital investment of CCS and its operating cost when added to the total generation cost of electricity could increase the electricity cost by as much as 40% (Figure 10.2). Considering how our society is able to adapt to steady rise in oil price, one can hope that consumers will be able to live with increased cost of GHG-free electricity.

The major problem is, however, the lead time for this technology. The CCS is not likely to have worldwide commercial implementation till 2030. At the present pace of rise in global inventory of $CO_2$, if the $CO_2$ emission is allowed to rise without any control, it is very much possible that earth's temperature might rise to alarming levels while waiting for CCS to be fully implemented. This underscores the need for biomass cofiring in the interim period to reduce the pace of $CO_2$ release to the atmosphere, and therefore the global warming, while waiting for CCS to arrive.

Figure 10.1 compares the cost of $CO_2$ abatement of different percentages of biomass cofiring with that of a CCS system for a landed cost of biomass $40/dry ton (19 MJ/kg) in a pulverized subbituminous coal-fired boiler. One notes from this figure that the extra cost for $CO_2$ abatement by CCS could be as high as 4.5–6 times that by cofiring.

Replacing an existing fossil fuel fired boiler with a new biomass-fired boiler or switching its fuel from coal to 100% biomass can achieve the above goal. The first option is exceptionally expensive while the second option of fuel switch could face major technical hurdles due to the large difference in

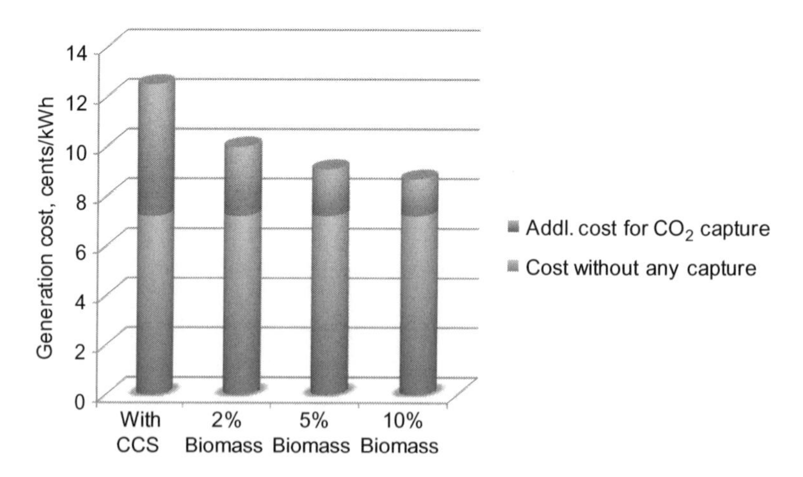


**FIGURE 10.2** The cost of electricity generation rises with $CO_2$ capture, but that due to cofiring is lower than that from carbon capture and storage (CCS). Source: *Plotted with data from Al Juaied and Whitmore (2008).*

the combustion and physical handling characteristics of biomass and coal. Such a fuel switch could reduce the output of an existing plant by as much as 40%. The major inhibiting factors for 100% fuel switch are therefore:

1. Large capital investment for replacing the entire coal-firing and handling system with biomass firing.
2. Long downtime of the existing plant for replacement of its firing system with loss of revenues.
3. Reduction in output of the plant and in some cases lower overall efficiency for operation of the steam cycle at reduced capacity.
4. The availability of large quantities of biomass at an affordable price could also be an issue.

The extent of the above problems could be reduced by partial fuel switch or cofiring such that the coal-fired plant can continue to fire coal along with a modest amount of biomass.

## 10.5 COFIRING OPTIONS

Biomass cofiring has been successfully demonstrated in the large number of installations worldwide for most combinations of fuels and boiler types. There are three major options for cofiring in coal-fired boilers (Figure 10.3):

1. Direct cofiring of raw or torrefied biomass
2. Gasification cofiring
3. Parallel cofiring

### 10.5.1 Direct Cofiring of Raw or Torrefied Biomass

In direct cofiring, the raw biomass and coal are fed directly into the boiler furnace using the same or separate set of mills and burners (Figure 10.3A). For this reason, it is simplest, least expensive, and most widely used. In direct cofiring, there are several options like below:

1. Dried or raw biomass is mixed with coal upstream of the coal feeders. They are then co-milled in the pulverizers.
2. Biomass is pulverized in separate mills and injected into the furnace through existing coal burners.
3. Separate burner, mills, and other feed preparation are used for biomass.

In the first option, the dried and sized biomass is mixed with coal upstream of the feeders. The mixture is then sent to the coal pulverizing mills, where it is pulverized along with coal. The mixture of pulverized biomass and coal is distributed among all coal burners. It is the simplest option requiring least capital investment, but it involves the highest risk of interference with the coal-firing capability of the boiler unit. Thus, it is applicable to a limited range of biomass

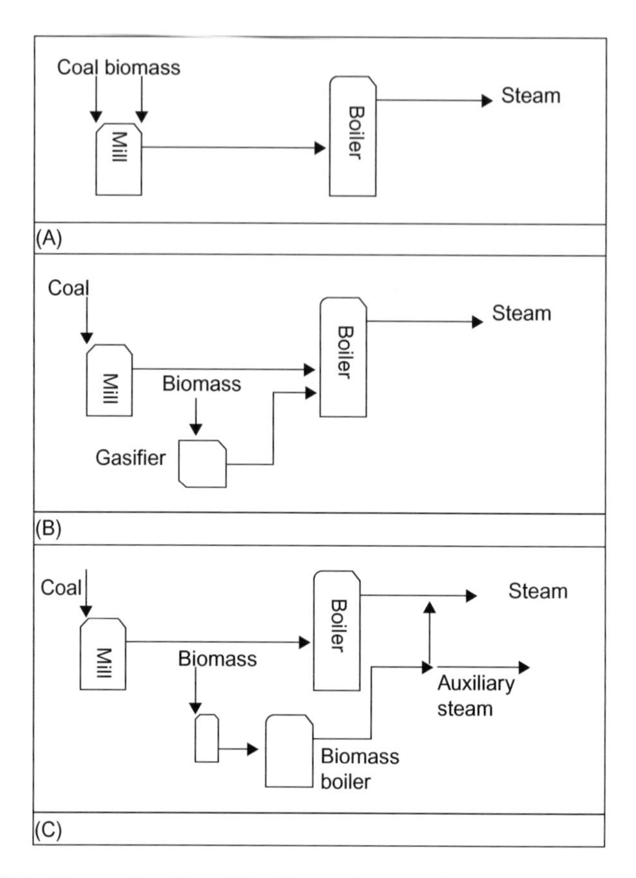


**FIGURE 10.3** Three options for cofiring biomass in a coal-fired boiler. (A) Direct cofiring—biomass fed directly into the coal pulverizing mill. (B) Indirect cofiring—biomass is gasified and gas is burnt in the boiler. (C) Parallel cofiring—biomass is fired in a separate boiler and steam is fed into the steam header.

types and to very low biomass to coal cofiring ratios that is typically less than 5% by mass (Tillman, 2000). The volumetric ratio of the cofiring would, however, be much higher. For example, 5% by mass switchgrass (80 kg/m$^3$) when mixed with coal (881 kg/m$^3$) could give a volume percentage of about 37%.

The second option involves separate handling, metering, and pulverizing the biomass in dedicated mills, and injecting it into the furnace through existing pulverized fuel pipework upstream of the burners. This option increases the pipeworks around the boiler, which may already be congested. It may also be difficult to control and to maintain the burner-operating characteristics over the normal boiler load curve.

The third option involves separate handling and pulverization of the biomass fuel. Here, pulverized biomass is injected through dedicated burners for burning. It involves highest capital investment but is of lowest risk for normal boiler operation. This option has an added benefit as the biomass can be used as a reburn fuel for superior $NO_x$ control. The dedicated burners are located downstream of the existing pulverized coal (PC) burner.

#### 10.5.1.1 Cofiring with Torrefied Biomass

All the above options of direct cofiring may also use torrefied biomass as the feedstock instead of raw one. Torrefaction can be carried out on-site upstream of the mill as shown in Figure 10.3A, or it could be torrefied at the biomass collection point such that the power plant receives biomass in a torrefied state instead of in raw form. The torrefied biomass could be mixed directly with coal and fed into the pulverizer. Owing to the similarity between torrefied wood and coal, there is reduced potential for operational problems, and the upper limit of the amount of biomass that can be fired can be relaxed.

### 10.5.2 Gasification Cofiring

Indirect cofiring through gasification involves gasification of solid biomass and subsequent combustion of its product gas (syngas) in the furnace along with coal (Figure 10.3B). The gasifier may be considered as a replacement for the pulverization equipment used for biomass preparation in the direct cofiring option. As the gasifier operates in tandem with the existing coal preparation stream, it does not interfere with the operation of the coal-firing system. So, this approach can offer a high degree of fuel flexibility.

One of the major problems of biomass cofiring comes from the evaporation of alkali and other compounds in the biomass that causes fouling and corrosion of boiler tubes. In gasification, one can strip the product gas of the alkali contents of the biomass prior to its combustion, and thereby minimize the impact of the impurities in biomass.

### 10.5.3 Parallel Cofiring

Parallel cofiring involves installation of a completely separate biomass-fired boiler to produce steam (Figure 10.3C). Low-grade (pressure and temperature) steam is generated in the biomass boiler and is utilized to meet the process demand of the coal-fired power plant instead of using high-pressure steam from the main boiler (Basu et al., 2011). This option has lowest risk with highest reliability as it runs independently in parallel to the existing boiler unit. This option avoids to a great extent the potential for biomass generated fouling or corrosion as the flue gas from biomass does not contact any heating surface of the boiler. It is, however, expensive though some cost can be offset by allowing the boiler to vent its flue gas into upstream of the

**TABLE 10.2** Economic Analysis of Three Major Biomass Cofiring Options

| | Direct Firing | External Firing | Gasification |
|---|---|---|---|
| Capital investment (m$) | 4.373[a] | 6.052 | 5.67[b] |
| Fuel cost savings (m$/yr) | 0.596 | 0.382 | 0.54 |
| Credit income (m$/yr) | 3.373 | 3.373 | 3.37 |
| Generation losses (m$/yr) | 0.657[c] | 0 | 0.33[d] |
| O&M cost (m$/yr) | 0.026[e] | 1.713 | 1.81 |
| Carrying charge (m$/yr) | 0.328 | 0.454 | 0.43 |
| Net after tax (m$/yr) | 2.957 | 1.588 | 1.35 |
| Internal rate of return (%) | 49.10 | 21.11 | 19.30 |

[a]*Scaled up from $279 per kW estimate of Cantwell (2003) using escalation factor 1.1084% per year.*
[b]*Cantwell (2003) took 25,000 per year using escalation factor 1.1084% per year.*
[c]*Capacity factor loss of 1%.*
[d]*Capacity factor loss of 0.5%.*
[e]*Antares (2003), Table A.1, estimate of $382 per $kW_e$.*
**Source:** From Basu et al. (2011).

existing particulate separator and fan system of the main boiler avoiding the cost of a dust separator, induced fan, and stack.

If one does not account for reliability or technological maturity, direct firing option of cofiring offers the highest return on investment (Basu et al., 2011). Table 10.2 shows a comparison of costs of three options for given plant with a given fuel and other scenario.

## 10.6 OPERATING PROBLEMS OF BIOMASS COFIRING

Cofiring of biomass poses some special problems due to the following basic differences between coal and biomass:

1. The elemental and proximate analysis of coal is much different from that of biomass. As one can see from the van Krevelen diagram (see Figure 3.10) biomass is on the top right corner with very high H/C and O/C ratio, while coal is near the lower left with low H/C and O/C ratio.
2. Unlike coal, the properties of biomass are highly variable and heterogeneous. Even different parts of a tree could have different composition.
3. Unlike coal, when stored for an extended period, biomass absorbs moisture and besides the adverse effect on thermal efficiency moisture could also lead to the development of harmful fungus.

4. Biomass is less brittle and more fibrous than coal which results in significantly different grinding characteristics.

The ash in biomass is much richer in compounds like K, Ca, and Si (Table 10.1). Waste biomass could additionally pick up chlorine, potassium, and heavy metals. All of these greatly increase the fouling, slagging, and corrosion potential in a coal-fired boiler. Table 10.1 shows the differences in ash constituents between biomass and coal that affect fouling, corrosion, and some results in operating problems in the cofired boiler.

### 10.6.1 Combustion Issues

Biomass particles are generally more reactive due to their higher volatile content and more porous structure compared to those of coal. So a biomass particle, dried to the same extent and ground to the same size as a coal particle, might burn faster. In this respect, it is not necessary to grind biomass as finely as PC, but its size should not be so large that it will drop into the furnace hopper unburnt when injected instead of being conveyed up the furnace in flames.

There is another issue related to the ignition of fuels. The relatively high-moisture content of biomass might delay the ignition of such particles. If the delay is significant, the flame could move further and further away from the burner extinguishing itself. This is countered to some extent by the fact that ignition temperature of dry biomass is much lower than that of coal.

Experience (Van Loo and Koppejan, 2008, p. 235) suggested that there is only negligible effect on the combustion efficiency for a modest level (3–5%) cofiring of relatively dry (<10% moisture) biomass. The efficiency, however, could be low for higher moisture in biomass. Batista et al. (1998) developed an empirical correlation to assess the impact of the extent of cofiring on the overall thermal efficiency of a coal-fired boiler:

$$\text{Boiler efficiency } (\%) = \text{TE} - A \cdot X - B \cdot Y - C \cdot Z \tag{10.3}$$

where

TE is the theoretical thermal efficiency (%)
$X$ is the oxygen concentration in flue gas at boiler economizer exit (%)
$Y$ is the sunburnt carbon in fly ash (%)
$Z$ is the biomass cofired as mass percent of coal fired
$A$, $B$, and $C$ are empirical constants

### 10.6.2 Fuel Preparation

The typical size of coal fed into a pulverized coal (PC) furnace is about 75 μm for suspension firing in burners. Biomass needs to be ground to a little larger (due to its lower density and higher reactivity) but to a comparable size for transportation through coal pipes and combustion in a PC burner. Raw biomass, due to its tenacious fibrous nature, is difficult to grind to such fine sizes. Grinding raw biomass is not only energy intensive but it is also difficult to obtain finer particle sizes in an adequate amount for a given energy input. Such high-energy requirement may be a result of the plastic behavior of biomass.

The pulverizing mills are designed for certain energy input. So, the output of such a mill reduces when called upon to grind biomass and that reduces with increase in several parameters like moisture content, degree of fineness, as well as tenacious nature of the material. For this reason, when raw biomass is fed through the coal feeder for pulverization in a mill designed for coal, the output of the mill reduces accordingly. If the plant has spare mill capacity, it can maintain the thermal input in the furnace but at the expense of additional energy consumption by the mills.

### 10.6.3 Storage

Raw biomass, agriculture waste in particular, tends to have high-moisture content. This characteristic makes transportation, handling, and storage of biomass difficult. If the local climate is dry and windy, open-air storage could help reduce the moisture of the biomass, but that is not the case for most plants. Some power plants go for expensive covered storage to prevent wetting of the biomass from rain or snow. Even drying of the biomass onsite will not prevent hygroscopic fresh biomass from future absorption of moisture from its surroundings while in storage. Additionally, the moisture within the biomass attracts fungal attacks and causes rotting while in storage. Such problems are not present to this extent for coal.

### 10.6.4 Fouling, Agglomeration, and Corrosion

The ash content of average biomass is lower than that of average coal. The ash in biomass is fine, while that from coal is in general coarse. For the low-ash mixed feed, one could expect a lower rate of ash deposition on boiler surfaces while cofiring biomass, but the reality is different. Many biomasses (herbaceous and agriculture waste in particular) contain high fraction of alkali and chlorine in their ash (Table 10.1). As a result, the fouling and corrosion rate of boiler heating surfaces could increase with the amount of biomass cofired. The alkali metals from biomass ash evaporates at combustion temperature and it subsequently condenses on cooler boiler surfaces downstream enriching alkali in the metal to ash interface. Fouling from such deposits could be hard enough to be beyond the cleaning capacity of standard soot blowers. The chlorine in biomass leads to hydrochloric acid in the flue gas that increases chlorine in metal-ash interface increasing the corrosion potential.

From a combustion standpoint, ash types in biomass may be divided into three classes, with their distinct fouling characteristics (Hiltunen et al., 2008):

1. Rich in calcium and potassium but lean in silica (e.g., woody biomass).
2. Rich in silica but lean in calcium and potassium (e.g., rice husk, straw, and bagasse).
3. Rich in phosphorous, calcium, and potassium.

Biomass ash is generally more reactive than that from coal (Hiltunen et al., 2003). Biomass with type 1 ash forms deposits of CaO, $CaSO_4$, and $K_2SO_4$ on high-temperature tubes of superheater and reheaters of a PC-fired boiler. These deposits harden with time. In fluidized bed boilers the Ca, K could react with quartz in bed material coating them with low softening temperature Ca or K silicate layers that causes agglomerates. Contrary to type 1 ash, type 2 ash that is rich in silica, separate sticky and molten ash particles are formed in a fluidized bed and cause agglomeration. Here, the quality of bed material does not affect the behavior of bed agglomeration in this case as was the case with type 1 ash. Type 3 ash being rich in phosphorous could form low-temperature melting potassium phosphate and make it high-fouling fuel. To avoid this, one can add limestone to fluidized-bed boilers to facilitate formation of high-temperature melting calcium phosphate, instead of low temperature potassium phosphate (Hiltunen et al., 2003).

On the basis of 10,000 fuel samples from 150 operating circulating fluidized-bed boilers, Foster Wheeler Corporation developed a probability index to give a quantitative assessment of how different biomass fuels ranks for fouling, agglomeration, and corrosion potential. They are shown in Table 10.3. The index is ranked as follows: 0–2 low probability, 2–4 medium probability, 4–5 high probability, and 5–10 very high probability.

## 10.6.5 Biomass Variability

Variability is a major issue in biomass supply. Unlike coal, it is difficult to get a consistent supply of large quantities of biomass of a given composition.

**TABLE 10.3** Probability for Agglomeration, Fouling, and Corrosion for Several Biomass Fuels as Developed by the Foster Wheeler Corporation

| Probability Index for | Rice Husk | Sawdust | Wood Chips | Sunflower Husk | Straw |
|---|---|---|---|---|---|
| Agglomeration | 0 | 1 | 1 | 4 | 5 |
| Fouling | 0 | 2 | 2 | 6 | 6 |
| Corrosion | 2 | 1 | 2 | 5 | 7 |

**Source:** Data from Janati et al. (2003).

A preferred feedstock for a cofired boiler is biomass pellet. The uniform size of pellets makes it convenient to handle the fuel and grind it in mills. It is, however, difficult to produce biomass pellets from a wide range of biomass feedstock (Bergman et al., 2005, p. 11). This is an important barrier in the wide-scale use of biomass cofiring, but this can be done using torrefaction pretreatment.

### 10.6.6 Capacity Reduction

As mentioned earlier, the output of a coal-fired power plant could reduce when biomass replaces coal in an existing boiler in spite of maintaining the energy input unchanged. Additionally, if the existing coal mills are used for biomass pulverization, the output of fuel would reduce due to higher moisture in biomass, low-energy density of biomass, and other factors.

### 10.6.7 Safety Issue

Cofiring biomass with coal raises some safety issues that involve potential for fire and explosion. Several incidents of fire and explosion in biomass underscore the need for this issue. There are three major hazards of handling coal in a cofired power plant. They are briefly described below:

1. *Combustible dust:* Coal, when moved through a belt conveyor or loaded into a bunker, produces dusts. Such dusts are an explosion hazard. The explosion potential depends on several factors including deflagration index. Higher the deflagration index larger is the explosion. Table 10.4 compares this index of several fuels.
2. *Spontaneous combustion:* Self-ignition causes spontaneous combustion of a fuel left in storage. Although very little data is available in published

**TABLE 10.4** Deflagration Index of Coal and Biomass

| Fuel | Petcoke | Bituminous coal | Powder river basin coal | Cellulose | Paper |
|---|---|---|---|---|---|
| Deflagration index | 47 | 150 | 225 | 229 | 200 |
| Fuel | Barley straw | Corn | Wood pellet | Wood bark | Sawdust |
| Deflagration index | 72 | 75 | 105 | 132 | 149 |

**Source:** From Power (2012), p. 22.

literature on biomass, it is easy to speculate that piled biomass runs the risk of self-ignition as much as the coal does. Several fires that occurred in biomass plants highlight this risk.

3. *Biomass-coal mixture:* Biomass being reactive, its mixing with coal dust adds another dimension to the explosion issue. Biomass is indeed more reactive than most coal, petcoke, or anthracite. Thus, explosion hazard for such mixtures has to be evaluated. Presently, very little data is available forcing the designer to assume the mixture to be all coal or all biomass in their assessment. The explosion behavior of the coal–biomass mixture is at the moment largely unexplored. This aspect is discussed further in Section 10.7.4.

## 10.7 COFIRING WITH TORREFIED WOOD

Torrefaction could offer some relief to the above problems and make biomass cofiring more viable. Such relief is due to the following intrinsic properties of torrefied biomass.

1. The fuel preparation for cofiring greatly benefits from the torrefaction of the biomass feed because this process makes biomass more brittle and less fibrous. The least cost option for cofiring uses the existing pulverization mills and feeds biomass directly into them along with coal. Though torrefaction cannot make biomass as grindable as coal, it makes significant improvement in the grindability of the biomass. As a result, the existing mills can grind the required amount of biomass without requiring additional energy. This allows the boiler to feed its burner with the required amount of coal and biomass to match the furnace heat input of the existing boiler. Furthermore, in cases where separate mills are used for coal and biomass, the improved grindability of torrefied wood allows the mills to produce particles of right size and in right quantity.
2. Capital investment for covered biomass storage could be a major component of the total cost of biomass cofiring upgrade of an existing plant. The carrying charge of that could tip the economic balance against cofiring. Even that may not prevent dried biomass from picking up additional moisture from the atmosphere and cause a health hazard due to fungal attack on biomass. Torrefied biomass, being relatively hydrophobic, does not pick up moisture even when stored outdoors and experience very little fungal attack. Thus, it obviates the need for expensive covered fuel storage allowing the plant to use parts of the existing coal yard to store the biomass.
3. Torrefaction acts like a quality leveller for multiple fuel feed. The difference between different biomass feedstock is reduced through torrefaction. Thus, while the quality of the delivered biomass supply is

variable, the actual variation in quality experienced by the burner is much less. Torrefaction helps reduce the difference in combustion characteristics and heating value of the biomass feed.

### 10.7.1 Effect of Cofiring on Plant Output

The capacity or thermal output of a boiler could reduce for two reasons: the boiler furnace is not able to generate the designed energy input and the available heating surfaces of the boiler are not able to absorb the required amount of heat. The volumetric energy density of biomass is much lower than that of coal. For example, volume energy density of raw wood is 5–8 $MJ/m^3$ while that for typical coal is 30-40 $MJ/m^3$ because of lower density (350–680 vs. 1100–1350 $kg/m^3$) and lower heating value (~17–21 vs. 24–33 MJ/kg dry basis) of the biomass (Table 4.3). When coal is replaced by an equivalent (by energy content) amount of biomass in a boiler plant, a significantly larger volume of biomass is to be handled by the existing feed preparation, feeder, and the burner system. In most cases, these components of a PC-fired boiler lack adequate spare capacity to handle such a large increase in volume throughput. Among these, the capacity of the pulverization mills is the major limitation.

There is another reason why capacity of a coal-fired boiler could reduce when cofired with biomass. For a given energy input, the amount of flue gas increases when one replaces coal with biomass. In Example 10.1, one can see that though biomass contains a significantly larger amount of oxygen, the air requirement per unit MJ heat input is about the same as that for coal, but the mass of flue gas produced by biomass is higher. So, biomass cofiring could place an extra load on the existing induced draft fan and downstream units of the boiler plant. This necessitates reductions in boiler output. For heat absorption, the only limitation may be on the flame emissivity due to higher $H_2O$ fraction in the flue gas. Since biomass may constitute only a small part of the total feed in normal cofiring, flue gas emissivity may not bring about major change in heat absorption. The only limitation could be the flame temperature if it is reduced due to lower heating value of the biomass fuel.

### 10.7.2 Feed Preparation

In a pulverized coal (PC)-fired boiler, pulverizing mills grind coal to about 75 μm size, and transport it pneumatically through pipes to burners for combustion in a flame. A fluidized-bed boiler, on the other hand, would require the fuel to be crushed to only less than 10 mm size and dropped under gravity into the furnace. Thus, cofiring of biomass in a fluidized-bed combustion boiler is a little easier than that in a PC-fired boiler because of the fuel flexible feature of fluidized-bed firing (Basu, 2006).

Raw biomass is highly fibrous in nature. Surface fibers of neighboring particles lock into each other making it difficult to flow smoothly. These along with the plastic behavior of biomass causes several problems such as:

1. Handling difficulties
2. Problem with pneumatic transportation in pipes
3. Difficult grinding and pulverization of fine sizes

Co-combustion of biomass would therefore require biomass to be ground to comparable sizes ($\sim 75\,\mu m$) and pneumatically conveyed through pipes. Because of its soft, nonbrittle characteristics, considerably more energy is required to grind untreated biomass to the above fineness. For example, to grind a ton of coal to a $d_{50}$ around $500\,\mu m$, about 7−36 kWh of grinding energy is required, while that for raw poplar and pine, the energy requirement would be 130 and 170 kWh, respectively (Esteban and Carrasco, 2006). There is thus nearly an order of magnitude increase in energy consumption for biomass grinding.

Additionally, the output (tons/h) of a given pulverizer is greatly reduced while grinding biomass along with coal for cofiring. Reduced mill output directly reduces the power production of the plant.

Torrefaction addresses the above problems to a great extent by making biomass particles more brittle, smoother, and less fibrous. An optical photomicrograph taken after torrefaction shows (Arias et al., 2008) an absence of fibrous exterior, sharp ends in the biomass. Thus, the friction created by the interlocking of the fibers during the handling of a pneumatic transportation is greatly reduced after torrefaction.

### 10.7.3 Effect of Torrefaction Parameters on Grinding

We note from above that torrefaction reduces the energy consumption for grinding biomass. This section discusses the grinding issue further. Torrefaction temperature is the most influential parameter affecting grinding. The higher the temperature at which torrefaction is performed the lower is the energy requirement for grinding, or for a given energy input, more amount of fine particles are obtained after grinding. Grinding of torrefied biomass gives smaller and uniform size distribution of the product (Phanphanich and Mani, 2011). The grinding energy requirement for a specified level of grinding decreases with torrefaction temperature. Here, one notes that the specific energy consumption reduced from about 250 kWh/ton for raw biomass to about 50 kWh/ton for that torrefied at 280°C.

#### 10.7.3.1 Grindability Index

Grindability index is a measure of the ease of grinding of a given feed. Utility industries use the Hardgrove grindability index (HGI) to express this parameter. In the direct cofired system, the existing mills designed for coal are used for grinding the cofired biomass. So, for a given mill, given rotation, and energy input, it is necessary to know how much biomass would be ground. HGI gives the comparative ease of grinding with reference to a standard coal. The higher the HGI the lower is the power requirements, and the finer the particle size. It represents a fuel that is easier to grind.

An HGI-measuring machine is a miniature ball mill type of pulverizer. Here, a standard mass (50 g) of coal is grounded for a given time in the mill subjecting the balls to a known force. The resulting product is sieved to measure amounts dropping below 75 μm size. This amount is compared against some specified standards to define the parameter, HGI. As the HGI ball mill works on the same principle as pulverizing mills, the index obtained from this could give a fair assessment of the grinding capability of torrefied biomass.

For biomass, one should use (Agus and Water, 1971; Joshi, 1978) a standard volume of sample instead of mass to compare the grinding ease to coal and torrefied biomass. Thus, an equivalent HGI was used to define the grinding ease of torrefied biomass. More details are given in Bridgeman et al. (2010).

The grindability index of torrefied biomass increases with torrefaction temperature (Boskovic, 2013). It also depends on the type of biomass.

### 10.7.4 Explosion and Fire

Dust explosion is a major problem in handling and conveying of fine dusts. So, special attention is paid in PC-fired power plants where coal dust is being conveyed or milled. Since pulverized biomass is being considered for cofiring, one needs to explore this potential to ensure that presence of fine dust of torrefied dust does not make the matter worse. In a typical explosive situation, the dust could be ignited by an energy source, and it is followed by rapid exothermic oxidation of the mixture. This leads to an increase in temperature that further increases the reaction and the gas expands rapidly. In a confined space like pipelines from the pulverizer to the burner, the pressure increases with temperature. The combustion rate increases with temperature and pressure, further aggravating the situation, which eventually leads to explosion that could burst the pipeline or its confinement. Large buildup of pressure leading to explosion is also possible even without a confinement.

The following factors favor dust explosion:

1. Fine particle size
2. High reactivity of dust materials
3. High concentration of dust in air

4. Low ignition temperature of dust particles
5. Proximity to a high-energy ignition source
6. Favorable oxidizing environment

Because of its low ignition temperature and high reactivity, torrefied biomass could potentially have worse risk for dust explosion. When the solid concentration in dust increases, the minimum temperature at which a dust-cloud ignites drops. Torrefied wood is more brittle in nature than biomass is; as such, it would have higher level of dust formation and greater potential for explosion. If one compares the above factors for coal and torrefied biomass, one could easily note that torrefied wood has greater potential of explosion.

Additionally, the ignition temperature of biomass is typically below that of coal. These make torrefied biomass particularly vulnerable to explosion and fire. Thus, care should be taken to reduce the risk of dust explosion in a cofired plant. Relatively low volatile content of torrefied biomass could, on the other hand, make it less explosive, but in the absence of any such data, it is only a speculation.

The intensity of explosion increases with increase in the combustibility of dust particles, So, depending on the combustibility of the torrefied wood, it may have a higher dust explosibility than raw wood, but this hypothesis is yet to be proved through experiments.

Torrefied biomass could produce more fines during handling than standard biomass would. Because of its low ignition temperature and high reactivity, chances of catching fire are real in a plant. Thus, particular attention needs to be paid to avoid fire in a plant cofired with torrefied biomass.

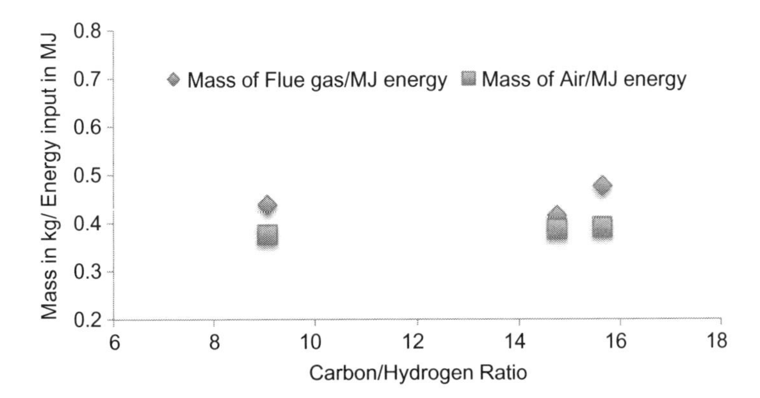


**FIGURE 10.4** Computation in Example 10.1 shows that when the C/H ratio of the fuel is changed the combustion air required per unit heat release does not change much but the flue gas produced changes.

**Example 10.1**

A PC-fired boiler is designed to fire bituminous coal with 20% excess air. It was decided to switch the boiler entirely into (i) torrefied biomass, (ii) raw biomass, or (iii) lignite firing.

Find the impact on combustion air requirement and flue gas production if the boiler is to retain the energy input into the furnace. Assume the air to contain 1.3% moisture. Composition and heating values of these fuels are given below:

| Composition | Bituminous Coal | Torrefied Biomass | Raw Biomass |
|---|---|---|---|
| C (%) | 67.36 | 54.70 | 38.49 |
| H (%) | 4.58 | 6.00 | 4.86 |
| O (%) | 5.69 | 36.40 | 37.19 |
| N (%) | 1.30 | 0.10 | 0.25 |
| S (%) | 2.08 | 0 | 0.00 |
| Ash (%) | 8.84 | 2.80 | 2.8 |
| Moisture (%) | 9.91 | 0.00 | 16.4 |
| HHV (MJ/kg) | 28.91 | 21.9 | 13.97 |

**Solution**

**a.** First we carry out the calculation for raw biomass:

Using Eq. (3.34), we calculate the theoretical dry air needed per kilogram raw biomass.

$$\begin{aligned} M_{da} &= 11.53 \times C + 34.34 \times [H - (O/8)] \\ &= 11.53 \times 38.49 + 34.34 \times (4.86 - 37.19/8) \\ &= 4.512 \text{ kg air/kg raw biomass} \end{aligned}$$

Actual air with 20% excess air $= 1.2 \times 4.512 = 5.41$ dry air/kg of raw biomass.

Moisture in air is 1.3%.

So, the wet air requirement is $M_{wa} = (1 + 0.013) \times 5.41 = 5.48$ wet air/kg of raw biomass.

Amount of flue gas product through complete combustion is found from Eq. (3.36).

$$\begin{aligned} \text{Flue gas mass } W_c &= M_{wa} - 0.2315 M_{da} + 3.66C + 9H + N + O + 2.5S \\ &= 5.48 - 0.2315 \times 4.512 + (3.66 \times 38.49 + 9 \times 4.86 \\ &\quad + 0.25 + 37.19)/100 \\ &= 6.66 \text{ kg/kg raw biomass} \end{aligned}$$

Heating value of biomass is 13.97 MJ/kg. So, for 1 MJ energy input the furnace needs (5.48/13.97) or 0.392 kg air and produce (6.66/13.97) or 0.476 kg flue gas.

**b.** Repeating the above computation for coal, we find that 1 kg coal needs 11.22 kg wet air and produces 12.03 kg flue gas and releases 28.91 MJ heat.

So, for 1 MJ energy input in the energy the furnace needs (11.22/28.91) or 0.39 kg air and produces (12.03/28.91) or 0.416 kg flue gas.

**c.** For torrefied biomass in similar way, we get that 1 kg torrefied biomass needs 8.27 kg wet air and produces 9.60 kg flue gas and releases 21.9 MJ heat.

So, for 1 MJ energy input in the energy the furnace needs (8.27/21.9) or 0.377 kg air and produces (9.6/21.9) or 0.438 kg flue gas. Figure 10.4 plots these data against C/H ratio of fuels.

Results are listed in the table below:

| | **Bituminous Coal** | **Torrefied Biomass** | **Raw Biomass** |
|---|---|---|---|
| Energy density of fuel (MJ/kg) | 28.91 | 21.9 | 13.97 |
| Mass of flue gas produced for per unit heat release (kg/MJ) | 0.416 | 0.438 | 0.476 |
| Wet air needed to burn unit mass of fuel (kg/kg) | 11.22 | 8.27 | 5.48 |
| Air required per unit energy release (kg/MJ) | 0.39 | 0.38 | 0.39 |
| Flue gas produced per unit fuel burnt (kg/kg) | 12.03 | 9.60 | 6.66 |
| Energy carried by unit mass of flue gas (MJ/kg) | 2.40 | 2.28 | 2.09 |

It is interesting to note that while wet air required for unit heat release is nearly independent on the fuel type, the amount of flue gas produced per unit energy input depends much on the fuel type. Raw biomass has higher flue gas weight than coal, but torrefied biomass produces mass of flue gas per unit heat release comparable to that of coal.

Chapter 11

# Production of Synthetic Fuels and Chemicals from Biomass

## 11.1 INTRODUCTION

Earlier chapters discussed the methods of converting biomass into convenient forms of gases, liquid, and solid products and their use in energy production. Besides energy production, such conversion processes have important applications in the production of chemicals and transport fuels. It is interesting to note that many of our daily necessities like plastic, resin, and fertilizer can potentially come from biomass.

Charcoal, a product of biomass carbonization, is not only a fuel but it is also an important reducing and adsorbing agent. Similarly, syngas, a mixture of $H_2$ and CO, is a fuel as well as a basic building block for many hydrocarbon chemicals. Transport fuel and a large number of chemicals are produced from different syntheses of CO and $H_2$. These products can be divided into three broad groups: (1) energy source (e.g., methane, carbon monoxide, and torrefied wood), (2) transportation fuels (e.g., biodiesel and biogas), and (3) chemical feedstock (e.g., methanol, ammonia, and charcoal).

Presently, syngas is produced not only from natural gas but also from any of the following feedstock:

- Biomass
- Solid fossil fuels (e.g., coal and petcoke)
- Liquid fuels (e.g., refinery wastes).

Interest in biomass as a chemical feedstock is rising given that it is renewable and carbon-neutral. There is a growing shift toward "green chemicals" and "green fuels," which are derived from carbon-neutral biomass. With the development of the chemical industry and new legislation concerning environment, the application of charcoal for purification of industrial wastes has increased markedly (FAO, 1985). Gasification and pyrolysis are effective and powerful ways to convert the biomass (or another fuel) into energy, chemicals, and transport fuels. Carbonization and torrefaction are important means of converting biomass into an effective adsorbent and reducing agent. This chapter discusses different ways to convert biomass-derived syngas into such useful products.

**Biomass Gasification, Pyrolysis and Torrefaction.**

## 11.2 SYNGAS

The following sections discuss syngas, its physical properties and uses, as well as its production and its cleaning and conversion.

### 11.2.1 What Is Syngas?

As mentioned earlier, syngas is a mixture of hydrogen ($H_2$) and carbon monoxide (CO). Syngas is an important feedstock for the chemical and energy industries. A large number of hydrocarbons traditionally produced from petroleum oil can also be produced from syngas gases. Natural gas is primarily made of methane gas ($CH_4$). Manufactured natural gas is called "synthetic (or substitute) natural gas" or SNG, which should not be confused with syngas.

Syngas may be produced from many hydrocarbons, including natural gas, coal, and petroleum coke, as well as from biomass. Syngas generated from biomass is called biosyngas so that it can be distinguished from that produced from fossil fuel. In this chapter, syngas implies that derived from biomass unless specified otherwise.

One of the major applications of syngas is the production of liquid transport fuel. For many years, South African Synthetic Oil Limited (SASOL) has been producing a large amount of liquid fuel from coal using Fischer–Tropsch synthesis (FTS) of syngas produced from the gasification of coal. The same liquid fuel may be produced from biomass-derived syngas. It is discussed further in Section 11.4.2.

The typical product gas of biomass gasification contains hydrogen, moisture, carbon monoxide, carbon dioxide, methane, aliphatic hydrocarbons, benzene, and toluene, as well as small amounts of ammonia, hydrochloric acid, and hydrogen sulfide. From this mixture, carbon monoxide and hydrogen are separated to produce syngas.

### 11.2.2 Applications of Syngas

As mentioned, syngas is an important source of valuable chemicals that include:

- Hydrogen, produced in refineries
- Diesel or gasoline, using FTS
- Fertilizer, through ammonia
- Methanol, for the chemical industry

It should be noted that a major fraction of the ammonia used for fertilizer production comes from syngas and nitrogen (Section 11.4.3).

### 11.2.3 Production of Syngas

Gasification is the preferred route for the production of syngas from coal or biomass. The current (2013) low price of natural gas is favorable to the production of syngas from methane, but the situation may change in future when the gas's price rises.

Steam reformation reaction that is widely used for bulk production of hydrogen is a popular method for production of syngas from natural gas.

#### 11.2.3.1 Steam Reforming of Methane

In the steam reforming method, natural gas ($CH_4$) reacts with steam at high temperatures (700–1100°C) in the presence of a metal-based catalyst (nickel).

$$CH_4 + H_2O \xrightarrow{\text{Catalyst}} CO + 3H_2 + 206\ \text{kJ/mol} \tag{11.1}$$

If hydrogen production is the main goal, the carbon monoxide produced is further subjected to the shift reaction (Eq. (11.3)) to produce additional hydrogen and carbon dioxide.

The ratio of hydrogen and carbon monoxide in the gasification product gas is a critical parameter in the synthesis of the reactant gases into desired products such as gasoline, methanol, and methane. The product desired determines that ratio. For example, gasoline may need the $H_2/CO$ ratio to be 0.5–1.0, while methanol may need it to be ~2.0 (Probstein and Hicks, 2006, p. 124). In a commercial gasifier, the $H_2/CO$ ratio of the product gas is typically less than 1.0, so the shift reaction (Eq. (11.3)) is necessary to increase this ratio by increasing the hydrogen content at the expense of CO. The shift reaction often takes place in a separate reactor, as the temperature and other conditions in the main gasifier may not be conducive to it.

#### 11.2.3.2 Partial Oxidation of Natural Gas

Steam reforming of natural gas is a highly endothermic reaction. An alternative approach for production of syngas from $CH_4$ is partial oxidation, which is slightly exothermic instead of being highly endothermic like the steam reforming reaction.

$$CH_4 + \tfrac{1}{2}O_2 = CO + 2H_2 - 22.1\ \text{kJ/mol} \tag{11.2}$$

A comparison between Eq. 11.1 and 11.2 shows that the latter produces less hydrogen. A lower $H_2/CO$ ratio (2:1) in the partial oxidation reaction is favorable for use of the syngas for FTS. The selectivities toward CO or $H_2$, however, are influenced by simultaneous occurrence of total combustion of methane and secondary oxidation reactions of CO and $H_2$. Potential catalysts for this reaction are Ni and Rh. Though Ni shows high conversion and selectivity, it suffers from catalyst deactivation (Smet, 2000).

### 11.2.4 Gasification for Syngas Production

The two main routes for production of syngas from biomass or fossil fuel are low-temperature ($\sim < 1000°C$) and high-temperature gasification ($\sim > 1200°C$).

Low-temperature gasification is typically carried out at temperatures below 1000°C. In most low-temperature gasifiers, the gasifying medium is air, which introduces undesired nitrogen in the gas. To avoid this, gasification can be carried out indirectly by one of the following means:

- An oxygen carrier (metal oxide) is used to transfer the oxygen from an air oxidizer to another reactor, where gasification takes place using the oxygen from the metal oxide.
- A certain amount of combustion in air is allowed in one reactor and heat-carrier solids carry the combustion heat to a second reactor, where the heat is used for endothermic gasification.
- Dilution of the product gas by nitrogen is avoided by using steam or oxygen instead of using air as the gasifying medium.

Low-temperature gasification produces a number of heavier hydrocarbons along with carbon monoxide and hydrogen. These heavier hydrocarbons are further cracked, separated, and used for other applications.

High-temperature gasification is carried out at temperatures above 1200°C, where biomass is converted mainly into hydrogen and carbon monoxide. Primary gasification is often followed by the shift reaction (Eq. (11.3)) to adjust the hydrogen-to-carbon monoxide ratio to suit the downstream application.

#### *11.2.4.1 Shift Reaction*

For a reaction like FTS that produces various gaseous and liquid hydrocarbons, a definite molar ratio of CO and $H_2$ in the syngas is necessary. This is adjusted through the shift reaction that converts excess carbon monoxide into hydrogen:

$$CO + H_2O \xrightarrow{\text{Catalyst}} CO_2 + H_2 - 41.1\ \text{kJ/mol} \tag{11.3}$$

The above reaction can be carried out either at higher temperatures (400–500°C) or at lower temperatures (200–400°C). At high temperatures, the shift reaction is often catalyzed using oxides of iron and chromium; it is equilibrium limited. At low temperatures, the shift reaction is kinetically limited; the catalyst is composed of copper, zinc oxide, and alumina, which help reduce the CO concentration down to about 1%.

### 11.2.5 Cleaning and Conditioning of Syngas

Whatever the gasification process is, the product gas must be cleaned before it is used for synthesis reactions. Special attention must be paid to clean the syngas of tar and other catalyst-poisoning elements, especially before it is used for FTS, which uses iron- or cobalt-based catalysts. So the gas must be cleaned of particulates and other contaminating gases. The raw syngas may contain three principal types of impurity: (1) solid particulates (e.g., unconverted char and ash), (2) inorganic impurities (e.g., halides, alkali, sulfur compounds, and nitrogen), and (3) organic impurities (e.g., tar, aromatics, and carbon dioxide).

The ash in the biomass appears as slag. At low temperatures, the ash remains in the product gas as dry ash. Cleaning has two aspects: removing undesired impurities and conditioning the gas to get the right ratio of $H_2$ and CO for the intended use. The end-use determines the level of cleaning and conditioning. Table 11.1 presents examples of product-gas specifications for different end uses.

**TABLE 11.1 Product-Gas Specifications for Various Applications**

| Specification | Hydrogen for Refinery Use | Ammonia Production | Methanol Synthesis | FTS |
|---|---|---|---|---|
| Hydrogen content | >98% | 75% | 71% | 60% |
| Carbon monoxide content | <10–50 ppm (v) | $[CO + CO_2]$ <20 ppm (v) | 19% | 30% |
| Carbon dioxide content | <10–50 ppm (v) | | 4–8% | |
| Nitrogen content | <2% | 25% | | |
| Other gases | $N_2$, Ar, $CH_4$ | Ar, $CH_4$ | $N_2$, Ar, $CH_4$ | $N_2$, Ar, $CH_4$, $CO_2$ |
| Balance | | As low as possible | As low as possible | Low |
| $H_2/N_2$ ratio | | ~3 | | |
| $H_2/CO$ ratio | | | | 0.6–2.0 |
| $H_2/[2CO + 3CO_2]$ ratio | | | 1.3–1.4 | |
| Process temperature | | 350–550°C | 300–400°C | 200–350°C |
| Process pressure | >50 bar | 100–250 bar | 50–300 bar | 15–60 bar |

**Source:** Adapted from Knoef (2005), p. 224.

#### *11.2.5.1 Cleanup Options*

For cleaning the gas of dust or particulates, there are four options: (1) cyclone, (2) fabric or other barrier filter, (3) electrostatic filter, and (4) solvent scrubber. Among organic impurities, tar is the least desirable. The three main options for tar removal are:

- Scrubbing with an organic liquid (e.g., methyl ester).
- Catalytic cracking by nickel-based catalysts or olivine sand.
- High-temperature cracking.

Inorganic impurities are best removed in sequence because some removal processes produce other components that need to be removed as well. In this sequence, first, water quenching removes char and ash particles. Next, hydrolysis removes COS and HCN by converting them into $H_2S$ and $NH_3$. The ammonia and halides can be washed with water, followed by adsorption of $H_2S$, which can be removed with the wash water. Solid or liquid adsorbents are used to remove carbon dioxide from the product gas.

## 11.3 BIO-OIL PRODUCTION

Bio-oil (or biofuel) is any liquid fuel derived from a recently living organism, such as plants and their residues or animal extracts. In view of its importance, a detailed discussion of bio-oil is presented next.

### 11.3.1 What Is Bio-Oil?

Bio-oil is the liquid fraction of the pyrolysis product of biomass. For example, a fast pyrolyzer typically produces 75% bio-oil, 12% char, and 13% gas. Bio-oil is a highly oxygenated, free-flowing, dark-brown (nearly black) organic liquid (Figure 11.1) that contains a large amount of water (~25%) that is partly the original moisture in the biomass and partly the reaction product. The composition of bio-oil depends on the biomass it is made from as well as on the process used.

Table 11.2 presents the composition of a typical bio-oil. It shows that water, lignin fragments, carboxylic acids, and carbohydrates constitute its major components. When it comes from the liquid yield of pyrolysis, bio-oil is called pyrolysis oil. Several other terms are often used to describe bio-oil or are associated with it, including:

- Tar or pyroligneous tar
- Biocrude
- Wood liquid or liquid wood
- Liquid smoke
- Biofuel oil
- Wood distillates

**FIGURE 11.1** Bio-oil is a thick black tarry liquid.

**TABLE 11.2 Composition of Bio-Oil**

| Major Group | Compounds | Mass (%) |
|---|---|---|
| Water | | 20–30 |
| Lignin fragments | Insoluble pyrolytic lignin | 15–30 |
| Aldehydes | Formaldehyde, acetaldehyde, hydroxyacetaldehyde, glyoxal, methylglyoxal | 10–20 |
| Carboxylic acids | Formic, acetic, propionic, butyric, pentanoic, hexanoic, glycolic | 10–15 |
| Carbohydrates | Cellobiosan, α-D-levoglucosan, oligosaccharides, 1.6 anhydroglucofuranose | 5–10 |
| Phenols | Phenol, cresols, guaiacols, syringols | 2–5 |
| Furfurals | | 1–4 |
| Alcohols | Methanol, ethanol | 2–5 |
| Ketones | Acetol (1-hydroxy-2-propanone), cyclopentanone | 1–5 |

**Source:** Adapted from Bridgwater et al. (2001), p. 989.

- Pyrolysis oil
- Pyroligneous acids.

There is an important difference between pyrolysis oil and biocrude. The former is obtained via pyrolysis; the latter can be obtained via other methods such as supercritical gasification.

Bio-oil may be seen as a two-phase micro-emulsion. In the continuous phase are the decomposition products of hollocellulose; in the discontinuous phase are the pyrolytic lignin macromolecules. Holocellulose is the fibrous residue that remains after the extractives, lignin, and ash-forming elements have been removed from the biomass. Similar to crude petroleum oil, which is extracted from the ground, pyrolysis liquid and biocrude contain tar as their heaviest component.

Bio-oil is a class-3 substance falling under the flammable liquid designation in the UN regulations for transport of dangerous goods (Peacocke and Bridgwater et al., 2001, p. 1485).

### 11.3.2 Physical Properties of Bio-Oil

As we observe from Figure 11.1, bio-oil is a free-flowing liquid. Its low viscosity is due to its high water content. Also, it has an acrid, smoky smell that can irritate eyes with long-term exposure. With a specific gravity of ~1.2, bio-oil is heavier than water or any oil derived from petroleum. A comparison of its physical and chemical properties with those of conventional fossil fuels is given in Table 11.3.

An important feature of bio-oil not reflected in Table 11.3 is that some of its properties change with time. For example, its viscosity increases and its volatility decreases (Mohan et al., 2006) with time. Some phase separation and deposition of gums may also occur with time, primarily because of polymerization, condensation, esterification, and etherification. This feature distinguishes bio-oil from mineral oils, the properties of which do not change with time.

Bio-oil is not soluble in water, although it contains a substantial amount of water. However, it is miscible in polar solvents, such as methanol and acetone but immiscible with petroleum-derived oils. Bio-oil can accept water up to a maximum limit of 50% (total moisture). Further addition of water could result in phase separation. Table 11.3 shows that bio-oil has a heating value nearly half that of conventional liquid fuels but has comparable flash and pour points.

### 11.3.3 Applications of Bio-Oil

Bio-oil is renewable and cleaner than nonrenewable mineral oil, a fossil fuel that is extracted from the ground (petroleum). Thus, it offers a "green"

**TABLE 11.3 Comparison of Physical and Chemical Properties of Bio-Oil and Three Liquid Fuels[a]**

| Property | Bio-Oil | Heating Oil | Gasoline | Diesel |
|---|---|---|---|---|
| Heating value (MJ/kg) | 18–20 | 45.5 | 44[b] | 42 |
| Density at 15°C (kg/m$^3$) | 1200 | 865 | 737[b] | 820–950[b] |
| Flash point (°C) | 48–55 | 38 | 40[b] | 42[b] |
| Pour point (°C) | −15 | −6 | −60 | −29[c] |
| Viscosity at 40°C (cP) | 40–100 (25% water)[d] | 1.8–3.4 per cSt | 0.37–0.44[d] | 2.4[d] |
| pH | 2.0–3.0 | – | | |
| Solids (% wt)[e] | 0.2–1.0 | – | 0 | 0 |
| **Elemental Analysis (% wt)** | | | | |
| Carbon | 42–47 | 86.4 | 84.9 | 87.4[f] |
| Hydrogen | 6.0–8.0 | 12.7 | 14.76 | 12.1[f] |
| Nitrogen | <0.1 | 0.006 | 0.08 | 392 ppm[f] |
| Sulfur | <0.02 | 0.2–0.7 | | 1.39[f] |
| Oxygen | 46–51 | 0.04 | | |
| Ash | <0.02 | <0.01 | | |

[a] *Except as indicated, all values are excerpted from www.dynamotive.com.*
[b] *http://www.engineeringtoolbox.com.*
[c] *Maples (2000).*
[d] *Bridgwater et al. (2001), p. 990.*
[e] *Mohan et al. (2006).*
[f] *Hughey and Henerickson (2001).*
cP, centipoise; cSt, centistokes. Values for gasoline and diesel are for a representative sample and can vary.

alternative in many applications where petro-oil is used. Bio-oil is mainly used not only as an energy source but also as a feedstock for the production of "green chemicals."

### *11.3.3.1 Energy Production*

Bio-oil may be fired in boilers and furnaces as a substitute for furnace oil. This could allow a rapid and easy switchover from fossil fuels to biofuels, as it does not call for complete replacement or any major renovation of the firing system as would be needed if raw biomass were to be fired in a furnace or boiler designed for furnace oil. The combustion performance of a bio-oil-fired furnace should be studied carefully before such a switchover is made because furnace oil and bio-oil have different combustion characteristics, including significant differences in ignition, viscosity, energy content, stability, pH, and emission level. In many cases, we can overcome these differences through proper design (Wagenaar et al., 2009).

#### *11.3.3.2 Chemical Feedstock Production*

Bio-oil is a hydrocarbon similar to petrocrude except that the former has more oxygen. Thus, most of the chemicals produced from petroleum can be derived from bio-oil. Such products include:

- Resins
- Food flavorings
- Agro-chemicals
- Fertilizers
- Levoglucosan
- Adhesives
- Preservatives
- Acetic acid
- Hydroxyacetaldehyde.

#### *11.3.3.3 Transport Fuel Production*

Bio-oil contains less hydrogen per carbon (H/C) atom than do conventional transport fuels like diesel and gasoline, but it can be hydrogenated (hydrogen added) to overcome this deficiency and thereby can produce transport fuels with a high H/C ratio. The hydrogen required for the hydrogenation reaction normally comes not only from an external source but it can also be supplied by reforming a part of the bio-oil into syngas. Dynamotive, a Canadian company, practices this method.

### 11.3.4 Production of Bio-Oil

Several options for the production of bio-oil are available. They fall under two broad groups: thermochemical or biochemical route. They are as follows:

- Gasification of biomass and synthesis of the product gases into liquid: thermochemical route.
- Production of biocrude using fast pyrolysis of biomass: thermochemical route.
- Production of biodiesel (fatty acid methyl ester or FAME) from vegetable oil or fats through transesterification: biochemical route.
- Production of ethanol from grains and cellulosic materials: biochemical route.

The common major steps in the production of bio-oil from biomass are as follows:

1. Receipt of biomass feedstock at the plant and storage.
2. Drying and sizing.
3. Reaction (e.g., pyrolysis, gasification, fermentation, and hydrolysis).
4. Separation of products into solids, vapor (liquid), and gases.
5. Collection of the vapor and its condensation into liquid.
6. Upgrading of the liquid to transport fuel or extraction of chemicals from it.

## 11.4 CONVERSION OF SYNGAS INTO CHEMICALS

As mentioned earlier, syngas is an important building block for a host of hydrocarbons. Commercially it finds use in two major areas: (1) alcohols (e.g., methanol and higher alcohols) and (2) chemicals (e.g., glycerol, fumaric acid, and ammonia). The following section briefly describes the production of some of these products.

### 11.4.1 Methanol Production

Methanol ($CH_3OH$) is an important feedstock for the production of transport fuels and many chemicals. The production of gasoline from methanol is an established commercial process. Methanol is produced through the synthesis of syngas (CO and $H_2$) in the presence of catalysts (Figure 11.2) (Higman and van der Burgt, 2008, p. 266):

$$CO + 2H_2 \xrightarrow{\text{Catalyst}} CH_3OH - 91\ \text{kJ/mol} \tag{11.4}$$

Methanol synthesis is an exothermic reaction influenced by both temperature and pressure. The equilibrium concentration of methanol in this reaction increases with pressure (in the 50–300 atm range) but decreases with

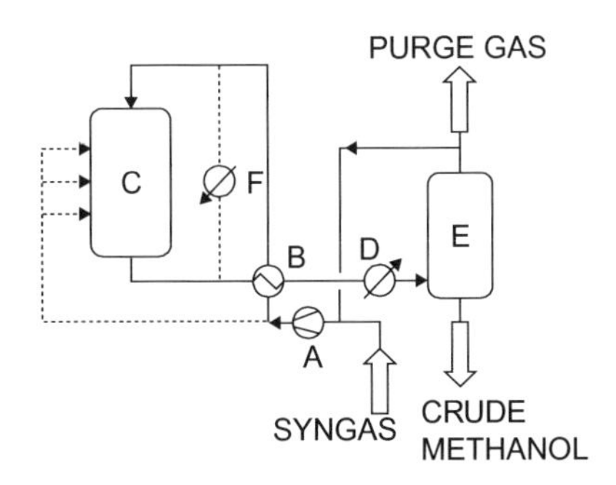


**FIGURE 11.2** Methanol production.

temperature (in the 240–400°C range). In the absence of a suitable catalyst, the actual yield is very low, so catalysts based on Zn, Cu, Al, and Cr are used.

Syngas, which is the feedstock for methanol production, can be produced from biomass through either thermal or hydrothermal gasification. One of the most commonly used commercial methods use natural gas ($CH_4$) as the feedstock. This process uses the steam reforming of methane as shown in the equation below:

$$CH_4 + H_2O \rightarrow CO + 3H_2 + 206\ \text{kJ/mol} \tag{11.5}$$

We note from this that for every mole of CO produced, 3 mol of $H_2$ are produced, but the methanol synthesis reaction (Eq. (11.4)) requires only 2 mol of hydrogen for every mole of carbon monoxide. Thus, there is an extra hydrogen molecule for every mole of methanol. In such a situation, carbon dioxide, if available, may be used in the following reaction to produce an additional methanol molecule utilizing the excess hydrogen:

$$CO_2 + 3H_2 \rightarrow CH_3OH + H_2O - 50\ \text{kJ/mol} \tag{11.6}$$

Methanol synthesis reaction (Reed, 2002, p. III-225) can take place at both high pressure (~30 MPa, 300–400°C) and low pressure (5–10 MPa, 220–350°C).

In the high-pressure process, the syngas is first compressed. The pressurized syngas is then fed into either a fixed- or a fluidized-bed reactor for synthesis in the presence of a catalyst at 300–350 atm and at 300–400°C. A fluidized bed has the advantage of continuous catalyst regeneration and efficient removal of the generated heat. The catalyst used is an oxide of Zn and Cr.

The product is next cooled to condense the methanol. Since the conversion is generally small, the unconverted syngas is recycled to the reactor to be further converted. Today, the most widely used catalyst is a mixture of copper, zinc oxide, and alumina.

The low-pressure process is similar to the high-pressure process, except that it uses low pressure and low temperature. In one of the several variations, a fixed bed of Cu/Zn/Al catalyst is used at 5–10 MPa and at 220–290°C (Reed, 2002, p. II-225).

Liquid-phase synthesis is another option, but it is in the development stage. This option could potentially give a much higher level of conversion (~90%) compared to 20% for the high-pressure process (Chu et al., 2002). Here, the syngas is fed into the slurry of the catalysts in an appropriate solvent. The compressed syngas is mixed with recycled gas and then heated in a heat exchanger to the desired reactor inlet temperature, which is usually about 220–230°C. In a cold-quench operation, only about two-thirds of the feed gas is preheated; the rest is used to cool the product gas between the individual catalyst layers.

**Example 11.1**

The production of methanol from syngas is given by the reaction:

$$CO + 2H_2 \rightarrow CH_3OH \quad (11.i)$$

The reaction heat at 25°C is −90.7 kJ/mol. Using Gibb's equation, calculate the equilibrium constant, K. Using the constant, K, find the fraction of the hydrogen in the syngas that will be converted into methanol at 1 atm at that temperature.

**Solution**

Let us assume that the reaction started with 1 mol of CO and 2 mol of $H_2$. If in the equilibrium state only $x$ moles of CO have been converted, it will have consumed $2x$ moles of $H_2$ and produced $x$ moles of $CH_3OH$ (as per Eq. (11.i)), leaving $(1 - x)$ moles of unreacted CO and $2(1 - x)$ moles of $H_2$. The total number of moles will comprise unreacted moles and the methanol produced. Hence, the total moles will be $1 - x + 2(1 - x) + x = 3 - 2x$.

Noting that partial pressure is proportional to mole fraction, the equilibrium constant is defined as:

$$K = \frac{P_{CH_3OH}}{P_{CO}(P_{H_2})^2} = \frac{xP}{(3-2x)} \times \frac{(3-2x)}{(1-x)P} \times \left[\frac{3-2x}{2(1-x)P}\right]^2 \quad (ii)$$

The equilibrium constant, K, is calculated from the Gibbs free energy using Eq. (7.56).

$$K = \exp\left(\frac{-\Delta G_T^0}{RT}\right) \quad (iii)$$

So, for $T = 25 + 273 = 298K$, we take the value of $\Delta G_T^0$ for methanol from Table 7.5:

$$\Delta G_{298}^0 = -161.6 \text{ kJ/mol}$$

The universal gas constant, $R$, is known to be 0.008314 kJ/mol K. Substituting these values in Eq. (11.iii) we get:

$$K = \exp(161.6/(0.008314 \times 298)) = 2.12 \times 10^{28}$$

Using this value in Eq. (11.i), we get a quadratic equation of $x$. Now, solving $x$, we get the following:

$$x = 1.0$$

So the equilibrium concentration of the product is:

$$CO = 1 - 1 = 0; \quad H_2 = 2\,(1 - 1) = 0; \quad \text{and} \quad CH_3OH = 1 \text{ mol}$$

At 25°C, the reaction will produce 2 mol of hydrogen and 1 mol of methanol.

### 11.4.2 Fischer–Tropsch Synthesis

Fischer-Tropsch or FTS synthesis is a highly successful method for the production of liquid hydrocarbons from syngas. The FTS process can produce high-quality diesel or gasoline from syngas derived from biomass or coal using reaction (11.7). The composition of such products is very similar to that of crude oil. Thus, it can be blended with petrodiesel/gasoline or used directly in an engine.

Some promising applications for the FTS process include biomass-to-liquid (BTL), coal-to-liquid (CTL), and gas-to-liquid (GTL) technologies. The BTL could provide a less carbon-intensive alternative fuel that could use not only agricultural feedstock but also waste biomass materials reducing dependence on carbon-rich fossil fuels. Additionally, the absence of sulfur and nitrogen in such biomass makes it superior to those derived from crude oil. A major commercial motivation of FTS is that it can turn natural gas into easily transportable liquid fuel. Such a GTL fuel could be an alternative to liquefied natural gas with an added advantage that it could transform the gas into other value-added chemical feedstock instead of using it as a fuel alone.

BTL conversion in FTS process may have several additional advantages because biomass's gasification product typically contains $H_2$:CO ratio of about unity, which is ideal for iron catalysts. The absence of sulfur in biomass is also favorable to most catalysts.

The most successful and well-known use of FTS is the production of liquid fuel from coal by SASOL in South Africa, where syngas is converted into petroleum products. Figure 11.3 shows a photograph of a SASOL plant in South Africa. The FTS process is also useful for conversion of biomass into liquid fuels and chemicals, but it is yet to be commercially utilized. FTS process requires large central facility. Collection, transportation, and preparation of such large amount of low bulk density biomass at a central gasification plant may have logistic problem. Additionally, large-scale gasification of biomass is still not in commercial use.

**FIGURE 11.3** Photograph of the SASOL plant in South Africa that is recognized as a leading use of FTS process (www.southafrica.info).

### *11.4.2.1 History*

Franz Fischer (1877–1947) and Hans Tropsch (1889–1935) (Figure 11.4) developed the FTS catalytic reaction in the 1920s at the Kaiser Wilhelm Institute for Coal Research (presently Max Planck Institute) at Mulhelm. It was prompted by the dire need of Germany, during the Second World War, which was rich in coal but had very little access to petroleum. So, many plants were built with this technology to produce liquid fuel from coal. After the war, these plants ceased to operate primarily due to increased availability and reduced price of crude oil.

The interest in FTS started rising again owing to its importance in the production of oil from biomass (Kreutz et al., 2008). Since then the original work in Germany's extensive research has been carried out on this synthesis reaction. More than 7500 references and citations are available in a Web site dedicated to this project (http://www.fischer-tropsch.org/).

### *11.4.2.2 Products*

FTS reaction produces a range of hydrocarbon for use as transport fuel like gasoline, diesel, and as chemical feedstock. Its products include some desirable products like olefins, paraffins, and alcohols, and some undesirable products like methane, aldehyde, acids, ketone, and carbon. Of these, olefins, paraffins, and alcohols are the most desirable products. For gasoline, the transportation fuel, desired products are in the olefins in hydrocarbon in carbon range $C_5$–$C_{10}$. (Probstein and Hicks, 2006, p. 128). Methane is an undesirable product of

**FIGURE 11.4** Photographs of two inventors of the FTS process, Franz Fischer and Hans Tropsch.

syngas produced from biomass or coal because it fetches much less value than the liquid yield of FTS. So, efforts are made to minimize its production. Water is a major by-product of FTS, and it has an important effect on the catalysts used for the synthesis (Dalai and Davis, 2008). Figure 11.5 shows the process schematically indicating two potential product streams.

There are four major components of a FTS plant:

1. Gasifier
2. Gas cleaning and conditioning unit
3. FTS reactor
4. Product upgrading units.

Of these, a gasifier is the most involved and expensive unit.

One can see from Figure 11.5 that there are two basic modes of FTS operation:

1. High-temperature Fischer–Tropsch (HTFT) process
2. Low-temperature Fischer–Tropsch (LTFT) process

The HTFT process uses iron-based catalysts at 300–350°C to produce gasoline and linear low molecular mass olefins (C3–C11). To maximize gasoline production, optimum combination is the use of a fluidized-bed reactor at 340°C with an iron catalyst (Dry 2002, p. 239). The straight run gasoline could be about 40%. Some of the remaining products could be oligomerized to gasoline. Other processes like hydrogenation and isomerization may be needed to get gasoline of the right octane number.

The LTFT process uses mainly cobalt-based catalysts in 200–240°C range for the production of high molecular weight linear wax that can be hydrogenated to diesel. A slurry bed is the preferred reactor for this process. The straight run yield of diesel for this process is about 20% (Dry, 2002, p. 240).

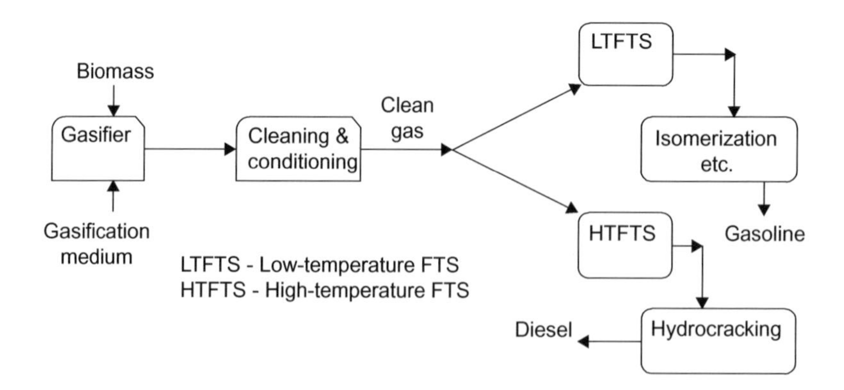


**FIGURE 11.5** The FTS process with its product stream.

The selectivity of the products of FTS reactions depends on several factors including:

- Process parameters (temperature, pressure, residence time).
- Catalyst (type, support).
- Choice of reactor type.
- Composition of the syngas.

Higher temperature, for example, results in faster reaction, but it leads to higher methane formation, faster carbon deposition on catalysts, and reduction in average chain length (Steynberg, 2004).

### 11.4.2.3 Reactions

The FTS reaction, which is typically carried out in the range 200–350°C and 20–300 atm (Reed, 2002, p. II-238), may be written in a generic form as:

$$nCO + 2nH_2 \xrightarrow{\text{Catalyst}} (CH_2)_n + nH_2O + \text{Heat} \tag{11.7}$$

where the hydrocarbon product is represented by the generic formula $(CH_2)n$.

The main reaction is (Dayton et al., 2011):

$$CO + 2H_2 = -CH_2 - + H_2O - 165\ \text{kJ/mol} \quad (227^\circ C) \tag{11.8}$$

For cobalt-based catalyst reaction, the required $H_2/CO$ molar ratio is 2.15. So, the reaction may be written as (Dry, 2002, p. 228):

$$CO + 2.15H_2 \xrightarrow{\text{Cobalt}} \text{hydrocarbon} + H_2O \tag{11.9}$$

When Fe catalyst is used, the water gas shift (WGS) reaction occurs simultaneously:

$$CO + H_2O \overset{\text{Fe}}{\rightarrow} H_2 + CO_2 - 41\ \text{kJ/mol} \tag{11.10}$$

The water produced in main FTS reaction (11.8) is converted into $H_2$ in WGS reaction (Eq. (11.10)). As this hydrogen could be utilized in the main reaction, the net hydrogen requirement of the FTS process is less when Fe catalyst is used. For low-temperature synthesis reaction, the overall $H_2/CO$ ratio is typically 1.7 (Dry, 2002, p. 229).

FTS produces a number of undesired (aldehyde, carbon etc.) and desired (paraffins, alcohol) products. For both desired and undesired products the reactions may be written as (Dalai and Davis, 2008) follows:

Desired Products:

$$\text{Paraffins: } nCO + (2n+1)H_2 \rightarrow C_nH_{2n+2} + nH_2O \tag{11.11}$$

$$\text{Olefins: } nCO + 2nH_2 \rightarrow C_nH_{2n} + nH_2O \tag{11.12}$$

$$\text{Alcohol: } nCO + 2nH_2 \rightarrow C_nH_{2n+1}OH + (n-1)H_2O \tag{11.13}$$

Undesired Products:

$$\text{Boudouard reaction: } 2\,CO \rightarrow C + CO_2 \quad (11.14)$$

$$\text{Water-gas shift reaction: } CO + H_2O \leftrightarrow CO_2 + H_2 \quad (11.15)$$

$$\text{Methanation reaction: } CO + 3H_2 \rightarrow CH_4 + H_2O \quad (11.16)$$

The FTS reaction is a highly exothermic polymerization reaction. It is kinetically controlled. It produces a wide spectrum of oxygenated compounds, including alcohols and aliphatic hydrocarbons ranging in carbon numbers from $C_1-C_3$ (gases) to $C_{35+}$ (solid waxes). For synthetic fuels, the desired products are olefinic hydrocarbons in the $C_5-C_{10}$ range (Probstein and Hicks, 2006, p. 128). The upper end of the range favors a gasoline product. The selectivity of different hydrocarbons may be predicted on the basis of a statistical distribution given by the Anderson-Schulz-Flory model. (Spath and Dayton 2003, p. 95):

$$W_n = n(1-\alpha)^2\alpha^{n-1} \quad (11.17)$$

where $W_n$ is the % weight of a product containing $n$ carbon atoms and $\alpha$ is the chain growth probability.

#### *11.4.2.4 Catalysts*

Catalysts play a pivotal role in the FTS process. Besides enhancing the yield of desired product or selectivity, it contributes to higher volumetric productivity of the process, which in turn reduces the volume of the reactor.

Group VIII metals have highest catalytic activity in FTS reaction. Besides the commonly used iron (Fe) and cobalt (Co) catalysts, several other catalysts are also used in FTS. Activities of these catalysts vary and they rank as below (Adesina, 1996):

$$\text{High activity} \quad Ru > Fe > Ni > Co > Rh > Pd > Pt \quad \text{Low activity} \quad (11.18)$$

It is apparent from here that ruthenium has the highest activity and high selectivity for products of large molecular mass at lower temperatures, but it is significantly more expensive (approx. 300,000 times) than iron. The other choice is Co which is also expensive but to a lesser extent (230 times).

Fe catalysts are less expensive and owing to its WGS activity requires lower hydrogen to carbon monoxide molar ratio, but it has a stronger tendency to form carbon that deposits on its surface deactivating the catalyst.

Cobalt (Co) catalysts, on the other hand, do not have WGS activity, which improves carbon conversion and it produces straight chain hydrocarbons at a high yield (Spath and Dayton, 2003). This catalyst has a longer life. The cobalt-based catalyst is only used in the low-temperature FTS as at higher temperature an excess of methane is produced (Dry, 2002, p. 233).

Promoters could greatly enhance the performance of catalysts. Potassium is a good promoter for iron. Copper is also a good promoter for iron catalysts (Spath and Dayton, 2003), but a Co catalyst is less sensitive to promoters.

Sulfur is a major poisoning element for both Fe and Co catalysts. Thus coal-based FTS needs an extensive sulfur removal process, while green biomass, being free from sulfur, enjoys a distinct advantage in this respect. Because a cobalt catalyst is more sensitive to poisoning from sulfur, it is not used in coal gasified syngas. In that respect, an iron catalyst is better suited for use with coal-derived syngas.

#### *11.4.2.5 Reactors*

Four main types of reactors have been used in FTS (Figure 11.6). They are:

1. Fixed-bed reactor (Figure 11.6A)
2. Slurry-bed reactor (Figure 11.6B)
3. Bubbling fluidized-bed (BFB) reactor (Figure 11.6C)
4. Circulating fluidized-bed (CFB) reactor (Figure 11.6D)

FTS is a highly exothermic reaction and is sensitive to the reactor temperature. Localized overheating could cause carbon deposition and production of methane at the expense of more valuable liquid products. Thus, it is critical to efficiently remove the large amount of heat produced. This criterion puts a fixed-bed reactor in a disadvantaged condition compared to a fluidized-bed reactor or slurry-bed reactor.

For low-temperature process (LTFT), slurry-bed reactors are preferred. Fluidized-bed reactors are preferred for high-temperature process. Initial designs of SASOL were based on CFB (Figure 11.6D) but present designs have moved to BFB (Figure 11.6B). The switchover from CFB was prompted by 40% lower cost, more space for heat exchange in reactor, the lowering of catalyst bulk density due to deposition of carbon in a BFB compared to those in CFB. Additionally, CFB had erosion issues from abrasive iron carbide catalysts in the narrow section of CFB (Dry, 2002, p. 232).

Multitubular fixed bed (Figure 11.6A) and slurry-bed reactors (Figure 11.6C) are well suited for LTFT where a large amount of wax is produced in liquid phase. The syngas enters from the top in the catalysts filled tubes in the multitubular reactor. The wax trickles down the vertical reactor tubes (Figure 11.6). In a slurry-bed reactor, the catalyst is suspended in molten wax product. Syngas bubbles through the slurry bed from the bottom. It is thus a three-phase reactor. Slurry-bed reactors are about 25% less expensive and has about fourfold reduction rate in catalyst loading and are more isothermal in nature (Dry, 2002, p. 233) compared to multitubular fixed-bed reactors.

Furthermore, biomass gasification products contain $CO_2$, which is beneficial for the production of liquid products (Reed, 2002, p. 242).

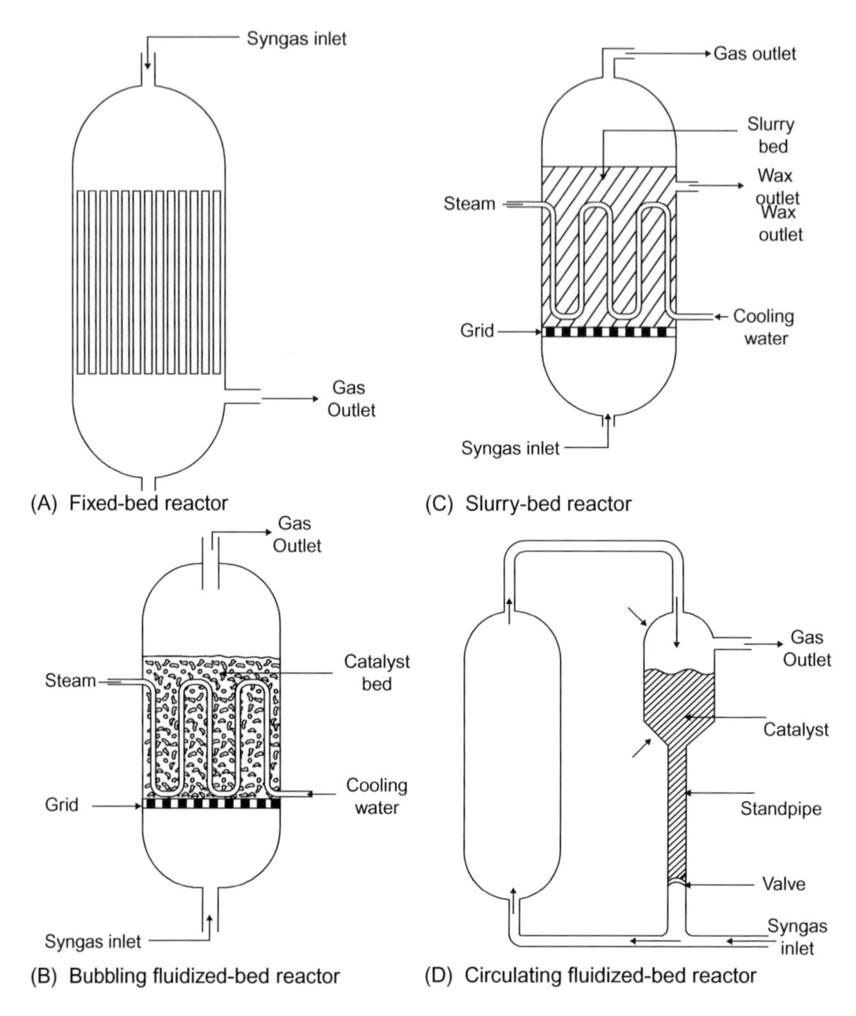


**FIGURE 11.6** Four main types of FTS reactors: (A) fixed-bed reactor, (B) bubbling fluidized-bed reactor, (C) slurry-bed reactor, and (D) circulating fluidized-bed reactor.

For all FTS catalysts, an increase in operating temperature results in a shift in selectivity toward lower carbon number products and to more hydrogenated products.

## 11.4.3 Ammonia Synthesis

Ammonia ($NH_3$) is an important chemical used for a large number of applications, including production of fertilizers, disinfectants, nitric acid, and refrigerants. It is produced by passing hydrogen and nitrogen over a bed of catalyst at high pressure but at moderate temperature. The hydrogen for this reaction can come from biomass gasification.

$$N_2 + 3H_2 \xrightarrow{\text{Catalyst}} 2\,NH_3 \tag{11.19}$$

Catalysts play an important role in this reaction. Iron catalysts (FeO, $Fe_2O_3$) with added promoters like oxides of aluminum and calcium, potassium, silicon, and magnesium are used (Reed, 2002, p. III-250).

Syngas contains both CO and $H_2$. So, for production of ammonia, the syngas must first be stripped of its CO through the shift reaction (Eq. (11.2)). As mentioned earlier, the shift conversion is aided by commercial catalysts, such as iron oxide and chromium oxide, that work in a high-temperature range (350–475°C); zinc oxide and copper oxide catalysts work well in a low-temperature range (200–250°C).

In a typical ammonia synthesis process, the syngas is first passed through the shift reactor, where CO is converted into $H_2$ and $CO_2$ following the shift reaction. Then the gas is passed through a $CO_2$ scrubber, where a scrubbing liquid absorbs the $CO_2$; this liquid is passed to a regenerator for regeneration by stripping the $CO_2$ from it. The cleaned gas then goes through a methanation reactor to remove any residual CO or $CO_2$ by converting it into $CH_4$. The pure mixture of hydrogen obtained is mixed with pure nitrogen and is then compressed to the required high pressure of the ammonia synthesis. The product, a blend of ammonia and unconverted gas, is condensed, and the unconverted syngas is recycled to the ammonia converter.

## 11.4.4 Glycerol Synthesis

Biodiesel, that is produced from fat or oil, generates a large amount (about 10%) of glycerol ($HOCH_2CH[OH]CH_2OH$) as a by-product. Large-scale commercial production of biodiesel can therefore bring a huge amount of glycerol into the market. For example, for every kilogram of biodiesel, 0.1 kg of glycerol is produced (i.e., 86% FAME, 9% glycerol, 4% alcohol, and 1% fertilizer) (www.biodiesel.org). If produced in the required purity ($>99\%$), glycerol may be sold for cosmetic and pharmaceutical production, but the market for them is not large enough to absorb it all. Therefore, alternative commercial uses need to be explored. They include:

- Catalytic conversion of glycerol into biogas ($C_8$–$C_{16}$ range) (Hoang et al., 2007).
- Liquid-phase or gas-phase reforming to produce hydrogen (Xu et al., 1996).

A large number of other chemicals may potentially come from glycerol. Zhou et al. (2008) reviewed several approaches for a range of chemicals and fuels. Through processes like oxidation, transesterification, esterification, hydrogenolysis, carboxylation, catalytic dehydration, pyrolysis, and gasification, many value-added chemicals can be produced from glycerol.

## 11.5 TRANSPORT FUELS FROM BIOMASS

Biodiesel, ethanol, and biogas are transport fuels produced from biomass that are used in the transportation industry. The composition of biodiesel and biogas may not be exactly the same as their equivalence from petroleum, but they perform the same task. Ethanol derived from biomass is either used as the sole fuel or mixed with gasoline in spark-ignition engines.

Two thermochemical routes are available for the production of diesel and gasoline from syngas:

1. Gasoline, through the methanol-to-gasoline (MTG) process.
2. Diesel, through the FTS process.

Similarly, there are two biochemical means for the production of ethanol and diesel:

- Diesel, through the transesterification of fatty acids.
- Ethanol, through the fermentation of sugar.

It may be noted that in both schemes, part of the syngas's energy content (30–50%) is lost during conversion into liquid transport fuel. It is apparent from Table 11.4 that this loss in conversion from biomass to methanol or ethanol can be as high as 50%, and further loss can occur when the methanol is converted into a transport fuel like gasoline. For this reason, when we consider the overall energy conversion efficiency of a car, running on biogas, and compare it with that of an electric car, the former shows a rather low fuel-to-wheel energy ratio.

### 11.5.1 Biochemical Ethanol Production

Ethanol is the most extensively used biofuel in the transportation industry. Ethanol can be mixed with gasoline (petroleum) or used alone for operating spark-ignition engines, just as biodiesel can be mixed with petrodiesel for

**TABLE 11.4** Energy Losses in Methanol Production

| Conversion Process | Energy Loss (%) |
|---|---|
| Biomass to methanol | 30–47 |
| Coal to methanol | 41–75 |

**Source:** Data compiled from Reed (2002), p. III-226.

operating compression-ignition engines. In most cases, engine modifications may not be needed for substitution of mineral oil with bio-oil-derived fuels. Ethanol is produced mainly from food crops, but, less commonly, it can also be produced from nonfood lignocellulosic biomass.

#### *11.5.1.1 Ethanol from Food Sources*

Ethanol ($C_2H_6O$) is presently produced primarily from glucose obtained from grains (e.g., corn and maize), sugar (sugarcane), and energy crops using the fermentation-based biochemical process. A typical process based on corn is shown in Figure 11.11. The process may comprise the following major steps:

**a.** Milling: Corn is ground to a fine powder called cornmeal.
**b.** Liquefying: A large amount of water is added to make the cornmeal into a solution.
**c.** Hydrolysis: Enzymes are added to the solution to break large carbohydrate molecules into shorter glucose molecules.
**d.** Fermentation: The glucose mixture is taken to the fermentation batch reactor, where yeast is added. The yeast converts the glucose into ethanol and carbon dioxide as represented by the equation:

$$C_6H_{12}O_6(\text{glucose}) + \text{yeast} \xrightarrow{\text{Fermentation}} 2C_2H_6O(\text{ethanol}) + 2CO_2 \qquad (11.20)$$

**e.** Distillation: The product of fermentation contains a large amount of water and some solids, so the water is removed through distillation. Distillation purifies ethanol to about 95–96%. The solids are pumped out and discarded as a protein-rich stock, which may be used only for animal feed.
**f.** Dehydration: The ethanol produced is good enough for car engines. So, countries like Brazil use it directly, but in some countries, further purification is needed if it has to be blended with mineral gasoline for ordinary cars. In this stage, a molecular sieve is used for dehydration. Small beads with pores large enough for water but not for ethanol absorb the water.

A large amount of energy is consumed in the distillation, dehydration, and other steps in this process. By one estimate, for the production of 1 liter of purified ethanol, about 12,350 kJ of energy is needed for processing, especially for dehydration. An additional 7440 kJ/l of energy consumed in harvesting the corn is required (Wang and Pantini, 2000). Although a liter of ethanol releases 21,200 kJ of energy when burnt, the farming and processing of the corn consume about 19,790 kJ of energy. The net energy production is therefore a meager 1410 kJ (21,200−19,790) per liter of ethanol.

Major criticism of this process is that it uses a valuable food source—indeed, a staple food in many countries. The search for an alternative is therefore ongoing.

### 11.5.1.2 Ethanol from Nonfood Sources

The conventional means of producing ethanol from food sources like corn and sugarcane is commercially highly successful. In contrast, the production of ethanol from nonfood biomass (lignocellulose), although feasible in principle, is not widely used. More processing is required to make the sugar monomers in lignocellulose feedstock available to the microorganisms that produce ethanol by fermentation. However, production from food sources, even though it strains the food supply and is wasteful, is widespread.

Consider that only 50% of the dry kernel mass is transformed into ethanol, while the remaining kernel and the entire stock of the corn plant, regardless that it is grown using cultivation energy and incurs expenses, remain unutilized. It is difficult to ferment this part, which contains lignocellulose mass, so it is discarded as waste. Alternative methods are being developed to convert the cellulosic components of biomass into ethanol so that they can also be utilized for transport fuel. This option is discussed further in Section 11.5.4.

## 11.5.2 Gasoline

Petrogasoline is a mixture of hydrocarbons having a carbon number (i.e., the number of carbon per hydrocarbon molecules) primarily in the range of 5–11. These hydrocarbons belong to the following groups:

- Paraffins or alkanes
- Aromatics
- Olefins or alkenes
- Cycloalkanes or naphthenes

### 11.5.2.1 Gasoline Production from Methanol

Methanol may be converted into gasoline using several processes. One of these, Exxon Mobil's MTG process, is well known (Figure 11.7). Methanol is converted into hydrocarbons consisting of mainly ($>75\%$) gasoline-grade materials ($C_5-C_{12}$) with a small amount of liquefied petroleum gas ($C_3-C_4$) and fuel gas ($C_1-C_2$). Mobil uses both fixed beds and fluidized beds of proprietary catalysts for this conversion. The reaction is carried out in two stages: the first stage is dehydration to produce dimethyl ether intermediate;

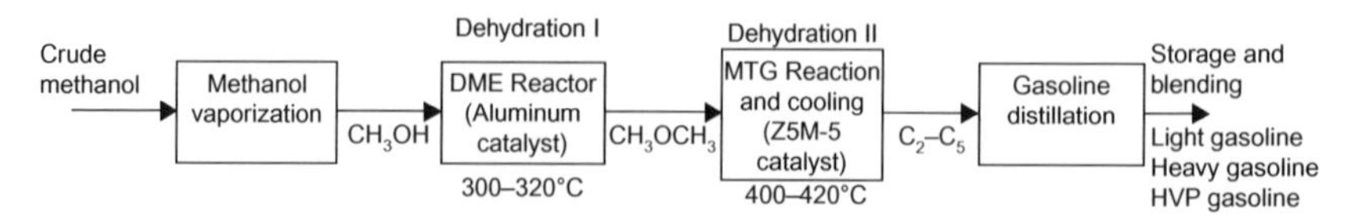


**FIGURE 11.7** Production of gasoline from methanol using MTG process.

the second stage is also dehydration, this time over a zeolite catalyst, ZSM-5, to give gasoline.

$$2CH_3OH \rightarrow (-H_2O) \rightarrow CH_3OCH_3 \rightarrow (-H_2O) \rightarrow C_2 - C_5 \rightarrow \text{paraffins} + \text{aromatics} + \text{cycloparaffins}$$

$$\begin{array}{ll} 300-320^{\circ}C & 400-420^{\circ}C \\ \text{alumina catalyst} & \text{ZSM catalyst} \end{array} \tag{11.21}$$

where $(-H_2O)$ represents the dehydration step.

The typical composition of the gasoline in weight percentage (see nzic.org.nz/ChemProcesses/energy/7D.pdf) is as follows:

- Highly branched alkanes: 53%
- Highly branched alkenes: 12%
- Napthenes: 7%
- Aromatics: 28%.

The dehydration process produces a large amount of water. For example, from 1000 kg of methanol, 387 kg of gasoline, 46 kg of liquefied petroleum gas, 7 kg of fuel gas, and 560 kg of water are produced (Adrian et al., 2007). Figure 11.7 shows a simplified scheme for the production of gasoline from methanol. This gasoline, sometimes referred to as MTG, is completely compatible with petrogasoline.

### 11.5.3 Diesel

Generally, the oil burnt in a diesel (compression-ignition) engine is called diesel. If produced from petroleum, it is called petrodiesel, and if produced from biomass, it is called biodiesel. Mineral diesel (or petrodiesel) is made up of a large number of saturated and aromatic hydrocarbons. The average chemical formula can be $C_{12}H_{23}$. Petrodiesel (also called fossil diesel) is produced from the fractional distillation of crude oil between 200°C and 350°C at atmospheric pressure, resulting in a mixture of carbon chains that typically contain between 8 and 21 carbon atoms per molecule (Collins, 2007).

According to the American Society for Testing and Materials (ASTM), biodiesel (B100) is defined as *a fuel comprised of mono-alkyl (methyl) esters of long chain fatty acids derived from vegetable oils or animal fats, and meeting the requirements of ASTM D 6751*. Biodiesel's characteristics are similar to those of petrodiesel but not identical. Biodiesel, which can be mixed with petrodiesel for burning in diesel engines, has several positive features for use in engines, and are as follows:

- Petrodiesel contains up to 20% polyaromatic hydrocarbon, while biodiesel contains none, making it safer for storage.
- Biodiesel has a higher flash point, making it safer to handle.

- Being oxygenated, biodiesel is a better lubricant than petrodiesel, and therefore gives longer engine life.
- Its higher oxygen content allows biodiesel to burn more completely.

#### 11.5.3.1 Biodiesel Production

Biodiesel is generally produced from vegetable oil and/or from animal fats with major constituents that are triglycerides. It is produced by transesterification of vegetable oil or fat in the presence of a catalyst. Biodiesel carries the name *fatty acid methyl (or ethyl) ester*, commonly abbreviated as "FAME." A popular production method involves mixing waste vegetable oil or fat with the catalyst and methanol (or ethanol) in appropriate proportion. A typical proportion is 87% oil, 1% NaOH catalyst, and 12% alcohol. Both acid and base catalysts can be used, but the base catalyst NaOH is most commonly used. Because NaOH is not recyclable, a "nongreen" feed is required to produce "green" biodiesel. Efforts are being made to produce recyclable catalysts and thereby make the product pure "green."

Figure 11.8 shows the reaction for the conversion of triglyceride into biodiesel FAME and its by-product, glycerol. Glycerol cannot be used as a transport fuel, and its disposal is a major issue.

An alternative noncatalytic conversion route for biodiesel is under development in which transesterification of triglycerides is by supercritical methanol (above 293°C, 8.1 MPa) without a catalyst (Kusdiana et al., 2006). The methanol can be recycled and reused, but the process for this must be carried out at high temperatures and pressures. Efforts are also being made to use woody biomass (lignocellulose) instead of fats or oil to produce biodiesel using the supercritical method (Minami and Saka, 2006). The reaction is carried out in a fixed or fluidized bed. The fluidized bed has the advantage of continuous catalyst regeneration and efficient removal of the heat of reaction.

### 11.5.4 Transport Fuel Production from Nonfood Biomass

Use of food cereals, such as wheat and corn for the production of biodiesel or ethanol, has been commercially successful; however, it has had a major impact on the world's food market, driving up prices and creating shortages.

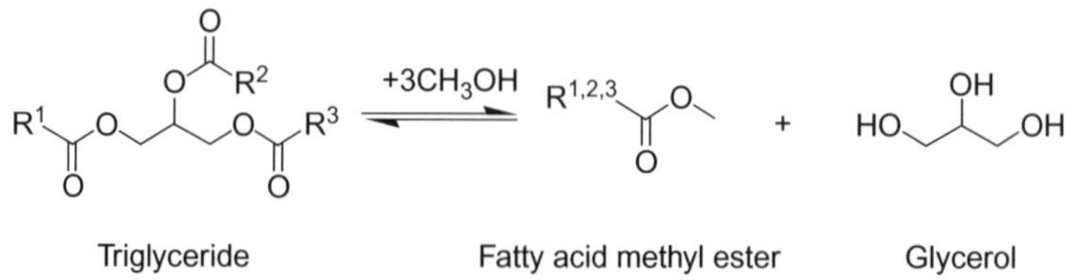

**FIGURE 11.8** Diesel (FAME) production from triglyceride.

Alternative sources of biodiesel are being researched. Instead of sugar beets or rapeseed, cellulosic biomass like wood may be used as the feedstock. With cellulosic materials, the industry can significantly increase the yield of fuel per unit of cultivated area.

There are two options for production of ethanol or gasoline from nonfood sources: thermal and biochemical.

### 11.5.4.1 Thermal Process

In the thermal process, cellulosic feedstock is subjected to fast pyrolysis (see Chapter 5). The liquid produced is refined and upgraded to gasoline or ethanol. Since cellulose is the feedstock, the ethanol from it is often referred to as cellulosic ethanol. An alternative thermal process involves gasification of the biomass to produce syngas and synthesis of the syngas into diesel oil using the FTS process. This process is described in Section 11.4.2 and is illustrated in Figure 11.9.

### 11.5.4.2 Biochemical Process

Figure 11.10 illustrates the biochemical process for production of ethanol from nonfood lignocellulosic biomass. To produce alcohol, the long-chain sugar molecules in the cellulose must be broken down into free sugar molecules. Only then can the sugar be fermented into alcohol (ethanol), as in the food-based process (refer to Figure 11.11). This extra step of breaking down

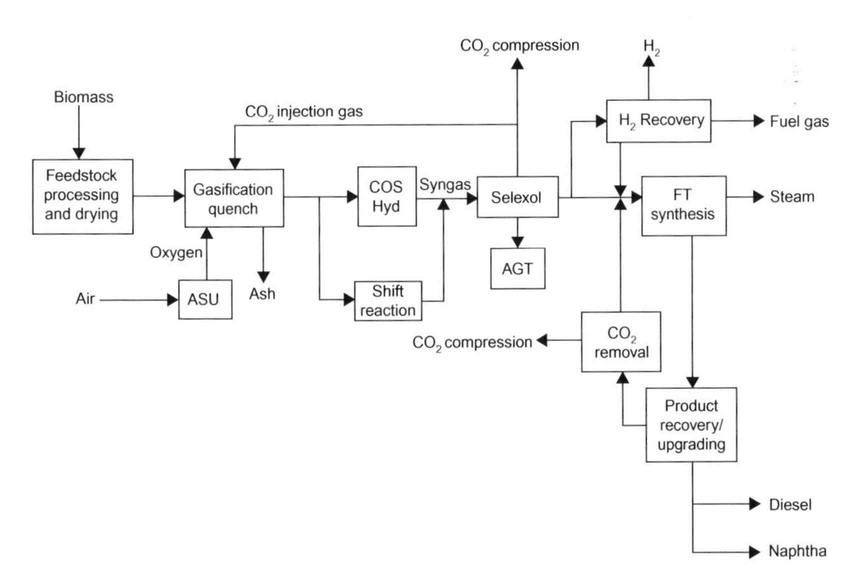


**FIGURE 11.9** Transport fuel production from coal and biomass using FTS Source: *Data from White et al. (2007).*

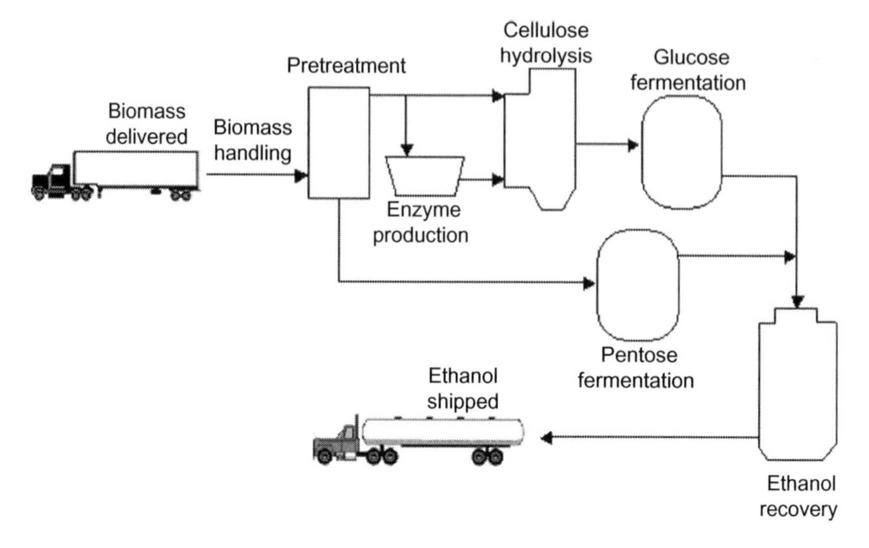


**FIGURE 11.10** Biochemical process for production of ethanol from lignocellulosic or nonfood feedstock.

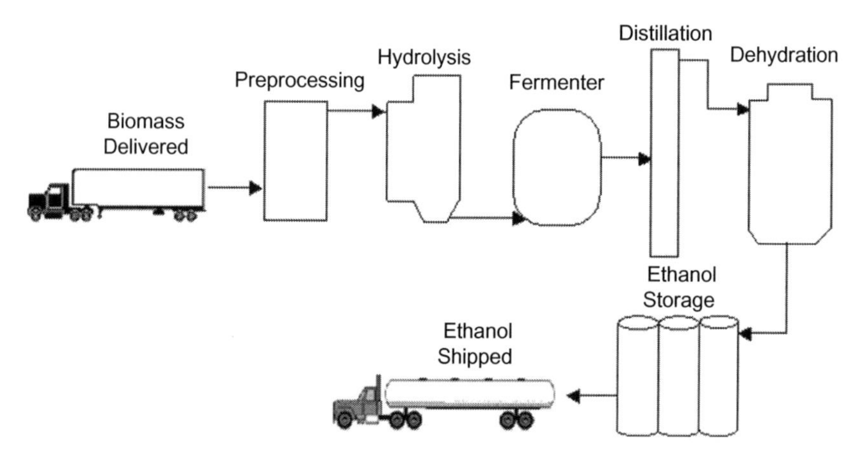


**FIGURE 11.11** Ethanol production from food cereal.

to free sugar molecules is not necessary in the latter process where the feedstock (e.g., corn and sugarcane) is already in sugar form.

The breakdown of cellulose into sugar can be carried out by either acid hydrolysis or enzymatic hydrolysis. The production of cellulosic ethanol typically takes five steps:

1. Feed preparation (i.e., mechanical cleaning and sizing, physicochemical preparation).

2. Hydrolysis (conversion into sugar).
3. Fermentation (conversion of the sugar into ethanol).
4. Distillation (removal of water and solids).
5. Dehydration (final drying).

The second step is different for the two biochemical processes. All other steps are the same.

### 11.5.4.3 Feed Preparation

This step prepares the biomass for processing. It involves cleaning and then pretreatment. Unlike food grain (e.g., corn and wheat), lignocellulose feedstock often comes mixed with dirt and debris. These must be cleaned from the delivered biomass, which is then shredded into small particles. In pretreatment, the hemicellulose/lignin sheath that surrounds the cellulose in plant material is disrupted. To make the cellulose more accessible to the hydrolysis process, one could adopt physical, chemical, or biological pretreatment as described below.

- Physical methods: grinding, milling, shearing (energy intensive), and steam explosion (to produce some inhibitory compounds).
- Chemical methods: treatment with acid (for pH neutralization and recovery of chemicals), treatment with alkalis (for pH adjustment and recycling of chemicals), and treatment with organic solvents (solvent removal and recycling is expensive).
- Biological method: enzymatic treatment of the cellulose (time consuming).

### 11.5.4.4 Hydrolysis

Acid hydrolysis uses dilute acid at high temperature and pressure. Concentrated acid at lower temperature and pressure may be used, but this produces a toxic by-product that inhibits fermentation and so must be removed.

In enzymatic hydrolysis, cellulose chains are broken into glucose molecules by cellulose enzymes, in a process similar to what occurs in the stomach of a cow to convert grass or fodder cellulose into sugar. Xylanose and hemicellulose enzymes can convert many cellulosic agricultural residues into fermentable sugars. These residues include corn stover, distiller grains, wheat straw, and sugarcane bagasse, as well as energy crops such as switchgrass. Lignin is difficult to convert into sugar, so it is discarded as waste. Figure 11.10 shows a process based on cellulose hydrolysis.

### 11.5.4.5 Fermentation of Hemicellulosic Sugars

Through a series of biochemical reactions, bacteria convert xylose and other hemicellulose and cellulose sugars into ethanol. The yeast or other microorganisms for the fermentation of cellulose and that for hemicellulose are not necessarily the same. In any case, they consume sugar molecules and produce ethanol and carbon dioxide.

#### *11.5.4.6 Distillation and Dehydration*

Dilute ethanol broth produced during the fermentation of hemicellulose and cellulose sugars is distilled to remove water and concentrate the ethanol. Solid residues containing lignin and microbial cells can be burned to produce heat or used to generate electricity consumed by ethanol production. Alternately, the solids can be converted to coproducts, such as animal feed and nutrients for crops. The last step in the process involves removal of the remaining water from the distilled ethanol.

Chapter 12

# Biomass Handling

## 12.1 INTRODUCTION

Handling of solid biomass poses some special challenges that are not present for liquid or gaseous fuels. Liquids and gases are relatively easy to handle, because they are fluid, which continuously deforms under a shear stress. Fluid easily takes the shape of any vessel they are kept in and flow easily under gravity, if they are heavier than air. For these reasons, storage, handling, and feeding of gases or liquids do not generally pose a major problem. However, solids can support shear stress without continuously deforming and, it thus does not flow freely. This problem is most evident when they are stored in a conical bin, and are withdrawn from its bottom. Because they do not deform under shear stress, solids can form a bridge over the cone and cease to flow.

Biomass is particularly notorious in this respect, because of its fibrous nature and nonspherical shape. The exceptionally poor flow characteristic of biomass poses a formidable challenge to both designers and operators of biomass plants. The cause of many shutdowns in these plants incidents can be traced to the failure of some parts of the biomass-handling system.

This chapter describes the design and operating issues involved in the flow of biomass through the system. It discusses options for the handling and feeding of biomass in a biomass conversion plant that include gasification, pyrolysis, and torrefaction plants.

## 12.2 DESIGN OF A BIOMASS ENERGY SYSTEM

A typical biomass energy system comprises farming, collection, transportation, preparation, storage, feeding, and conversion. This is followed by transmission of the energy produced to the point of use. The concern here is with the handling of biomass upstream of a conversion system. The production of biomass through biomass farming is a subject by itself and is beyond the scope of this chapter.

Biomass has two major (Figure 12.1) applications: (1) energy production and (2) production of chemicals and fiber-based items (e.g., paper). In either

**Biomass Gasification, Pyrolysis and Torrefaction.**

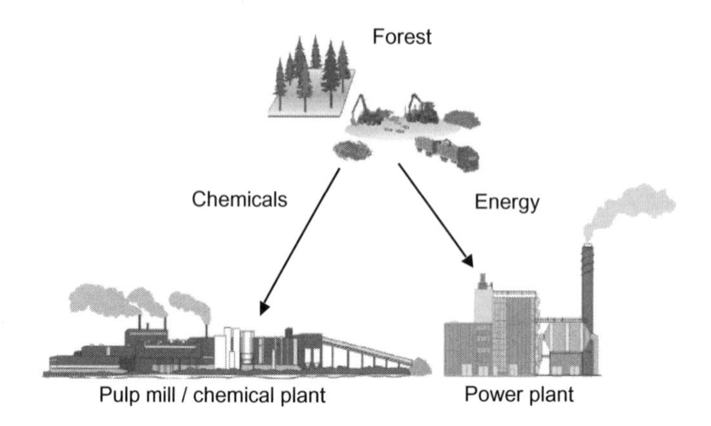


**FIGURE 12.1** Biomass is used for energy production or other commercial products like paper or chemicals.

case, biomass has to be procured. It can be procured from the following sources:

1. Energy crop (farming) or forestry
2. Lignocellulosic wastes that are from forestry, agriculture, wood, or other industries
3. Carbohydrates such as fat, oil, and other wastes.

The collection methods for biomass vary depending on its type and source. Forest residues are a typical lignocellulosic biomass used in biomass conversion plants. They are collected by various pieces of equipment and transported to the conversion plant by special trucks (or rail cars in some cases). There, the biomass is received, temporarily stored, and pretreated as needed. Sometimes the plant owner purchases prepared biomass to avoid the cost of on-site pretreatment. The treated biomass is placed in storage bins, located in line with the feeder, which feeds it into the gasifier at the required rate.

Biomass typically contains only a small amount of ash, but it is often mixed with undesirable foreign materials. These materials require an elaborate system for separation. If the plant uses oxygen for gasification, it needs an air separation unit for oxygen production. If it uses steam, a steam generator is necessary. Thus, a biomass plant could involve several auxiliary units. The capacity of each of these units and the selection of equipment depend on a large number of factors. These are beyond the scope of this chapter.

Forestry and agriculture are two major sources of biomass. In forestry, large trees are cut, logged, and transported to the market. The logging process involves delimbing, and taking out the large-diameter tree trunks as logs. The processes involved in biomass harvesting, such as delimbing, deburking, and chipping, produce a large amount of woody residue, all of which constitutes a major part of the forest residue. The entire operation involves chopping the tree into chips and then using those chips to make fuels or feedstock for pulp industries.

## 12.3 BIOMASS-HANDLING SYSTEM

A typical biomass conversion plant comprises a large number of process units, of which the biomass-handling unit is the most important. Unlike coal-fired boiler plants, an ash-handling system is not a major component of a biomass plant because biomass contains a relatively small amount of ash. Also a biomass plant does not produce a large volume of spent catalysts or sorbents like in a coal-based plant. The transportation and handling of biomass are the main focus of a biomass plant; as transportation, feed preparation, and feeding are more important for biomass than they are for coal- or oil-gas-fired units. There is, however, a major challenge in the design of a biomass conversion plant due to the large variability of bulk density of biomass. It varies from species to species. For example, loosely piled straw has bulk density of 32–48 kg/$m^3$, loose bagasse has 112–160 kg/$m^3$, while that of sawdust is 256–577 kg/$m^3$ (Susawa, 1989). Delivered form (chopped vs. loose, pelletized vs. dust) could also affect the bulk density.

The biomass-handling system could typically comprise of the following units or components:

- Receiving unit
- Storage and screening unit
- Feed preparation unit
- Conveying system
- Feeding system

While the biomass conversion unit may vary widely depending on the system used, the design of the biomass-handling system is very similar, and is equal to that of a biomass-fired steam plant. Figure 12.2 illustrates the layout of a typical plant showing receiving, screening, storage, and conveying.

Major considerations for the design of feeding and handling systems are transportation, sealing, and injection of the feed into the reactor. The feed should be transported smoothly from the temporary storage to the feed system, which must be sealed against the conversion unit's pressure and temperature, and is then injected into the reactor. Metering or measurement of the fuel feed rate is an important aspect of the feed system, as it is a key parameter in the control of the entire process.

The following subsections discuss the individual components of a solids-handling system for biomass. They assume the biomass to be solid, although some biomass, such as sewage sludge, is in slurry or semisolid form.

### 12.3.1 Receiving

Biomass is brought to the plant typically by truck, barge, or by rail car. For large biomass plants, unloading from the truck or rail car is a major task. Manual unloading can be strenuous and uneconomical in terms of manpower

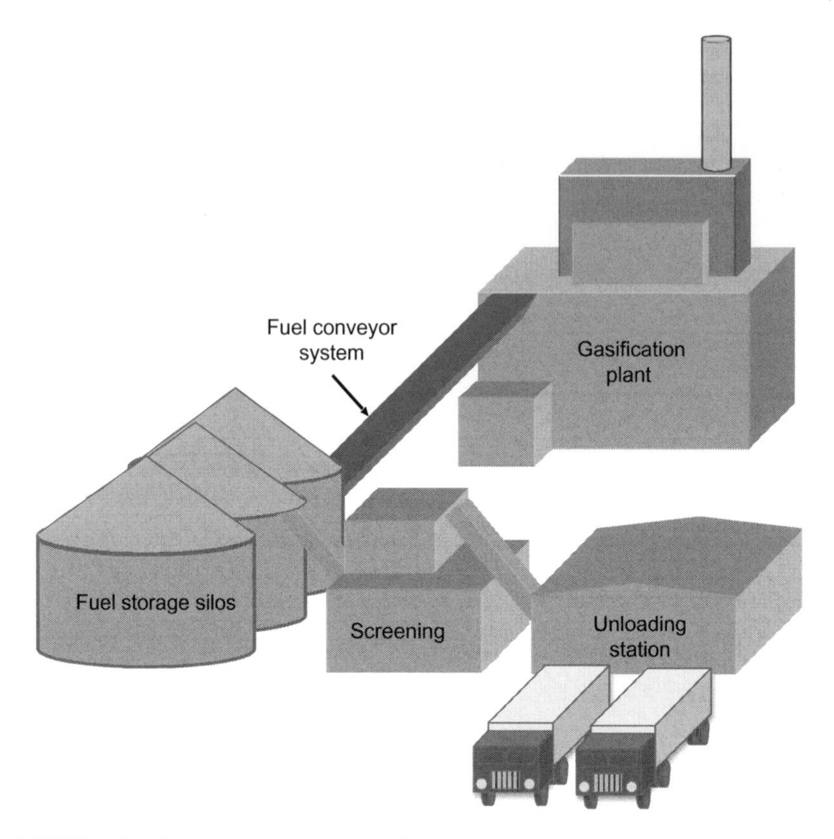


**FIGURE 12.2** Plant layout for biomass gasification. Fuel received by trucks are cleaned of foreign materials before it is stored in silos.

except for small plants. A front loader is used for manual unloading. This is why large plants use truck hoisters, wagon tipplers, or bottom-discharge wagons. Figure 12.3 shows a typical system where a truck hoister unloads the biomass. The truck drives onto the hoist platform or unloading bay where it is clamped down. The hoister tilts to a sharp angle, allowing the entire load to drop into the receiving chute under gravity. This method is fast and economical.

A bottom-discharge wagon may be used for rail cars. The wagon drops its load into a large bin located below the rail. An alternative is a standard open-top wagon and a tippler to rotate it 180° to empty its contents into a bin underneath. Such units are procured from the suppliers of various bulk material-handling equipment. Their capacities depend on a number of factors, including plant throughput and frequency of truck and/or rail arrival.

Climate and local regulations could impact the design of the receiving system. For example, some places allow large trucks to enter during

**FIGURE 12.3** Biomass carrying truck is tilted to unload biomass in the plant. Source: *Photograph by the author.*

weekdays and within specified hours. So the receiving system needs to have spare capacity to receive feed for the full week within few hours. A plant designed for a cold country like Canada may not be the same as that in a hot country like India.

## 12.3.2 Storage

The primary purpose of storage is to retain the biomass in a good condition and in a position convenient for easy transfer to the next stage of operation, such as drying or feeding into the biomass conversion (i.e., combustor, gasifier, torrefier, or pyrolyzer) unit. For this reason, the stored biomass should be protected from rain, snow, and infiltration of groundwater.

Once unloaded, belt conveyers move the biomass to the storage yard, where it is stored in piles according to usage patterns. If the biomass is from several sources and is to be mixed before use, the piles are arranged in such a way that they can be mixed conveniently into the desired proportions and withdrawn on first in first out basis. Because of the large volume of biomass, indoor storage may not be always economical. Open-air storage is most common, though it can cause absorption of additional moisture from rain or snow and produce dust pollution. Storage can be of two types: above ground, for large-volumc biomass, or enclosed in a silo or bunker.

Figure 12.2 shows the general arrangement of the solids-handling system in a typical biomass conversion plant. A truck-receiving station unloads into an underground hopper (Figure 12.3) from which a belt conveyor takes the biomass to a screening or scalping station. After removal of undesired foreign materials, the biomass is crushed and screened to the desired size range and then transported into silos for covered storage. From there, it is reclaimed and taken to the plant as required. Figure 12.4 shows a photograph of receiving, size-screening, and aboveground outdoor storage.

**FIGURE 12.4** Biomass is conveyed to the storage pile, from where the scraper collects it when needed and transfers it to conveyors to the fuel preparation plant. Source: *Photograph by the author.*

Underground bunker storage is very convenient from a fuel delivery and dust pollution standpoint. Also it protects the biomass from rain and snow. However, because it needs good ventilation and drainage for safety and environmental protection, its capital cost is higher than that of aboveground storage. The hygroscopic nature of biomass is a major issue, as it causes the prepared biomass to absorb moisture even if stored indoors. Moreover, long-term storage can cause physical and chemical changes in the biomass that might adversely affect its flow and product qualities. For these reasons, it is desirable to occasionally turn the biomass. A simple and practical way of doing this is to draw it at a rate higher than that required and return the excess to the top of the pile.

An important issue especially with a tall storage silo is that static pressure could increase from the top to the bottom of the silo. For some loosely packed biomass like silage, the increased static pressure could increase the bulk density of the feed. For example, Otis and Tomroy (1957) noticed that the bulk density of Alfalfa silage increased from 270 to 825 kg/m$^3$ as the depth in the silo increased from 0.6 to 11 m.

Moving or retrieving the biomass from the storage piles to the gasifier plant requires careful design, because interruptions or delays can have a major effect on the operation of the plant. Generally, it is desirable to withdraw biomass from the bottom of the pile such that the first in first out principle is followed to allow a relatively uniform shelf life.

The properties of the biomass determine the ease with which it is retrieved or handled. Oversized materials, frozen chunks (in cold countries), and compaction can lead to poor or interrupted fuel flow. If the fuel bin is not filled uniformly, erratic operation can result, creating problems for hydraulic scrapers and bridging over the unloaders. Sticks, wires, and gloves in the feed, for example, can jam augers. Mobile loaders normally achieve uniformity in aboveground storage buildings or in live-bottom unloaders and augers in bins and silos. For large plants, a scraper connected to a conveyor, as shown in Figure 12.4, is more efficient for reclamation.

The following are some common methods for retrieval of biomass from storage:

- Simple gravity feed or chute
- Screw-type auger feed
- Conveyor belt
- Pneumatic blower
- Pumped flow
- Bucket conveyor
- Front loader
- Bucket grab

Walking beams are sometimes used on the floors of large bunkers or storage buildings, to facilitate the movement of biomass to the discharge end of the storage.

#### 12.3.2.1 Outdoor Aboveground Storage

In large-scale plants, aboveground outdoor storage is the only option (Figure 12.4). Indoor storage is usually too expensive. Biomass needs to be piled in patterns that allow maximum flexibility in retrieval as well as in delivery. Furthermore, it is necessary to ensure the first in first out principle. In some cases, an emergency or strategic reserve is kept separate from the regular flow of biomass. This is a special consideration for long-term storage.

Good ventilation is important in storage design. Biomass absorbs moisture. Ventilation prevents condensation of moisture and the formation of mold (a fungal species) that can pose serious health hazards. It also prevents composting (formation of methane), which not only reduces the energy content of the biomass, but also run the risk of fire. Because tall storage piles are difficult to ventilate, the maximum height of a wood chip storage pile should not exceed 8–10 m (Biomass Energy Centre, 2009). For an indoor facility, water or moisture accumulation may occur inadvertently. Unless moved periodically, the biomass may form fungi and cause a health hazard. Drainage is an important issue, especially for outdoor storage.

#### 12.3.2.2 Silos and Bins for Storage of Biomass

Improper storage not only makes retrieval difficult, but it also can adversely affect the quality of the biomass. Retrieval or reclamation from storage is equally important, if not more so. It represents one of the most trouble-prone areas of biomass plant operation. The handling system and its individual components must be designed to ensure uninterrupted flow to the conversion unit at a measured rate.

Bunkers, silos, and bins provide temporary storage in a protective environment. Bunkers are a type of large-scale storage. Although the term *bunker* is generally associated with underground shelter, here it refers to the indoor storage of fuel in power or process plants that is not necessarily underground. Silos could be fairly large in diameter (4–10 m), and are very tall, which is good for storing grain-type biomass. For example, Figure 12.5 shows a tower silo for cattle feed. Bins are for smaller capacity temporary storage.

### *12.3.2.3 Hopper Design*

Hoppers or chutes facilitate withdrawal of biomass or other solids from temporary storage such as a silo. Major issues in their design include: (1) mode of solids flow, (2) slope angle of discharge, and (3) size of discharge end.

There are several modes of solids flow in a hopper: funnel flow, mass flow, etc. *Funnel flow* would have an annular zone of stationary solids and a moving core of solids at the center. In this case, the solids flow primarily through the core of the hopper. Solids in the periphery either remain stationary (Figure 12.6C, left) or move very slowly (Figure 12.6C, right). Fine particles tend to move through the core while coarser particles stay preferentially in the annulus. The particles from the top surface can flow into the funnel, thus violating the design norm of "first in first out." When that does not happen, a stationary annulus is formed and the discharge stops, causing a rat hole to form through the hopper that becomes void and stops the flow. Remaining solids in the hopper stay in the annulus (Figure 12.6A, right), which prevents the hopper from emptying completely. The only positive thing about a funnel-flow hopper is that it requires a lower height.

**FIGURE 12.5** Typical grain silo used for storing cattle feed. Source: *Photograph by the author.*

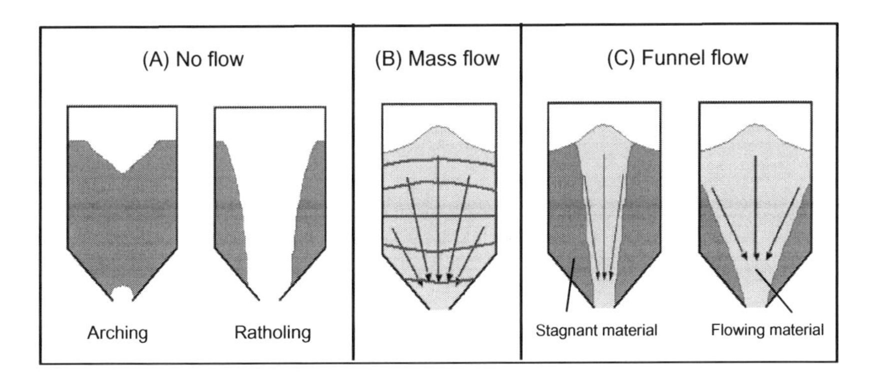


**FIGURE 12.6** Schematic representation of three types of solids flow through a hopper.

*Mass flow* (Figure 12.6B) is the preferred flow mode because here the solids flow across the entire hopper cross section. Though there may be some difference in velocity, this allows an uninterrupted and consistent flow with very little radial size segregation, which permits the hopper to effectively follow the first in first out norm. However, because of the solids' plug-flow behavior, there can be more wear on the hopper walls with abrasive solids. Therefore, the required height of a mass-flow hopper must be greater than that of a funnel-flow hopper. A steeper cone angle of a hopper improves the probability of a "mass flow" mode of solids flow through it. Some common operating problems with hoppers are:

- Ratholing
- Funnel flow
- Arching
- Flushing
- Insufficient flow and incomplete emptying
- Caking

Two of the most common problems experienced in an improperly designed silo or bin (hereafter referred to as *silo*) are no-flow and erratic flow. No-flow from a silo can be due to either arching or ratholing (Figure 12.6A).

Ratholing (Figure 12.6A, right) most often happens in the flow of biomass with particles that are cohesive, have low shape factor, or rough surface texture. This is a serious problem in hoppers. To facilitate solids flow, the rat hole must be collapsed by proper aeration in the hopper or by vibrations on the hopper wall.

Arching occurs when rough or cohesive particles form an obstruction over the exit (Figure 12.6A, left), usually in the shape of an arch or a bridge above the hopper outlet that prevents further discharge. The arch can be interlocking, with the particles mechanically locking to form the obstruction, or it can be held simply by particle cohesion. Coarse particles can also form an arch while competing for an exit, as a traffic jam results from a large number of automobiles trying to pass through a narrow road in an unregulated manner. By making the outlet size at least 8–12 times the size of the largest particle, this type of arching can be avoided (Jacob, 2000).

Flushing results in uncontrolled flow of fine solids like Geldart's group A or group C particles (Basu, 2006, p. 443) through the exit hole. It is uncommon for relatively coarse biomass, but it can happen if the hopper is improperly aerated in an attempt to collapse a rat hole.

Another problem influenced by hopper design is inadequate emptying. This can happen if the base of the hopper is improperly sloped, causing some solids to remain on the floor that cannot flow by themselves.

Erratic flow from an inappropriately designed hopper often results from alternating between an arch and a rat hole. Interestingly, a rat hole could collapse because of an external force of a flow-aid device such as an air cannon or vibrator, or even vibrations created by a passing train or a plant pulverizer (mill). Some biomass discharges as the rat hole collapses, but the falling material can compact over the outlet and form an arch. The arch may break because of a similar external force, and the material flow will resume until the flow channel is emptied and a rat hole is once again formed (Hossfeld and Barnum, 2007).

Material discharge problems can also occur if the biomass stays in the bunker for a very long time, forming cakes because of humidity, pressure, and temperature. This easily results in arching or rat holes. To avoid this, renewal of solids in the hopper is necessary.

Besides solids flow there are some special problems in fuel-handling systems. For example, spontaneous ignition of biomass can occur if fine biomass particles stay stagnant in a bunker for too long. Even in an operating silo, a stagnant region can be a problem for fuels like biomass or coal, which are prone to spontaneous combustion. Dust explosion that could occur in fine dust in the silo is another problem.

If the fuel flows through a channel in the silo, the fuel outside the channel remains stagnant for a long time. The residence time of such fuels in the silo should be reduced by emptying the silo frequently or by using a first in first out mass-flow pattern (Figure 12.6B), where all of the material is in motion whenever the fuel is discharged.

### *12.3.2.4 Achieving Mass Flow*

To achieve mass flow, the following conditions are to be met:

- The hopper wall must be sufficiently smooth for mass flow.

- The hopper angle should be adequately steep to force solids to flow at the walls.
- The hopper outlet must be large enough to prevent arching.
- The hopper outlet must be adequately large to achieve the maximum discharge rate.

The required smoothness and sloping angle for mass flow in a hopper depends on the friction between the particles and the hopper surface. This friction can be measured in a laboratory using a standard test (ASTM, 2000). Factors that affect wall friction for a given fuel are:

- Wall material
- Surface texture or roughness of the wall
- Moisture content and variations in solids composition and particle size
- Length of time solids remain unmoved
- Corrosion of wall material due to reaction with solids
- Scratching of wall material caused by abrasive materials.

To enhance the smoothness of the surface, sometimes the hopper is coated or a smooth lining is applied to it. Lining materials that can be used include polyurethane sheets, TIVAR-88, ultra high molecular weight polyethylene plastic, and krypton polyurethane.

Mass flow can be adversely affected by the narrowness of the hopper outlet. A too-narrow outlet (compared to particle size) permits the particles to interlock when exiting and form an arch over the outlet. The probability of this happening increases when:

- The particles are large compared to the outlet width.
- There is high moisture in the solids.
- The particles are of a low shape factor and have a rough surface texture.
- The particles are cohesive.

Wedge-shaped hoppers require a smaller width than conical hoppers do, in order to prevent bridging. Slotted outlets must be at least three times as long as they are wide.

Negative angle of outlet walls could also help avoid solids flow problem through a hopper. The edges of the hopper could be flared (positive angle) on one side while it is tapered (negative angle) on other side.

#### 12.3.2.5 Hopper Design for Mass Flow

The design of the hopper outlet significantly affects the flow of solids. When solids flow through the hopper, air or gas enters, dilating the particles. It is essential for powder solids to flow freely through the outlet. Air drag, which is proportional to surface area, must be balanced by gravitational force that is equal to the weight of the particle. Fine particles have a lower ratio of weight to surface area compared to coarser particles. So, for fine particles, this force balance becomes an important consideration. For such particles, the following expression is used (Carleton, 1972):

$$\frac{4V_0^2\sin\theta}{B} + 15\frac{(\rho_a \mu^2 V_0^4)^{1/3}}{\rho_p d_p^{5/3}} = g \quad \text{for} \quad d_p < 500\ \mu\text{m} \tag{12.1}$$

where

$V_0$ is the average solid velocity through the outlet, m/s
$\rho_a$ and $\rho_p$ are the densities of the air and solids, respectively, kg/m$^3$
$d_p$ is the particle size, m
$\mu$ is the viscosity of the air, kg/m s
$\theta$ is the semi-included angle of the hopper
$g$ is the acceleration due to gravity, 9.81 m/s$^2$
$B$ is the parameter

The mass-flow rate, $m$, is given in terms of the bulk solid density, $\rho_b$, and the outlet area, $A$:

$$m = \rho_b A V_0 \tag{12.2}$$

For coarse particles (>500 μm), an alternative relation is used (Johanson, 1965):

$$m = \rho_b A \sqrt{\frac{Bg}{2(1+C)\tan\theta}}\ \text{kg/s} \quad \text{for} \quad d_p > 500\ \mu\text{m} \tag{12.3}$$

Values of the parameters $A$, $B$, and $C$ are given as:

| Parameter | Conical Outlet | Symmetric Slot |
|---|---|---|
| $B$ | Outlet diameter, $D$ | Slot width, $W$ |
| $A$ | $(\pi/4)D^2$ | Width × breadth |
| $C$ | 1.0 | 0 |

### *12.3.2.6 Design Steps*

Hopper design involves determining particle properties, such as interparticle friction, particle-to-wall friction, and particle compressibility or permeability. With these properties known, the outlet size, hopper angle, and discharge rate are found.

Dedicated experiments like shear tests are carried out to determine the interparticle friction. Particle–wall friction should also be measured by purpose-designed experiments. Parameters, such as angle of repose, have little value in hopper design, as it simply gives the heap angle when solids are poured in.

The stress distribution on the silo wall is important, especially for a tall unit. Figure 12.7 compares the wall pressure in a biomass-filled silo with that of a liquid-filled silo. As we can see, the wall pressure in a solid-filled silo does not vary linearly with height, but it does in a liquid-filled silo. In the former case, the pressure increases with depth, reaching an asymptotic value that depends on the diameter of the hopper rather than on the height. Because there is no further increase in wall stress with height, large silos are made smaller in diameter but taller.

To find the stress in the barrel, or the vertical wall section, of a hopper, we consider the equilibrium of forces on a differential element, d$h$, in a straight-sided silo (Figure 12.8):

- Vertical force due to pressure acting from above: $P_v A$
- Weight of material in element: $\rho A g \, dh$
- Vertical force due to pressure acting from below: $(P_v + dP_v)A$
- Solid friction on the wall acting upward: $\tau \pi D \, dh$

The force balance on the elemental solid cross section gives

$$(P_v + dP_v)A + \tau \pi D \, dh = P_v A + \rho A g \, dh \tag{12.4}$$

The wall friction is equal to the particle–wall friction coefficient, $k_f$, times the normal pressure on wall, $P_w$:

$$\tau = k_f P_w \tag{12.5}$$

Janssen (1895) assumed the lateral pressure to be proportional to the vertical pressure, as shown in the following equation:

$$P_w = K P_v \tag{12.6}$$

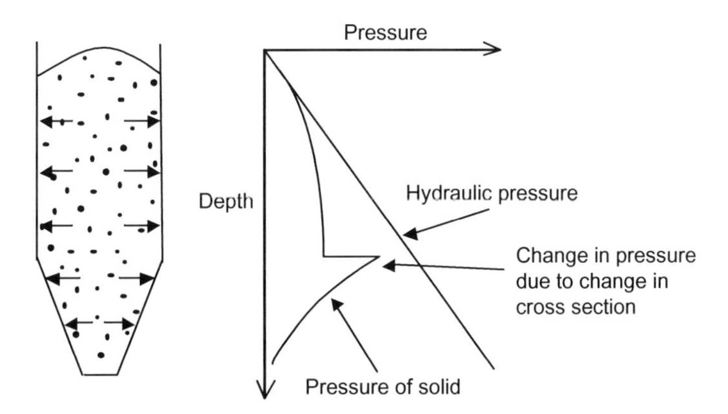


**FIGURE 12.7** Wall pressure distribution in a hopper filled with solids. The pressure profile changes in the inclined section.

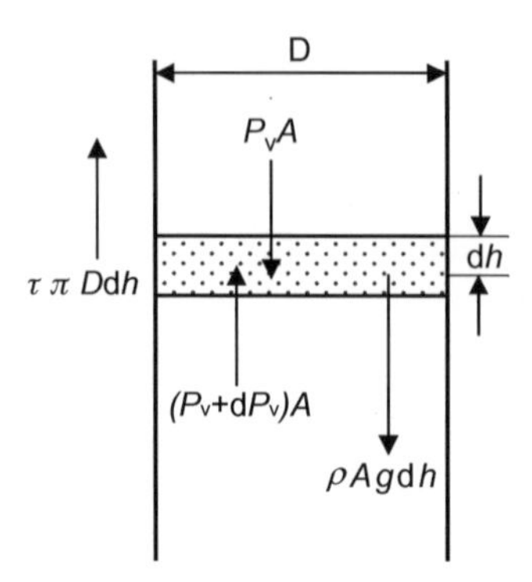


**FIGURE 12.8** Force balance on an element of storage silo.

where $K$ is the Janssen coefficient. For liquids, the pressure is uniform in all directions, so $K$ is 1.0. This relation is not strictly valid for all solids, but for engineering approximations we can start with this assumption.

Substituting Eqs. (12.5) and (12.6) in Eq. (12.4), we get:

$$A\,\mathrm{d}P_v = \rho A g\,\mathrm{d}h - k_f K P_v \pi D\,\mathrm{d}h \tag{12.7}$$

Boundary conditions for this equation are $h=0$, $P_v=0$; $h=H$, $P_v=P_0$. With this, Eq. (12.7) is integrated from $h=0$ to $h=H$ to get the pressure at the base of the silo's vertical section, $P_0$. Substituting

$$A = \frac{\pi D^2}{4}$$

we get

$$P_0 = \frac{\rho D g}{4 k_f K}\left(1 - \exp\left(-\frac{4 H k_f K}{D}\right)\right) \tag{12.8}$$

This is known as the Janssen equation.

Figure 12.7 illustrated the pressure distribution along the height of a silo. The straight line shows the pressure we expect if the stored substance is a liquid; the discontinuous exponential curve is the one predicted for solids. There is a sharp increase in pressure at the beginning of the inclined wall. The pressure decreases with height (Figure 12.7).

The stress on the inclined section is different from that calculated from the preceding. To calculate this, we use the Jenike equation, which states that the radial pressure is proportional to the distance of the element from the hopper apex, which is the point where inclined surfaces would meet if they were extended (Jenike, 1964). It can be seen that the magnitude of stress at the hopper exit is the lowest, although this is the lowest point in the hopper.

**Example 12.1**

Find the wall stress at the bottom of a large silo, 4.0 m in diameter and 20 m in height, that uses a flat bottom for its discharge. Compare the stress when the silo is filled with wood chips (bulk density 300 kg/m$^3$) with that when it is filled with water.

Given that the wall-to-wood chip friction coefficient, $k_f$, is 0.37, assume the Janssen coefficient, $K$, to be 0.4.

**Solution**

We use Eq. (12.8) to calculate the vertical pressure, $P_0$, in the silo. Data given are as follows:

- The bulk density of the wood chips, $\rho_b$ is 300 kg/m$^3$.
- The wall–solid friction coefficient, $k_f$, is 0.37.
- The diameter, $D$, is 4.0 m.
- The height, $H$, is 20 m.
- The Janssen coefficient, $K$, is 0.4.

$$P_0 = \frac{\rho D g}{4k_f K}\left(1 - \exp\left(-\frac{4Hk_f K}{D}\right)\right)$$

$$= \frac{300 \times 4 \times 9.81}{4 \times 0.37 \times 0.4}\left[1 - \exp\left(-\frac{4 \times 20 \times 0.37 \times 0.4}{4}\right)\right] = 18,854 \text{ Pa}$$

Since the lateral pressure, $P_w$, is proportional to the vertical pressure, $P_v$,

$$P_w = KP_0 = 0.4 \times 18,854 = 7542 \text{ Pa}$$

For water, the vertical pressure is the weight of the liquid column:

$$P_0 = \rho g H$$

Because the lateral and vertical pressures are the same (i.e., $K = 1.0$), we can write:

$$P_w = P_0 = 1000 \times 9.81 \times 20 = 196,200 \text{ Pa}$$

The lateral pressure for water is therefore (196,200/7542) or 26 times greater than that for wood chips.

### *12.3.2.7 Chute Design*

In a silo, the solids are withdrawn through chutes at the bottom. Previous discussions examined solids flow through the silo. Now, we will look at the flow out of the silo through the chute, which connects the silo to the feeder. A proper chute design ensures uninterrupted flow from storage to feeder. Improper design results in nonuniform flow. Figure 12.9 illustrates the problem, showing partial solids flow with a uniform-area chute and full solids flow with a properly designed chute. As the solids accumulate on the belt, their uniform flow through the hopper prevents them from accumulating at

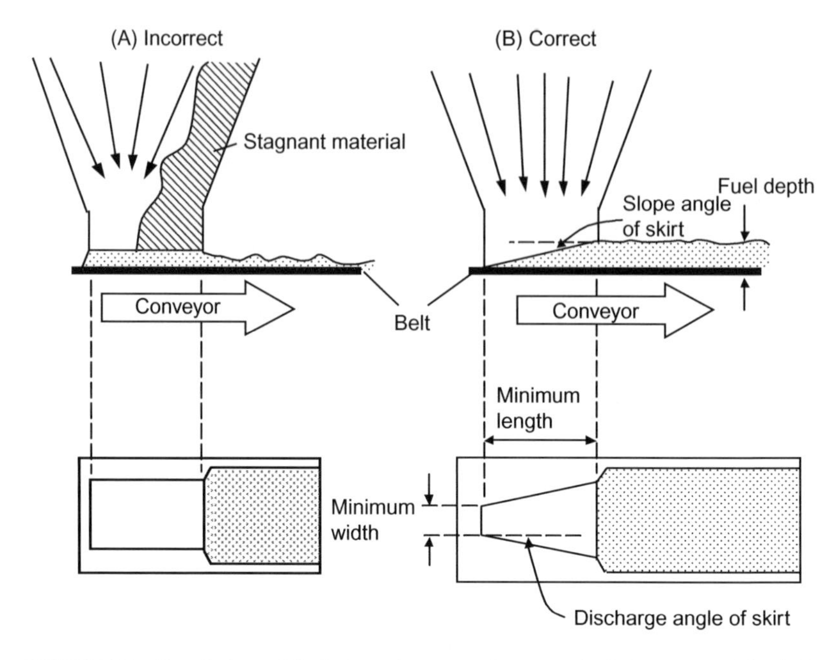


**FIGURE 12.9** Two designs of feed chutes between a hopper and a belt conveyor. The design of (A) is simpler but causes partial flow while (B) gives complete flow.

the chute's downstream section. The chute's expanded and lifted opening helps the solids spread well, allowing uniform withdrawal. For this reason, the modified design of Figure 12.9B shows the skirt on the chute to be lifted and expanded (in plan view) to facilitate uniform solid discharge from the hopper. These angles (slope and discharge) should be in the range of 3–5°.

Figure 12.10 is another illustration of this phenomenon, this time with a rotary feeder. Here, the design of Figure 12.10A is without the short vertical section like that of Figure 12.10B. Solids are compressed in the direction of rotation and pushed up through the hopper. The design on the right uses a short vertical chute that limits this backflow only to the chute height, giving a relatively steady flow.

The two key requirements for chute design are: (1) the entire cross section of the outlet must be active, permitting the flow of solids; and (2) the maximum discharge rate of the chute must be higher than the maximum handling rate of the feeder to which it is connected.

A restricted outlet, caused by a partially open slide gate, results in funnel flow with a small active flow channel regardless of hopper design. A rectangular outlet ensures that feeder capacity increases in the direction of the flow. With a belt feeder, the increase in capacity is achieved by a tapered

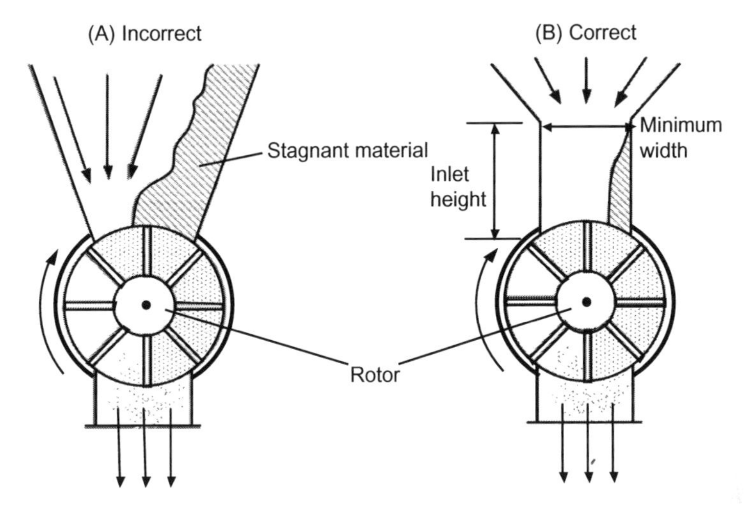


**FIGURE 12.10** A small vertical section above a rotary feeder as shown on the right gives better flow.

interface. The capacity increase along the feeder length is achieved by the increase in height and width of the interface above the belt.

Poor design of a feeder is a common cause of flow problems, as it prevents smooth withdrawal of solids. If the discharge rate of the chute is lower than the maximum designed feeding rate of the feeder, the feeder can be starved of solids and its flow control will be affected.

### 12.3.3 Conveying

There are several options for conveying biomass from one point to the other. Some of the popular ones include:

1. Belt conveyor
2. Chain conveyor
3. Pneumatic conveyor

A belt conveyor is less expensive but have a larger footprint due to its relatively low slope angle. The belt speed should be kept low below 2 m/s (Janze 2010). It allows a magnet hanging from the top to remove magnetic materials and other devices to remove scrap materials and oversize feed as the biomass moves along the belt. A chain conveyor, however, requires lower space as it can operate at a steeper slope, but the chain conveyor is more costly. Pneumatic conveying is the least costly option, but it has a high power requirement. The fuel should necessarily be small for pneumatic conveying. One could use dilute phase conveying. Installations also use dense phase conveying.

### 12.3.4 Feed Preparation

Biomass received from its source cannot be fed directly into the gasifier for the following reasons:

- Presence of foreign materials (e.g., rocks and metals)
- Unacceptable level of moisture
- Too large (or uneven in size)

Such undesirable conditions not only affect the flow of solids through the feeder, but also operation of the gasifier. It is thus necessary to eliminate them and prepare the collected biomass appropriately for feeding. Foreign materials pose a major problem in biomass-fired plants. They jam feeders, form arches in silos, and thus affect the gasifier operation, so it is vitally important to remove them as much as possible. The three main types of foreign materials are: (1) stones, (2) ferrous metals (e.g., iron), and (3) nonferrous metals (e.g., aluminum).

Some of the equipment used to remove foreign materials from the collected biomass are as follows:

*De-stoner.* The basic purpose of a de-stoner is the separation of heavier-than-biomass materials such as glass, stones, and metals. Typical de-stoners use vibration in tandem with suitable airflow to stratify heavy materials according to their specific gravity.

*Nonferrous metal separators.* Separation of nonferrous metals like aluminum has always been a challenge. One solution is an eddy current separator—essentially a rotor with magnetic blocks that, depending on the application, is made of either standard ferrite ceramic or a more powerful rare earth magnet. The rotors are *spun at high revolutions* (more than 3000 rpm) to produce an "eddy current," which reacts differently with different metals according to their specific mass and resistivity, to create a repelling force on the charged particle. If a metal is light yet conductive, such as aluminum, it is easily levitated and ejected from the normal flow of the product stream, making separation possible (Figure 12.11). Separation of stainless steel is also possible by eddy current, but it is more difficult and depend on its gravity. Particles from material flows can be sorted down to a minimum size of 3/32 in. (2 mm) in diameter. Eddy current separators are crucial in the recycling industry because of their ability to separate nonmagnetic materials.

*Magnetic metal separation.* The use of powerful magnets to separate iron and other magnetic materials from the feed is a standard procedure in many plants. Magnets are located at several places along the feed stream. They are generally suspended above the belt, to attract magnetic materials, which are then discharged away.

#### 12.3.4.1 Size Reducers

Biomass comes from different sources, so the presence of oversized solids or trash is very common in the fuel delivered. Woody biomass may be sized

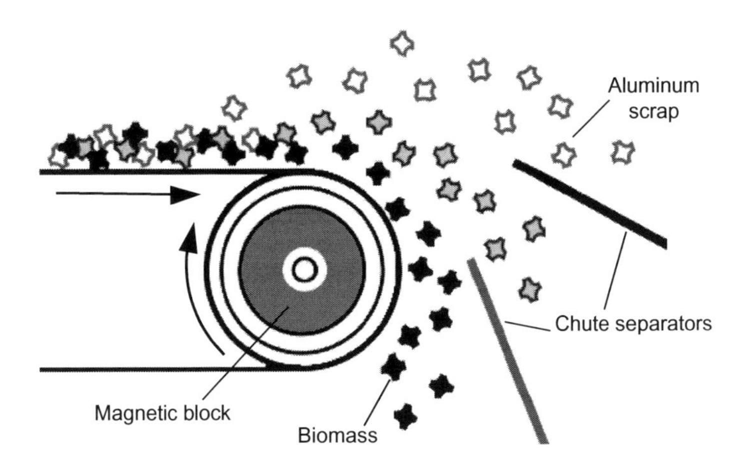


**FIGURE 12.11** Separation of nonferrous metals from biomass using eddy current separation.

and classified at the source or at the plant. The following list contains some of the equipment used for its preliminary sizing along with the typical sizes produced (Van Loo and Koppejan, 2008, p. 64):

- Chunker: 250 to 50 mm
- Chipper: 50 to 5 mm
- Grinder: <80 mm
- Pulverizer: <100 μm (dust)

Different types of equipment are necessary for sizing biomass. One example is a chunker with multiple blades. Another is a spiral chunker with a helical cutter mounted on a shaft; as the wood is fed into the machine, the cutter draws it in and slices it into chunks. The power consumption of a chunker is relatively low.

Chippers are used to break wood into small pieces. Disc and drum chippers are two common types. In a disc chipper, the wood is fed from the side, meeting a large disc with several rotating knives. In a drum chipper, several knives are embedded in grooves (Figure 12.12); as the drum rotates, wood fed at one end is chipped. The chips, which are now uniform in size, are carried away and thrown to the other end by the grooves.

Size-reducing machines consume energy in proportion to the reduction in size. Chipping typically consumes energy equivalent to 1–3% of the energy content of the wood (Van Loo and Koppejan, 2008, p. 65).

Grinding is needed when a finer size (<80 mm) of biomass particle is needed. Hammer mills may be used for this purpose. Where wood is thrown to the wall of the mill and crushed by hammers. A conventional biomass combustor or gasifier does not require biomass to be ground to such a fine size. However, direct cofiring in pulverized coal-fired boilers and the use of

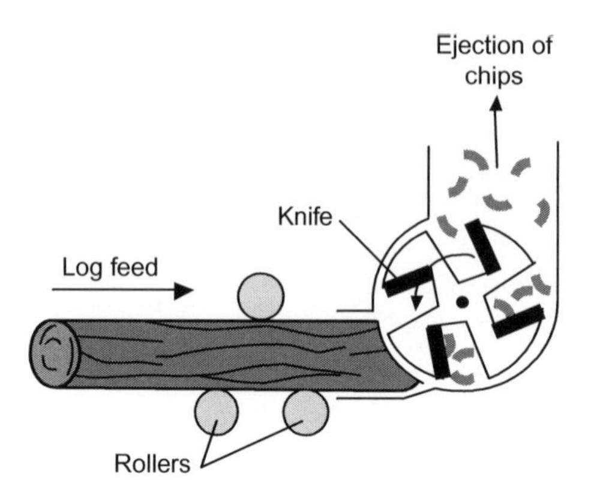


**FIGURE 12.12** Drum chipper.

entrained beds for gasification require the biomass to be ground extremely fine so that the particles can be conveyed like pulverized coal.

### *12.3.4.2 Size Classification*

Oversized materials often cause major problems in a biomass plant. They jam belts, bunkers, and other components. Sometimes trummels are used in the fuel yard to separate the oversize pieces before feeding to the plant. A trummel (Figure 12.13) is a rotating drum, with holes of various sizes, that separates the smaller and larger feed.

A scalping screen is recommended for removal of large oversized solids. Disc screens are good for removing stones larger than the screen opening.

### *12.3.4.3 Drying*

Freshly cut biomass can contain 40–60% surface moisture when harvested, but thermal gasification typically requires a moisture content of less than 10–15%—this moisture is inherent in the biomass. Furthermore, biomass is hygroscopic, so even after drying it can still absorb moisture from the atmosphere; only after torrefaction does the biomass stop absorbing moisture (see Chapter 4). This could happen even when the dried biomass is stored in a shed. Because biomass is bulky, with low energy density, a very large storage space is necessary for the typical fuel inventories required in an energy conversion (boiler or gasifier) plant. For this reason, the biomass is often stored outdoors, though it could absorb additional moisture from rain and snow. Leaving freshly harvested biomass outdoors can at times have some positive effect. For example, straw is sometimes left in the field for a few days or weeks to lose moisture and to leach away K and Cl before it is put in bales (Van Loo and Koppejan, 2008). Leaving wood logs outside over the

**FIGURE 12.13** Portable trummel used in the fuel yard for size classification.

summer can reduce moisture by as much as 20% (Van Loo and Koppejan, 2008, p. 70).

The moisture in biomass must be reduced before use because it represents a large drain on a plant's deliverable energy. Every kilogram of moisture needs about 2300 kJ of heat to vaporize and an additional 1500 kJ to be raised to a typical gasifier temperature of 700°C. This large amount of energy (3800 kJ/kg) has to come from the energy released by the gasifier's exothermic reactions. Therefore, lower moisture makes a higher amount of heat available in the product gas.

Outdoor storage may not work well because of rain and snow, but precipitation can have a beneficial effect on some herbaceous biomass, such as straw, since it leaches water-soluble agglomerates and corrosion-causing elements such as chlorine and potassium. The three types of moisture in a biomass gasifier are: (1) surface moisture, (2) chemical moisture, and (3) moisture in air or steam used for gasification.

While the chemical (also called inherent) moisture cannot be reduced, it is possible to reduce the surface moisture by drying, using the sensible heat in the gasifier product gas, the flue gas of the combustor, or heat from other external sources. A surface moisture less than 10–20% is desirable for most gasifier types (Cummers and Brown, 2002).

The temperature of the hot gas used for drying the biomass is a critical design parameter. Generally, it is in the range of 50–60°C. If much hotter gas is used, it can heat the biomass above 100°C, and pyrolysis can set in on the outer surface of the biomass before the heat reaches the interior. Besides contributing to an energy loss though, such a hot gas can cause volatile organic compounds to be released from the biomass that are potentially hazardous. They are detected by a "blue haze" in the exhaust gas (Cummers and Brown, 2002). The presence of excessive oxygen in the dryer can also lead to ignition of fuel dust in the dryer, resulting in a potential explosion. Therefore, oxygen concentration in the dryer should be kept below 10% to avoid this risk (Brammer and Bridgwater, 1999).

It allows a magnet hanging from the top to remove magnetic materials and other devices to remove oversized solids and other scrap materials as the biomass moves along the belt. The biomass that remains is fed into a silo for temporary storage.

### 12.3.5 Feeding

Feeding is the last step in the feedstock-handling stream. A feed system should include a weighing scale and a tramp metal magnet. A final scalping screen at this point to remove oversize could provide additional reliability to the system. Many types of feeder are used depending on biomass type and other process parameters. This topic is discussed next.

## 12.4 BIOMASS FEEDERS

Based on the type of biomass, feeders can be divided into two broad groups: (1) those for harvested biomass and (2) those for nonharvested biomass.

Harvested fuels include long and slender plants like straw, grass, and bagasse, which carry considerable amounts of moisture. Examples of nonharvested fuels are wood chips, rice husk, shells, barks, and pruning. These fuels are not as long or as slender as harvested fuels, and some of them are actually granular in shape.

### 12.4.1 Feeding Systems for Harvested Fuel

Harvested biomass, such as straw and nonharvested hay, is pressed into bales in the field, and sometimes the bales are left in the field to dry (Figure 12.14). Baling facilitates transportation and handling (Figure 12.15). Cranes are used to load the bales at a certain rate depending on the rate of fuel consumption. The bales are brought to the boiler house from storage by chain conveyors.

Whole bales are fed into a bale shredder and a rotary cutter chopper to reduce the straw to sizes adequate for feeding into a fluidized-bed gasifier or combustor. In the final leg, the chopped straw is fed into the furnace by one of the several feeder types. Figure 12.15 shows a ram feeder which pushes the straw into the furnace. In some cases, the straw falls into a double-screw stoker, which presses it into the furnace through a water-cooled tunnel.

### 12.4.2 Feeding Systems for Nonharvested Fuels

Wood and by-products from food-processing industries are generally granular in shape. Wood chips and bark may not be of the right size when delivered to the plant, so they need to be shredded to the desired size in a

**FIGURE 12.14** Tall grass are cut in the field. They are made into bales and left in the field for drying. Source: *Photograph taken by the author in a countryside in Nova Scotia, Canada.*

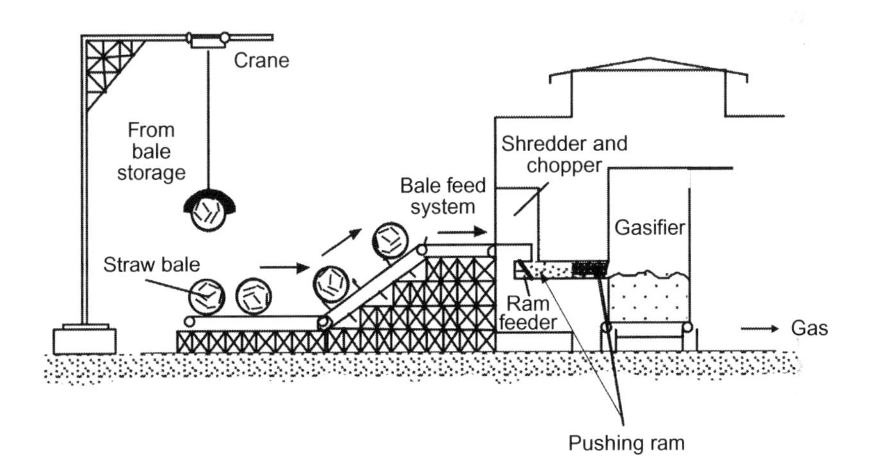


**FIGURE 12.15** Straw bale handling system for feeding into a straw gasifier.

chopper. However, fuels like rice husk and coffee beans are of a fixed granular size and so do not need further chopping. Rice husk, a widely used biomass, is flaky and 2–10 mm × 1–3 mm in size. As such, it can be fed as it comes from the source, but it can be easily entrained in a fluidized bed. For this reason, one can press it into pellets using either heat or a nominal binder in a press.

Feeders for nonharvested fuels are similar to those for conventional fuels like coal. Speed-controlled feeders take the fuel from the silo and drop measured amounts of it into several conveyors. Each conveyor takes the fuel to an air-swept spout that feeds it into the furnace. If the moisture in the fuel is too high, augers are used to push the fuel into the furnace.

### 12.4.3 Feeder Types

The six main feeder types for biomass are: (1) gravity chute, (2) screw conveyor, (3) pneumatic injection, (4) rotary spreader, (5) moving-hole feeder, and (6) belt feeder. These are broadly classified as traction, nontraction, and other types as shown in Figure 12.16. In the traction type, there is linear motion of the surface carrying the fuel, as with a belt feeder or a moving-hole or drag-chain feeder. In the nontraction type, the motion is rotating and oscillatory screw feeders and rotary feeders belong to this group. Oscillatory feeders are of the vibratory or ram type. Other feeder types move the fuel by gravity or air pressure.

#### *12.4.3.1 Gravity Chute*

A gravity chute is a simple device in which fuel particles are dropped into the bed with the help of gravity. The pressure in the furnace needs to be at least slightly lower than the atmospheric pressure; otherwise, hot gas will blow back into the chute, creating operational hazards and possible choking of the feeder due to coking near its mouth.

In spite of the excellent mixing capabilities of a fluidized bed, a fuel-rich zone is often created near the outlet of a chute feeder that is subjected to severe corrosion. Since the fuel is not well dispersed in gravity chute feeding, much of the volatile matter is released near the feeder outlet, which causes a reducing environment. To reduce this problem, the chute can be extended into the furnace. However, the extension needs insulation and some cooling air to avoid premature devolatilization of the feed passing through it. Additionally, a pressure surge might blow fine fuel particles back into the chute, while reducing conditions might encourage corrosion. An air jet can help disperse the fine particles away from the fuel-rich zone.

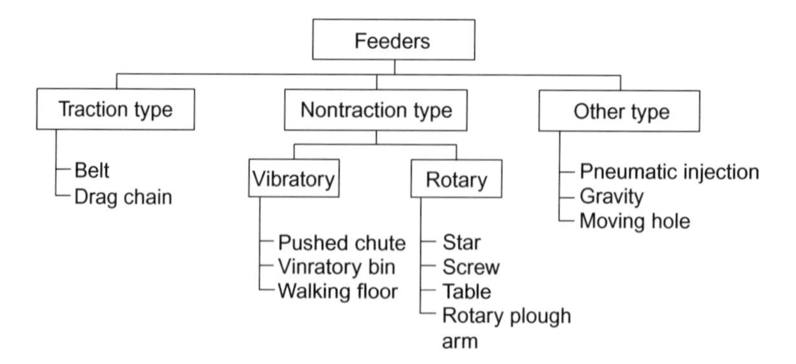


**FIGURE 12.16** Types of feeders used in biomass plants.

A gravity feeder is not a metering device. It can neither control nor measure the feed rate of the fuel. For this, a separate metering device such as a screw feeder is required upstream of the chute.

### 12.4.3.2 Screw Feeder

A screw feeder is a positive displacement device. Not only can it move solid particles from a low-pressure zone to a high-pressure zone with a pressure seal, but it can also measure the amount of fuel fed into the bed. By varying the speed of its drive, a screw feeder can easily control the feed rate. As with a gravity chute, the fuel coming out of a screw does not have any means for dispersion. An air dispersion jet employed under the screw feeder can serve this purpose.

Plugging of the screw is a common problem. Solids in the screw flights are compressed as they move downstream; sometimes they are packed so hard that they do not fall off the screw. Compaction against the sealed end of the trough carrying the screw is even worse, often leading to jamming of the screw. Plugging and jamming can be avoided by one of the following:

- Variable pitch screw (Figure 12.17A)
- Variable diameter to avoid compression of fuels toward the feeder's discharge end (Figure 12.17B)
- Wire screw
- Multiple screws (Figure 12.18)

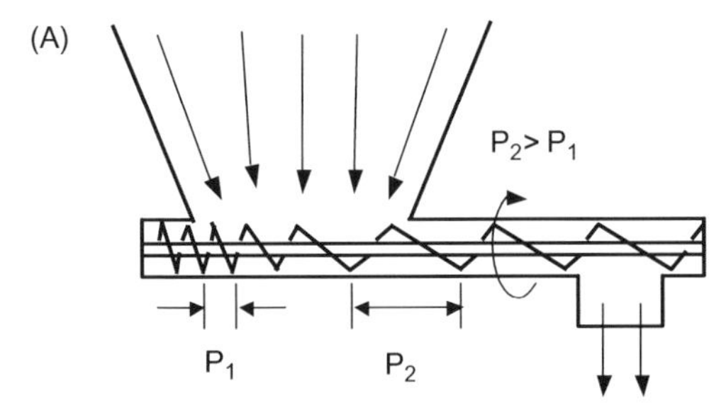


(B)

**FIGURE 12.17** Two types of screw used for trouble-free feeding of biomass. Uniform flow by (A) variable pitch screw and (B) variable diameter screw. Source: *Photograph by the author.*

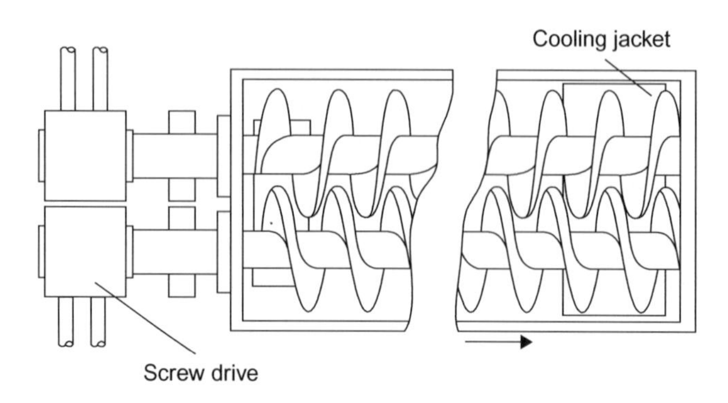


**FIGURE 12.18** Double-screw feeders help uniform flow of biomass.

A wire screw is suitable for a highly fibrous biomass. It is made of a helical spring like wire with no central shaft or blades. Because there is minimum metal-feed contact, there is less chance of feed buildup even if the feed is cohesive.

Multiple screws are effective especially for large-biomass fuels. Figure 12.18 shows a feeder with two screws. Some feed systems use three, four, or more.

The hopper outlet, to which the inlet of a feeder is connected, needs careful design. Figure 12.9 showed two designs. The first (Figure 12.9A) has a tapered wall hopper. It develops a large stagnant layer on the hopper's downstream wall. The second (Figure 12.9B) is a vertical hopper wall toward the discharge end. This is superior to the traditionally inclined wall because it develops a smaller stagnant layer and thus avoids formation of rat holes.

A screw feeder typically serves 3 $m^2$ or less area of a bubbling fluidized bed, so several feeders are needed for a large bed. A major and very common operational problem arises when the fuel contains high moisture. It has to be dried first before it enters the screw conveyor to avoid plugging.

Dai and Grace (2008) developed a theoretical model to determine the mechanism of solids flow through a screw feeder. They noted that the torque required by the screw is proportional to the vertical stress exerted on the hopper outlet by the bulk material in the hopper; it also depends strongly on screw diameter. The choke section (the part of the screw extending beyond the hopper exit) accounts for more than half of the total torque required to feed the biomass, especially with compressible particles. The torque, $T$, required by a screw of diameter, $D_0$, rotating in a shaft of diameter, $D_c$, is given as:

$$T = K_i \sigma_v D_0^3 \tag{12.9}$$

where $\sigma_v$ is the vertical stress for the flow and $D_0$ is the screw diameter. The constant, $K_i$, depends on the ratios $P/D_0$ and $D_c/D_0$ (normal stress/axial stress) and the wall friction, where $D_c$ is the shaft diameter and $P$ is the pitch of the screw.

### 12.4.3.3 Spreader

For a wide dispersion of fuel over the bed, spreader wheels are used (Figure 12.19). The spreader throws the fuel received from a screw or other type of metering feeder over a large area of the bed surface. Typically, it comprises a pair of blades rotating at high speed; slightly opposite orientation of the blades helps throw the fuel over a larger lateral area. This is not a metering device. A major problem with the spreader is that it encourages segregation of particles in the bed.

### 12.4.3.4 Pneumatic Injection Feeder

A pneumatic injection feeder is not a metering device; rather, it helps feed already metered biomass into the reactor. This works well for gravity feeding, and it is especially suitable for fine solids. Pneumatic injection is preferred for less reactive fuels, which must reside in a gasifier bed longer, for complete conversion. It transports dry fuel particles in an air stream at a velocity higher than their settling velocity. The fuel is typically fed from underneath a bubbling fluidized bed. The maximum velocity of air in the fuel transport lines may not exceed 11–15 m/s to avoid line erosion. The air for transporting constitutes part of the air for gasification.

Splitting of the fuel–air mixture into multiple fuel lines is a major problem with pneumatic injection. A specially designed feed splitter, like pneumatic or fluidized splitters (Basu, 2006, p. 355), can be used.

In an underbed pneumatic system, air jets that carry solid particles with high momentum to enter into the fluidized bed, form a plume that could punch through the bed. To avoid this, a cap sits at the top of the exit of each feeder nozzle. This cap reduces the momentum of the jets breaking into the

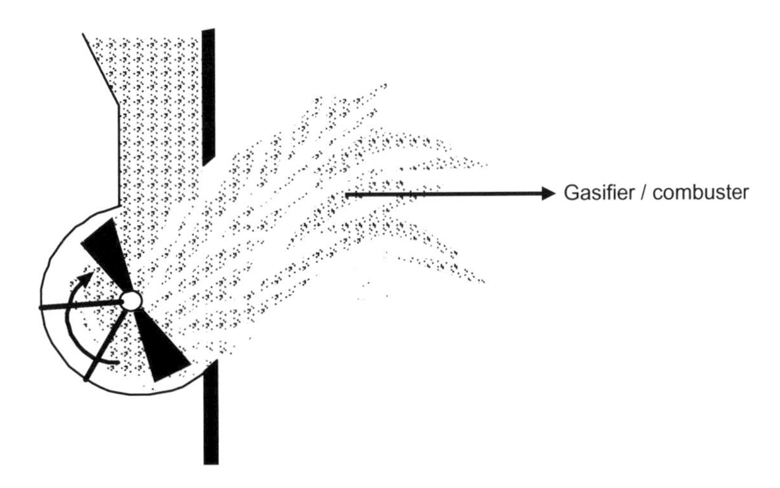


**FIGURE 12.19** Rotary spreader for spreading the fuel over a large bed area.

bed of a bubbling fluidized bed. A highly erosive zone may be formed, near each outlet nozzle of the feeders, which might corrode the tubes nearby.

Another innovative, but one that is less common, feed system uses pulsed air. Controlled-air pulses push the biomass into the gasifier, avoiding pyrolysis of feed in the gasifier feed line. A very small amount of air minimizes dilution of the product gas with nitrogen.

#### *12.4.3.5 Moving-Hole Feeder*

A moving-hole feeder is particularly useful for fluffy biomass or solids, with flakes, which are not free-flowing. Such solids can cause excessive packing in the hopper and screw feeder. Unlike other types, moving-hole feeders do not draw solids from one particular point in the silo.

A moving-hole feeder essentially consists of slots that traverse back and forth with no friction between the stored material and the feeder deck. At a desired rate, a moving hole or aperture slides under the hopper. The solids drop by gravity into the trough or belt that carry the feed at that rate.

With a moving-hole feeder there is no compaction of solids that are typically seen in screw feeders. Rat holes are also avoided by using vertical instead of sloped walls in the hopper the only stipulation is that the size of the hole must be sufficiently large to avoid arching of a given biomass.

#### *12.4.3.6 Fuel Auger*

A metering device such as a screw is used to meter biomass like hog fuel and feed it onto the main fuel belt. The belt carries the fuel to the gasifier front, where the fuel stream is divided into several 50%-capacity fuel trains. Each train consists of a surge bin with a metering bottom and a fuel auger to deliver the fuel into the furnace. The auger is cantilevered and driven at a constant speed through a gear reducer. The bearing of the auger shaft is located away from the heat of the gasifier. Cooling air is provided to cool the auger's inner trough as well as to propel the fuel toward the bed.

#### *12.4.3.7 Ram Feeder for Refuse-Derived Fuel*

A ram feeder is essentially a hydraulic pusher. Refuse-derived fuel (RDF) is at times too fibrous or sticky to be handled by any of the aforementioned feeders. In this case, a ram-type feeder can be effective in forcing them into the gasifier. A fuel auger can convey the solids into a hopper at the bottom of which is the ram feeder. The ram pushes the RDF onto a sloped apron-type feeder that feeds the fuel chute (Figure 12.15). From the fuel chute, the RDF drops into the fuel spout, where sweep air transports it into the furnace. The air also prevents any backflow of hot gases. The RDF stored in the inlet hopper provides a seal against positive furnace pressure. The apron feeders are driven by a variable-speed drive for controlling the amount of fuel going into the system.

### *12.4.3.8 Belt Feeder*

Belt feeders are very effective for feeding nonfree-flowing biomass that is cohesive, fibrous, friable, coarse, elastic, sticky, or bulky. However, they are not recommended for fine or granular solids. Typically, a moving belt is located directly under the outlet chute of the fuel hopper. The belt is supported on rollers that can be mounted on load cells to directly measure the fuel feed rate. Such feeders are referred to as *belt-weigh feeders.*

The width and speed of the belt depend on the density and size of the feed material. A narrow belt with a high design speed may be the most economical, but it is limited by other considerations such as dust generation and hopper width. Most manufacturers provide data on available belt widths, permissible speed, feed density, and recommended spacing of idlers supporting the belt. Such data can be used for the design of the belt feeder and the feed system.

## 12.4.4 Fuel Feed into the Reactor

Biomass feed into the reactor of biomass conversion unit needs special considerations that are discussed in the following sections. For bubbling fluidized beds, we have the choice of two types of feed systems: (1) overbed and (2) underbed (Figure. 12.20).

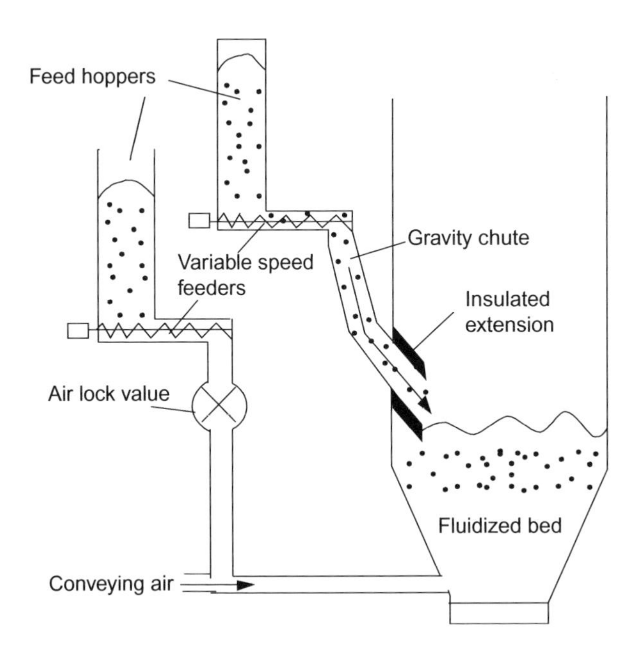


**FIGURE 12.20** Position of over- and underbed feeders in a bubbling fluidized bed.

Gasification and torrefaction are relatively slow processes, compared to combustion, so the rapid mixing of fuels is not as critical as it is in a combustor. Table 12.1 compares the characteristics of the two types of feeder as used in a combustor. Such a comparison may be valid for fluidized-bed gasifiers but only on a qualitative basis.

Overbed feeders can handle coarse particles; underbed feeders need fine sizes with less moisture. An underbed system consists of crushers, bunkers, gravimetric feeders, air pumps, a splitter, and small fuel-transporting lines. An overbed feed system, however, consists of crushers, bunkers, gravimetric feeders, small storage bins, a belt conveyor, and spreaders.

#### *12.4.4.1 Overbed System*

The overbed system (Figure 12.20) is simple, reliable, and economical, but it causes a loss of fine biomass particles through entrainment. In this system, the top size of the fuel particles is coarser than that in an underbed system, making fuel preparation simpler and less expensive. However, the feed can contain a large amount of fines with a terminal velocity that is higher than the superficial velocity in the freeboard. When the terminal velocity is lower than the superficial velocity of the fluidized bed, the particles are elutriated before they completely gasify, resulting in a large carbon loss. This represents most of the carbon loss in a fluidized-bed gasifier.

In an overbed feed system, biomass particles are crushed to sizes less than 20 mm, which is usually coarser than the particle size used in the underbed system. In a typical setup, the fuel passes through bunkers, gravimetric feeders, and a belt conveyor, and is then dropped into a feed hopper.

Fewer feed points is an important characteristic of an overbed feed system when used in a fluidized-bed unit. A typical unit will receive the biomass from a metering feeder. The chute will need a knife or isolation valve for safety. A seal-like rotary air lock could prevent hot gas from the combustor or gasifier to be transmitted into the fuel chute. Thereafter, the fuel is spread over the bed of the reactor. An air jet is often used to facilitate the flow of biomass through the fuel chute.

A rotary spreader throws the fuel particles over the bed surface. The coarser particles travel deeper into the gasifier while the finer particles drop closer to the feeder. The bed thus receives particles of a nonuniform size distribution. The maximum throwing distance of a typical spreader is around 4–5 m. The location of the spreaders is dependent on the dimensions of the bubbling bed. When the width is less than the depth, the spreaders are located on the side walls; when the depth is less than the width, they are located on the front wall. When both width and depth are greater than 4.5 m, the spreaders can be located on both the side walls. Sometimes air is used to assist the throw of fuel by spreaders. An air jet also helps gravity into a CFB unit.

**TABLE 12.1** Feed Points for Some Commercial Bubbling Fluidized-Bed Boilers

| Boilers | Boiler Rating (MWe) | Bed Area ($m^2$) | Feed Points | MWth per Feed | $m^2$ per Feeder | Feed Type | Fuel Type | HHV of Fuel (MJ/kg) |
|---|---|---|---|---|---|---|---|---|
| Shell | 43 (MWth) | 23.6 | 2 | 21.7 | 11.8 | OB | Bituminous | |
| Black dog | 130 | 93.44 | 12 | 31.0 | 7.8 | OB | Bituminous | 19.5–34.9 |
| TVA | 160 | 234 | 120 | 3.8 | 2.0 | UB | Bituminous | 24–25 |
| Wakamatsu | 50 | 99 | 86 | 1.7 | 1.2 | UB | Bituminous | 25.8 |
| Stork | 90 | 61 | 36 | 2.8 | 1.7 | UB | Lignite | 25 |

*Note:* OB = overbed spreader feeder; UB = underbed pneumatic feed.

#### 12.4.4.2 Underbed System

In an underbed feed system (Figure 12.20), the fuel particles are crushed into sizes smaller than 8–10 mm. Introduced in Section 12.4.2, as pneumatic feeding, this system is relatively expensive, complicated, and less reliable than the overbed system (especially with moist fuels), but it does achieve high char conversion efficiency.

Fuel entering at a feed point disperses over a much smaller area, than it does in overbed feeding, so the feed points are more numerous and more closely spaced. Spacing greatly affects gasification. Because a deeper bed allows wider dispersion of the fuel and hence works with wider feed-point spacing; increased spacing with no sacrifice of char conversion efficiency can be achieved, but only if it is compensated by a corresponding increase in bed height. A decrease in bed height must be matched by increased feed-point spacing; otherwise, the conversion efficiency can drop. Coarser particles take longer to gasify and are less prone to entrainment. Therefore, wider spacing is preferred for them; finer particles require closer spacing.

The freeboard can provide room for further reaction of particles entrained from the bed. Freeboard design is important, especially when wide feed-point spacing is used.

#### 12.4.4.3 Feed-Point Allocation

The excellent solids–solids mixing in a fluidized bed helps disperse the fuel over the bed. A single underbed fuel injection point is adequate for a small bed having a cross-sectional size of less than 2 $m^2$. A much larger area is served by one overbed fuel feeder. A circulating fluidized bed would require even a smaller number of feeder per unit bed area because of its superior mixing. Larger beds need multiple feeders. The number required for a given bed depends on factors such as quality of fuel, type of feeding system, amount of fuel input, and bed area. Table 12.1 shows feed-point allocation for some overbed and underbed fluidized-bed combustors.

Highly reactive fuels with high volatiles need a larger number of feed injection points because they react relatively fast; less reactive fuels on the other hand require fewer feed points.

Industrial designs often call for redundancy. For example, if a reactor needs two overbed feeders, designers will provide three, each with a capacity that is at least 50% of the design feed rate. In this way, if one feeder is out of service, the plant can still maintain full output on the other two. The number of redundant feeders depends on the capacity and reliability required of the plant.

## 12.5 COST OF BIOMASS-HANDLING SYSTEM

Material handling is a major item in the capital cost of a biomass conversion plant. So it is worth discussing this aspect of biomass-handling system

**TABLE 12.2 Relative Cost of Biomass-Handling System**

| Capacity on as-Received Basis | 100 Tons/Day | 680 Tons/Day |
|---|---|---|
| System type | Manual | Automated |
| Truck tipper | 11% | 5% |
| Conveyor to wood pile | – | 1% |
| Radial stacker, adder | – | 4% |
| Front end loaders, adder | 5% | – |
| Reclaim feeder | – | 5% |
| Conveyor | – | 3% |
| Metal separator | 2% | 1% |
| Screener | 1% | 5% |
| Grinder | 12% | 12% |
| Buffer storage | 3% | 5% |
| Fuel metering | 12% | 1% |
| Controls | 5% | 4% |
| Equipment installation | 24% | 5% |
| Civil/structural work | 18% | 1% |
| Electrical work | 8% | 4% |
| Total direct cost in 2003 dollars | $2,102,000 | $4,857,000 |
| Operating personnel required | 5 | 4 |

**Source**: Prepared based on data from www.epa.gov_chp_documents_chp_catalog_part4.pdf.

design. Table 12.2 gives an example of relative costs of different elements of a typical biomass-handling plant. This table developed for biomass-fired combined heat and power plant gives both relative capital cost of the biomass-handling plant for a small 100 tons/day with manual loading of biomass with front loader and for a 680 tons/day automated loading plant. The manual plant requires a larger number of operating personnel. The unit cost of dollar per ton per day is higher for lower cost and it is lower for larger capacity unit. This analysis found the capital cost values as a function of 0.85 power (EPA, 2007) of the capacity of the plant.

$$\text{Unit cost} \sim \text{capacity}^{0.85} \tag{12.10}$$

## SYMBOLS AND NOMENCLATURE

| | |
|---|---|
| $A$ | area of the cross-sectional area of silo ($m^2$) |
| $B$ | parameter in Eq. (12.1) and (12.3) |
| $C$ | parameter in Eq. (12.3) |
| $d_p$ | particle size (m) |
| $D$ | diameter of the silo (m) |
| $D_0$ | diameter of the screw (m) |
| $D_C$ | shaft diameter (m) |
| $dh$ | height of a differential element in the silo (m) |
| $g$ | acceleration due to gravity (9.81 $m/s^2$) |
| $H$ | height of the silo (m) |
| $k_f$ | wall friction coefficient (−) |
| $K$ | Janssen coefficient (−) |
| $K_i$ | a constant in Eq. (12.9), depending on $D_c/D_0$ or $P/D_0$ (−) |
| $m$ | mass-flow rate (kg/s) |
| $P$ | pitch of the screw (m) |
| $P_w$ | normal pressure on the wall in the silo (Pa) |
| $P_v$ | vertical pressure on the biomass in the silo (Pa) |
| $P_0$ | pressure at the base of the silo (Pa) |
| $T$ | torque of the screw (Nm) |
| $V_0$ | average solid velocity through outlet (m/s) |
| $\tau$ | wall friction (Pa) |
| $\rho$, $\rho_p$ | density of solids ($kg/m^3$) |
| $\rho_a$ | density of air ($kg/m^3$) |
| $\sigma_v$ | vertical stress for the flow (Pa) |
| $\mu$ | viscosity of air (kg/m s) |
| $\theta$ | semi-included angle of hopper (°) |

Chapter 13

# Analytical Techniques

Feedstock analysis is a vital and important part of a process. It gives the critical information on biomass that is needed for a rational design or better understanding of a process. This chapter discusses the methods used to determine the composition of biomass as a whole with specific reference to its cell walls and its thermal and other properties. A typical wood comprises the followings:

Wood = Extractives + Holocellulose + Lignin + Ash
Holocellulose = Hemicellulose + Cellulose

The analysis of biomass starts with reduction in the sample size from a large representative one. Then a sequence of tests as below could be carried out to determine different thermophysical properties of the biomass.

1. Ultimate analysis
2. Proximate analysis
   a. Moisture content
   b. Ash content
   c. Volatile content
   d. Fixed carbon
3. Extractive: polar (water) and nonpolar (organic)
4. Holocellulose
5. Hemicellulose
6. Lignin
7. Other analyses: chromatographic, spectroscopic, microscopic, and thermal analysis.

The following sections present a brief description of some of the analytical processes.

## 13.1 COMPOSITION OF BIOMASS

### 13.1.1 Ultimate (Elemental) Analysis

Ultimate analysis gives the elemental composition of a fuel. Its determination is relatively difficult and expensive compared to proximate analysis.

**Biomass Gasification, Pyrolysis and Torrefaction.**

The following ASTM standards are available for determination of the ultimate analysis of biomass components:

- Carbon, hydrogen: E-777 for refuse-derived fuels (RDF)
- Nitrogen: E-778 for RDF
- Sulfur: E-775 for RDF
- Moisture: E-871 for wood fuels
- Ash: D-1102 for wood fuels.

Although no standard for other biomass fuels is specified, we can use the RDF standard with a reasonable degree of confidence. For determination of the carbon, hydrogen, and nitrogen components of the ultimate analysis of coal, we may use the ASTM standard D-5373-08. Table 13.1 lists standard methods of ultimate analysis for biomass materials.

### 13.1.2 Proximate Analysis

Proximate analysis gives the gross composition of the biomass and hence it is relatively easy to measure. One can do this without any elaborate set up or expensive analytical equipment. For wood fuels, we can use standard E-870-06. Separate ASTM standards are applicable for determination of the individual components of biomass:

- Volatile matter: E-872 for wood fuels
- Ash: D-1102 for wood fuels
- Moisture: E-871 for wood fuels
- Fixed carbon: determined by difference.

The moisture and ash determined in proximate analysis refer to the same moisture and ash determined in ultimate analysis. However, the fixed carbon

**TABLE 13.1** Standard Methods for Biomass Compositional Analysis

| Biomass Constituent | Standard Methods |
|---|---|
| Carbon | ASTM E 777 for RDF |
| Hydrogen | ASTM E 777 for RDF |
| Nitrogen | ASTM E 778 for RDF |
| Oxygen | By difference |
| Ash | ASTM D 1102 for wood, E 1755 for biomass, D 3174 for coal |
| Moisture | ASTM E 871 for wood, E 949 for RDF, D 3173 for coal |

in proximate analysis is different from the carbon in ultimate analysis. In proximate analysis, it does not include the carbon in the volatile matter and is often referred to as the char yield after devolatilization.

#### 13.1.2.1 Volatile Matter

For the determination of volatile matter, the fuel is heated to a standard temperature and at a standard rate in a controlled environment. The applicable ASTM standard for determination of volatile matter is E-872 for wood fuels and D-3175-07 for coal and coke.

Standard E-872 specifies that 50 g of test sample be taken out of no less than a 10 kg representative sample of biomass using the ASTM D-2013 sample reduction protocol. This sample is ground to less than 1 mm in size through cutting or shearing, and 1 g is taken from it. The sample is put in a covered crucible, so as to avoid contact with air, during devolatilization. The covered crucible is placed in a furnace maintained at 950 ± 20°C. The volatiles released are detected by luminous flame observed from the outside. The crucible is heated for 7 min. After 7 min, the crucible is taken out, cooled in a desiccator, and weighed as soon as possible to determine the weight loss due to devolatilization.

For nonsparking coal or coke, Standard D-3175-07 is used which, follows a similar process except that it requires a 1.0 g sample ground to 250 micron size (as per D-346 protocol). The rest of the procedure is the same as above. For sparking samples, it should be slowly heated to 600°C in 6 min and then heated at 950°C exactly for 6 min.

#### 13.1.2.2 Ash

The ash content of fuel is determined by ASTM test protocol D-1102 for wood, E-1755-01 for other biomass, and D-3174 for coal.

Standard D-1102 specifies a 2.0 g sample of wood (sized below 475 μm) dried in a standard condition and placed in a muffle furnace with the lid of the crucible removed. The temperature of the furnace is raised slowly to 580–600°C to avoid flaming. When all the carbon is burnt, the sample is cooled and weighed. Standard E-1755-01 specifies 0.5–1.0 g of dried biomass to be heated for 3 h at 575 ± 25°C. After that the sample is cooled and weighed.

For coal or coke, standard D-3174-04 may be used. Here a 1.0 g sample (pulverized below 250 μm) is dried under standard conditions and heated to 450–500°C for the first 1 h and then to 700–750°C (950°C for coke) for the second 1 h. The sample is heated for 2 h or longer at that temperature to ensure that the carbon is completely burnt. It is then removed from the furnace, cooled, and weighed.

#### *13.1.2.3 Moisture*

Moisture content (M) is determined by the test protocol given in ASTM standards D-871-82 for wood, D-1348-94 for cellulose, D-1762-84 for wood charcoal, and E-949-88 for RDF (total moisture). For equilibrium moisture in coal one could use D-1412-07. In these protocols, a weighed sample of the fuel is heated in an air oven at 103°C and weighed after cooling. To ensure complete drying of the sample, the process is repeated until its weight remains unchanged. The difference in weight between a dry and a fresh sample gives the moisture content in the fuel.

Standard E-871-82, for example, specifies that a 50 g wood sample be dried at 103°C for 30 min. It is left in the oven at that temperature for 16 h before it is removed and weighed. The weight loss gives the moisture (M) of the proximate analysis.

Standard E-1358-97 provides an alternative means of measurement of moisture using microwave. However, this alternative represents only the physically bound moisture; moisture released through chemical reactions during pyrolysis constitutes volatile matter.

Klass (1998) proposed an alternative means of measuring the proximate composition of a fuel using thermogravimetry (TG). In these techniques, a small sample of the fuel is heated in a specified atmosphere at the desired rate in an electronic microbalance. This gives a continuous record of the weight change of the fuel sample in a TG apparatus. The differential thermogravimetry apparatus gives the rate of change in the weight of the fuel sample continuously. Thus, from the measured weight loss-versus-time graphs, we can determine the fuel's moisture, volatile matter, and ash content. The fixed carbon can be found from Eq. (3.23). This method, though not an industry standard, can quickly provide information regarding the thermochemical conversion of a fuel.

TG analysis provides additional information on reaction mechanisms, kinetic parameters, thermal stability, and heat of reaction. A detailed database of thermal analysis is given in Gaur and Reed (1995).

### 13.1.3 Analysis of Polymeric Components of Biomass

Thermochemical conversion of biomass (torrefaction in particular) greatly depends on the polymeric composition of biomass. As explained in Chapter 3, a typical biomass primarily include:

- Ash
- Extractives
- Cellulose
- Hemicellulose
- Lignin.

Several techniques are available for the determination of the above components of biomass. Determination of some of the components is

**TABLE 13.2 Some Standards Used for the Determination of Different Components of Biomass**

| | | | |
|---|---|---|---|
| Ash | ASTM D 1102-84 | TAPI T 211 om-85 | ASTM E 830-87 |
| | Ash in wood | Ash in wood and pulp | Ash in RDF |
| Moisture | ASTM D 2016-74 (1983) (withdrawn 1988) | E 871 for wood | NF B 51-004-85 |
| | Method for determination of moisture content in wood | E 949 for RDF | Woods: determination of moisture content |
| | | E 3173 for coal | |
| Extractives | ASTM D 1108-84 | TAPPI T 204 om-88 | CPPA G.20 |
| | Standard method for preparation of extractive-free wood | Alcohol–benzene and dichloromethane soluble in wood | Solvent extractives in wood |
| Hemicellulose | | TAPPI 223 | |
| | | Determination of pentosans in wood | |
| Lignin | ASTM D 1106-83 | TAPPI T 222 om-88 | CPPA G.8 |
| | Acid insoluble lignin in wood | Acid insoluble lignin in wood | Acid insoluble lignin or "Klason lignin" in wood |
| | E 1721-01 | | |
| Cellulose | ASTM D 1104-56 (1978) (discontinued) | TAPPI T 17 wd-70 (withdrawn) | JIS P 8007-76 (1984) |
| | Determines holocellulose (hemicellulose plus cellulose) in wood | | Testing method for cellulose in wood for pulp |
| Sample preparation | ISO 3129-75 | TAPPI T 264 om-88 and T 257 cm-85 | CPPA G.31P |
| | Wood sampling methods and requirements for physical and mechanical tests | Preparation of wood for chemical analysis | Preparation of wood for chemical analysis |

ASTM—American Society for Testing and Materials.
CPPA—Canadian Pulp and Paper Association.
ISO—International Organization for Standardization.
JIS—Japanese Industrial Standards.
TAPPI—Technical Association of the Pulp and Paper Industry.

covered by specific ASTM standards (Table 13.2). The following is a brief description of methods that extracts the constituents of the cell walls of wood for their determination. The methods described follow those described by Rowell et al. (2005).

### 13.1.3.1 Sample Preparation

The biomass sample should be taken such that it truly represents the stock it is taken from. The sample is to be prepared such that it is free from foreign elements. ASTM standard, E 1757 describes a standard method for the preparation of a sample for the analysis. The sample is dried at 105°C for several hours to free it from moisture and is then ground to below 40 mesh size.

### 13.1.3.2 Extractive Components of Biomass

Extractives are the natural chemical products of biomass that are capable of being extracted by some solvents. Based on the solvent used in extraction process, extractives can be classified as water soluble, toluene–ethanol, and ether soluble extractives (Rowell, 2005). Major chemicals in an extractive of biomass are fats, fatty acids, fatty alcohols, phenols, terpenes, steroids, resin acids, rosin, waxes, and other organic compounds. Extractives are nonstructured nonpolymer composition of biomass that could affect the analysis of polymer compositions. Therefore, it needs to be removed prior to downstream analysis of the biomass sample (NREL, 2008).

ASTM standard D 1105-96 is available to determine the extractive components of wood in which ethanol–benzene and hot water are used as the solvents. Ethanol–benzene is used to extract waxes, fats, resins, and a portion of wood gums whereas hot water is used to remove tannins, gums, sugars, starches, and color producing chemicals (TAPPI, 2007). On the other hand, ASTM standard test methods E 1690-08/95 are also used for extractive measurement in biomass. These standards are applicable for wider biomass materials such as for both hard and soft barkless woods, herbaceous materials (switchgrass and sericea), agriculture residues (corn stover, wheat straw, and bagasse), and waste papers (office waste, boxboard, and newsprint). Extractives can be removed from wood by using neutral solvents, water, toluene or ethanol, or combinations of solvents. The nature of solvents may change according to the type of chemicals present as an extractive in different biomass.

The following methods as per Rowell (2005) may be used for determination of the extractives. A fresh sample is recommended to avoid the errors due to fungal attack. It is also desirable to peel off the bark from the stem. To prepare, the sample is dried at 105°C in an oven for 24 h prior to milling. It is then ground to an average size of 0.40 mm using a Wiley mill. Major apparatus required for this method include Buchner funnel, extraction thimbles, extraction apparatus, extraction flask (500 ml), Soxhlet extraction tube, heating mantle, boiling chips for taming boiling action, chemical fume hood, and vacuum oven

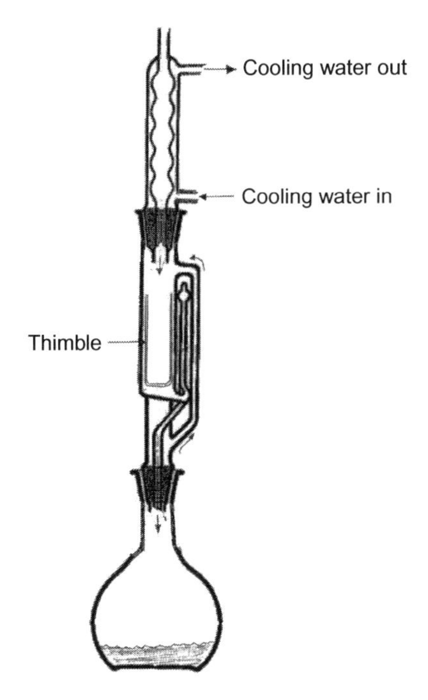


**FIGURE 13.1** Soxhlet extractor for the determination of extractives.

for drying (Figure 13.1). Ethanol 200 proof, toluene (reagent grade), and toluene–ethanol mixture (1:1 volume basis) are required as solvents.

The test Procedures:

**a.** Weigh the extraction thimbles and cover without sample and weigh thimbles by adding 2–3 g of sample.
**b.** Dry thimbles in vacuum oven not exceeding 45°C for 24 h or to a constant weight.
**c.** Thimbles are then cooled down to room temperature in a desiccator for an hour and weighed.
**d.** Cooled thimbles are then kept inside the Soxhelt extraction units (Figure 13.1).
**e.** Take 200 ml of ethanol–toluene mixture in a 500 ml round bottom flask with several boiling chips to prevent bumping.
**f.** Carry out the extraction with well-ventilated chemical fume hood for 2 h, keeping the liquid boiling. Ensure that the siphoning from the extractor is no less than four times per hour.
**g.** After the extraction, take the thimbles out from the extractors to drain the excess solvent.
**h.** Then wash the sample using ethanol and dry in the vacuum oven over night at a temperature not exceeding 45°C for 24 h.

**i.** Dried sample is removed and cooled down in a desiccator for an hour.
**j.** Take weight of the sample; the sample is called *extractive-free sample*.

### 13.1.3.3 Holocellulose

Holocellulose is a water-insoluble carbohydrate fraction of wood materials. It can be extracted by the chlorination method by getting rid of the lignin.

A 2.5 g of extractive-free dry sample is placed in a 250 ml Erlenmeyer flask. Then 80 ml hot distilled water, 0.5 ml acetone, and 1 g of sodium chlorite ($NaClO_2$) are added to it. The mixture is heated in a water bath at 70°C for 1 h. After this, another dose of 0.5 ml acetone, 1 g $NaClO_2$ are added and heated further for 1 h. This process is repeated for 6–8 h until the lignin is completely removed. The mixture is left for 24 h and then it is filtered through a tarred and fritted disk glass thimble (Rowell, 2005, p. 63). The residue is washed with acetone and left in a vacuum oven to dry at 105°C for 24 h.

ASTM E 1721 uses 72% sulfuric acid instead to hydrolyze the sample instead of sodium chlorite used in the above process.

The solid whitish residue left on the filter gives the weight of the lignin-free holocellulose.

### 13.1.3.4 Hemicellulose

The extractive- and lignin-free holocellulose as obtained from above is used for the determination of hemicellulose. The sample is treated with sodium hydroxide (NaOH) and acetic acid to get cellulose as a solid residue and hemicellulose as the filtrate. This filtrate could be run on high performance liquid chromatograph (HPLC) to determine the concentration of each different monomer, for example, glucose, galactose, mannose (hexose), xylose, and arabinose (pentose).

A dry sample of holocellulose (2 g) is taken in a 250 ml flask. Then 10 ml of 17.5% NaOH is added to the flask and a lid put on it. The flask is kept in a water bath at 20°C and the mixture is stirred. After 5 min another 5 ml of the same 17.5% sodium hydroxide is added to it. The process is repeated and continued for 15 min. After that it is left for 30 min and 33 ml of water is added and kept at 20°C. The solid residue containing cellulose is filtered through a tarred alkali resistance fritted glass filter. It is washed again in 100 ml of 8.3% NaOH solution. The cellulose is subjected to acid treatment in 15 ml of 10% acetic acid. Thereafter wash it in distilled water and dry the residue containing cellulose before weighing it.

### 13.1.3.5 Lignin

ASTM D-1166-84 gives a method for determination of acid insoluble lignin also called Klason lignin by dissolving a dry extractive-free sample in 72% sulfuric acid followed by secondary hydrolysis in fucose. ASTM E 1721-1 is based on hydrolysis in 72% sulfuric acid and water alone.

### 13.1.3.6 Cellulose

Cellulose is distinguished from extractives by its insolubility in water or organic solvents, from hemicellulose by its insolubility in aqueous alkaline solutions, and from lignin by its relative resistance to oxidizing agents and susceptibility to hydrolysis by acids (Browning, 1967, p. 387; Fengel and Wegener, 1989).

Cellulose can be measured by isolating it from other components of biomass like, extractive, lignin, and hemicellulose. The following method determines the amount of holocellulose, which comprises cellulose and hemicellulose.

An extractive-free dry sample (2 g) is taken in an extraction thimble. It is then extracted successively with ethanol–benzene for 4 h and with 95% ethanol for 4 h in a Soxhlet extractor. The extracted sample is further extracted with hot water for 3 h in a flask. It is filtered on a fritted glass crucible and washed with hot water followed by cold water. Suction is applied to the bottom of the crucible to remove excess air. It is then chlorinated by passing chlorine gas from the inverted funnel on the crucible with fritted glass filter. The crucible is kept in an ice water bath. After 5 min the suction is released and cold water is drained. Then the sample is treated twice with hot solution (75°C) of monoethanolamine (3% in 95% ethanol) for 2 min each time. The solvent is removed and the wood is washed with ethanol followed by cold water.

The process of chlorination (3 min) and extraction (2 min) are repeated until the residue remains white after chlorination and it is no longer colored by the addition of hot monoethanolamine. The residual substance is weighed to give the mass of hollocellulose.

The residue is finally washed with alcohol, cold water, and ether and dried for 2.5 h at 105°C. It is weighed and left in a bottle.

### 13.1.3.7 Ash

A crucible is cleaned and left in a muffle furnace for a required time to burn any combustibles. It is then cooled and left in a desiccator. The weighted mass of the sample is left in the crucible. The crucible is placed in a muffle furnace at 575°C for several hours to burn off the carbon. After this, the crucible is cooled and weighed. The difference in weight would determine the amount of ash in the sample.

## 13.2 HEATING VALUE

The higher heating value (HHV) can be measured in a bomb calorimeter using ASTM standard D-2015 (withdrawn by ASTM 2000, and not replaced).

The bomb calorimeter consists of pressurized oxygen "bomb" (30 bar), which houses the fuel. A 10 cm fuse wire connected to two electrodes is kept in contact with the fuel inside the bomb. The oxygen bomb is placed in a container filled with 2 l of deionized water. The temperature of the water is measured by means of a precision thermocouple. A stirrer stirs the water continuously. Initially, the temperature change would be small (Figure 13.2) as the only heat generated would be from the stirring of the water molecules. After the temperature is stabilized, the sample is fired, meaning a high voltage is sent across the electrodes and through the fuse wire. The electric current passing through the fuse wire would almost instantly ignite and combust the fuel sample in oxygen. The water absorbs the heat, released by the combustion of the sample, resulting in a sharp rise in the water temperature (Figure 13.2). The temperature continues to rise for sometime before leveling off. The water temperature is continuously recorded till the temperature readings are stable. Knowing the heat capacity of the bomb calorimeter material, water, and of the fuse wire, one can calculate the exact amount of heat released by combustion of the sample.

By knowing the initial mass of the fuel sample, the heating value of the sample can be calculated by dividing the heat released by the mass of the sample. As the product of combustion is cooled below the condensation temperature of water, this technique gives the HHV of the fuel.

## 13.3 DIFFERENTIAL SCANNING CALORIMETRY

Heat capacity, glass transition temperature, crystallization temperature, and melting point are some important parameters of a fuel undergoing

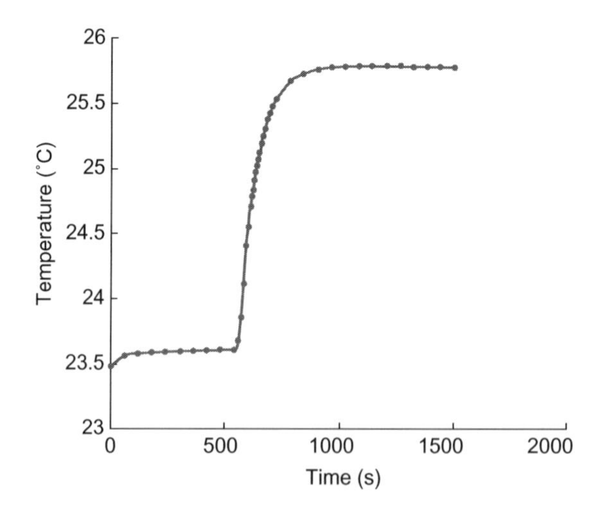


**FIGURE 13.2** Temperature profile from a bomb calorimeter experiment.

thermochemical processing. The differential scanning calorimetry (DSC) may be used to determine these parameters of a substance.

The unit essentially consists of two solid pans resting on top of heaters. The sample to be tested is kept on one of the pans while the other is kept empty (reference pan). A computer program accurately controls the rate of temperature rise, and records the exact amount of heat supplied to each pan. The amount of heat supplied to the reference pan is lower than that supplied to the pan with the solid sample. An additional heat is required exclusively to heat the sample to maintain the same rate of temperature rise.

The difference in heat given out by two heaters to maintain an identical temperature rise is plotted in the *y*-axis while the temperature is plotted in the *x*-axis (Figure 13.3). Different temperatures, like glass transition temperature $T_g$, crystallization temperature $T_c$, and melting point $T_m$, are denoted in the graph shown, which is the typical result from a DSC experiment.

One can determine the heat capacity $C_p$ by dividing the heat flow-rate by the temperature change rate.

$$C_p = \frac{(dQ/dt)}{(dT/dt)}$$

When the sample reaches its glass transition temperature, the heat capacity does not remain constant anymore. It requires more heat to raise the temperature. Because of the change in heat capacity, the heat input rate to the sample rises shifting the graph upward. Beyond this range, the graph flattens. The glass transition temperature ($T_g$) at the middle of the incline (Figure 13.3).

Certain molecules crystallize at a temperature $T_c$. Since crystallization is an exothermic process during which heat is released, this process can be characterized by a dip in the plot.

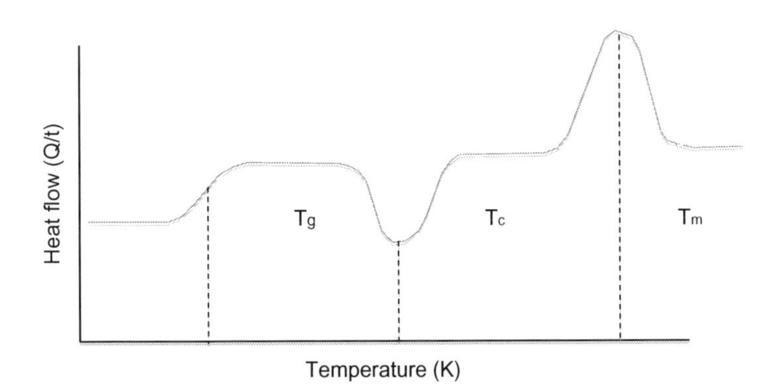


**FIGURE 13.3** Heat flow plot in a DSC.

Further heating melts the substance, when the temperature does not rise with heat due to phase change. So the melting point ($T_m$) of the substance is given by the temperature where $dQ/dt = 0$.

## 13.4 REACTIVITY MEASUREMENTS

This section presents a brief description of derivation of kinetic parameters from measured data.

### 13.4.1 Thermogravimetric Analysis

Thermogravimetric analysis (TGA) instruments can measure a host of parameters like moisture loss, decarboxylation, pyrolysis, loss of solvent, loss of plasticizer, oxidation, and decomposition for biomass or other substances. It could also give vital information of the torrefaction of biomass. It also finds application to determine the carbon content, to compare two similar products, as a quality control tool and for analysis of nanomaterial.

The working of this apparatus is based on the change in mass of the sample with change in temperature or time (Figure 13.4). A typical TGA instrument consists of a pan that rests on a sensitive analytical balance. The test sample is placed on the pan and is heated externally. A purge gas that may be reactive or inert (depending on test requirements) is passed over the sample and exits through the exhaust. The heating rate of the sample can be controlled, and the mass change over the entire period is monitored continuously.

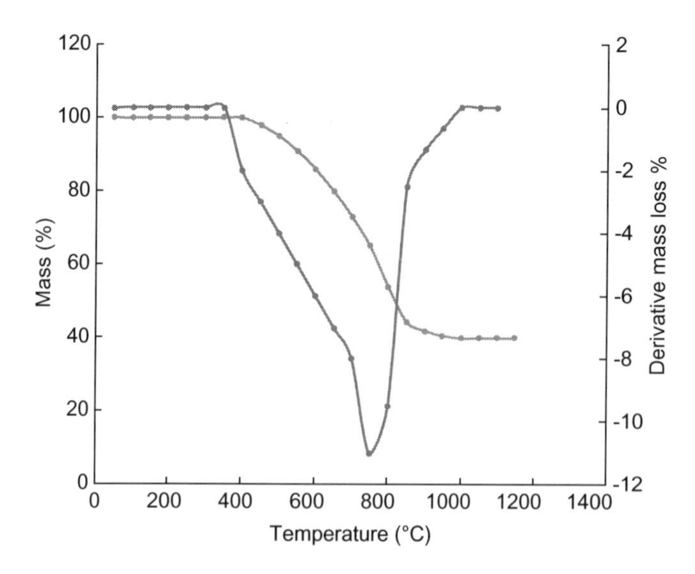


**FIGURE 13.4** A TGA curve shows the mass loss percentage and derivative mass loss.

TGA instruments need very accurate measure of mass. Certain models have an under hanging pan which hangs down from the balance. All models use computers to accurately record the change in mass.

The temperature scanning rate and the purge gas flow-rate can be changed. Certain experiments are conducted isothermally, while some samples are cooled. Software programs are available which plots the first derivative curve which is an essential tool to determine the point of greatest change on the mass loss curve.

### 13.4.2 Differential Thermal Analyzer

Kinetic data can be derived from a differential thermal analyzer. A procedure is described in Gawz and Reed (1995).

### 13.4.3 Quartz Wool Matrix Apparatus

This is a simple universal reactor for a wide range of process and reaction. It works on the same gravimetric principle, as TGA, with an important additional feature that it can study the influence of particle size, shape, and hydrodynamics to some extent. In a typical TGA, the sample is generally ground and placed on a pan as a fixed bed. It impedes uniform access of gases to all particles, and the thermal or concentration gradient around one particle affects those around another. In a quartz wool matrix (QWM) reactor, particles are dispersed widely on a highly expanded matrix of high temperature inert wool. This allows individual particles to have equal access to gas and temperature field. For this reason, it is possible to study the effect of particle size, and shape if any on a reaction. Furthermore, such a highly expanded bed better simulates the hydrodynamics of a fluidized bed or entrained bed. It is thus able to study the effect of mass transfer on a certain reaction.

A typical QWM reactor, originally developed by Chi et al., (1994) is made of tube that is heated externally by an electric heater. A tubular furnace controls the furnace temperature. Reactant gases like oxygen, nitrogen, carbon dioxide, sulfur dioxide, or others are mixed at desired proportion using a precision electronic flow meter (Figure 13.5). The mixture, which simulates the gaseous environment of a reaction, is preheated to the reaction temperature and is passed into the reactor at desired flow-rates. The volumetric flow-rates of the individual gases are calculated based on the cross-sectional area available for flow in the reactor.

The sample fuel ground to the desired size and shape is dispersed on the inert matrix of quartz wool. The wool is supported on a wire basket, which in turn is hanged from a microbalance, vertically on the top but outside the reactor (Figure 13.5). The dispersed location of the sorbent particles closely resembles that in an actual fluidized bed. Solid samples dispersed on the matrix allow free access of gas to all sides of the samples. The fuel sample

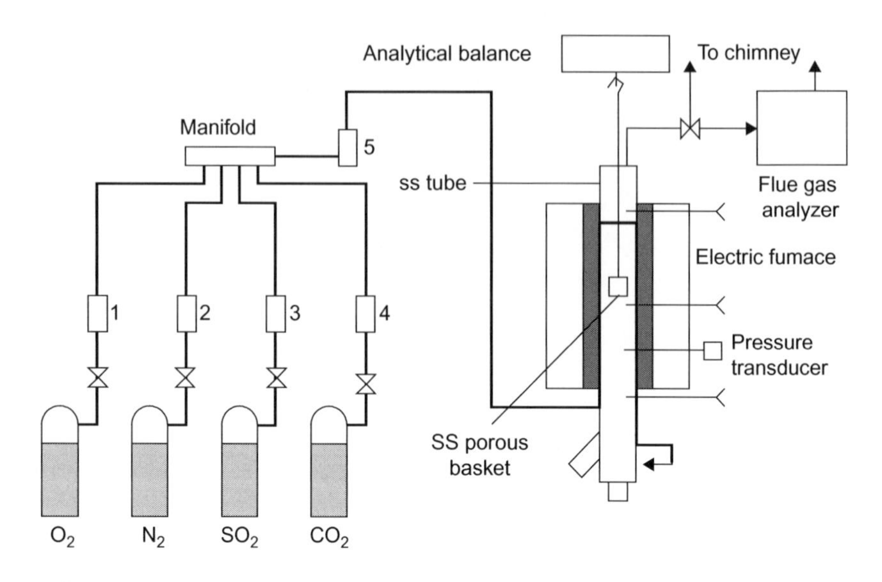


**FIGURE 13.5** Schematic of a QWM apparatus.

in the porous basket undergoes reaction, during which period the mass of the sample is recorded continuously by an analytical balance, which is interfaced with a computer. Gas temperatures are measured at various points by thermocouples. Part of the gases escaping from the chamber is sent to a continuous flue gas analyzer, which determines the percentages and composition of the flue gas; the weight loss of the fuel is plotted as a function of time.

QWM helps to determine the reaction rate, reactivity, and study changes in the physicochemical condition of the biomass or other samples. The following is an example of how reaction rate constant can be determined using the QWM apparatus. Here we take the example of a calcination reaction to illustrate its use.

Calcination reaction: $CaCO_3 \rightarrow CaO + CO_2$.

We define rate constant, $K$ ($s^{-1}$), in a first-order reaction of kinetic model to examine the kinetics of calcination reaction.

$$\frac{dX}{dt} = K\frac{(1-X)(P_{eq}-P_{CO_2})}{P_{eq}} \tag{13.1}$$

$$K = k_0 e^{-\frac{E_a}{RT}}$$

$$X = \left(\frac{W_0 - W_t}{W_0}\right) \times \left(\frac{100}{44}\right)$$

where $X$ = conversion (−), $P_{eq}$ = equilibrium decomposition pressure (atm), $P_{CO_2}$ = partial pressure of $CO_2$ (atm), $k_0$ = reaction rate constant ($s^{-1}$),

$E_a$ = activation energy (kJ/mol), $R$ = universal gas constant (kJ/mol K), $T$ = temperature (K), $W_0$ = initial weight of calcium carbonate (g), $W_t$ = weight of calcium carbonate after time $t$ (g).

The equilibrium partial pressure given by (Stanmore and Gilot, 2005):

$$P_{eq} = 4.137 \times 10^7 \ e^{-(20474/T)}$$

Using QWM, we can continuously monitor the change in weight with time. So using this data we can calculate d$X$/d$t$ as well as the conversion $X$.

---

**Example 16.1**

While calcination is done in presence of $CO_2$ at 900°C for 1590 s, the conversion obtained was 20%. So $dX/dt = 0.000126\ s^{-1}$.

For the same condition, the partial pressure of $CO_2$ = 1.010 atm while the equilibrium pressure for calcination reaction at 900°C is 1.087 atm.

Step 1: Substituting the values of $X$, $dX/dt$, $P_{CO_2}$, and $P_{eq}$ in Eq. (13.1), one can calculate the value of $k$.

$$k = 0.002212\ s^{-1}$$

Step 2: Now $k$ is the function of temperature so to find the reaction rate constant and activation energy, the above step has to be repeated at different temperatures and calculate $k$ for each temperature.

| Temperature (°C) | 900 | 950 | 1000 |
|---|---|---|---|
| $k$ ($s^{-1}$) | 0.00221 | 0.00310 | 0.00958 |

Step 3: Arrhenius plot and identifying the reaction rate constant ($k_0$) and the activation energy ($E_a$).

The Arrhenius plot is the plot of ln $k_0$ versus $1/T$. It is plotted from the data above. (Figure 13.6). So the intercept on $y$-axis of the plot will give the value of ln $k_0$ and the slope will give $-E_a/R$. Thus from this one can calculate the value of $k_0$ and $-E_a$.

From the above graph: $E_a/R = 21{,}717$

$$E_a = 21{,}717 \times 0.008314 = 180.56\ \text{kJ/mol}$$

And $\ln k_0 = 12.265$

$$k_0 = 212{,}139.64\ s^{-1}$$

So the final kinetic equation becomes:

$$\frac{dX}{dt} = 0.21 \times 10^6 \times e^{-\frac{180.56}{RT}} \times (1 - X) \times \frac{P_{eq} - P_{CO_2}}{P_{eq}}$$

---

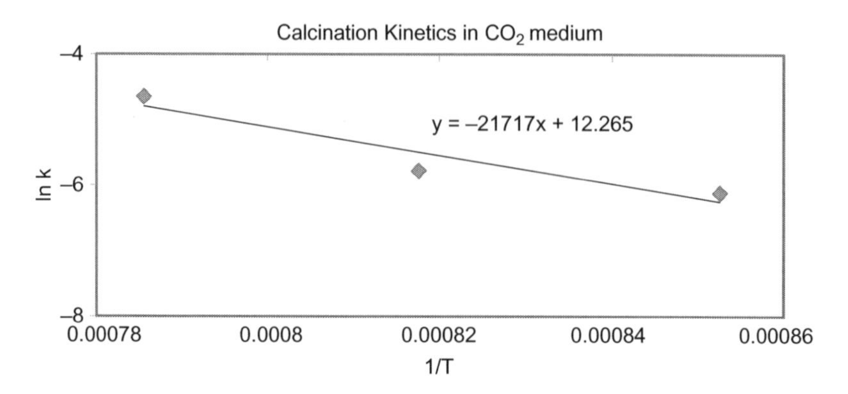


**FIGURE 13.6** Arrhenius plot to identify the reaction rate and activation energy.

## 13.5 PYROLYSIS-GAS CHROMATOGRAPHY/MASS SPECTROMETRY

Pyrolysis-gas chromatography/mass spectrometry (Py-GC/MS) is an analytical method to identify compounds. It involves heating the sample to high temperatures, where they are decomposed to smaller molecules, which are separated by gas chromatography and identified by mass spectrometry.

As described in Chapter 5, pyrolysis is a thermal decomposition of materials, which occurs at high temperatures (above 600°C) in the absence of oxygen. Usually, the sample is contacted with a platinum fuse wire or placed in a quartz tube. A high temperature of 600–1000°C or even hotter ambient is employed. Heating the sample is very rapid and temperatures of 700°C are reached in about 10 s. Resistive heating, isothermal furnace, and inductive heating methods are commonly used in such equipment. Large molecules breakdown (cleave) and produce more volatile fragments. A methylating reagent sometimes aids the production of the volatile fragments. The pyrolysis and gas chromatography/mass spectrometry (GC/MS) could take place in either one instrument or pyrolysis is separately performed before sending it to a GC/MS.

Once the sample is decomposed into smaller molecules, a small fraction of the volatile produced is injected into a GC at 300°C, along with an inert carrier gas like helium. The molecules are then carried into a 30 m GC column. The GC column is housed in an oven that maintains a temperature of 40–320°C. The inside of the column is coated with a special polymer. The mixture is separated depending on their volatility. Higher volatile particles travel faster through the column and lower volatile particles travel slower.

The volatile molecules are then ionized using an electric charge. The charged ions are then sent to an electromagnetic field that filters the ions based on their mass. The user can define the range of mass, through the filter, which continuously scans through the range of masses. An ion detector

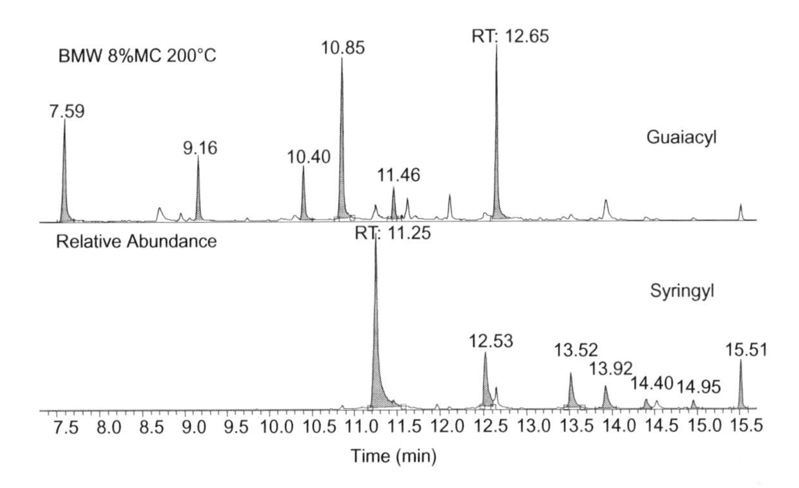


**FIGURE 13.7** Pyrogram ball milled wood hot-pressed poplar (200°C at 8% moisture content). Source: *From Osman (2010).*

then detects the number of ions with a specific mass. A mass spectrum is created using this information. It is essentially a graph of the number of ions with different masses that passed the filter. The graph from a Py-GC/MS contains the mass on the *x*-axis and its abundance on the *y*-axis (Figure 13.7).

Users can compare the mass spectrum of an unknown compound with a database of known compounds to identify them. This, however, needs extra care because the identification of a compound could be misleading or difficult.

Appendix A

# Definition of Biomass

Although it is generally agreed that biomass is formed from living species like plants and animals that is now alive or was alive a short time ago, its legal definition is less straightforward. A legal definition is necessary in some countries where special financial incentive or provisions are made for biomass-based product or energy.

**1.** In the United States, the definition of *biomass* has been hotly debated. Currently, the generally accepted definition can be found in the American Clean Energy and Security Act of 2009, H.R. 2454, excerpted as follows:

   *The term "renewable biomass" means any of the following:*

   **(A)** *Plant material, including waste material, harvested or collected from actively managed agricultural land that was in cultivation, cleared, or fallow and non-forested on the date of enactment of this section;*

   **(B)** *Plant material, including waste material, harvested or collected from pastureland that was non-forested on such date of enactment;*

   **(C)** *Nonhazardous vegetative matter derived from waste, including separated yard waste, landscape right-of-way trimmings, construction and demolition debris or food waste (but not municipal solid waste, recyclable waste paper, painted, treated or pressurized wood, or wood contaminated with plastic or metals);*

   **(D)** *Animal waste or animal byproducts, including products of animal waste digesters;*

   **(E)** *Algae;*

   **(F)** *Trees, brush, slash, residues, or any other vegetative matter removed from within 600 feet of any building, campground, or route designated for evacuation by a public official with responsibility for emergency preparedness, or from within 300 feet of a paved road, electric transmission line, utility tower, or water supply line;*

   **(G)** *Residues from or byproducts of milled logs;*

   **(H)** *Any of the following removed from forested land that is not Federal and is not high conservation priority land:*

      **(i)** *Trees, brush, slash, residues, inter-planted energy crops, or any other vegetative matter removed from an actively managed tree plantation established—*

**(I)** *Prior to the date of enactment of this section; or*
**(II)** *On land that, as of the date of enactment of this section, was cultivated or fallow and non-forested.*

**(ii)** *Trees, logging residue, thinnings, cull trees, pulpwood, and brush removed from naturally regenerated forests or other non-plantation forests, including for the purposes of hazardous fuel reduction or preventative treatment for reducing or containing insect or disease infestation.*

**(iii)** *Logging residue, thinnings, cull trees, pulpwood, brush and species that are non-native and noxious, from stands that were planted and managed after the date of enactment of this section to restore or maintain native forest types.*

**(iv)** *Dead or severely damaged trees removed within 5 years of fire, blow down, or other natural disaster, and badly infested trees:*

**(I)** *Materials, pre-commercial thinnings, or removed invasive species from National Forest System land and public lands (as defined in section 103 of the Federal Land Policy and Management Act of 1976 (43 U.S.C. 1702)), including those that are byproducts of preventive treatments (such as trees, wood, brush, thinnings, chips, and slash), that are removed as part of a federally recognized timber sale, or that are removed to reduce hazardous fuels, to reduce or contain disease or insect infestation, or to restore ecosystem health, and that are—*

**(i)** *Not from components of the National Wilderness Preservation System, Wilderness Study Areas, inventoried road-less areas, old growth or mature forest stands, components of the National Landscape Conservation System, National Monuments, National Conservation Areas, Designated Primitive Areas, or Wild and Scenic Rivers corridors;*
**(ii)** *Harvested in environmentally sustainable quantities, as determined by the appropriate Federal land manager; and*
**(iii)** *Harvested in accordance with Federal and State law and applicable land management plans.*

**2.** Another accepted definition is that of the Ontario Corporations Tax Act, excerpted as follows:

*The term "biomass resource" means*

**(a)** *organic matter that is derived from a plant and available on a renewable basis, including organic matter derived from dedicated energy crops, dedicated trees, agricultural food and feed crops, or*

**(b)** *waste organic material from harvesting or processing agricultural products, including animal waste and rendered animal fat, forestry products, including wood waste, and sewage.*

**3.** United Nations Framework Convention on Climate Change (UNFCCC) offers an alternative definition of biomass that must be used for calculation of tradable carbon credit under clean development mechanism (CDM). As per Annex 8 of EB 20[1] Report it is defined as:

*Biomass means non-fossilized and biodegradable organic material originating from plants, animals and micro-organisms. This shall also include products, by-products, residues and wastes from agriculture, forestry and related industries as well as the non-fossilized and biodegradable organic fractions of industrial and municipal wastes. Biomass also includes gases and liquids recovered from the decomposition of non-fossilized and biodegradable organic material.*

It is apparent from above that the first definition is most restrictive and aimed at one specific need. The third definition of UNFCCC is more general and scientific.

[1] UNFCCC, Annex 8, Clarifications on definition of biomass and consideration of changes in carbon pools due to a CDM project activity, EB 20 Report, Annex 8, p. 1. *cdm.unfccc.int/Reference/Guidclarif/mclbiocarbon.pdf*

Appendix B

# Physical Constants and Unit Conversions

## B1 PHYSICAL CONSTANTS

Atmospheric pressure
- 101.325/N/$m^2$
- 101.325/kPa
- 1.013/bar

Avogadro's number
- $6.022 \times 10^{23}$/mol

Boltzmann's constant
- $1.380 \times 10^{-23}$/J/K

Gravitational acceleration (sea level), $g$
- 9.807/m/$s^2$

Planck's constant
- $6.625 \times 10^{-34}$/J s

Speed of light in vacuum
- $2.998 \times 10^8$/m/s

Stefan–Boltzmann constant
- $5.670 \times 10^{-8}$/W/$m^2$ $K^4$

Universal gas constant, $R$
- $8.205 \times 10^{-2}$/$m^3$ atm/kmol K = $8.314 \times 10^{-2}$/$m^3$ bar/kmol K
- = 8.314/kJ/kmol K
- = 282/N m/kg K
- = 8.314/kPa $m^3$/kmol K
- = 1.98/kCal/kmol K

## B2 SUMMARY OF COMMON CONVERSION UNITS

**Length:** 1 m = $10^{-3}$ km = $10^{10}$ Angstrom units = $10^6$ micron = 39.370 inch = 3.28084 ft = 4.971 links = 1.0936 yd = 0.54681 fathoms = 0.04971 chain = $4.97097 \times 10^{-3}$ furlong = $5.3961 \times 10^{-4}$ UK nautical miles = $5.3996 \times 10^{-4}$ US nautical miles = $6.2137 \times 10^{-4}$ miles

**Area:** 1 $m^2$ = 1550.0 $inch^2$ = 10.7639 $ft^2$ = 1.19599 $yd^2$ = 2.47105 × $10^{-4}$ acre = 1 × $10^{-4}$ hectare = $10^{-6}$ $km^2$ = 3.8610 × $10^{-7}$ $miles^2$
**Density:** 1 $kg/m^3$ = $10^{-3}$ $g/cm^3$ = 0.06243 $lb_m/ft^3$ = 0.01002 lbm/UK gallons = 8.3454 × $10^{-3}$ $lb_m$/US gallons = 1.9403 × $10^{-3}$ $slug/ft^3$
**Energy:** 1 kJ = 238.85 cal = 2.7778 × $10^{-4}$ kW h = 737.56 ft $lb_f$ = 0.94782 Btu = 3.7251 × $10^{-4}$ hp h
**Heat transfer co-efficient:** 1 W/($m^2$ K) = 0.8598 kcal/($m^2$ h °C) = $10^{-4}$ W/($cm^2$ K) = 0.2388 × $10^{-4}$ cal/($cm^2$ s °C) = 0.1761 Btu/($ft^2$ h °F)
**Mass:** 1 kg = $10^{-3}$ tonne = 1.1023 × $10^{-3}$ US ton = 0.98421 × $10^{-3}$ UK ton = 2.20462 $lb_m$ = 0.06852 slug
**Mass flow rate:** 1 kg/s = 2.20462 lb/s = 132.28 lb/min = 7936.64 lb/h = 3.54314 long ton/h = 3.96832 short ton/h
**Power:** 1 W = 1 J/s = $10^{-3}$ kW = $10^{-6}$ MW = 0.23885 cal/s = 0.8598 kcal/h = 44.2537 ft $lb_f$/min = 3.41214 Btu/h = 0.73756 ft $lb_f$/s
**Pressure:** 1 bar = $10^5$ $N/m^2$ = $10^5$ Pa = 0.1 MPa = 1.01972 $kg/cm^2$ = 750.06 mmHg = 750.06 Torr = 10197 $mmH_2O$ = 401.47 $inchH_2O$ = 29.530 inchHg = 14.504 psi = 0.98692 atm = 0.0145 $kip/in^2$
**Specific energy:** 1 kJ/kg = 0.2388 cal/g = 0.2388 kcal/kg = 334.55 ft $lb_f$/$lb_m$ = 0.4299 Btu/$lb_m$
**Specific heat:** 1 kJ/(kg K) = 0.23885 cal/(g °C) = 0.23885 kcal/(kg °C) = 0.23885 Btu/($lb_m$ °F)
**Surface tension:** 1 N/m = 5.71015 × $10^{-3}$ $lb_f$/inch
**Temperature:** $T$(K) = $T$(°C) + 273.15 = [$T$(F) + 459.67]/1.8 = $T$(R)/1.8
**Thermal conductivity:** 1 W/(m K) = 0.8598 kcal/(m h °C) = 0.01 W/(cm K) = 0.01 W/(cm K) = 2.390 × $10^{-3}$ cal/(cm s °C) = 0.5782 Btu/(ft h °F)
**Torque:** 1 N m = 141.61 oz inch = 8.85073 $lb_f$ inch = 0.73756 $lb_f$ ft = 0.10197 $kg_f$ m
**Velocity:** 1 m/s = 100 cm/s = 196.85 ft/min = 3.28084 ft/s = 2.23694 mile/h = 2.23694 mph = 3.6 km/h = 1.94260 UK knot = 1.94384 Int. knot
**Viscosity (dynamic):** 1 kg/(m s) = 1 (N s)/$m^2$ = 1 Pa s = 10 Poise = 2419.1 $lb_m$/(ft h) = $10^3$ centipoise = 75.188 slug/(ft h) = 0.6720 $lb_m$/(ft s) = 0.02089 ($lb_f$ s)/$ft^2$
**Viscosity (kinematic):** 1 $m^2$/s = 3600 $m^2$/h = 38,750 $ft^2$/h = 10.764 $ft^2$/s
**Volume:** 1 $m^3$ = 61,024 $inch^3$ = 1000 liters = 219.97 UK gallons = 264.17 US gallons = 35.3147 $ft^3$ = 1.30795 $yd^3$ = 1 stere = 0.81071 × $10^{-3}$ acre-foot = 6.289 barrel (oil) = 8.648 US dry barrel
**Volume flow rate:** 1 $m^3$/s = 35.3147 $ft^3$/s = 2118.9 $ft^3$/min = 13,198 UK gallons/min = 791,891 UK gallons/h = 15,850 gallons/min = 951,019 US gallon/h

**TABLE B.3 Detailed Conversion Factors**

| Multiply | By | To Obtain |
|---|---|---|
| acre | 4046.86 | square meter |
| ampere/centimeter | 2.54000 | ampere/inch |
| ampere/inch | 39.3701 | ampere/meter |
| ampere/pound (mass | 2.20462 | ampere/kilogram |
| ampere/square foot | 10.7639 | ampere/square meter |
| ampere/square inch | 1550.00 | ampere/square meter |
| ampere/square meter | 0.092903 | ampere/square foot |
| ampere/volt | 1.00000 | siemens |
| ampere/volt inch | 39.3701 | siemens/meter |
| ampere/weber | 1.00000 | unit/henry |
| ampere turn | 1.25664 | gilbert |
| ampere turn/inch | 39.3701 | ampere turn/meter |
| ampere turn/meter | 0.012566 | Oersted |
| atmosphere (kilogram (force)/ square centimeter) | 98.0665 | kilopascal |
| atmosphere (760 Torr) | 101.325 | kilopascal |
| bar | 100.000 | kilopascal |
| barrel (42 US gallons) | 0.158987 | cubic meter |
| barrel/ton (UK) | 0.156476 | cubic meter/metric ton |
| barrel/ton (US) | 0.175254 | cubic meter/metric ton |
| barrel/hour | 0.044163 | cubic decimeter/second |
| barrel/million standard cubic feet | 0.133010 | cubic decimeters/kilomol |
| British thermal unit | 0.251996 | kilo calorie |
| Btu (mean) | 1.05587 | kilojoule |
| Btu (thermochemical) | 1.05435 | kilojoule |
| Btu (39°F) | 1.05967 | kilojoule |
| Btu (60°F) | 1.05468 | kilojoule |
| Btu (IT) | 1.05506 | kilojoule |

(*Continued*)

**TABLE B.3 (Continued)**

| Multiply | By | To Obtain |
|---|---|---|
| Btu (IT)/brake horsepower hour | 0.000393 | kilowatt/kilowatt |
| Btu (IT)/cubic foot | 37.2589 | kilojoule/cubic meter |
| Btu (IT)/hour | 0.293017 | watt |
| Btu (IT)/hour cubic foot | 0.010349 | kilowatt/cubic meter |
| Btu (IT)/hour cubic foot °F | 0.018629 | kilowatt/cubic meter kelvin |
| Btu (IT)/hour square foot | 3.15459 | watt/square meter |
| Btu (IT)/hour square foot °F | 5.67826 | watt/square meter kelvin |
| Btu (IT)/hour square foot °F/foot | 1.73074 | watt/meter kelvin |
| Btu (IT)/minute | 0.017581 | kilowatt |
| Btu (IT)/pound mol | 2.32600 | joule/mol |
| Btu (IT)/pound mol °F | 4.18680 | kilojoule/kilomol kelvin |
| Btu (IT)/pound (mass) | 0.555555 | kilocalorie/kilogram |
| Btu (IT)/pound (mass) | 2.32600 | kilojoule/kilogram |
| Btu (IT)/pound (mass) °F | 4.18680 | kilojoule/kilogram kelvin |
| Btu (IT)/second | 1.05487 | kilowatt |
| Btu (IT)/second cubic foot | 37.2590 | kilowatt/cubic meter |
| Btu (IT)/second cubic foot °F | 67.0661 | kilowatt/cubic meter kelvin |
| Btu (IT)/second square foot | 11.3565 | kilowatt/square meter |
| Btu (IT)/second square foot °F | 20.4418 | kilowatt/square meter kelvin |
| Btu (IT)/gallon (UK) | 232.080 | kilojoules/cubic meter |
| Btu (IT)/gallon (US) | 278.716 | kilojoules/cubic meter |
| calorie (IT) | 4.18680 | joule |
| calorie (mean) | 4.19002 | joule |
| calorie (15°C) | 4.18580 | joule |
| calorie (TC) | 4.18400 | joule |
| calorie (TC) | 0.003966 | Btu (IT) |
| calorie (20°C) | 4.18190 | joule |
| calorie (TC)/gram kelvin | 4.18400 | kilojoule/kilogram kelvin |

(*Continued*)

**TABLE B.3 (Continued)**

| Multiply | By | To Obtain |
|---|---|---|
| calorie (TC)/hour cubic centimeter | 1.16222 | kilowatt/cubic meter |
| calorie (TC)/hour square centimeter | 0.011622 | kilowatt/square meter |
| calorie (TC)/milliliter | 4.18400 | megajoule/cubic meter |
| calorie (TC)/pound (mass) | 9.22414 | joule/kilogram |
| calorific heat hour | 2.64778/ | megajoule |
| calorific value | 0.795500 | kilowatt |
| calorific heat unit | 1.89910 | kilojoules |
| candela/square meter | 0.291864 | foot lambert |
| candela/square meter | 0.000314 | lambert |
| centimeter water (4°C) | 0.098064 | kilopascals |
| centipoises | 0.001000 | pascal second |
| centistokes | 1.00000 | square millimeter/second |
| chain | 20.1168 | meter |
| coulomb/cubic foot | 35.3146 | coulomb/cubic meter |
| coulomb/foot | 3.28084 | coulomb/meter |
| coulomb/inch | 39.3701 | coulomb/meter |
| coulomb/meter | 0.025400 | coulomb/inch |
| coulomb/square foot | 10.7639 | coulomb/square meter |
| coulomb/square meter | 0.092930 | coulomb/square foot |
| cubic centimeter | 0.035195 | ounce fluid (UK) |
| cubic centimeter | 0.033814 | ounce fluid (US) |
| cubic centimeter/cubic meter | 0.034972 | gallon (UK)/1000 oil barrels |
| cubic centimeter/cubic meter | 1.00000 | volume parts/million |
| cubic decimeter/second | 2.11888 | cubic foot/minute |
| cubic decimeter/second | 0.035315 | cubic foot/second |
| cubic decimeter/metric ton | 0.005706 | oil barrel/ton (US) |
| cubic decimeter/metric ton | 0.006391 | barrel/ton (UK) |

*(Continued)*

**TABLE B.3 (Continued)**

| Multiply | By | To Obtain |
|---|---|---|
| cubic decimeter/metric ton | 0.268411 | gallon (US)/ton (UK) |
| cubic decimeter/metric ton | 0.239653 | gallon (US)/ton (US) |
| cubic foot | 0.028317 | cubic meter |
| cubic foot | 28.3169 | cubic decimeter |
| cubic foot/foot | 0.092903 | cubic meter/meter |
| cubic foot/hour | 0.007866 | cubic decimeter/second |
| cubic foot/minute | 0.471947 | cubic decimeter/second |
| cubic foot/minute square foot | 0.005080 | cubic meter/second square meter |
| cubic foot/pound (mass) | 62.4280 | cubic decimeter/kilogram |
| cubic foot/pound (mass) | 0.062428 | cubic meter/kilogram |
| cubic foot/second | 28.3169 | cubic decimeter/second |
| cubic inch | 0.016387 | cubic decimeter |
| cubic kilometer | 0.239913 | cubic mile |
| cubic meter | 6.28976 | barrel (42 US gallons) |
| cubic meter | 35.3147 | cubic foot |
| cubic meter | 1.30795 | cubic yard |
| cubic meter | 219.969 | gallon (UK) |
| cubic meter | 264.172 | gallon (US) |
| cubic meter/kilogram | 16.0185 | cubic foot/pound (mass) |
| cubic meter/meter | 10.7639 | cubic foot/foot |
| cubic meter/meter | 80.5196 | gallon (US)/foot |
| cubic meter/second/meter | 4022.80 | gallon (UK)/minute/foot |
| cubic meter/second/meter | 4831.18 | gallon (US)/minute/foot |
| cubic meter/second/meter | 20,114.0 | gallon (UK)/hour/inch |
| cubic meter/second/meter | 24,155.9 | gallon (US)/hour/inch |
| cubic meter/second/square meter | 88,352.6 | gallon (US)/hour/square foot |
| cubic meter/square meter | 3.28084 | cubic foot/second/square foot |
| cubic meter/second/ square meter | 196.850 | cubic foot/minute square foot |

*(Continued)*

**TABLE B.3 (Continued)**

| Multiply | By | To Obtain |
|---|---|---|
| cubic meter/second/ square meter | 510.895 | gallon (UK)/hour square inch |
| cubic meter/square meter | 613.560 | gallon (US)/hour square inch |
| cubic meter/metric ton | 5.70602 | oil barrel/ton (US) |
| cubic meter/metric ton | 6.39074 | oil barrel/ton (UK) |
| cubic mile | 4.16818 | cubic kilometer |
| cubic yard | 0.764555 | cubic meter |
| degree Celsius (difference) | (9/5) | degree Fahrenheit (difference) |
| degree Celsius (traditional) | (9/5)°C + 32 | degree Fahrenheit (traditional) |
| degree Fahrenheit/100 feet | 0.018227 | kelvin/meter |
| degree Fahrenheit (difference) | (5/9) | degree Celsius (difference) |
| degree Fahrenheit (traditional) | (5/9)°F − 32 | degree Celsius (traditional) |
| degree Rankine | (5/9) | kelvin |
| degree (angle) | 0.017453 | radian |
| dyne | 0.000010 | newton |
| dyne/square centimeter | 0.100000 | pascal |
| dyne second/square centimeter | 0.100000 | pascal second |
| farad/inch | 39.3701 | farad/meter |
| farad/meter | 0.025400 | farad/inch |
| fathom (US) | 1.82880 | meter |
| foot | 304.800 | millimeter |
| foot lambert | 3.42626 | candel/square meter |
| foot degree F | 0.548640 | meter/kelvin |
| foot/gallon (US) | 80.5196 | meter/cubic meter |
| foot/barrel (oil barrel) | 1.91713 | meter/cubic meter |
| foot/cubic foot | 10.7639 | meter/cubic meter |
| foot/day | 0.003528 | millimeter/second |
| foot/hour | 0.084667 | millimeter/second |
| foot/mile | 0.189394 | meter/kilometer |

(*Continued*)

**TABLE B.3 (Continued)**

| Multiply | By | To Obtain |
|---|---|---|
| foot/minute | 0.005080 | meter/second |
| foot/second | 0.304800 | meter/second |
| foot poundal | 0.042140 | joule |
| foot pound (force) | 1.35582 | joule |
| foot pound (force)/gallon (US) | 0.3358169 | kilojoule/cubic meter |
| foot pound (force)/second | 1.35582 | watt |
| foot pound (force)/square inch | 0.210152 | joule/square centimeter |
| footcandle | 10.7639 | lux |
| gallon (UK) | 0.004546 | cubic meter |
| gallon (UK)/hour foot | $4.14306 \times 10^{-6}$ | cubic meter/second meter |
| gallon (UK)/hour square foot | $1.35927 \times 10^{-5}$ | cubic meter/second square meter |
| gallon (UK)/minute | 0.075768 | cubic decimeter/second |
| gallon (UK)/minute foot | 0.000249 | cubic meter/second meter |
| gallon (UK)/minute square foot | 0.000816 | cubic meter/second square meter |
| gallon (UK)/pound (mass) | 10.0224 | cubic decimeter/kilogram |
| gallon (UK)/1000 barrels | 28.5940 | cubic centimeter/cubic meter |
| gallon (US) | 0.003785 | cubic meter |
| gallon (US)/cubic foot | 133.681 | cubic decimeter/cubic meter |
| gallon (US)/foot | 0.012419 | cubic meter/meter |
| gallon (US)/hour foot | $3.44981 \times 10^{-6}$ | cubic meter/second meter |
| gallon (US)/hour square foot | $1.13183 \times 10^{-5}$ | cubic meter/second square meter |
| gallon (US)/minute | 0.063090 | cubic decimeter/second |
| gallon (US)/minute foot | 0.000207 | cubic meter/second meter |
| gallon (US)/minute square foot | 0.000679 | cubic meter/second square meter |
| gallon (US)/pound (mass) | 8.34540 | cubic decimeter/kilogram |
| gallon (US)/ton (UK) | 3.72563 | cubic decimeter/metric ton |
| gallon (US)/1000 barrels | 23.8095 | cubic centimeter/cubic meter |
| gauss | 0.000100 | telsa |
| gauss/Oersted | $1.25664 \times 10^{-6}$ | henry/meter |

(*Continued*)

**TABLE B.3 (Continued)**

| Multiply | By | To Obtain |
|---|---|---|
| gilbert | 0.795775 | ampere turn |
| gilbert/maxwell | $7.95575 \times 10^7$ | unit/henry |
| grain | 64.7989 | milligram |
| grain/cubic foot | 2.28835 | milligram/cubic decimeter |
| grain/100 cubic feet | 22.8835 | milligram/cubic meter |
| gram | 0.035274 | ounce (avoirdupois) |
| gram | 0.032151 | ounce (troy) |
| gram mol | 0.001000 | kilomol |
| gram/cubic meter | 3.78541 | milligram/gallon (US) |
| gram/cubic meter | 0.058418 | grains/gallon (US) |
| gram/cubic meter | 0.350507 | pound (mass)/1000 barrels |
| gram/cubic meter | 0.008345 | pound (mass)/1000 gallons (US) |
| grams/cubic meter | 0.010022 | pound (mass)/1000 gallons (UK) |
| grams/gallon (UK) | 0.219969 | kilogram/cubic meter |
| grams/gallon (US) | 0.264172 | kilogram/cubic meter |
| gray | 100.000 | rad |
| henry | $7.95775 \times 10^7$ | maxwell/gilbert |
| henry | 1.00000 | weber/ampere |
| henry | $1.00000 \times 10^8$ | line/ampere |
| henry/meter | 795,775 | gauss/Oersted |
| henry/meter | $2.54000 \times 10^6$ | lines/ampere inch |
| horsepower (electric) | 0.746000 | kilowatt |
| horsepower (hydraulic) | 0.745700 | kilowatt |
| horsepower (US) | 0.745702 | kilowatt |
| horsepower (US) | 42.4150 | Btu/minute |
| horsepower hour (US) | 2.68452 | megajoule |
| horsepower hour (US) | 2544.433 | Btu (IT) |
| horsepower/cubic foot | 26.3341 | kilowatt/cubic meter |
| hundred weight (UK) | 50.8024 | kilogram |

*(Continued)*

**TABLE B.3 (Continued)**

| Multiply | By | To Obtain |
|---|---|---|
| hundred weight (US) | 45.3592 | kilogram |
| inch | 25.4000 | millimeter |
| inch water (39.2°F) | 0.249082 | kilopascal |
| inch mercury (32°F) | 3.38639 | kilopascal |
| inches/minute | 0.423333 | millimeter/second |
| inches/second | 25.4000 | millimeter/second |
| joule | 0.737562 | foot pound (force) |
| joule | 23.7304 | foot poundal |
| joule | 1.00000 | watt second |
| joule | 0.239126 | calorie (20°C) |
| joule | 0.238903 | calorie (15°C) |
| joule | 0.238662 | calorie (mean) |
| joule | 0.238846 | calorie (IT) |
| joule | 0.239006 | calorie (TC) |
| joule/kilogram | 0.108411 | calorie (TC)/pound (mass) |
| joule/mol | 0.429923 | Btu (IT)/pound mol |
| joule/square centimeter | 4.75846 | foot pound (force)/square inch |
| joule/square centimeter | 0.101972 | kilogram meter/square centimeter |
| kelvin (degree) | (9/5) | degree Rankine |
| kelvin (degree) (minus) | −273.16 | degree centigrade |
| kilocalorie (TC) | 4.18400 | kilojoule |
| kilocalorie (TC)/hour | 1.16222 | watt |
| kilocalorie (TC)/hour square meter °C | 1.16222 | watt/square meter °K |
| kilocalorie (TC)/kilogram °C | 4.18400 | kilojoule/kilogram °K |
| kilogram | 0.196841 | hundred weight (UK) |
| kilogram | 0.220462 | hundred weight (US) |
| kilogram | 2.20462 | pound (avoirdupois) |
| kilogram meter/second | 7.23301 | pound (mass) foot/second |

(*Continued*)

**TABLE B.3 (Continued)**

| Multiply | By | To Obtain |
|---|---|---|
| kilogram meter/square centimeter | 9.80665 | joule/square centimeter |
| kilogram/cubic decimeter | 8.34541 | pound (mass)/gallon (US) |
| kilogram/cubic decimeter | 10.0224 | pound (mass)/gallon (UK) |
| kilogram/cubic meter | 0.062428 | pound (mass)/cubic foot |
| kilogram/cubic meter | 0.350507 | pound (mass)/barrel |
| kilogram/cubic meter | 3.78541 | grams/gallon (US) |
| kilogram/cubic meter | 4.54609 | grams/gallon (UK) |
| kilogram/meter | 0.671969 | pound (mass)/foot |
| kilogram/mol | 2.20462 | pound (mass)/mol |
| kilogram/second | 7936.64 | pound (mass)/hour |
| kilogram/second | 2.20462 | pound (mass)/second |
| kilogram/second | 0.059052 | ton (mass)(UK)/minute |
| kilogram/second | 0.066139 | ton (mass)(US)/minute |
| kilogram/second | 3.54314 | ton (mass)(UK)/hour |
| kilogram/second | 3.96832 | ton (mass)(US)/hour |
| kilogram/second | 31,058.5 | ton (mass)(UK)/year |
| kilogram/second | 34,762.5 | ton (mass)(US)/year |
| kilogram/second meter | 0.671969 | pound (mass)/second foot |
| kilogram/second meter | 2419.09 | pound (mass)/hour foot |
| kilogram/second square meter | 0.204816 | pound (mass)/second square foot |
| kilogram/second square meter | 737.338 | pound (mass)/hour square foot |
| kilogram/square meter | 0.204816 | pound (mass)/square foot |
| kilojoule | 0.947817 | Btu (IT) |
| kilojoule | 0.943690 | Btu (39°F) |
| kilojoule | 0.948155 | Btu (60°F) |
| kilojoule | 0.947086 | Btu (mean) |
| kilojoule | 0.948452 | Btu (TC) |
| kilojoule/cubic meter | 0.026839 | Btu (IT)/cubic foot |

(*Continued*)

**TABLE B.3 (Continued)**

| Multiply | By | To Obtain |
|---|---|---|
| kilojoule/cubic meter | 0.004309 | Btu (IT)/gallon (UK) |
| kilojoule/cubic meter | 0.003588 | Btu (IT)/gallon (US) |
| kilojoule/cubic meter | 2.79198 | footpound (force)/gallon (US) |
| kilojoule/kilogram | 0.429923 | Btu (IT)/pound (mass) |
| kilojoule/kilogram kelvin | 0238846 | Btu (IT)/pound (mass) °F |
| kilojoule/kilogram kelvin | 0.238846 | Btu (IT)/pound mol °F |
| kilojoule/kilogram kelvin | 0.239006 | calorie (TC)/gram kelvin |
| kilojoule/kilogram kelvin | 0.239006 | calorie (TC)/gram mol °C |
| kilojoule/kilogram kelvin | 0.239006 | kilocalorie (TC)/kilogram °C |
| kilojoule/kilogram kelvin | 0.000278 | kilowatt hour/kilogram °C |
| kilojoule/mol | 0.239006 | kilocalorie (TC)/gram mol |
| kilometer | 0.621371 | mile |
| kilometer | 0.539957 | nautical mile |
| kilometer/cubic decimeter | 2.35215 | mile/gallon (US) |
| kilometer/hour | 0.539957 | knot |
| kilometer/hour | 0.621371 | miles/hour |
| kilomol | 1000.00 | gram mol |
| kilomol | 2.20462 | pound mol |
| kilomol | 836.610 | standard cubic foot (60°F, 1 atmosphere) |
| kilomol | 22.4136 | standard cubic meter (0°C, 1 atmosphere) |
| kilomol | 23.6445 | standard cubic meter (15°C, 1 atmosphere) |
| kilomol/cubic meter | 0.0624280 | pound mol/cubic foot |
| kilomol/cubic meter | 0.010022 | pound mol/gallon (UK) |
| kilomol/cubic meter | 0.008345 | pound mol/gallon (US) |
| kilomol/cubic meter | 133.010 | standard cubic foot/barrel (60°F, 1 atmosphere) |
| kilomol/second | 2.20462 | pound mol/second |

(*Continued*)

**TABLE B.3 (Continued)**

| Multiply | By | To Obtain |
|---|---|---|
| kilomol/second | 7936.64 | pound mol/hour |
| kilonewton | 0.224809 | kip (1000 foot pound) |
| kilonewton | 0.100361 | ton (force) (UK) |
| kilonewton | 0.112405 | ton (force) (US) |
| kilonewton meter | 0.368781 | ton force (US) foot |
| kilopascal | 0.010197 | kilogram/square centimeter |
| kilopascal | 0.009869 | atmosphere (760 Torr) |
| kilopascal | 0.01000 | bar |
| kilopascal | 10.1974 | centimeter water (4°C) |
| kilopascal | 4.01474 | inch water (39.2°F) |
| kilopascal | 0.295300 | inch mercury (32°F) |
| kilopascal | 0.296134 | inch mercury (60°F) |
| kilopascal | 7.50062 | millimeter mercury (0°C) |
| kilopascal | 20.8854 | pound (force)/square foot |
| kilopascal | 0.145038 | pound (force)/square inch |
| kilopascal/meter | 0.044208 | pound (force)/square inch/foot |
| kilopascal second | 0.145038 | pound (force)/square inch |
| kilowatt | 56.8690 | Btu (IT)/minute |
| kilowatt | 0.947817 | Btu (IT)/second |
| kilowatt | 1.35962 | calorific value |
| kilowatt | 1.34048 | horsepower (electric) |
| kilowatt | 1.34102 | horsepower (550 foot pound/ second) |
| kilowatt | 1.34102 | horsepower (hydraulic) |
| kilowatt | 0.284345 | ton of refrigeration |
| kilowatt hour | 3.60000 | megajoule |
| kilowatt hour/kilogram °C | 3600.00 | kilojoule/kilogram kelvin |
| kilowatt/cubic meter | 96.6211 | Btu (IT)/hour cubic foot |
| kilowatt/cubic meter | 0.026839 | Btu (IT)/second cubic foot |

*(Continued)*

**TABLE B.3 (Continued)**

| Multiply | By | To Obtain |
|---|---|---|
| kilowatt/cubic meter | 0.860421 | calorie (TC)/hour cubic centimeter |
| kilowatt/cubic meter | 0.037974 | horsepower/cubic foot |
| kilowatt/cubic meter kelvin | 53.6784 | Btu (IT)/hour cubic foot °F |
| kilowatt/cubic meter kelvin | 0.014911 | Btu (IT)/second cubic foot °F |
| kilowatt/kilowatt | 2544.43 | Btu (IT)/brake horsepower hour |
| kilowatt/square meter | 0.088055 | Btu (IT)/second square foot |
| kilowatt/square meter | 86.0421 | calorie (TC)/hour cubic centimeter |
| kilowatt/square meter kelvin | 0.048919 | Btu (IT)/second square foot °F |
| kip (1000 foot pounds) | 4.44822 | kilonewton |
| kip/square inch | 6.89476 | megapascal |
| knot | 1.85200 | kilometer/hour |
| lambert | 3183.10 | candela/square meter |
| line | 1.00000 | maxwell |
| line | $1.00000 \times 10^{-8}$ | weber |
| lines/ampere | $1.00000 \times 10^{-8}$ | henry |
| lines/ampere inch | $3.93701 \times 10^{-7}$ | henry/meter |
| lines/square inch | 1550.0031 | telsa |
| link | 0.201168 | meter |
| lumen/square foot | 10.7639 | lux |
| lumen/square inch | 1550.00 | lux |
| lux | 0.092903 | footcandle |
| lux | 0.092903 | lumen/square foot |
| lux | 0.000645 | lumen/square inch |
| lux second | 0.092903 | foot candle second |
| maxwell | 1.00000 | line |
| maxwell | $1.00000 \times 10^{-8}$ | weber |
| maxwell/gilbert | $7{,}95775 \times 10^{7}$ | henry |
| megagram | 1.00000 | ton (mass) (metric) |
| megagram | 0.984206 | ton (mass) (UK) |

(*Continued*)

**TABLE B.3 (Continued)**

| Multiply | By | To Obtain |
|---|---|---|
| megagram | 1.10231 | ton (mass) (US) |
| megagram/square meter | 0.102408 | ton (mass) (US)/square foot |
| megajoule | 947.817 | Btu (IT) |
| megajoule | 0.377675 | calorific value hour |
| megajoule | 0.372506 | horsepower hour |
| megajoule | 0.277778 | kilowatt hour |
| megajoule | 0.009478 | therm |
| megajoule | 0.102408 | ton (mass) (US) mile |
| megajoule/cubic meter | 4.30886 | Btu (IT)/gallon (UK) |
| megajoule/cubic meter | 3.58788 | Btu (IT)/gallon (US) |
| megajoule/cubic meter | 0.239006 | calorie (TC)/milliliter |
| megajoule/meter | 0.021289 | ton (force) (US) mile/foot |
| megapascal | 0.145038 | kip/square inch |
| megapascal | 145.038 | pound/square inch |
| megapascal | 10.4427 | ton (force) (US)/square foot |
| megapascal | 0.072519 | ton (force) (US)/square inch |
| megawatt | 3.41214 | million Btu (IT)/hour |
| meter | 0.049710 | chain |
| meter | 0.546807 | fathom |
| meter | 3.28084 | feet |
| meter | 4.97097 | link |
| meter | 0.198839 | rod |
| meter | 1.09361 | yard |
| meter/cubic meter | 0.521612 | foot/barrel |
| meter/cubic meter | 0.092903 | foot/cubic foot |
| meter/cubic meter | 0.012419 | foot/gallon (US) |
| meter/kelvin | 1.82269 | foot/°F |
| meter/kilometer | 5.28000 | foot/mile |
| meter/second | 3.28084 | foot/second |

*(Continued)*

**TABLE B.3 (Continued)**

| Multiply | By | To Obtain |
|---|---|---|
| meter/second | 196.850 | foot/minute |
| microbar | 0.100000 | pascal |
| micrometer | 0.039370 | mil |
| micrometer | 1.00000 | micron |
| micron | 1.00000 | micrometer |
| microsecond/foot | 3.28084 | microsecond/meter |
| microsecond/meter | 0.304800 | microsecond/foot |
| mil | 25.4000 | micrometer |
| mile | 5280.00 | foot |
| mile | 1.60934 | kilometer |
| mile/gallon (US) | 0.425144 | kilometer/cubic decimeter |
| mile/hour | 1.60934 | kilometer/hour |
| milligram | 0.015432 | grain |
| milligram/cubic decimeter | 0.436996 | grain/cubic foot |
| milligram/cubic meter | 0.043700 | grain/100 cubic foot |
| milligram/gallon (US) | 0.264172 | gram/cubic meter |
| millimeter | 0.039370 | inch |
| millimeter | 0.003281 | foot |
| millimeter mercury (0°C) | 133.322 | pascal |
| millimeter mercury (0°C) | 0.133322 | kilopascal |
| millimeter/second | 283.465 | foot/day |
| millimeter/second | 11.8110 | foot/hour |
| millimeter/second | 2.36221 | inch/minute |
| millimeter/second | 0.039370 | inch/second |
| million Btu (IT)/hour | 0.293071 | megawatt |
| million electron volt | 0.160218 | picojoule |
| million pound (mass)/year | 0.014374 | kilogram/second |
| minute (angle) | 0.000291 | radian |
| mol/foot | 3.28084 | mol/meter |

(*Continued*)

**TABLE B.3 (Continued)**

| Multiply | By | To Obtain |
|---|---|---|
| mol/kilogram | 0.453592 | mol/pound (mass) |
| mol/meter | 0.304800 | mol/foot |
| mol/pound (mass) | 2.20462 | mol/kilogram |
| mol/square foot | 10.7639 | mol/square meter |
| mol/square meter | 0.092903 | mol/square foot |
| nautical mile | 1.85200 | kilometer |
| newton | $1.00000 \times 10^{5}$ | dyne |
| newton | 0.224809 | pound (force) |
| newton | 7.23301 | poundal |
| newton meter | 0.737562 | pound (force) foot |
| newton meter | 8.85075 | pound (force) inch |
| newton meter | 23.7304 | poundal foot |
| newton meter/meter | 0.018734 | pound (force) foot/inch |
| newton meter/meter | 0.224809 | pound (force) inch/inch |
| newton/meter | 0.068522 | pound (force)/foot |
| newton/meter | 0.005710 | pound (force)/inch |
| Oersted | 79.5775 | ampere turn/meter |
| ohm circular mil/foot | $1.66243 \times 10^{-9}$ | ohm square meter/meter |
| ohm foot | 0.304800 | ohm square meter/meter |
| ohm inch | 0.025400 | ohm square meter/meter |
| ohm square meter/meter | $6.01531 \times 10^{8}$ | ohm circular mil/foot |
| ohm square meter/meter | 3.28084 | ohm foot |
| ohm square meter/meter | 39.3701 | ohm inch |
| ounce (avoirdupois) | 28.3495 | gram |
| ounce (troy) | 31.1035 | gram |
| ounce (fluid) (UK) | 28.4131 | cubic centimeter |
| ounce (fluid) (US) | 29.5735 | cubic centimeter |
| pascal | 1 | newton/square meter |
| pascal | 10.0000 | dyne/square centimeter |

*(Continued)*

**TABLE B.3 (Continued)**

| Multiply | By | To Obtain |
|---|---|---|
| pascal | 10.0000 | microbar |
| pascal | 0.007501 | millimeter mercury (0°C) |
| pascal second | 1000.00 | centipoise |
| pascal second | 10.0000 | dyne second/square centimeter |
| pascal second | 0.020885 | pound (force) second/square foot |
| pascal second | 2419.09 | pound (mass)/foot hour |
| pascal second | 0.671969 | pound (mass)/foot second |
| picojoule | 6.241509 | million electron volt |
| pint (liquid) (UK) | 0.568262 | cubic decimeter |
| pint (liquid) (US) | 0.473167 | cubic decimeter |
| pint (UK)/1000 barrels | 3.57425 | cubic decimeter/cubic meter |
| pound mol | 0.453592 | kilomol |
| pound mol/cubic foot | 16.0185 | kilomol/cubic meter |
| pound mol/gallon (UK) | 99.7763 | kilomol/cubic meter |
| pound mol/gallon (US) | 119.826 | kilomol/cubic meter |
| pound mol/hour | 0.000126 | kilomol/second |
| pound mol/second | 0.453592 | kilomol/second |
| poundal | 0.138255 | newton |
| poundal | 0.031081 | pound (force) |
| poundal foot | 0.042140 | newton meter |
| pound (force) | 4.44822 | newton |
| pound (force) foot | 1.35582 | newton meter |
| pound (force) foot/inch | 53.3787 | newton meter/meter |
| pound (force) inch | 0.112985 | newton meter |
| pound (force)/foot | 14.5939 | newton/meter |
| pound (force)/inch | 175.127 | newton/meter |
| pound (force)/square foot | 0.047880 | kilopascal |
| pound (force)/square inch | 6.89476 | kilopascal |
| pound (force)/square inch/foot | 22.6206 | kilopascal/meter |

*(Continued)*

**TABLE B.3 (Continued)**

| Multiply | By | To Obtain |
|---|---|---|
| pound (force)/square foot | 47.8803 | pascal second |
| pound (mass) | 32.1719 | poundal |
| pound (mass) | 0.453592 | kilogram |
| pound (mass) | 1.21528 | pound (troy) |
| pound (mass)/barrel | 2.85301 | kilogram/cubic meter |
| pound (mass)/Btu | 1.80018 | kilogram/kilogram calorie |
| pound (mass)/cubic foot | 16.0185 | kilogram/cubic meter |
| pound (mass)/foot | 1.48816 | kilogram/meter |
| pound (mass)/foot hour | 0.000413 | pascal second |
| pound (mass)/foot second | 1.48816 | pascal second |
| pound (mass)/gallon (UK) | 0.099776 | kilogram/cubic decimeter |
| pound (mass)/gallon (US) | 0.119826 | kilogram/cubic decimeter |
| pound (mass)/1000 gallons (UK) | 99.7763 | gram/cubic meter |
| pound (mass)/1000 gallons (US) | 119.826 | gram/cubic meter |
| pound (mass)/hour | 0.000126 | kilogram/second |
| pound (mass)/hour foot | 0.000413 | kilogram/second meter |
| pound (mass)/hour square foot | 0.001356 | kilogram/second square meter |
| pound (mass)/minute | 0.007560 | kilogram/second |
| pound (mass)/mol | 0.453592 | kilogram/mol |
| pound (mass)/second foot | 1.48816 | kilogram/second meter |
| pound (mass)/second square foot | 4.88243 | kilogram/second square meter |
| pound (mass)/square foot | 4.88243 | kilogram/square meter |
| pound (mass)/second/second | 0.138255 | kilogram meter/second |
| pound (mass)/second/ square foot | 0.042140 | kilogram square meter |
| quart (dry) (UK) | 0.968939 | quart (dry) (US) |
| quart (liquid) (UK) | 1.136523 | cubic decimeter |
| quart (liquid) (UK) | 1.136523 | liter |

*(Continued)*

**TABLE B.3 (Continued)**

| Multiply | By | To Obtain |
|---|---|---|
| quart (liquid) (UK) | 1.20030 | quart (liquid) (US) |
| quart (liquid) (US) | 0.946353 | liter |
| quart (liquid) (US) | 1.163647 | quart (dry) (US) |
| rad | 0.010000 | gray |
| radian | $2.06265 \times 10^5$ | second (angle) |
| radian | 3437.75 | minute (angle) |
| radian | 57.2958 | degree (angle) |
| radian/second | 0.159155 | revolutions/second |
| radian/second | 9.54930 | revolutions/minute |
| radian/second squared | 0.159155 | revolutions/second squared |
| radian/second squared | 572.958 | revolutions/minute squared |
| revolutions/minute | 0.104720 | radian/second |
| revolutions/minute squared | 0.001745 | radian/second squared |
| revolutions/second | 6.28319 | radian/second |
| revolutions/second squared | 6.28319 | radian/second squared |
| rod | 5.02920 | meter |
| second (angle) | $4.84814 \times 10^{-6}$ | radian |
| section | 2.58999 | square kilometer |
| siemens | 1.00000 | ampere/volt |
| siemens/meter | 0.025400 | ampere/volt inch |
| slug | 14.59 | 3903 kilogram/meter |
| slug/cubic foot | 515.379 | kilogram/cubic meter |
| square foot | 0.092903 | square meter |
| square foot/cubic inch | 5669.29 | square meter/cubic meter |
| square foot/hour | 25.8064 | square millimeter/second |
| square foot/pound (mass) | 0.204816 | square meter/kilogram |
| square foot/second | 92903.04 | square millimeter/second |
| square inch | 645.160 | square millimeter |
| square kilometer | 0.386102 | section |

*(Continued)*

**TABLE B.3 (Continued)**

| Multiply | By | To Obtain |
|---|---|---|
| square kilometer | 0.386102 | square mile |
| square meter | 10.7639 | square foot |
| square meter | 0.000247 | acre |
| square meter | 1.19599 | square yard |
| square meter/cubic meter | 0.000176 | square foot/cubic inch |
| square meter/kilogram | 4.88243 | square foot/pound (mass) |
| square mile | 2.58999 | square kilometer |
| square millimeter | 0.001550 | square inch |
| square millimeter/second | $1.07639 \times 10^{-5}$ | square foot/second |
| square millimeter/second | 0.038750 | square foot/hour |
| square millimeter/second | 1.00000 | centistoke |
| square yard | 0.836127 | square meter |
| tesla | 10,000.0 | gauss |
| tesla | 64,516.0 | lines/square inch |
| therm | 105.506 | megajoule |
| ton (force) (UK) | 9.96402 | kilonewton |
| ton (force) (US) | 8.89644 | kilonewton |
| ton (force) (US) foot | 2.71164 | kilonewton meter |
| ton (force) (US) mile | 14.3174 | megajoule |
| ton (force) (US) mile/foot | 46.9732 | megajoule/meter |
| ton (force) (US)/square foot | 0.095761 | megapascal |
| ton (force) (US)/square inch | 13.7895 | megapascal |
| ton (mass) (UK) | 1.01605 | megagram |
| ton (mass) (UK) | 1.01605 | metric ton |
| ton (mass) (UK) | 1.12000 | ton (mass) (US) |
| ton (mass) (US) | 0.907185 | megagram |
| ton (mass) (US) | 0.907185 | metric ton |
| ton (mass) (US) | 0.892857 | ton (mass) (UK) |
| ton (metric) | 1.00000 | megagram |

(*Continued*)

**TABLE B.3 (Continued)**

| Multiply | By | To Obtain |
|---|---|---|
| ton (mass) (UK)/day | 0.011760 | kilogram/second |
| ton (mass) (US)/day | 0.010500 | kilogram/second |
| ton (mass) (UK)/hour | 0.282235 | kilogram/second |
| ton (mass) (US)/hour | 0.251996 | kilogram/second |
| ton (mass) (UK)/minute | 16.9341 | kilogram/second |
| ton (mass) (US)/minute | 15.1197 | kilogram/second |
| ton (mass) (US)/square foot | 9.76486 | megagram/square meter |
| ton refrigeration | 3.51685 | kilowatt |
| unit/foot | 3.28084 | unit/meter |
| unit/henry | 1.00000 | ampere/weber |
| unit/henry | $1.25664 \times 10^{-8}$ | gilbert/maxwell |
| unit/meter | 3.28084 | volt/meter |
| volt/foot | 3.28084 | volt/meter |
| volt/inch | 39.3701 | volt/meter |
| volume parts per million | 1.00000 | cubic centimeter/cubic meter |
| watt | 3.412142 | Btu (IT)/hour |
| watt | 44.2537 | foot pound (force)/minute |
| watt | 0.737562 | foot pound (force)/second |
| watt | 0.860421 | kilocalorie (TC)/hour |
| watt hour | 3.60000 | kilojoule |
| watt/inch | 39.3701 | watt/meter |
| watt/meter | 0.025400 | watt/inch |
| watt/meter kelvin | 0.577789 | Btu (IT)/hour square foot °F/foot |
| watt/meter kelvin | 6.93347 | Btu (IT)/hour square foot °F/inch |
| watt/meter kelvin | 8.60421 | calorie (TC)/hour square centimeter °C/centimeter |
| watt/meter kelvin | 0.002390 | calorie (TC)/second square centimeter °C/centimeter |
| watt/square meter | 0.316998 | Btu (IT)/hour square foot |
| watt/square meter kelvin | 0.176110 | Btu (IT)/hour square foot °F |

(*Continued*)

**TABLE B.3 (Continued)**

| Multiply | By | To Obtain |
|---|---|---|
| watt/square meter kelvin | $0.860421 \times 10^{-4}$ | kilocalorie (TC)/hour square centimeter °C |
| watt second | 1.00000 | joule |
| weber | $1.00000 \times 10^{8}$ | lines |
| weber | $1.00000 \times 10^{8}$ | maxwell |
| weber/ampere | 1.00000 | henry |
| yard | 0.914402 | meter |

*TC—Thermochemical.*

Appendix C

# Selected Design Data Tables

**TABLE C.1** Fusibility of Biomass Ash

| Type of Biomass | Temperature (°C) | | | |
|---|---|---|---|---|
| | Initial Deformation | Softening | Hemispherical | Fluid |
| Corn cob[a] | 900 | | 1020 | |
| Corn stalk[a] | 820 | | 1091 | |
| Grape pruning (oxidizing)[b] | 1313 | 1368 | 1374 | 1424 |
| Grape pruning (reducing)[b] | 1310 | 1360 | 1371 | 1382 |
| Olive pit[a] | 850 | | 1480 | |
| RDF pellet[a] | 890 | | 1130 | |
| RDF (oxidizing)[c] | 1065 | 1092 | 1131 | 1193 |
| RDF (reducing)[c] | 1024 | 1063 | 1097 | 1182 |
| Rice hulls | 1439 | | >1650 | |
| Rice straw[a] | 1060 | | 1250 | |
| Walnut shell[a] | 820 | | 1225 | |

[a] *Osman and Goss (1983).*
[b] *Rossi (1984, pp. 69–99).*
[c] *Alter and Campbell (1979, pp. 127–142).*

**TABLE C.2** Volumetric Heating Values and Other Properties of Constituents of Product Gas from Biomass Gasification

| Gases | $H_2$ | CO | $CO_2$ | $CH_4$ | $C_2H_6$ | $C_2H_4$ | $C_2H_2$ | $C_3H_8$ | $N_2$ |
|---|---|---|---|---|---|---|---|---|---|
| HHV ($MJ/Nm^3$)[a] | 12.74 | 12.63 | | 39.82 | 70.29 | 63.41 | 58.06 | 101.24 | |
| LHV ($MJ/Nm^3$)[a] | 10.78 | 12.63 | | 35.88 | 64.34 | 59.45 | 56.07 | 99.09 | |
| Viscosity[b] ($\mu$P) | 90 | 182 | 150 | 112 | 94 | 103 | 104 | 82 | 180 |
| Thermal conductivity[b] (W/m K) | 0.1820 | 0.0251 | 0.0166 | 0.0343 | 0.0218 | 0.0214 | 0.0213 | 0.0183 | 0.026 |
| Specific heat[b] (kJ/kg K) | 3.467 | 1.05 | 0.85 | 2.226 | 1.926 | 1.691 | 1.775 | 1.708 | 1.05 |

| | $C_3H_6$ | i-$C_4H_8$ | i-$C_4H_{10}$ | n-$C_4H_{10}$ | $C_6H_6$ | $NH_3$ | $H_2S$ |
|---|---|---|---|---|---|---|---|
| HHV ($MJ/Nm^3$)[a] | 93.57 | 125.08 | 133.12 | 134.06 | 142.89 | 13.07 | 25.10 |
| LHV ($MJ/Nm^3$)[a] | 87.57 | 116.93 | 122.91 | 123.81 | 141.41 | 10.13 | 23.14 |

[a] *Data compiled from Waldheim and Nilsson (2001).*
[b] *Data compiled from Jenkins (1989, p. 887).*

**TABLE C.3** Composition of Standard Dry Air at Atmospheric Pressure

| Gas | Volume (%) | Weight (%) | Molecular Weight |
|---|---|---|---|
| Nitrogen | 78.09 | 75.47 | 28.02 |
| Oxygen | 20.95 | 23.2 | 32 |
| Argon | 0.933 | 1.28 | 39.94 |
| Carbon dioxide | 0.03 | 0.046 | 44.01 |

**TABLE C.4** Temperature Dependence of Molar Specific Heat of Some Gases

| Gas | Molecular Weight | Specific Heat (kJ/kmol K) at Temperature, *T* (K) | Range of Validity (K) |
|---|---|---|---|
| $H_2S$ | 34 | $30.139 + 0.015T$ | 300–600 |
| $H_2O_{steam}$ | 18 | $34.4 + 0.000628T + 0.0000052T^2$ | 300–2500 |
| $H_2$ | 2 | $27.71 + 0.0034T$ | 273–2500 |
| $CH_4$ | 16 | $22.35 + 0.048T$ | 273–1200 |
| CO | 28 | $27.62 + 0.005T$ | 273–2500 |
| $CO_2$ | 44 | $43.28 + 0.0114T - 818{,}363/T^2$ | 273–1200 |
| $O_2$ | 32 | $34.62 + 0.00108T - 785{,}712/T^2$ | 300–5000 |
| $N_2$ | 28 | $27.21 + 0.0042T$ | 300–5000 |

**Source:** Adapted from Perry and Green (1997).

**TABLE C.5** Thermo-Physical Properties of Air at Various Temperatures

| Temperature (K) | Density (kg/m$^3$) | Dynamic Viscosity, $\mu \times 10^7$ (N s/m$^2$) | Kinematic Viscosity, $\gamma \times 10^6$ (m$^2$/s) | Thermal Conductivity, $K_g \times 10^3$ (W/m K) | Thermal Diffusivity, $\alpha \times 10^6$ (m$^2$/s) | Prandtl Number |
|---|---|---|---|---|---|---|
| 100 | 3.5562 | 71.1 | 2.00 | 9.34 | 2.54 | 0.786 |
| 150 | 2.3364 | 103.4 | 4.426 | 13.8 | 5.84 | 0.758 |
| 200 | 1.7458 | 132.5 | 7.590 | 18.1 | 10.3 | 0.737 |
| 250 | 1.3947 | 159.6 | 11.44 | 22.3 | 15.9 | 0.720 |
| 300 | 1.1614 | 184.6 | 15.89 | 26.3 | 22.5 | 0.707 |
| 350 | 0.9950 | 208.2 | 20.92 | 30.0 | 29.9 | 0.700 |
| 400 | 0.8711 | 230.1 | 26.41 | 33.8 | 38.3 | 0.690 |
| 450 | 0.7740 | 250.7 | 32.39 | 37.3 | 47.2 | 0.686 |
| 500 | 0.6964 | 270.1 | 38.79 | 40.7 | 56.7 | 0.684 |
| 550 | 0.6329 | 288.4 | 45.57 | 43.9 | 66.7 | 0.683 |
| 600 | 0.5804 | 305.8 | 52.69 | 46.9 | 76.9 | 0.685 |
| 650 | 0.5356 | 322.5 | 60.21 | 49.7 | 87.3 | 0.690 |
| 700 | 0.4975 | 338.8 | 68.10 | 52.4 | 98.0 | 0.695 |
| 750 | 0.4643 | 354.6 | 79.63 | 54.9 | 109 | 0.702 |
| 800 | 0.4354 | 369.8 | 84.93 | 57.3 | 120 | 0.709 |
| 850 | 0.4097 | 384.3 | 93.80 | 59.6 | 131 | 0.716 |
| 900 | 0.3868 | 398.1 | 102.9 | 62.0 | 143 | 0.720 |
| 950 | 0.3666 | 411.3 | 112.2 | 64.3 | 155 | 0.723 |
| 1000 | 0.3482 | 424.4 | 121.9 | 66.7 | 168 | 0.726 |
| 1100 | 0.3166 | 449.0 | 141.8 | 71.5 | 195 | 0.728 |
| 1200 | 0.2902 | 473.0 | 162.9 | 76.3 | 224 | 0.728 |
| 1300 | 0.2679 | 496.0 | 185.1 | 82 | 238 | 0.719 |
| 1400 | 0.2488 | 530 | 213 | 91 | 303 | 0.703 |
| 1500 | 0.2322 | 557 | 240 | 100 | 350 | 0.685 |
| 1600 | 0.2177 | 584 | 268 | 106 | 390 | 0.688 |

**TABLE C.6 Heat of Formation of Some Elements and Compounds at Standard Condition 25°C and 1 Bar Pressure**

| Substance | Heat of Formation $\Delta H_f^\circ$ (kJ/mol) | Absolute Entropy, $S^\circ$ (J/K mol) | Gibbs Function of Formation, $\Delta G_f^\circ$ (kJ/mol) |
|---|---|---|---|
| $C_{(s)}$(graphite) | 0 | 5.7 | 0 |
| $C_{(s)}$(diamond) | 1.9 | 2.38 | 2.90 |
| $CH_{4(g)}$ | −74.8 | 186.3 | −50.7 |
| $C_2H_{2(g)}$ | 226.7 | 200.9 | 209.2 |
| $C_2H_{4(g)}$ | 52.3 | 219.6 | 68.2 |
| $C_2H_{6(g)}$ | −84.7 | 229.6 | −32.8 |
| $C_3H_{8(g)}$ | −103.8 | 269.9 | −23.5 |
| $C_6H_{6(l)}$ | 49.0 | 172.8 | 124.5 |
| $CH_3OH_{(l)}$ | −238.7 | 126.8 | −166.3 |
| $C_2H_5OH_{(l)}$ | −277.7 | 160.7 | −178.8 |
| $CH_3CO_2H_{(l)}$ | −484.5 | 159.8 | −389.9 |
| $CO_{(g)}$ | −110.5 | 197.7 | −137.2 |
| $CO_{2(g)}$ | −393.5 | 213.7 | −394.4 |
| $He_{(g)}$ | 0 | 126.0 | 0 |
| $H_{2(g)}$ | 0 | 130.7 | 0 |
| $H_2O_{(l)}$ | −285.8 | 69.9 | −237.1 |
| $H_2O_{(g)}$ | −241.8 | 188.8 | −228.6 |
| $H_2O_{2(l)}$ | −187.8 | 109.6 | −120.4 |
| $N_{2(g)}$ | 0 | 191.6 | 0 |
| $NH_{3(g)}$ | −46.1 | 192.5 | −16.5 |
| $NO_{(g)}$ | 90.3 | 210.8 | 86.6 |
| $NO_{2(g)}$ | 33.2 | 240.1 | 51.3 |
| $N_2O_{(g)}$ | 82.1 | 219.9 | 104.2 |
| $N_2O_{4(g)}$ | 9.2 | 304.3 | 97.9 |
| $HNO_{3(l)}$ | −174.1 | 155.6 | −80.7 |
| $O_{(g)}$ | 249.2 | 161.1 | 231.7 |
| $O_{2(g)}$ | 0 | 205.1 | 0 |

(*Continued*)

**TABLE C.6** (Continued)

| Substance | Heat of Formation $\Delta H_f^\circ$ (kJ/mol) | Absolute Entropy, $S^\circ$ (J/K mol) | Gibbs Function of Formation, $\Delta G_f^\circ$ (kJ/mol) |
|---|---|---|---|
| $O_{3(g)}$ | 142.7 | 238.9 | 163.2 |
| $K_{(s)}$ | 0 | 64.2 | 0 |

(g) gaseous state; (l) liquid state; (s) solid state.
**Source:** Taken from University of Saskatoon chemistry web site.

**TABLE C.7** Equilibrium Constants for the Water–Gas, Boudouard, and Methane Formation Reactions (JANAF Thermochemical Tables)

| | Water Gas Reaction | Boudouard Reaction | Methanation Reaction |
|---|---|---|---|
| Temperature (K) | $K_{e2}(= P_{H_2} P_{CO}/P_{H_2O})$ | $K_{e1}(= P_{CO}^2/P_{CO_2})$ | $K_{e3}(= P_{CH_4}/P_{H_2}^2)$ |
| 400 | $7.709 \times 10^{-11}$ | $5.225 \times 10^{-14}$ | $9.481 \times 10^{4}$ |
| 600 | $5.058 \times 10^{-5}$ | $1.870 \times 10^{-6}$ | $8.291 \times 10^{2}$ |
| 800 | $4.406 \times 10^{-2}$ | $1.090 \times 10^{-2}$ | $5.246 \times 10^{0}$ |
| 1000 | 2.617 | 1.900 | $2.727 \times 10^{-2}$ |
| 1500 | $6.081 \times 10^{2}$ | $1.622 \times 10^{3}$ | $3.762 \times 10^{-8}$ |

See Eqs. 7.70, 7.71 and 7.72 for definitions.

**TABLE C.8** Specific Heat of Solid Biomass and Related Materials

| Type | Specific Heat (kJ/kg K) | Temperature (K) |
|---|---|---|
| Carbon[a] | 0.70 | 299–349 |
| | 1.60 | 329–1723 |
| Cellulose[b] | 1.34 | |
| Graphite | 0.84[c] | 273–373 |
| | 1.62[a] | 329–1723 |
| Wood (Oven dry, avg. 20 species) | 1.37[d] | 273–379 |
| Wood charcoal | 0.84[c] | 273–273 |

[a]*Perry et al. (1984).*
[b]*Kollman and Cote (1968).*
[c]*Baumeister (1967).*
[d]*Dunlap (1912).*

**TABLE C.9 Bulk Densities of Some Biomass Materials**

| Biomass | Conditions of Feed | Bulk Density, kg/m$^3$ |
|---|---|---|
| Rice | Hulled | 770–840 |
| | Rough | 540–570 |
| Wheat | | 710–800 |
| Corn | Ear | 449 |
| | Shelled | 718–721 |
| Oats | | 410–417 |
| Soybean | Seed | 721–801 |
| Peanut | With shell | 218–320 |
| | Without shell | 561–721 |
| Potato | Loose | 625–771 |
| Sugarcane | Long bundle | 882–1042 |
| | Cut | 240–289 |
| Bagasse | Loose | 112–160 |
| Wood | Bark | 160–321[a] |
| | Chips | 160–481 |
| | Sawdust | 256–577 |
| Straw | Bales in field | 107–128 |
| | Chopped | 64–80 |
| | Piled loose | 59–64 |

[a]*Revised food manufacturing process picture book 1984. Kogaku Kogyo Sha, Tokyo, pp. 539–542.*

**TABLE C 10** Thermal Conductivity of Some Biomass Species

| | Apparent Density ($kg/m^3$) | Thermal Conductivity (W/m K) | Temperature (K) | References |
|---|---|---|---|---|
| Softwood (balsam fir, white spruce, and black spruce) | 360 | 0.0986–0.1114 | 310–341 | Gupta et al. (2003) |
| Douglas fir | 512 | 0.110 | | Dinwoodie et al. (1989) |
| Norway spruce | 350 | 0.100 | | Dinwoodie et al. (1989) |
| Fir | 540 | 0.138 | | Kanury Murty (1970) |
| Hardwood birch | 680 | 0.214–0.250 | 295–373 | Suleiman et al. (1999) |

# Glossary

**Carbohydrates** Organic compounds of carbon, hydrogen, and oxygen, with the general formula $C_m(H_2O)_n$. The carbon and hydrogen are in the 2:1 atom ratio. Carbohydrates can be viewed as hydrates of carbon. They include sugars, starches, cellulose, and other cellular products.

**Cellulose** The main constituent of cell walls with the generic formula ($C_6H_{10}O_5$).

**Cracking** The breaking up of large complex organic molecules into smaller molecules using pressure and temperature with or without a catalyst. The product of cracking depends on temperature, pressure, and catalyst used. No single unique reaction takes place in the cracker. The hydrocarbon molecules are broken up in a fairly random way to produce mixtures of smaller hydrocarbons.

**Depolymerization** The decomposition of a polymer into smaller fragments, or the breakdown of macromolecular compounds into relatively simple compounds.

**Esterification** The chemical process for making esters, which are compounds of the chemical structure R—COOR′ in which R and R′ are either alkyl or aryl groups. The most common method for preparing esters is to heat a carboxylic acid, R—CO—OH, with an alcohol, R'—OH, and remove the water that is formed.

**Esters** Any chemical compounds derived by reacting an oxoacid (it contains an oxo group, X = O) with a hydroxyl compound such as an alcohol or a phenol.

**Ethanol** A popular alcohol ($C_2H_5OH$) used in spark-ignition engines, either alone or blended with petroleum-derived gasoline.

**Gasoline** In the United States and Canada, the petroleum-derived oil that runs normal spark-ignition car engines is called gasoline. In many other places it is called petrol. Gasoline is a mixture of a large number of hydrocarbons containing 4—12 carbon atoms per molecule in proportions that can vary depending on the crude oil and the user's specification. In the United States, gasoline is usually a blend of straight-run gasoline, reformate, alkylate, and some butane. The approximate composition is 15% C4—C8 straight-chain alkanes, 25—40% C4—C10 branched alkanes, 10% cycloalkanes, $<25\%$ aromatics ($<1.0\%$ benzene), and 10% straight-chain and cyclic alkenes (ACS, 2005). The average heating value of gasoline is 44.4 MJ kg and its specific gravity is 0.67—0.77. Its average molecular weight is $\sim 108$ (Ritter, 2005).

**Hemicellulose** An important component of plant cell walls that can be any of several heteropolymers present in almost all plant cell walls along with cellulose.

**Hydrocracking** A cracking process that uses a catalyst and occurs at high hydrogen partial pressure to "crack" the fractions into smaller molecules, to produce high-octane gasoline and other good quality, stable distillates.

**Hydrolysis** The breaking of hydrogen bonds in long-chained organic molecules.

**Hydrotreating** A process used in a refinery in which the feedstock is treated with hydrogen at elevated temperature and pressure in the presence of appropriate catalysts to remove contaminants such as sulfur, nitrogen, metals, and condensed-ring aromatics or metals.

**Lignin** A component of the cell wall of wood. It is a complex polymer that binds cellulose cells in biomass.

**Methanol** An important alcohol ($CH_3OH$) that serves as a feedstock for a host of chemicals and liquid transportation fuels.

**Polymerization** The building up of larger molecules by combining smaller molecules that is necessarily similar.

**Producer gas** Primarily a mixture of carbon monoxide, hydrogen, and nitrogen produced by blowing air and steam through the fuel bed.

**Protein** Any organic compound made up of amino acids arranged in a linear chain and folded into a globular form. These high-molecular-weight compounds of carbon, hydrogen, oxygen, and nitrogen are synthesized by plants and animals.

**Reforming** The structural manipulation of a molecule to improve its product quality. The process does not always involve major change in the molar mass. The steam reforming of methane is a widely used method of producing hydrogen.

**Starch** A polysaccharide carbohydrate consisting of a large number of glucose units joined together by glycosidic bonds, with the generic formula $(C_6H_{10}O_5)_n$. All green plants produce starch for energy storage.

**Steam reforming** A method for producing hydrogen from methane. Steam reacts with methane to produce hydrogen and carbon monoxide when heated to very high temperatures in the presence of a metal-based catalyst.

**Substituent** An atom or a group of bonded atoms that can be considered to have replaced a hydrogen atom in a parent molecular entity.

**Sugar** Any monosaccharide or disaccharide used especially by organisms to store energy. Glucose ($C_6H_{12}O_6$), a monosaccharide, is the simple sugar that stores chemical energy that biological cells convert to other types of energy.

**Syngas** A mixture of carbon monoxide and hydrogen.

**Synthesis** The building up of larger molecules from smaller ones. The molecules being synthesized need not be similar.

**Synthetic natural gas (SNG)** A methane gas artificially produced from other gaseous, solid, or liquid fuels using various methods.

**Triglycerides** A glyceride in which the glycerol is esterified with three fatty acids; the main constituent of vegetable oil and animal fats. The chemical formula is RCOO—$CH_2$CH (—OOCR′) $CH_2$—OOCR″, where R, R′, and R″ are longer alkyl chains.

**Torrefaction** Dictionary meaning is roasting, but here it refers to thermal decomposition within 200–300°C temperature in an oxygen-starved environment.

# References


Abdulsalam, P., 2005. A comparative study of hydrodynamics and gasification performance of two types of spouted bed reactor designs. Ph.D. Thesis. Asian Institute of Technology, School of Environment, Resources and Development, Bangkok.

Acharya, B., Dutta, A., Basu, P., 2009. Chemical looping gasification of biomass for hydrogen enriched gas production with in-process carbon-dioxide capture. Energ. Fuels 23 (10), 5077–5083.

Adesina, A.A., 1996. Hydrocarbon synthesis via Fischer–Tropsch reaction: travails and triumph. Appl. Catal. A Gen. 13 (2), 345–367.

Adrian, M.H., Brandl, A., Hooper, M., Zhao, X., Tabak, S.A., 2007. An alternative route for coal to liquids: methanol-to-gasoline (MTG) technology. Gasification Technologies Conference, San Francisco, CA, USA.

Agus, F., Waters, P.L., 1971. Determination of the grindability of coal, shale and other minerals by a modified Hardgrove machine method. Fuel 50, 405–431.

Al Juaied, M., Whitmore, A., 2008. Realistic cost of carbon capture. Harvard Kennedy School. Discussion paper BCSIA 2009-08, p-9.

Altafini, C.R., Wander, P.R., Barreto, R.M., 2003. Prediction of the working parameters of a wood waste gasifier through an equilibrium model. Energ. Convers. Manage. 44 (17), 2763–2777.

Alter, H., Campbell, J.A., 1979. The preparation and properties of densified refuse-derived fuel. In: Jones, J.L., Radding, S.B. (Eds.), Thermochemica Conversion of Solid Waste. ACS Symposium Series, 130, Washington, pp. 127–142.

Amin, S., Reid, R.C., Modell, M., 1975. Reforming and decomposition of glucose in an aqueous phase. ASME 75-ENAs-21, Intersociety Conference on Environmental System, San Francisco, CA.

Antal, M.J., Gronli, M., 2003. The art, science, and technology of charcoal production. Ind. Eng. Chem. Res. 42, 1619–1640.

Antal, M.J., Allen, S.G., Schulman, D., Xu, X., Divilio, R.J., 2000. Biomass Gasification in Supercritical Water. Ind. Eng. Chem. Res. 39, 4040–4053.

Antares Group, Inc., Biomass CHP catalog, EPA, combined heat and power partnership, US environmental protection agency, <http://www.epa.gov/CHP/documents/biomass_chp_catalog.pdf/> (accessed 22.07.08).

Appel, B.S., Adams, T.N., Roberts, M.J., Lange, W.F., Freiss, J.H., Einfeldt, C.T., et al., 2004. Process for conversion of organic, waste, or low-value materials into useful products. US Patent 2004/0192980 A1, September.

Arias, B., Pevida, C., Fermoso, J., Plaza, M., Rubiera, F., Pis, J., 2008. Influence of torrefaction on the grindability and reactivity of woody biomass. Fuel Process. Technol. 89, 169–175.

Arthur, J.R., 1951. Reactions between carbon and oxygen. Trans. Faraday Soc. 47, 164–178.

ASTM, 2000. Standard test method for shear testing of bulk solids using the Jenike shear cell. Am. Soc. Test. Mater.D-6128–00.

Aznar, M.P., Delgado, J., Corella, J., Borque, J.A., Campos, I.J., 1997. Steam gasification in fluidized bed of a synthetic refuse containing chlorine with a catalytic gas cleaning at high temperature. In: Bridgwater, A.V., Boocock, D.G.B. (Eds.), Developments in Thermochemical Biomass Conversion. Blackie Academic and Professional, London, UK, pp. 1194–1208. ISBN 0 7514 0350 4.

Babu, B.V., Chaurasia, A.S., 2004a. Parametric study of thermal and thermodynamic properties on pyrolysis of biomass in thermally thick regime. Energ. Convers. Manage. 45 (1), 53–72.

Babu, B.V., Chaurasia, A.S., 2004b. Pyrolysis of biomass: improved models for simultaneous kinetics and transport of heat, mass and momentum. Energ. Convers. Manage. 45 (9–10), 1297–1327.

Baker, E.G., Brown, M.D., Elliott, D.C., Mudge, L.K., 1988. Characterization and Treatment of Tars from Biomass Gasifiers. AIChE 1988 Summer National Meeting, Denver, CO, pp. 1–11.

Barea, A.G., 2009. Personal communication.

Barooah, J.N., Long, V.D., 1976. Rates of thermal decomposition of some carbonaceous materials in a fluidized bed. Fuel 55, 116–120.

Barrio, M., Hustad, J.E., 2001. $CO_2$ gasification of birch char and the effect of CO inhibition on the calculation of chemical kinetics. In: Bridgwater, A.V. (Ed.), Progress in Thermochemical Biomass Conversion, vol. 1. Blackwell Science, Oxford, pp. 47–60.

Barrio, M., Gøbel, B., Risnes, H., Henriksen, U., Hustad, J.E., Sørensen, L.H., 2001. Steam gasification of wood char and the effect of hydrogen inhibition on the chemical kinetics. In: Bridgwater, A.V. (Ed.), Progress in Thermochemical Biomass Conversion, vol. 1. Blackwell Science, Oxford, pp. 32–46.

Basu, P., 1977. Burning rate of carbon in fluidized beds. Fuel 56 (4), 390–392.

Basu, P., 2006. Combustion and Gasification in Fluidized Beds. CRC Press, Taylor & Francis, New York, NY.

Basu, P., Wu, S., 1993. Surface Reaction Rate of Coarse Sub-bituminous Char Particles in the Temperature Range 600-1340 K, Fuel, October.

Basu, P., Kefa, C., Jestin, L., 2000. Boilers and Burners – Design and Theory. Springer, New York, p. 25.

Basu, P., Butler, J., Leon, M.A., 2011. Biomass co-firing options on the emission reduction and electricity generation costs in coal-fired power plants. Renew. Energ. 36, 282–288.

Basu, P., Rao, S., Dhungana, A., 2013a. An investigation into the effect of biomass particle size on its torrefaction. Can. J. Chem. Eng. June 94, 466–474.

Basu, P., Rao, S., Acharya, B., Dhungana, A., 2013b. Effect of torrefaction on the density and volume changes of coarse biomass particles. Can. J. Chem. Eng. Available from: http://dx.doi.org/10.1002/c3ce-2817.

Basu, P., Dhungana, A., Rao, S., Acharya, B., 2013c. Effect of oxygen presence in torrefier. J. Energy Inst 128.3r in press. 10.1179/.00000000060.

Basu, P., Dhungana, A., Dutta, A., 2013d. Effects of reactor design on the torrefaction of biomass. ASME J. Energ. Res. 134 (4), 11–1062. Available from: http://dx.doi.org/10.1115/1.4007u84.

Batista, J.D., Tilaman, D., Hughers, E., 1998. Cofiring wood waste with coal in a wall fired boiler: initiating a three year demonstration program. Proceedings Ionergy'98, 4–8 October, Madison, WI, pp. 243–250.

Baumeister, T. (Ed.), 1967. Standard Handbook for Mechanical Engineers. McGraw-Hill.

Behrendt, F., Neubauer, Y., Overmann, M., Wilmes, B., Zobel, N., 2008. Direct liquefaction of biomass. Chem. Eng. Technol. 31 (5), 667–677.

Bergman, P.C.A., Kiel, J.H.A., 2005. Torrefaction for biomass upgrading. Fourteenth European Biomass Conference and Exhibition, Paris. ECN Report ECN-RX-05-180.

Bergman, P.C.A., Boersma, A., Zwart, R., Kiel, J., 2005a. Torrefaction for biomass co-firing in existing coal-fired power stations. Report ECN-C–05-013, ECN.

Bergman, P.C.A., Boersma, A.R., Kiel, J.H.A., Prins, M.J., Ptasinski, K.J., Janssen, F.J.J.G., 2005b. Torrefaction for entrained-flow gasification of biomass. Energy Research Centre of The Netherlands, ECN Report ECN-C-05-067.

Bergman, P.C.A., Prins, M.J., Boersma, A.R., Ptasinski, K.J., Kiel, J.H.A., Janssen F.J.J.G., 2005c. Combined Torrefaction and Pellestisation. Energy Research Centre of The Netherlands. ECN Report ECN-05-013.

Bergman, P.C.A., Boersma, A.R., Zwart, R.W.H., Kiel, J.H.A., 2005d, Development of torrefaction biomass co-firing in existing coal-fired power stations. Energy Research Centre of The Netherlands. ECN Report ECN-C-05-013.

Bingyan, X., Zengfan, L., Chungzhi, W., Haitao, H., Xiguang, Z., 1994. Circulating fluidized bed gasifier for biomass. In: Nan, L., Best, G., Coelho, C., de Carvalho, O. (Eds.), Integrated Energy Systems in China—The Cold Northeastern Region Experience. FAO, The United Nations, Rome.

Biomass Energy Centre. <www.biomassenergycentre.org.uk> (accessed 07.09).

Blasi, C.D., 1993. Modeling and simulation of combustion processes of charring and non-charring solid fuels. Prog. Energ. Combust. Sci. 19 (1), 71–104.

Blasi, C.D., 2009. Combustion and gasification rates of lignocellulosic chars. Prog. Energ. Combust. Sci. 35 (2), 121–140.

Blasi, C.D., Branca, C., Galgano, A., Meier, D., Brodzinski, I., Malmros, O., 2007. Supercritical gasification of wastewater from updraft wood gasifiers. Biomass Bioenergy 31 (11–12), 802–811.

Boerrigter, H., Rauch, R., 2005. Syngas production and utilization. In: Knoef, H.A.M. (Ed.), Biomass Gasification Handbook. Biomass Technology Group (BTG), Enschede, Chapter 10.

Boley, C.C., Landers, W.S., 1969. Entrainment drying and carbonization of wood waste. Bureau of Mines, Report of Investigations 7282, Washington, DC.

Boskovic, A., 2013. Physical processing of torrefied biomass. M.Sc. Thesis. Mechanical Engineering, Dalhousie University, Halifax, Canada.

Boskovic, A., 2013. Studies on the effects of torrefaction on grindability and explosion potential of biomass. M.Sc Thesis, Mechanical Engineering, Dalhousie University, Halifax.

Boukis, N., Franz, G., Habicht, W., Dinjus, E., 2001. Corrosion resistant materials for SCWO applications: experimental results from long time experiments. CORROSION 2001, March. 11–16, Paper 01353. Houston, Texas.

Boukis N., Galla U., D'Jesus P., Muller H., Dinjus E., 2005. Gasification of wet biomass in supercritical water: results of pilot plant experiments. Fourteenth European Biomass Conference, Paris, pp. 964–967.

Boukis, I.P., Grammelis, P., Bezergianni, S., Bridgwater, A.V., 2007. CFB air-blown flash pyrolysis. Part I: Engineering design and cold model performance. Fuel 86, 1372–1386.

Boutin, O., Lédé, J., 2001. Use of a concentrated radiation for the determination of cellulose thermal decomposition mechanisms. In: Bridgwater, A.V. (Ed.), Progress in Thermochemical Biomass Conversion. Blackwell Science, Oxford, UK, pp. 1034–1045. chapter 84.

Bowerman, F.R., 1969. Introductory chapter to principles and practices of incineration. In: Corey, R.C. (Ed.), John-Wiley and Sons, New York, P. 7.

Bradbury, A.G.W., Sakai, Y., Shafizadeh, F., 1979. A kinetic model for pyrolysis of cellulose. J. Appl. Polym. Sci. 23 (11), 3271–3280.

Brammer, J.G., Bridgwater, A.V., 1999. Drying technologies for an integrated gasification bio-energy plant. Renew. Sust. Energ. Rev. 3, 243–289.

Bridgeman, T., Jones, J.M., Shield, I., Williams, P.T., 2008. Torrefaction of reed canary grass, wheat straw and willow to enhance solid fuel qualities and combustion properties. Fuel 87, 844–856.

Bridgeman, T.G., Jones, J.M., Williams, A., Waldron, D.J., 2010. An investigation of the grindability of two torrefied energy crops. Fuel 89, 3911–3918.

Bridgwater, A.V., 1995. The technical and economic feasibility of biomass gasification for power generation. Fuel 74 (5), 633–653.

Bridgwater, A.V., 1999. Principles and practice of biomass fast pyrolysis processes for liquids. J. Anal. Appl. Phys. 51 (1–2), 3–22.

Bridgwater, A.V., 2002. Fast Pyrolysis of Biomass: A Handbook, vol. 2. CPL Press.

Bridgwater, A.V., Czernik, S., Piskorz, J., 2001. An overview of fast pyrolysis. In: Bridgwater, A.V. (Ed.), Progress in Thermochemical Biomass Conversion. Blackwell Science, pp. 977–997.

Bridgwater, A.V., Toft, A.J., Brammer, J.G., 2002. A techno-economic comparison of power production by biomass fast pyrolysis with gasification and combustion. Renew. Sust. Energ. Rev. 6 (3), 181–246.

Browning, B.L., 1967. Methods of Wood Chemistry, vol. II. Interscience publishers, New York, NY, pp. 387–388.

Bui, T., Loof, R., Bhattacharya, S.C., 1994. Multi-stage reactor for thermal gasification of wood. Energy 19 (4), 397–404.

Callaghan, C.A., 2006. Kinetics and catalysis of the water-gas-shift reaction: a microkinetic and graph theoretic approach. Ph.D. Thesis. Chemical Engineering Department, Worcester Polytechnic Institute, pp. 116.

Cantwell. K., 2002. Economic value of biomass for aging power plants, MBA Thesis, Saint Mary's University, Halifax, Canada.

Carleton, A.J., 1972. The effect of fluid-drag forces on the discharge of free-flowing solids from hoppers. Powder Technol. 6 (2), 91–96.

Carlos, L., 2005. High temperature air/steam gasification of biomass in an updraft fixed batch type gasifier. Ph.D. Thesis. Royal Institute of Technology, Energy Furnace and Technology, Stockholm.

Cetin, E., Moghtaderi, B., Gupta, R., Wall, T.F., 2005. Biomass gasification kinetics: influence of pressure and char structure. Combust. Sci. Technol. 177 (4), 765–791.

Chakraverty, A., Mujumdar, A.S., Raghavan, G.S.V., Ramaswamy, H.S., 2003. Handbook of Post-Harvest Technology. Taylor & Francis.

Channiwala, S.A., Parikh, P.P., 2002. A unified correlation for estimating HHV of solid, liquid and gaseous fuels. Fuel 81, 1051–1063.

Chase, M.W., 1998. NIST JANAF thermochemical tables. Am. Chem. Soc.

Chembukulam, S.K., Dandge, A.S., Kovilur, N.L., Seshagiri, R.K., Vaidyeswaran, R., 1981. Smokeless fuel from carbonized sawdust. Ind. Eng. Chem. Prod. Res. Dev. 20 (4), 714–719.

Chen, W.H., Kuo, P.C., 2010. A study on torrefaction of vaious biomass materials and its impact on lignocellulosic structure simulated by a thermogravimetry. Energy 35, 2580–2586.

Chen, W.H., Kuo, P.C., 2011. Torrefaction and co-torrefaction characterization of hemicellulose, cellulose and lignin as well as torrefaction of some basic constituents in biomass. Energy 36 (2), 803–811.

Chen, P., Zhao, Z., Wu, C., Zhu, J., Chen, Y., 2005. Biomass gasification in circulating fluidized bed. In: Cen, K. (Ed.), Circulating Fluidized Bed Technology, vol. 8. International Academic Publishers, pp. 507–514.

Chi, Y., Basu, P., Kefa, C., 1994. A simplified technique for measurement of reactivity of sorbent for use in circulating fluidized bed combustors. Fuel 73, 117–122.

Chowdhury, R., Bhattacharya, P., Chakravarty, M., 1994. Modeling and simulation of a down draft rice husk gasifier. Int. J. Energ. Res. 18 (6), 581–594.

Chu, W., Zhang, T., He, C., Wu, Y., 2002. Low-temperature methanol synthesis in liquid phase on novel copper based catalysts. Catal. Lett. 79 (1–4), 129–132.

Cielkosz, D., Wallace, R., 2011. Review: torrefaction for bioenergy feedstock production. Biofuels Bioprod. Biorefin. 5 (3), 317–329.

Ciferno, J.P., Marano, J.J., 2002. Benchmarking Biomass Gasification Technologies for Fuels, Chemicals and Hydrogen Production. U.S. Department of Energy, National Energy Technology Laboratory, <http://seca.doe.gov/technologies/coalpower/gasification/pubs/pdf/BMassGasFinal.pdf>.

Collins, C.D., 2007. Implementing phytoremediation of petroleum hydrocarbons, Methods in Biotechnology, vol. 23. Humana Press, pp. 99–108.

Corella, J., Sanz, A., 2005. Modeling circulating fluidized bed biomass gasifiers: a pseudo-rigorous model for stationary state. Fuel Process. Technol. 86 (9), 1021–1053.

Corella, J., Toledo, J.M., Molina, G., 2007. A review on dual fluidized-bed biomass gasifiers. Ind. Eng. Chem. Res. 46 (21), 6831–6839.

Cruz, D.C., Baddour, C.E., Ferrante, L., Berruti, F., Briens, C., 2012. Biocoal production from torrefied woody biomass. In: Arena, U., et al. (Eds.), Proceedings of Twelfth International Conference on Fluidized Bed Combustion. pp. 843–850.

Cummers, K.R., Brown, R.C., 2002. Ancillary equipment for biomass gasification. Biomass Bioenergy 23 (2), 113–128.

Dai, J., Grace, J.R., 2008. A model for biomass screw feeding. Powder Technol. 186 (1), 40–55.

Dalai, A.K., Davis, B.H., 2008. Fischer–Tropsch synthesis: a review of water effects on the performances of unsupported and supported Co catalysts. Appl. Catal. A Gen. 348, 1–15.

Darton, R.C., LaNauze, R.D., Davidson, J.F., Harrison, D., 1977. Bubble growth due to coalescence in fluidized beds. Trans. Inst. Chem. Eng. 55 (4), 274–280.

Dayton, D., 2002. A Review of the Literature on Catalytic Biomass Tar Destruction. National Renewable Energy Laboratory, NREL Report TP-510-32815.

Dayton, D.C., Turk, B., Gupta, R., 2011. Syngas cleanup, conditioning and utilization. In: Brown, R.C. (Ed.), Thermochemical Processing of Biomass. Wiley, p. 99.

de Smet, C.R.H., 2000. Partial oxidation of methane to synthesis gas: reaction kinetics and reactor modelling. Ph.D. Thesis. Technical University of Eindhoven.

Debdoubi, A., el Amarti, A., Colacio, E., Blesa, M.J., Hajjaj, L.H., 2006. The effect of heating rate on yields and compositions of oil products from esparto pyrolysis. Int. J. Energ. Res. 30 (15), 1243–1250.

Degroot, W.F., Shafidzadeh, F., 1984. Kinetics of gasification of Douglas fir and cottonwood chars by carbon dioxide. Fuel 63, 210.

Delgado, J., Aznar, M.P., Corella, J., 1996. Calcined dolomite, magnesite, and calcite for cleaning hot gas from a fluidized bed biomass gasifier with steam: life and usefulness. Ind. Eng. Chem. Res. 35 (10), 3637–3643.

Demirbas, A., 2000. Mechanisms of liquefaction and pyrolysis reactions of biomass. Energ. Convers. Manage. 41 (6), 633–646.

Demirbas, A., 2001. Biomass resources facilities and biomass conversion processing for fuels and chemicals. Energ. Convers. Manage. 42 (11), 1357–1378.

Demirbas, A., 2009. Biorefineries: Current activities and future developments. Energy Convers. Manage. 50, 2782–2801.

Department of Energy, 1978. Coal Conversion Systems Technical Data Book. USDOE, Contract no. EX–76–C–0102286, HCP/T2286–01, pp. IIIA-52–IIIA-53.

Desch, H.E., Dinwoodie, J.M., 1981. Timber: Its Structure, Properties and Utilization, sixth ed. Macmillan Press.

Devi, L., Ptasinski, K.J., Janssen, F.J.J.G., 2003. A review of the primary measures for tar elimination in biomass gasification processes. Biomass Bioenergy 24 (2), 125–140.

Di Blasi, C., 1993. Modeling and simulation of combustion processes of charring and non-charring solid fuels. Progr. Energy Combust. Sci. 19 (1), 71–104.

Di Blasi, C., 2008. Modeling chemical and physical processes of wood and biomass pyrolysis. Prog. Energ. Combust. Sci. 34 (1), 47–90.

Diebold, J.P., Bridgwater, A.V., 1997. Overview of fast pyrolysis of biomass for the production of liquid fuels. In: Bridgwater, A.V., Boocock, D.G.B. (Eds.), Developments in Thermochemical Biomass Conversion. Blackie Academic & Professional, pp. 5–27.

Diebold, J.P., Power, A., 1988. Engineering aspects of the vortex pyrolysis reactor to produce primary pyrolysis oil vapors for use in resins and adhesives. In: Bridgwater, A.V., Kuester, J.L. (Eds.), Research in Thermochemical Biomass Conversion. Elsevier Applied Science, pp. 609–628.

Diebold, J.P., Milne, T.A., Czernik, S., Osamaa, A., Bridgwater, A.V., Cuevas, A., et al., 1997. Proposed specification for various grades of pyrolysis oils. In: Bridgwater, A.V., Boocock, D.G.B. (Eds.), Developments in Thermochemical Biomass Conversion. Blackie Academic & Professional.

Dinjus, E., Kruse, A., 2004. Hot compressed water—A suitable and sustainable solvent and reaction medium. J. Phys. Condens. Matter 16, S1161–S1169.

Dinwoodie, J.M., 1989. Wood: Nature's Cellular, Polymeric Fiber-Composite. The Institute of Metals, London, pp. 27–34 (Chapters 3 and 4).

Dry, M.E., 2002. The Fischer–Tropsch process: 1950–2000. Catal. Today 71, 227–241.

Dryer, F.L., Glassman, I., 1973. High temperature oxidation of CO and $CH_4$. Fourteenth Symposium of Combustion. The Combustion Institute, Pittsburgh, pp. 987–1003.

Dunlap, F., 1912. Specific heat of wood, USDA Forest Service Bulletin 110, Washington, DC.

Ebasco Services Inc., 1981. Supplemental Studies for Anthracite Coal Gasification to Produce, Fuels and Chemicals, Nepgas Project—vol. II. Supplemental Tasks. Energy Development and Resource Corporation, Nanticoke, PA (Contract number FG01–79RA20221).

Electric Power Research Institute, 1991. Technical Assessment Guide, TR–100281, 3, 7–6 December.

Encinar, J.M., Beltran, F.J., Bernalte, A., Ramiro, A., Gonzalez, J.F., 1996. Pyrolysis of two agricultural residues: olive and grape bagasse. Influence of particle size and temperature. Biomass Bioenergy 11 (5), 397–409.

Encinar, J.M., Gonzalez, J.F., Rodriguez, J.J., Ramiro, M.J., 2001. Catalysed and uncatalysed steam gasification of eucalyptus char: influence of variables and kinetic study. Fuel 80 (14), 2025–2036.

EPA, 2007. Biomass Combined Heat and Power Catalog of Technologies. EPA Combined Heat and Power Partnership (Chapter 4). www.epa.gov_chp_documents_biomass_chp_catalog_part4.pdf (accessed 2012).

Esteban, L.S., Carrasco, J.E., 2006. Evaluation of different strategies for pulverization of forest biomasses. Powder Technol. 166, 139–151.

EUBIONET, 2003. Biomass Cofiring—An Efficient Way to Reduce Greenhouse Gas Emissions. European Bioenergy Network, p. 21. ec.europa.eu/energy/renewables/. . ./2003_cofiring_eu_bionet.pdf.

Evans, R.J., Milne, T.A., 1997. Chemistry of tar formation and maturation in the thermochemical conversion of biomass. In: Bridgwater, A.V., Boocock, D.G.B. (Eds.), Developments in Thermochemical Biomass Conversion, vol. 2. Blackie Academic & Professional, pp. 803–816.

FAO, 1983. Use of charcoal in blast furnace. <http://www.fao.org/docrep/03500e/03500e07.htm>.

FAO, 1985. Charcoal utilization and marketing.. ISSN 0259-2800. (Forestry Paper, Chapter 6). <www.fao.org/docrep/X5555E/x5555e07> (accessed 2012).

FAO, 1985. Industrial charcoal making. Wood Carbonization and the Product it Makes. FAO (Document repository, Chapter 2) www.fao.org/docrep/x5555e/x5555e03.htm.

FAO TCP 3101, 2008. Industrial Charcoal Production.

Felfli, F.F., Luengo, C.A., Suarez, J.A., Beaton, P.A., 2005. Wood briquette torrefaction. Energ. Sust. Dev. IX (3), 19–22.

Feng, W., Hedzer, J., Kooi, V.D., Arons, J.S., 2004a. Biomass conversions in subcritical and supercritical water: driving force, phase equilibria, and thermodynamic analysis. Chem. Eng. J. 43, 1459–1467.

Feng, W., Hedzer, J., Kooi, V.D., Arons, J.S., 2004b. Phase equilibria for biomass conversion processes in subcritical and supercritical water. Chem. Eng. J. 98, 105–113.

Fengel, D., Wegener, G., 1989. Wood: Chemistry, Ultrastructure and Reactions. ISBN: 3-11-00841-3.

Florin, N.H., Harris, A.T., 2008. Enhanced hydrogen production from biomass with in situ carbon dioxide capture using calcium oxide sorbents. Chem. Eng. Sci. 63 (2), 287–316.

Font, R., Marcilla, A., Verdu, E., Devesa, J., 1990. Kinetics of the pyrolysis of almond shells and almond shells impregnated with cobalt dichloride in a fluidized bed reactor and in a pyroprobe 100. Ind. Eng. Chem. Res. 29 (9), 1846–1855.

Freda, C., Canneto, G., Mariotti, P., Fanelli, E., Molino, A., Braccio, G., 2008. Cold model testing of an internal circulating fluid bed gasifier. Gasification Workshop, Marmara Research Centre, Turkey.

Friedrich, C., Kritzer, P.N., Boukis, N., Franz, G., Dinjus, E., 1999. The corrosion of tantalum in oxidizing sub- and supercritical aqueous solutions of HCl, $H_2SO_4$ and $H_3PO_4$. J. Mater. Sci. 34 (13), 3137–3141.

Gao, K., Wu, J., Wang, Y., Zhang, D., 2006. Bubble dynamics and its effect on the performance of a jet fluidized bed gasifier simulated using CFD. Fuel 85 (9), 1221–1231.

García, X.A., Alarcón, N.A., Gordon, A.L., 1999. Steam gasification of tars using a CaO catalyst source. Fuel Process. Technol. 58 (2), 83–102.

Gasafi, E., Reinecke, M., Kruse, A., Schebek, L., 2008. Economic analysis of sewage sludge gasification in supercritical water for hydrogen production. Biomass Bioenergy 32, 1085–1096.

Gaur, S., Reed, T.B. (1995), An atlas of thermal data for biomass and other fuels. National Renewable Energy Laboratory report no, NREL/TP-433-7965.UC, category: 1310. DE95009212, chapter 10.

Gaur, S., Reed, T.B., 1995. An Atlas of Thermal Data for Biomass and Other Fuels. National Renewable Energy Laboratory, NREL no. NREL/TP-433-7965.

Gerrard, A.M. (Ed.), 2000. A Guide to Capital Cost Estimating. fourth ed. Institution of Chemical Engineers, UK.

Gesner, A., 1861. A practical treatise on coal petroleum and other distilled oils (n.p.).

Gil, J., Corella, J., Aznar, M.P., Caballero, M.A., 1999. Biomass gasification in atmospheric and bubbling fluidized bed: effect of the type of gasifying agent on the product distribution. Biomass Bioenergy 17 (5), 389–403.

Gilbert, P., Ryu, C., Sharifi, V., Swithenbank, J., 2009. Effect of process parameters on pelletization of herbaceous crops. Fuel 88, 1491–1497.

Gomez-Barea, A., Leckner, B., Campoy, M., 2008. Conversion of char in CFB gasifiers. In: Werther, J., Nowak, W., Wirth, K., Hartge, E. (Eds.), Circulating Fluidized Bed Technology, vol. 10. Tu-Tech Innovation GmbH, Hamburg, pp. 727–732.

Graham, R.G., Bain, R., 1993. Biomass gasification: hot gas clean up. International Energy Agency, Biomass Gasification Working Group.

Grezin, A.K., Zakharov, N.D., 1988. Thermodynamic analysis of air separation equipment with a throttling refrigerating cycle. Chem. Petroleum Eng. 24 (5), 223–227 (translated from Russian Khimicheskoe i Neftyanoe Mashinostroenie).

Grotkjær, T., Da-Johansen, K., Jensen, A.D., Glarborg, P., 2003. An experimental study of biomass ignition. Fuel 82 (1), 825–833.

Guo, B., Shen, Y., Li, D., Zhao, F., 1997. Modelling coal gasification with a hybrid neural network. Fuel 76 (12), 1159–1164.

Guo, B., Li, D., Cheng, C., Lu, Z., Shen, Y., 2001. Simulation of biomass gasification with a hybrid neural network model. Bioresource Technol. 76 (2), 77–83.

Guo, L.J., Lu, Y.J., Zhang, X.M., Ji, C.M., Guan, Y., Pei, A.X., 2007. Hydrogen production by biomass gasification in supercritical water—A systematic experimental and analytical study. Catal. Today 129 (3–4), 275–286.

Gupta, M., Yang, J., Roy, C., 2003. Specific heat and thermal conductivity of softwood bark and softwood char particles. Fuel 82, 919–927.

Haar, L., Gallagher, J.S., Kell, G.S., 1984. NBS/NRC Steam Tables. Hemisphere Publishing.

Hajek, M., Judd, M.R., 1995. Use of neural networks in modeling the interactions between gas analysers at coal gasifiers. Fuel 74 (9), 1347–1351.

Han, J., Kim, H., 2008. The reduction and control technology of tar during biomass gasification/pyrolysis: an overview. Renew. Sust. Energ. Rev. 12 (2), 397–416.

Hao, X.H., Guo, L.J., Mao, X., Zhang, X.M., Chen, X.J., 2003. Hydrogen production from glucose used as a model compound of biomass gasified in supercritical water. Int. J. Hydrogen Energ. 28, 55–64.

Hasler, P., Nussbaumer, T., 1999. Gas cleaning for IC engine applications from fixed bed biomass gasification. Biomass Bioenergy 16, 385–395.

Hemati, M., Laguerie, C., 1987. The kinetic study of the pyrolysis of sawdust in a thermobalance: kinetic approach to the pyrolysis of oak sawdust. Chem. Eng. J. 35 (3), 147–156.

Hemati, M., Laguerie, C., 1988. Determination of the kinetics of the wood sawdust steam–gasification of charcoal in a thermobalance. Entropie 142, 29–40.

Herguido, J., Corella, J., Gonzalez-Saiz, J., 1992. Steam gasification of lignocellulosic residues in a fluidized bed at a small pilot scale: effect of the type of feedstock. Ind. Eng. Chem. Res. 31 (5), 1274–1282.

Higman, C., Burgt, M.V.D., 2003. Gasification. Elsevier Science, USA.

Higman, C., Burgt, M.V.D., 2008. Gasification, second ed. Gulf Professional Publishing/ Elsevier.

Higman, C., Burgt, M.V.D., 2008. Gasification, second ed. Gulf Professional Publishing, Elsevier, Oxford, p. 2, p. 351.

Hiltunen, M., Barisic, V., Zabetta, C., 2008. Combustion of different types of biomass in CFB boilers. Sixteenth European Biomass Conference, Valencia, Spain, 2–6 June.

Hoang, T., Ballantyne, G., Danuthai, T., Lobban, L.L., Resasco, D., Mallinson, R.G., 2007. Glycerol to gasoline conversion. Spring National Meeting, AIChE.

Hodge, B.K., 2010. Biomass. Alternative Energy Systems & Applications. Wiley (Chapter 12), pp. 296–329.

Hofbauer, H., 2002. Biomass CHP–plant Güssing: a success story. Paper presented at Pyrolysis and Gasification of Biomass and Waste Expert Meeting, Strasbourg.

Holgate, H.R., Meyer, J.C., Tester, W.J., 1995. Glucose hydrolysis and oxidation in supercritical water. AIChE J. 41, 637–648.

Hossfeld, R.J., Barnum, R.A., 2007. How to avoid flow stoppages during storage and handling. Power Eng. 111 (10), 42–50.

Hsu, I.C., 1979. Heat transfer to isobutene flowing inside a horizontal tube at supercritical pressure. Ph.D. Thesis. University of California, Lawrence Berkeley Laboratory, CA.

Hughey, C.A., Henerickson, C.L., 2001. Elemental composition analysis of processed and unprocessed diesel fuel by electrospray ionization Fourier transform ion cyclotron resonance mass spectrometry. Energ. Fuels 15 (5), 1186–1193.

Hulet, C., Briens, C., Berruti, F., Chan, E.W., 2005. A review of short residence time cracking processes. Int. J. Chem. Reactor Eng. 3, 1–71.

Humphreys, K.K., Wellman, P., 1996. Basic Cost Engineering, third ed. Marcel Dekker, New York, NY.

IFP, 2007. Potential biomass mobilization for biofuel production worldwide. In: Europe and in France, Innovation Energy Environment, Panaroma. <www.ifp.fr/>.

International Workshop on Biomass Torrefaction for Energy, 2012. France, 10–11 May, paper K3.

Jacob, K., 2000. Bin and Hopper Design. Dow Chemical Company, www.uakron.edu.

Jager, W., Espinoz, V.H.S., Hurtado, A., 2011. Review and proposal for heat transfer predictions at supercritical water conditions using existing correlations and experiments. Nucl. Eng. Des. 241, 2184–2203.

Janati, T., Sarkki, J., Lampenius, H., 2003. The utilization of CFB technology for large-scale biomass firing power plants. Power-Gen Europe 2010 Conference. 8–10 June, Foster Wheeler TP-CFB-10_03, Amsterdam.

Janssen, H.A., 1895. Versuche über getreidedruck in silozellen. Zeitschrift des Verein Deutscher Ingenieure 39, 1045–1049.

Janze, P., 2010, Move your biomass, Canadian Biomass Magazine, May-June, <www.canadianbiomassmagazine.ca/> (accessed April 2013).

Jarungthammachote, S., Dutta, A., 2008. Equilibrium modeling of gasification: Gibbs free energy minimization approach and its application to spouted bed and spout-fluid bed gasifiers. Energ. Convers. Manage. 49 (6), 1345–1356.

Jenike, A.W., 1964. Storage and flow of solids. Bulletin of the University of Utah, vol. 53, no. 26, Bulletin no. 123. University of Utah Engineering Experiment Station.

Jenkins, B.M., 1989. Physical properties of biomass. In: Kitani, O., Hall, C.W. (Eds.), Biomass Handbook. Gordon & Breach Science Publishers, Amsterdam.

Jenkins, B.M., Ebeling J.M., 1985. Correlations of physical and chemical properties of terrestrial biomass with conversion. Proceedings of Energy from Biomass and Wastes. IX Institute of Gas Technology, Chicago, pp. 371–400.

Jensen, A.A., Hoffman, L., Moller, B.T., Schmidt, A., 1997. Life cycle assessment. Environmental Issues Series no. 8. European Environment Agency.

Ji, P., Feng, W., Chen, B., Yuan, Q., 2006. Finding appropriate operating conditions for hydrogen purification and recovery in supercritical water gasification of biomass. Chem. Eng. J. 124, 7–13.

Johanson, J.R., 1965. Method of calculating rate of discharge from hoppers and bins. Trans. Soc. Mining Eng. 232 (3), 69–80.

Jones, J.M., Nawaz, M., Darvell, L.I., Ross, A.B., Pourkashanian, M., Williams, A., 2006. Towards biomass classification for energy applications. In: Bridgwater, A.V., Boocock, D.G.B. (Eds.), Science in Thermal and Chemical Biomass Conversion, vol. 1. CPL Press, pp. 331–339.

Joshi, N.R., 1979. Relative grindability of bituminous coals on volume basis, Fuel, Letter to editor. 58.

Kabyemela, B.M., Adschiri, T., Malaluan, R.M., Arai, K., 1997. Kinetics of glucose epimerization and decomposition in subcritical and supercritical water. Ind. Eng. Chem. Res. 36, 1552–1558.

Kalogirou, S.A., 2001. Artificial neural networks in renewable energy systems applications: a review. Renew. Sust. Energ. Rev. 5 (4), 373–401.

Kalogirou, S.A., Panteliou, S., Dentsoras, A., 1999. Artificial neural networks used for the performance prediction of a thermosiphon solar water heater. Renew. Energ. 18 (1), 87–99.

Kanury Murty, A., Blackshear Perry, L., 1970. Combust. Sci. Technol. 1, 339.

Kasman, H., Berg, M., 2006. Ash related problems in woodfired boilers and effect of additives. Workshop on Ash Deposition and Corrosion, Glasgow, September, Vattenfall, AB.

Kaushal, P., Pröll, T., Hofbauer, H., 2008. Model for biomass char combustion in the riser of a dual fluidized bed gasification unit: Part 1—Model development and sensitivity analysis. Fuel Process. Technol. 89 (7), 651–659.

Kersten, S.R.A., 2002. Biomass gasification in circulating fluidized beds. Dissertation. Twente University, Twente University Press, Enschede, The Netherlands.

Kinoshita, C.M., Wang, Y., Zhou, J., 1994. Tar formation under different biomass gasification conditions. J. Anal. Appl. Pyrol. 29 (2), 169–181.

Kitani, O., Hall, C.W., 1989. Biomass Handbook. Gordon & Breach Science Publishers.

Kitto, J.B., Stultz, S.C., 2005. Thermodynamics of steam, Steam: Its Generation and Use, forty-first ed. The Babcock and Wilcox Company, Barberton, OH, pp. 10–10.

Klass, D.L., 1998. Biomass for Renewable Energy, Fuels, and Chemicals. Academic Press, pp. 30, 276–277, 233, 239.

Kleinschmidt, C.P., 2010. Overview of international developments in torrefaction. <www.kema.com>.

Kleinschmidt, C.P., 2010 Overview of International Developments in Torrefaction. <www.kema.com/>.

Klose, W., Wolki, M., 2005. On the intrinsic reaction rate of biomass char gasification with carbon dioxide and steam. Fuel 84 (7–8), 885–892.

Knight, R.A., 2000. Experience with raw gas analysis from pressurized gasification of biomass. Biomass Bioenergy 18 (1), 67–77.

Knoef, H.A.M. (Ed.), 2005. Handbook Biomass Gasification. BTG Publisher, Enschede, The Netherlands, pp. 32, 239–241.

Kobe Steel. <www.admin@kobelco.co.jp> (accessed 06.10).

Kollman, F.F.P., Cote, W.A., 1968. Principle of Wood Science and Technology, vol. I. Solid Wood. Springer-Verlag.

Koltz, I.M., Rosenberg, R.M., 2008. Chemical Thermodynamics, seventh ed. Wiley, p. 237.

Krammer, P., Mittelstadt, S., Vogel, H., 1999. Investigating the synthesis potential in supercritical water. Chem. Eng. Technol. 22, 126–130.

Kreutz, T.G., Larson, E.D., Liu, G., Williams, R.H., 2008. Fischer–Tropsch Fuels from Coal and Biomass. Twenty-fifth Annual International Pittsburgh Coal Conference, October.

Kritz, P., 2004. Corrosion in high-temperature and supercritical water and aqueous solutions: a review. J. Supercrit. Fluids 29, 1–29.

Kruse, A., Abel, J., Dinjus, E., Kluth, M., Petrich, G., Schacht, M., et al., 1999. Gasification of biomass and model compounds in hot compressed water. International Meeting of the GVC-Fachausschub Hochdruckverfahrenstechnik, Karlsruhe, Germany.

Kruse, A., Henningsen, T., Sinag, A., Pfeiffer, J.J., 2003. Biomass gasification in supercritical water: influence of the dry matter content and the formation of phenols. Ind. Eng. Chem. Res. 42 (16), 3711–3717.

Kudo, K., Yoshida, E., 1957. The decomposition process of wood constituents in the course of carbonization I: The decomposition of carbohydrate and lignin in Mizunara. J. Jpn. Wood Res. Soc. 3 (4), 125–127.

Kumar, J.V., Pratt, B.C., 1996. Compositional analysis of some renewable biofuels. Am. Lab. 28 (8), 15–20.

Kunii, D., Levenspiel, O., 1991. Fluidization Engineering. Butterworth-Heinemann.

Kusdiana, D., Minami, E., Saka, S., 2006. Non-catalytic biodiesel fuel production by supercritical methanol pre-treatment. In: Bridgwater, A.V., Boocock, D.G.B. (Eds.), Science in Thermal and Chemical Biomass Conversion. CPL Press, pp. 424–435.

Lammars, G., Beenackers, A.A.C.M., Corella, J., 1997. Catalytic tar removal from biomass producer gas with secondary air. In: Bridgwater, A.V., Boocock, D.G.B. (Eds.), Developments in Thermochemical Biomass Conversion. Blackie Academic & Professional.

Lee, I., Kim, M.S., Ihm, S.K., 2002. Gasification of glucose in supercritical water. Ind. Eng. Chem. Res. 41, 1182–1188.

Lehman, J., Gaunt, J., Rondon, M., 2006. Bio-char sequestration in terrestrial ecosystem—A review. Mitig. Adapt. Strat. Glob. Change 11, 403–427.

Lewellen, P.C., Peters, W.A., Howard, J.B., 1977. Cellulose pyrolysis kinetics and char formation mechanism, 16th International Symposium on Combustion, (1), 1471–1480.

Li, X.T., Grace, J.R., Watkinson, A.P., Lim, C.J., Ergüdenler, A., 2001. Equilibrium modeling of gasification: a free energy minimization approach and its application to a circulating fluidized bed coal gasifier. Fuel 80 (2), 195–207.

Li, B., Kado, S., Mukainakano, Y., Miyazawa, T., Miyao, T., Naito, S., et al., 2007. Surface modification of Ni catalysts with trace Pt for oxidative steam reforming of methane. J. Catal. 245, 144–155.

Li, H., Liu, X., Legros, R., Bi, X.T., Lim, J., Sokhansanj, S., 2012. Torrefaction of sawdust in a fluidized bed reactor. Bioresour. Technol. 103, 453–458.

Liden, A.G., Berruti, F., Scott, D.S., 1988. A kinetic model for the production of liquids from the flash pyrolysis of biomass. Chem. Eng. Commun. 65, 207–221.

Liliedahl, T., Sjostrom, K., 1997. Modeling of char-gas reaction kinetics. Fuel 76 (1), 29–37.

Lim, M.T., Alimuddin, Z., 2008. Bubbling fluidized bed gasification—Performance, process findings and energy analysis. Renew. Energ. 33 (10), 2339–2343.

Liska, A.J., Yang, H.S., Bremer, V.R., Klopfenstein, T.J., Walters, D.T., Erickson, G.E., Cassman, K.G., 2009. Improvements in life cycle energy efficiency and greenhouse gas emissions of corn-ethanol. J. Ind. Ecol. 13 (1), 58–74.

Littlewood, K., 1977. Gasification: theory and application. Prog. Energ. Combust. Sci. 3 (1), 35–71.

Liu, H., Li, S., Zhang, S., Chen, L., Zhou, G., Wang, J., et al., 2008. Catalytic performance of monolithic foam Ni/SiC catalyst in carbon dioxide reforming of methane to synthesis gas. Catal. Lett. 120 (1–2), 111–115.

Loppinet-Serani, A., Aymonier, C., Cansell, F., 2008. Current and foreseeable applications of supercritical water for energy and the environment. ChemSusChem 1, 486–503.

Lu, Y.J., Guo, L.J., Ji, C.M., Zhang, X.M., Hao, X.H., Yan, Q.H., 2006. Hydrogen production by biomass gasification in supercritical water—A parametric study. Int. J. Hydrogen Energ. 31, 822–831.

MacLean, J.D., 1941. Thermal conductivity of wood. Trans. Am. Soc. Heating Ventilating Eng. 47, 323–354.

Malhotra, A., 2006. Thermodynamic Properties of Supercritical Steam. www.steamcenter.com, Jaipur, <www.lulu.com/items/volume_13/254000/254766/1/preview/preview-thpb.doc>.

Mani, S., Sokhansanj, S., Bi, X., Turhollow, A., 2006. Economics of producing fuel pellets from biomass. Appl. Eng. Agric. 22 (3), 421–442.

Maniatis, K., 2001. Progress in biomass gasification: an overview. In: Bridgwater, A.S. (Ed.), Progress in Thermochemical Biomass Conversion, vol. I. Blackwell Science, pp. 1–31.

Maples, R.E., 2000. Petroleum Refinery Process Economics, second ed. PennWell.

Marrone, P.A., Hong, G.T., 2008. Corrosion control methods in supercritical water oxidation and gasification processes, NACE Conference, New Orleans, LA.

Marrone, P.A., Hong, G.T., Spritzer, M.H., 2007. Developments in supercritical water as a medium for oxidation, reforming, and synthesis. J. Adv. Oxid. Technol. 10 (1), 157–168.

Mastellone, M.L., Arena, U., 2008. Olivine as a tar removal catalyst during fluidized bed gasification of plastic waste. AIChE J. 54 (6), 1656–1667.

Matsumura, Y., 2002. Evaluation of supercritical water gasification and biomethanation for wet biomass utilization in Japan. Energ. Convers. Manage. 43, 1301–1310.

Matsumura, Y., Minowa, T., 2004. Fundamental design of a continuous biomass gasification process using a supercritical water fluidized bed. Int. J. Hydrogen Energ. 29, 701–707.

Matsumura, Y., Harada, M., Nagata, K., Kikuchi, Y., 2006. Effect of heating rate of biomass feedstock on carbon gasification efficiency in supercritical water gasification. Chem. Eng. Comm. 193, 649–659.

Matsumura, Y., Minowa, T., Xu, X., Nuessle, F.Q., Adschiri, T., Antal, M.J., 1997. High pressure $CO_2$ removal in supercritical water gasification of biomass. In: Bridgewater, A.V., Boocock, D.G.B. (Eds.), Developments in Thermochemical Biomass Conversion. Blackie Academic & Professional, pp. 864–877.

Matsumura, Y., Minowa, T., Potic, B., Kerstein, S.R.A., Prins, W., van Swaiij, W.P.M., et al., 2005. Biomass gasification in near and supercritical water-status and prospect. Biomass Bioenergy 29 (4), 269–292.

McKendry, P., 2002. Energy production from biomass. Part 1—Overview of biomass. Bioresour. Technol. 83 (1), 37–46.

Medic, D., Darr, M., Shah, A., Rahn, S., 2012. Effect of torrefaction on water vapor adsorption properties and resistance to microbial degradation of corn stover. Energ. Fuels. doi:10.1021/ef3000449.

Mettanant, V., Basu, P., Leon, M.A., 2009. Gasification of rice husk in supercritical water. Eighth World Conference on Chemical Engineering, Montreal, August, paper # 971.

Mettanant, V., Basu, P., Butler, J., 2009. Agglomeration of biomass fed fluidized bed gasifier and combustor. Can. J. Chem. Eng. 87 (Oct), 656–684.

Miao, Q., Zhu, J., Yin, X.L., Wu, C.Z., 2008. Modeling of biomass gasification in circulating fluidized beds. In: Werther, J., Nowak, W., Wirth, K., Hartge, E. (Eds.), Circulating Fluidized Bed Technology, vol. 9. Tu-Tech Innovation GmbH, Hamburg, pp. 685–690.

Miller, R.B., 1999. Structure of wood. Wood Handbook—Wood as an Engineering Material. USDA Forest Service, Forest Products Laboratory, Madison, WI, Technical report FPL-FTR-113.

Milne, T., 2002. Pyrolysis: the thermal behaviour of biomass below 600°C. In: Reed, T.B. (Ed.), Encyclopedia of Biomass Thermal Conversion, third ed. Biomass Energy Foundation Press, pp. II-96–II-131. (Chapter 5).

Milne, T.A., Evans, R.J., Abatzoglou, N., 1998. Biomass Gasifier Tars: Their Nature, Formation, and Conversion. NREL/TP-570-25357.

Min, K., 1977. Vapor-phase thermal analysis of pyrolysis products from cellulosic materials. Combust. Flame 30, 285–294.

Minami, E., Saka, S., 2006. Chemical conversion of woody biomass and supercritical methanol to liquid fuels and chemicals. In: Bridgwater, A.V., Boocock, D.G.B. (Eds.), Science in Thermal and Chemical Biomass Conversion. CPL Press, pp. 1028–1037.

Minchener, A.J., 2005. Coal gasification for advanced power generation. Fuel 23 (17), 2222–2235.

Minowa, T., Zhen, F., Ogi, T., 1998. Cellulose decomposition in hot-compressed water with alkali or nickel catalyst. J. Supercrit. Fluids 13, 253–259.

Mohan, D., Pittman, C.U., Steele, P.H., 2006. Pyrolysis of wood/biomass for bio-oil: a critical review. Energ. Fuels 20 (3), 848–889.

Mokry, S., Pioro, I., Farah, A., King, K., Gupta, S., Peiman, W., Kirillov, P., 2011. Development of supercritical water heat-transfer correlation for vertical bare tubes. Nucl. Eng. Des. 241, 1126–1136.

Mozaffarian, M., Deurwaarder, E.P., Kerste, S.R.A., 2004. Green gas (SNG) production by supercritical gasification of biomass. The Netherlands Energy Research Foundation report ECN-C-04-081. <www.ecn.nl>.

Muhlen, H.J., Sowa, F., 1995. Factors influencing the ignition of coal particles- studies with a pressurized heated-grid apparatus. Fuel 74 (11), 1551–1554.

Mullins, E.J., McKnight, T.S., 1981. Canadian Wood—Its Properties and Uses. Ministry of Supply and Services, Canada.

Narváez, I., Orío, A., Aznar, M.P., Corella, J., 1996. Biomass gasification with air in an atmospheric bubbling fluidized bed. Effect of six operational variables on the quality of the produced raw gas. Ind. Eng. Chem. Res. 35 (7), 2110–2120.

National Non-Food Crops Centre, 2009. Miscanthus. <http://www.nnfcc.co.uk/metadot/index.pl?id=2416;isa=DBRow;op=show;dbview_id=2329> (accessed 12.08.09).

National Renewable Energy Laboratory (NREL), 2008. Determination of extractives in biomass. Laboratory Analytical Procedure (LAP). Technical Report NREL/TP-510-42619.

Neeft, J.P.A., Knoef, H.A.M., Zielke, U., Sjöström, K., Hasler, P., Simell, P.A., et al., 1999. Guideline for sampling an analysis of tar and particles in biomass producer gas. Version 3.1. Energy project EEN5-1999-00507 (tar protocol).

NETL, 2011. Cost and Performance Baseline for Fossil Energy Plants—vol. 3b: Low Rank Coal to Electricity: Combustion Cases, March 2011. DOE/NETL-2011/1463.

Nikitin, 1966. Chemistry of Cellulose and Wood. Academy of Sciences of the USSR, Institute of High Molecular Compounds, Moscow, Leningrad (translated in 1966 from Russian).

Nolan, P.F., Brown, D.J., Rothwell, E., 1973. Fourteenth Symposium (Intl.) on Combustion. The Combustion Institute, pp. 1143.

Norton, T., Sun, D.W., Grant, J., Fallon, R., Dodd, V., 2007. Applications of computational fluid dynamics (CFD) in the modeling and design of ventilation systems in the agricultural industry: a review. Bioresource Technol. 98 (12), 2386–2414.

Ontario Power Generation, 2010. Torrefaction treatment process, Internal presentation, November 25.

Ortiz, D.S., Curtiright, A.E., Samaras, C., Litovitz, A., Burger, N., 2011. Near-term opportunities for integrating biomass into the US electricity supply. Technical Report. Rand Corporation for NREL, pp. xvii.

Osman, E.A., Goss, J.R. 1982. Ash chemical composition, deformation and fusion temperatures for wood and agricultural residues Paper no. 83–3549, American Society of Agricultural Engineers, St. Joseph, MI.

Osman, N.B., 2010. Chemistry of hot-pressing hybrid poplar wood. PhD Dissertation. University of Idaho, p. 140.

Otis, P.K., Tomroy, J.H., 1957. Density: a tool in silo research. Agric. Eng. 38, 806–807; 860–863.

Overend, R.P., 1982. Wood Gasification: Review of Recent Canadian Experience. National Research Council of Canada, Ottawa, ON, Report no. 20094.

Overend, R.P., 2004. Thermo-chemical gasification technology and products. Presentation at Global Climate and Energy Project, April, Stanford University.

Overmann, M., Gerber, S., Behrendt, F., 2008. Euler–Euler and Euler–Lagrange modelling of wood gasification in fluidized beds. In: Werther, J., Nowak, W., Wirth, K., Hartge, E. (Eds.), Circulating Fluidized Bed Technology, vol. 9. Tu-Tech Innovation GmbH, Hamburg, pp. 733–738.

Paasen, S.V.B., 2004. Tar removal with wet ESP: parametric study. The Second World Conference and Technology Exhibition on Biomass for Energy, Industry and Climate Protection, pp. 205–215.

Pastorova, I., Arisz, P.W., Boon, J.J., 1993. Preservation of D-glucose oligosaccharides in cellulose chars. Carbohyd. Res. 248, 151–165.

Peacocke, G.V.C., Bridgwater, A.V., 2001. Transport, handling and storage of biomass derived fast pyrolysis liquid. In: Bridgwater, A.V. (Ed.), Progress in Thermochemical Biomass Conversion, vol. 2. Blackwell Science, Oxford, pp. 1482–1499.

Peng, J.H., Bi, H.T., Sokhasanj, S., Lim, J.C., Melin, S., 2011. An economical and market analysis of Canadian wood pellets. Int. J. Green Energ. 7 (2), 128–142.

Perry, R.H., Green, D.W., 1997. Perry's Chemical Engineer's Handbook, seventh ed. McGraw-Hill, pp. 2-161–2-169.

Perry, R.H., Green, D.W., Maloney, J.O. (Eds.), 1984. Chemical Engineer's Handbook. McGraw-Hill.

Petersen, L., Werther, J., 2005. Experimental investigation and modeling of gasification of sewage sludge in the circulating fluidized bed. Chem. Eng. Proc. 44 (7), 717–736.

Peterson, A.A., Vogel, F., Lachance, R.P., Froling, M., Antal, M.J., Tester, J.W., 2008. Thermochemical biofuel production in hydrothermal media: a review of sub- and supercritical water technologies. Energ. Environ. Sci. 1, 32–65.

Pfeifer, C.M., Pourmodjib, M., Rauch, R., Hofbauer, H., 2005. Bed material and additive variations for a dual fluidized bed steam gasifier in laboratory, pilot and demonstration scale. In: Cen, K. (Ed.), Circulating Fluidized Bed Technology VIII. International Academic Publishers, pp. 482–489.

Phanphanich, M., Mani, A., 2011. Impact of torrefaction on the grindability and fuel characteristics of forest biomass. Bioresour. Technol. 102, 1246–1253.

Pimchuai, A., Dutta, A., Basu, P., 2009. Torrefaction of agriculture residue to enhance combustible properties. Eighth World Conference in Chemical Engineering, Montreal, paper 1815.

Pimchuai, A., Dutta, A., Basu, P., 2010. Torrefaction of agriculture residue to enhance combustible properties. Energ. Fuels 24 (9), 4638–4645.

Piskorz, J., Scott, D.S., Radlien, D., 1988. Composition of oils obtained by fast pyrolysis of different woods. In: Soltes, J., Milne, T.A. (Eds.), Pyrolysis Oils from Biomass: Producing, Analyzing and Upgrading. American Chemical Society, pp. 167–178. (Chapter 16).

Power Magazine, 2012. July, p. 22.

Prins, M.J., 2005. Thermodynamic analysis of biomass gasification and torrefaction. Technizche Universiteit Eindhoven, Proefscrift. ISBN: 90-386-2886-2.

Prins, W., Kersten, S.R.A., Pennington, J.M.L., van de Beld, L., 2005. Gasification of wet biomass in supercritical water. In: Knoeff, H.A.M. (Ed.), Handbook of Biomass Gasification. BTG Biomass Technology Group, Entschede, p. 234.

Prins, M.J., Ptasinski, K.J., Janssen, F.J.J.G., 2006a. Torrefaction of wood. Part 2. Analysis of products. J. Anal. Appl. Pyrol. 77, 35–40.

Prins, M.J., Krzysztof, J.P., Janssen, F.J.J.G., 2006. More efficient biomass gasification via torrefaction. Energy 31 (15), 3458–3470.

Prins, M.J., Ptasinski, K.J., Janssen, F.J.J.G., 2006. Torrefaction of wood. Part 1. Weight loss kinetics. J. Anal. Appl. Pyrol. 77, 28–34.

Probstein, R.F., Hicks, R.E., 1982. Synthetic Fuels. McGraw-Hill, New York, NY.0070509085

Probstein, R.F., Hicks, R.E., 2006. Synthetic Fuels. Dover Publications, pp. 63, 98–99.

Pyle, D.L., Zaror, C.A., 1984. Heat transfer and kinetics in the low temperature pyrolysis of solids. Chem. Eng. Sci. 39 (1), 147–158.

Quaak, P., Knoef, H.A.M., Stassen, H., 1999. Energy from biomass: a review of combustion and gasification technologies. Energy Series, The World Bank, Technical Paper no. 422, pp. 56–57.

Ragland, K.W., Aerts, D.J., Baker, A.J., 1991. Properties of wood for combustion analysis. Bioresource Technol. 37, 161–168.

Rao, M.S., Singha, S.P., Sodha, M.S., Dubey, A.K., Shyam, M., 2004. Stoichiometric, mass, energy and energy balance analysis of countercurrent fixed-bed gasification of post-consumer residues. Biomass Bioenergy 27 (2), 155–171.

Rapagnà, S., Jand, N., Kiennemann, A., Foscolo, P.U., 2000. Steam-gasification of biomass in a fluidized-bed of olivine particles. Biomass Bioenergy 19 (3), 187–197.

Razvigorova, M., Goranova, M., Minkova, V., Cerny, J., 1994. On the composition of volatiles evolved during the production of carbon adsorbents from vegetable wastes. Fuel 73 (11), 1718–1722.

Reed, T.B., 2002. Kinetics of char gasification reactions above 500°C, Encyclopedia of Biomass Thermal Conversion, third ed. Biomass Energy Foundation Press (Chapter 7), pp. II-289. II-51, III-5.

Reed, T.B., Das, A., 1988. Handbook of Biomass Downdraft Gasifier for Engine System. Solar Energy Research Institute, SERUSP-271-3022, NTIS.

Ren, S., Lei, H., Wang, L., Bu, Q., Wei, Y., Liang, J., et al., 2012. Microwave torrefaction of douglas fir sawdust pellets. Energ. Fuels. doi:10.1021/ef300633c.

Rezaiyan, J., Cheremisinoff, N.P., 2005. Gasification Technologies- a primer for engineers and technologists. Taylor and Francis Group, CRC Press, p. 15.

Richey, C.B., et al., 1961. Agricultural Engineers Handbook. McGraw-Hill, New York, NY, p. 683.

Risnes, H., Sørensen, L.H., Hustad, J.E., 2001. $CO_2$ reactivity of chars from wheat, spruce, and coal. In: Bridgwater, A.V. (Ed.), Progress in Thermochemical Biomass Conversion, vol. 1. Blackwell Science, pp. 61–72.

Roberts, K.G., Gloy, B.A., Joseph, S., Scott, N.R., Lehmann, J., 2010. Life cycle assessment of biochar system: estimating the energetic, economic and climate change potential. Environ. Sci. Technol. 44 (2), 827–833.

Rocha, J.D., Brown, S.D., Love, G.D., Snape, C.E., 1997. Hydropyrolysis: a versatile technique for solid fuel liquefaction, sulphur speciation and biomarker release. J. Anal. Appl. Pyrol. 40–41, 91–103.

Ross, D., Noda, R., Adachi, M., Nishi, K., Tanaka, N., Horio, M., 2005. Biomass gasification with clay minerals using a multi-bed reactor system. In: Cen, K. (Ed.), Circulating Fluidized Bed Technology VIII. International Academic Publishers, pp. 468–473.

Rossi, A., 1984. Fuel characteristics of wood and nonwood biomass fuels. In: Tilaman, D.A., Jahn, E.C. (Eds.), Progress in Biomass Conversion, vol. 5. Academic Press, pp. 6–69.

Rouset, P., Macedo, L., Commandre, J.M., Moreira, A., 2012. Biomass torrefaction under different oxygen concentrations and its effect on the composition of the solid by-product. J. Anal. Appl. Pyrolysis 96, 86–91.

Rowell, R.M., 2005. Handbook of Wood Chemistry and Wood Composites. Taylor and Francis Group, CRC Press, FL, USA.

Rowell, R.M., Pettersen, R., Han, J.S., Rowell, J.S., Tshabala, M.A., 2005. Cell wall chemistry. In: Rowell, R.M. (Ed.), Handbook of Wood Chemistry and Wood Composites. CRC Press, Boca Raton, FL, (Chapter 3).

Ryall, E., 2012, The diversity of torrefaction feedstock. International Workshop on Biomass Torrefaction for Energy, Ecole des Mines d'Albi, 10–11 May, paper K3.

Sadak, S., Negi, S., 2009. Improvements of biomass physical and thermochemical characteristics via torrefaction. Proc. Environ. Progr. Sustainable Energy, AIChE 28 (3), 427–434.

Sanner, W.S., Ortuglio, C., Walters, J.G., Wolfson, D.E., 1970. Conversion of municipal and industrial refuse into useful materials by pyrolysis. U.S. Bureau of Mines, RI 7428.

Schmieder, H., Abeln, J., Boukis, E., Dinjus, A., Kruse, M., Kluth, G., et al., 2000. Hydrothermal gasification of biomass and organic wastes. J. Supercrit. Fluids 17, 145–153.

Schulz, H., 1999. Short history and present trends of Fischer–Tropsch synthesis. Appl. Catal. A Gen. 186, 3–12.

Scott, D.S., Piskorz, J., 1984. The continuous flash pyrolysis of biomass. Can. J. Chem. Eng. 62 (3), 404–412.

Seebauer, V., Petek, J., Staudinger, G., 1997. Effects of particle size, heating rate and pressure on measurement of pyrolysis kinetics by thermogravimetric analysis. Fuel 76 (13), 1277–1282.

Sensoz, S., Angin, D., 2008. Pyrolysis of safflower seed press cake: Part 1. The effect of pyrolysis parameters on the product yields. Bioresource Technol. 99, 5492–5497.

Serani, A.L., Aymonier, C., Cansell, F., 2008. Current and foreseeable applications of supercritical water for energy and the environment. ChemSusChem, 486–503. doi:10.1002/cssc.2007001671.

Shafidazeh, F., 1984. Chemistry of solid wood. Advances in Chemistry Series No. 207. American Chemical Society.

Shafidazeh, F., McGinnis, G.D., 1971. Chemical composition and thermal analysis of cottonwood. Carbohyd. Res. 16, 273–277.

Shafizadeh, F., 1984. Chemistry of solid wood. Advances in Chemistry Series No. 207. American Chemical Society, Washington DC, Chapter 13, pp. 489–529.

Shang, Z., Yao, Y., Chen, S., 2008. Numerical investigation of system pressure effect on heat transfer of supercritical water flows in a horizontal round tube. Chem Eng Sci 63, 4150–4158.

Shaw, R.W., Dahmen, N., 2000. Destruction of toxic organic materials using supercritical water oxidation: current state of the technology. In: Kiran, E., Debenedetti, P.G., Peters, C.J. (Eds.), Supercritical Fluids—Fundamentals and Applications. NATO Science Series E, vol. 366. Kluwer Academic Publishers, pp. 425–438.

Shen, J., Wand, X., Perez, M., Mourant, D., Rhodes, M., Li, C., 2009. Effects of particle size on the fast pyrolysis of oil mallee woody biomass. Fuel 88 (10), 1810–1817.

Simpson, W., Tenwolde, A., 1999. Physical properties and moisture relations of wood. In: Wood Handbook: Wood as an Engineering Material. Madison, WI: USDA Forest Service, Forest Products Laboratory, 1999. General technical report FPL, pp. 3–17.

Sinag, A., Kruse, A., Rathert, J., 2004. Influence of the heating rate and the type of catalyst on the formation of key intermediates and on the generation of gases during hydropyrolysis of glucose in supercritical water in a batch reactor. Ind. Eng. Chem. Res. 43, 502–508.

Sit, S.P., Grace, J.R., 1978. Interphase mass transfer in an aggregative fluidized bed. Chem. Eng. Sci. 33 (8), 1115–1122.

Sjostrom, E., 1993. Wood Chemistry. Fundamentals and Applications, second ed. Academic press, San Diego, CA, p. 292.

Smith, E.B., 2005. Basic Chemical Thermodynamics. Imperial College Press, pp. 49–57.

Smoot, L.D., Smith, P.J., 1985. Coal combustion and gasification (The Plenum Chemical Engineering Series), pp. 88–89.

Soltes, E.J., Elder, T.J., 1981. Pyrolysis. In: Goldstein, I.S. (Ed.), Organic Chemicals from Biomass. CRC Press.

Souza-Santos, M.L.D., 2004. Solid Fuels Combustion and Gasification. CRC Press.

Spath, P.L., Dayton, D.C., 2003. Preliminary screening – technical and economic assessment of synthesis gas to fuels and chemicals with emphasis on the potential for biomass-derived syngas, Chapter 7, Technical Report of National Renewable Energy Laboratory, NREL/TP-510-34929, pp. 90–110.

Stanmore, B.R., Gilot, P., 2005. Review—calcination and carbonation of limestone during thermal cycling for $CO_2$ sequestration. Fuel Proc. Technol. 86, 1707–1743.

Stelt, M.J.C., Gerhauser, H., Kiel, J.H.A., Ptasinski, K.J., 2011. Biomass upgrading by torrefaction for the production of biofuels: a review. Biomass Bioenergy 35, 3748–3762.

Stevens, D.J., 2001. Hot gas conditioning: recent progress with larger scale biomass gasification systems. National Renewable Energy Laboratory, Report NREL/SR-510-29952.

Steynberg, A.P., 2004. Introduction to Fischer–Tropsch technology. In: Steynberg, A., Dry, M. (Eds.), Studies in Surface Science and Catalysis 152. Elsevier B.V., pp. 1–62. (Chapter 1).

Ståhl, K., Neergaard, M., Nieminen, J., 2001. Final report: Värnamo demonstration programme. In: Bridgwater, A.V. (Ed.), Progress in Thermochemical Biomass Conversion. Blackwell, pp. 549–563.

Stiegel, G.J., 2005. Overview of Gasification Technologies. In: Global Energy and Energy Project (GCEP) Advanced Coal Workshop, March 15–16, Provo, Utah, USA.

Sudo, S., Takahasi, F., Takeuchi, M., 1989. Chemical properties of biomass. In: Kitani, O., Hall, C. W. (Eds.), Biomass Handbook. Gordon and Breach Science Publishers, p. 899. (Chapter 5.3).

Suleiman, B.M., Larfeldt, J., Leckner, B., Gustavsson, M., 1999. Thermal conductivity and diffusivity of wood. Wood Sci. Technol. 33, 465.

Sun, H., Song, B., Yang, Y., Kim, S., Li, H., Cheng, J., 2007. The characteristics of steam gasification of biomass and water filter carbon. Korean J. Chem. Eng. 24 (2), 341–346.

Susanto, H., Beenackers, A.A.C.M., 1996. Moving bed gasifier with internal recycle of pyrolysis gas. Fuel 75 (11), 1339–1347.

Susawa, K., 1989. In: Kitani, O., Hall, C.,.W. (Eds.), Biomass Handbook. Gordon and Breach Science Publications, New York, NY, p. 498. (Chapter 2.6.4).

Sutton, D., Kelleher, B., Ross, J.R.H., 2001. Review of literature on catalysts for biomass gasification. Fuel Process. Technol. 73 (3), 155–173.

Syred, N., Kurniawan, K., Griffith, T., Gralton, T., Ray, R., 2007. Development of fragmentation models for solid fuel combustion and gasification as subroutines for inclusion in CFD codes. Fuel 86 (14), 2221–2231.

Technical Association of Pulp and Paper Industry (TAPPI), 2007. Standard test method for preparation of extractive-free wood. ASTM Designation: D1105-96.

Termuehlen, H., Emsperger, W., 2003. Clean and Efficient Coal-Fired Power Plants: Development Toward Advanced Technologies. ASME Press, New York, p. 23.

Tester, J.W., Holgate, H.R., Armellini, F.A., Webley, P.A., Killilea, W.R., et al., 1993. Supercritical water oxidation technology process development and fundamental research. In: Teddler, W.D., Pohland, F.G. (Eds.), Emerging Technologies for Hazardous Waste Management III, vol. 518. American Chemical Society Symposium Series, Washington, pp. 35–76.

Thunman, H., Leckner, B., 2002. Thermal conductivity of wood—Models for different stages of combustion. Biomass Bioenergy 23 (1), 47–54.

Thurner, F., Mann, U., 1981. Kinetic investigation of wood pyrolysis. Ind. Eng. Chem. 20 (3), 482–488.

Tillman, D.A., 1978. Wood as an Energy Resource. Academic Press.

Tillman, D.A., 2000. Biomass cofiring: the technology, the experience, the combustion consequences. Biomass Bioenergy 19, 365–394.

Tillman, D.A., Jahn, E.C., 1984. Progress in Biomass Conversion, vol. 5. Academic Press, Orlando, FL, pp. 69–99.

Tumuluru, J.S., Sokhansanj, S., Hess, J.R., Wright, C.J., Baardman, R.D., 2011a. A review on biomass torrefaction process and product properties for energy applications. Ind. Biotechnol. 7 (5), 384–491.

Tumuluru, J.S., Sokhansanj, S., Wright, C.T., Boradman, R.D., Yancey, N.A., 2011b. A review on biomass classification and composition, co-firing issues and pretreatment methods. ASABE Annual International Meeting, INL/CON-11-22458, Paper no. 1110458, Idaho National Laboratory.

Uden, A.G., Berruti, F., Scott, D.S., 1988. A kinetic model for the production of liquids from the flash pyrolysis of biomass. Chem. Eng. Commun. 65 (1), 207–221.

Uematsu, M., Franck, E.U., 1980. Static dielectric constant of water and steam. J. Phys. Chem. Ref. Data 9 (4), 1291–1306.

Uemura, Y., Omar, W., Othman, N.A., Yusup, S., Tsutsui, T., 2011. Torrefaction of oil palm EFB in the presence of oxygen. Fuel 90 (8), 2585–2591.

UNFCCC, 2005. Clarifications of definition of biomass and consideration of changes in carbon pools due to a CDM project activity, EB-20, Appendix 8, July.

Uslu, A., Faaij, A.P.C., Bergman, P.C.A., 2008. Pre-treatment technologies, and their effect on international bioenergy supply chain logistics, techno-economic evaluation of torrefaction, fast pyrolysis and pelletisation. Energy 33, 1206–1223.

Vamvuka, D., Woodburn, E.T., Senior, P.R., 1995. Modelling of an entrained flow coal gasifier 1. Development of the model and general predictions. Fuel 74 (10), 1452–1460.

van den Enden, P.J., Lora, E.S., 2004. Design approach for a biomass fed fluidized bed gasifier using the simulation software CSFB. Biomass Bioenergy 26, 281–287.

van Heek, K.H., Muhlen, H.J., 1990. Chemical kinetics of carbon and char gasification. In: Lahaye, J., Ehrburger, P. (Eds.), Fundamental Issues in Control of Carbon Gasification Reactivity. Kluwer Academic Publisher, pp. 1–34.

Van Loo, S., Koppejan, J., 2003. Handbook on Biomass Combustion and Cofiring. Task 32. International Energy Agency (IEA), Twente University Press, Enschede, The Netherlands.

Van Loo, S., Koppejan, J., 2008. The Handbook of Biomass Combustion and Co-firing. Earthscan, London, p. 1, p. 295.

Van Loo, S., Koppejan, J., 2008. The Handbook of Biomass Combustion and Co-firing. Task 32. Earthscan, London.

Van Swaaij, W.P.M., 2003. Technical feasibility of biomass gasification in a fluidized bed with supercritical water, Report GP-01. University of Twente, p. 160.

Verhoeff, F., Pels, J.R., Boersma, A.R., Zwart, R.W.R., Kiel, J.H.A., 2011. ECN torrefaction technology heading for demonstration. ECN report ECN-M-11-079. Presented at nineteenth European Biomass conference and Exhibition, Berlin.

Vilienskii, T.V., Hezmalian, D.M., 1978. Dynamics of the Combustion of Pulverized Fuel (Dinamika Gorenia Prilevidnovo Topliva). Energia, Moscow.

Wagenaar, B.M., Florijn, J.H., Gansekoele, E., Venderbosch, R.H., 2009. Bio-Oil as Natural Gas Substitute in a 350 MW Power Station. Biomass Technology Group, <www.btgworld.com>.

Waldheim, L. and Nilson, T. 2001. Heating value of gases from biomass gasification, Report prepared for IEA Bioenergy Agreement, Task 20 – Thermal gasification of biomass. TPS-01/16, May.

Walker, P.L., Rusinko, F., Austin, L.G., 1959. Gas reactions of carbon. Adv. Catal. 11, 133–221.

Wang, Y., Kinoshita, C.M., 1993. Kinetic model of biomass gasification. Sol. Energ. 51 (1), 19–25.

Wang, S., Lu, G.Q., 1996. Carbon dioxide reforming of methane to produce synthesis gas over metal-supported catalysts: state of the art. Energ. Fuels 10, 896–904.

Wang, M., Pantini, D., 2000. Corn Based Ethanol Does Indeed Achieve Energy Benefits. Center for Transportation Research, Argonne National Laboratory.

Wang, Y., Yan, L., 2008. CFD studies on biomass thermochemical conversion. Int. J. Mol. Sci. 9 (6), 1108–1130.

Watanabe, M., Inomata, H., Osada, M., Sato, T., Adschiri, T., Arai, K., 2003. Catalytic effects of NaOH and $ZrO_2$ for partial oxidative gasification of n-hexadecane and lignin in supercritical water. Fuel 82, 545–552.

Weigner, P., Martens, F., Uhlenberg, J., Wolff, J., 2002. Increased flexibility of Shell gasification plant. IChemE Conference on Gasification: The Clean Choice for Carbon Management. Noordwijk, The Netherlands.

Weisser, D., 2007. A guide to life-cycle greenhouse gas (GHG) emissions from electric supply technologies. Energy 32 (9), 1543–1559, September.

Wen, C.Y., Bailie, R.C., Lin, C.Y., O'Brien, W.S., 1974. Production of low BTU gas involving coal pyrolysis and gasification, Advances in Chemistry Series, vol. 131. American Chemical Society, Washington, DC.

Werpy, T., Petersen, G., 2004. Top value added chemicals from biomass. Natl. Renew. Res. Lab. USA 1, 11–13, August.

Westbrook, C.K., Dryer, F.K., 1981. Simplified reaction mechanisms for the oxidation of hydrocarbon fuels in flames. Combust. Sci. Technol. 27, 31–43.

White, R.H., Dietenberger, M.A., 2001. Wood products: thermal degradation and fire. In: Buschow, K.H.J., Cahn, R.W., Flemings, M.C., Ilschner, B., Kramer, E.J., Mahajan, S. (Eds.), The Encyclopedia of Materials: Science and Technology. Elsevier, Amsterdam, pp. 9712–9716.

White, C., Gray, D., Tomlinson, G., 2007. Co-conversion of coal and biomass to produce Fischer–Tropsch transportation fuels. Gasification Technologies Conference, San Francisco, CA.

Williams, A., Pourkashanian, M., Jones, J.M., Skorupska, N., 2000. The Combustion and Gasification of Coal. Taylor and Francis, London.

World Energy Council, 2010. 2010-Survey of energy resources, London, UK, p. 360. ISBN 9780946121021.

Xia, B., Sun, D.W., 2002. Applications of computational fluid dynamics (CFD) in the food industry: a review. Comput. Electron. Agric. 34 (1–3), 5–24.

Xiao, G., Ni, M.N., Chi, Y., Jin, B.S., Xiao, R., Zhong, Z.P., et al., 2009. Gasification characteristics of MSW and an ANN prediction model. Waste Manage. 29 (1), 240–244.

Xu, X., Matsumura, Y., Stenberg, J., Antal, M.J., 1996. Carbon-catalyzed gasification of organic feed-stocks in supercritical water. Ind. Eng. Chem. Res. 35 (8), 2522–2530.

Yamagata, K., Nishikawa, K., Hasegawa, S., Fyjii, T., 1972. Forced convective heat transfer to supercritical water flowing in tubes. Int. J. Heat Mass Trans. 15, 2575–3593.

Yan, W., Acharjee, T.C., Coronella, C.J., Vasquez, V.R., 2009. Thermal pretreatment of lignocellulosic biomass. Environ. Prog. Sustain. Energ. 28 (3), 435–440.

Yan, W., Hastings, J., Acharjee, T., Coronella, C., Vasquez, V., 2010. Mass and energy balances of wet torrefaction of lignocellulosic biomass. Energ. Fuels 24, 4738–4742.

Yoshida, Y., Dowaki, K., Matsumura, Y., Matsuhashi, R., Li, D., Ishitani, H., et al., 2003. Comprehensive comparison of efficiency and $CO_2$ emissions between biomass energy conversion technologies—Position of supercritical water gasification biomass technologies. Biomass Bioenergy 25, 257–272.

Yu, D.M., Aihara, M., Antal, M.J., 1993. Hydrogen production by steam reforming of glucose in supercritical water. Energ. Fuels 7, 574–577.

Zeleznik, F.J., Gordon, S., 1968. Calculation of complex chemical equilibria. Ind. Eng. Chem. 60 (6), 27–57.

Zen, H.C. (Ed.), 2006. Coal Gasification Technology (in Chinese), Chemical Industry Press, Beijing, pp. 140, 154.

Zeng, H., Jin, F., Guo, J., 2004. Removal of elemental mercury from coal combustion flue gas by chloride-impregnated activated carbon. Fuel 83 (1), 143–146.

Zhou, C., Beltraminin, J.N., Fan, Y.X., Lu, F.Q., 2008. Chemoselective catalytic conversion of glycerol as a biorenewable source to valuable commodity chemicals. Chem. Soc. Rev. 37 (3), 527–554.

Zhu, X., Venderbosch, R., 2005. A correlation between stoichiometrical ratio of fuel and its higher heating value. Fuel 84, 1007–1010.

Zimont, V.I., Trushin, Y.M., 1969. Total combustion kinetics of hydrocarbon fuels. Combust. Explo. Shock Waves 5 (4), 567–573.

Zolin, A., Jensen, A., Jensen, P.A., Frandsen, F., Johansen, K.D., 2001. The influence of inorganic materials on the thermal deactivation of fuel chars. Energ. Fuels 15 (5), 1110–1122.

Zwart, R.W.R., Boerrigter, H., Drift, A.V.D., 2006. The impact of biomass pretreatment on the feasibility of overseas biomass conversion to Fischer-Tropsch products. Energy Fuels 20, 2192.

Available from: www.biodiesel.org

Available from: www.kamengo.com

# Index

*Note*: Page numbers with "*f*" and "*t*" refer to figures and tables, respectively.

## C

# G

## I



## J



## K



## L



## M

# Q



# R



# S

## T

Printed and bound by CPI Group (UK) Ltd, Croydon, CR0 4YY
16/05/2025
01874020-0001